Distribution Dependent Stochastic Differential Equations

World Scientific Series on Probability Theory and Its Applications

Print ISSN: 2737-4467
Online ISSN: 2737-4475

Series Editors: Zenghu Li *(Beijing Normal University, China)*
Yimin Xiao *(Michigan State University, USA)*

Published:

Vol. 5 *Distribution Dependent Stochastic Differential Equations*
by Feng-Yu Wang (Tianjin University, China) and Panpan Ren (City University of Hong Kong, Hong Kong SAR, China)

Vol. 4 *Stationary Stochastic Models: An Introduction*
by Riccardo Gatto (University of Bern, Switzerland)

Vol. 3 *Introduction to Probability Theory: A First Course on the Measure-Theoretic Approach*
by Nima Moshayedi (University of California, Berkeley, USA)

Vol. 2 *Introduction to Stochastic Processes*
by Mu-Fa Chen (Beijing Normal University, China) and Yong-Hua Mao (Beijing Normal University, China)

Vol. 1 *Random Matrices and Random Partitions: Normal Convergence*
by Zhonggen Su (Zhejiang University, China)

World Scientific Series on
Probability Theory and Its Applications

Volume 5

Distribution Dependent Stochastic Differential Equations

Feng-Yu Wang
Tianjin University, China

Panpan Ren
City University of Hong Kong, Hong Kong SAR, China

World Scientific

NEW JERSEY · LONDON · SINGAPORE · BEIJING · SHANGHAI · HONG KONG · TAIPEI · CHENNAI · TOKYO

Published by
World Scientific Publishing Co. Pte. Ltd.
5 Toh Tuck Link, Singapore 596224
USA office: 27 Warren Street, Suite 401-402, Hackensack, NJ 07601
UK office: 57 Shelton Street, Covent Garden, London WC2H 9HE

Library of Congress Control Number: 2024040471

British Library Cataloguing-in-Publication Data
A catalogue record for this book is available from the British Library.

World Scientific Series on Probability Theory and Its Applications — Vol. 5
DISTRIBUTION DEPENDENT STOCHASTIC DIFFERENTIAL EQUATIONS

Copyright © 2025 by World Scientific Publishing Co. Pte. Ltd.

All rights reserved. This book, or parts thereof, may not be reproduced in any form or by any means, electronic or mechanical, including photocopying, recording or any information storage and retrieval system now known or to be invented, without written permission from the publisher.

For photocopying of material in this volume, please pay a copying fee through the Copyright Clearance Center, Inc., 222 Rosewood Drive, Danvers, MA 01923, USA. In this case permission to photocopy is not required from the publisher.

ISBN 978-981-12-8014-6 (hardcover)
ISBN 978-981-12-8015-3 (ebook for institutions)
ISBN 978-981-12-8016-0 (ebook for individuals)

For any available supplementary material, please visit
https://www.worldscientific.com/worldscibooks/10.1142/13512#t=suppl

Preface

As an extension to Itô's stochastic differential equations (SDEs) describing linear parabolic equations, distribution dependent SDEs (DDSDEs) characterize nonlinear Fokker-Planck equations. This type of SDEs are named after McKean-Vlasov due to the pioneering work of H. P. McKean (1966) where an expectation dependent SDE is proposed to study nonlinear PDEs for Maxwellian gas. Moreover, according to the propagation of chaos, a DDSDE is characterized as the limit of the equation for a single particle in the corresponding mean field particle systems, when the number of particles goes to infinity. So, DDSDEs are also called mean-field SDEs. To restrict a DDSDE in a domain, we consider the reflection boundary by following the line of A. V. Skorohod (1961), or kill the solution at the hitting time of the boundary. This book aims to provide a self-contained account on singular SDEs and DDSDEs in $\mathbb{R}^d$, and the reflected or killed equations in a domain which might be unbounded and non-convex.

In Chapter 1, we study singular SDEs with coefficients satisfying local integrability conditions, which allow the drift to be unbounded in bounded domains. The main idea is to kill the singular drift using a strong enough noise, for which Krylov's and Khasminskii's estimates as well as Girsanov's and Zvonkin's transforms are used. The study goes back to A. J. Veretennikov (1979) who proved the well-posedness of non-degenerate SDEs with bounded drifts, and by now has led to a relatively complete theory on singular SDEs. Besides the well-posedness, we also characterize the regularity (Harnack inequalities, gradient estimates, Bismut formulas) and the exponential ergodicity.

In Chapter 2, we investigate singular reflected SDEs in a domain. A key point is to construct Zvonkin's transform in the domain, for which we extend a PDE from the domain to the whole space such that estimates presented in Chapter 1 apply to the reflecting setting.

Chapter 3 presents criteria on the well-posedness of DDSDEs in several different situations including the monotone case, the time-spatially singular case with distribution independent noise, and the time-spatial singular case with distribution dependent noise.

Chapter 4 devotes to derivative estimates on the law of singular DDSDEs with respect to the initial distribution. We first establish the log-Harnack and power Harnack inequalities, then introduce the intrinsic and L derivatives for functions of probability measures, and finally present Bismut type formulas for these derivatives of the DDSDEs in the initial distribution.

Chapter 5 focuses on long time behaviors of DDSDEs. The uniform ergodicity is proved for a class of singular equations, the exponential convergence in entropy and Wasserstein distance is derived for the dissipative case, the exponential ergodicity in weighted Wasserstein distances is presented for non-dissipative equations, and the Donsker-Varadhan large deviation principle is established by comparing DDSDEs with the corresponding stationary SDEs.

In Chapters 6 and 7 we study reflected and killed DDSDEs in a domain which may be unbounded and non-convex, such that results presented in Chapters 3–5 are extended to the domain case.

For readers' references, some remarks and further results are presented at the end of most chapters. Results and techniques introduced in the book are mainly organized from recent papers worked out by the authors and collaborators. We have to indicate that due to the limitation of space and our own interests, many related results and references are not included.

We would like to thank our colleagues for helpful comments and fruitful collaborations, among many others we would like to mention Jianhai Bao, Xing Huang and Yulin Song. Most materials of the book have been reported for summer schools and advanced courses organized during 2020–2022 by Professor Xiaoyue Li in Northeast Normal University, Professor Runzhang Xu in Habin Engineering University, Professor Rongchan Zhu in Beijing Institute of Technology, Professors Xicheng Zhang and Wei Liu in Wuhan University, and Professor Jing Wu in Sun Yat-Sen University. A part of the contents has been presented online as a TCC course for universities in the United Kingdom. We would like to thank the organizers and participants for their interests. Thanks are also given to anonymous reviewers for valuable comments, and to Hongbo Han, Wei Hong, Shanshan Hu, Shen Wang, Bingyao Wu, and Xiaoyu Zhao for typo corrections. Last but not least, we would like to thank the executive editor Ms. Lai Fun Kwong for very efficient work.

We gratefully acknowledge the support from the National Key R&D Program of China (2022YFA1006000, 2020YFA0712900), the National Nature Science Foundation of China (11831014, 11921001) and Alexander von Humboldt Foundation (Germany).

Feng-Yu Wang & Panpan Ren

Contents

Preface   v

1. Singular Stochastic Differential Equations   1
   - 1.1 Itô's SDE and linear Fokker-Planck equation   1
   - 1.2 Krylov's and Khasminskii's estimates   4
   - 1.3 Well-posedness   13
     - 1.3.1 The main result   13
     - 1.3.2 Some lemmas   16
     - 1.3.3 Proof of Theorem 1.3.1   19
   - 1.4 Bismut formula   25
     - 1.4.1 Malliavin calculus   26
     - 1.4.2 The main result   27
   - 1.5 Dimension-free Harnack inequality   34
     - 1.5.1 Log-Harnack inequality   35
     - 1.5.2 Power Harnack inequality   37
   - 1.6 Exponential ergodicity   44
     - 1.6.1 Main results   44
     - 1.6.2 Ergodic theorems   47
     - 1.6.3 Elliptic equations   49
     - 1.6.4 Proofs of Theorem 1.6.1 and Corollary 1.6.2 . . .   51
   - 1.7 Notes and further results   55
     - 1.7.1 Singular SDEs on Heisenberg groups   56
     - 1.7.2 Singular stochastic Hamiltonian systems   58
     - 1.7.3 Singular SDEs with Kato and critical drifts . . .   60

2. Singular Reflected SDEs — 61

    2.1 Reflected SDE and Neumann problem 61
    2.2 Krylov's and Khasminskii's estimates 68
    2.3 Weak well-posedness 77
        2.3.1 Preparations 78
        2.3.2 Proof of Theorem 2.3.1 83
    2.4 Strong well-posedness and gradient estimates 85
        2.4.1 Proof of Theorem 2.4.1 86
        2.4.2 Proof of Theorem 2.4.2 95
    2.5 Power Harnack inequality 98
    2.6 Exponential ergodicity 99

3. DDSDEs: Well-posedness — 101

    3.1 DDSDE and nonlinear Fokker-Planck equation 101
    3.2 Fixed point in distribution and Yamada-Watanabe principle 106
    3.3 The monotone case 109
    3.4 Singular case: $\|\cdot\|_V$-Lipschitz in distribution 112
    3.5 Singular case: $(\|\cdot\|_{k,var} + \mathbb{W}_k)$-Lipschitz in distribution . 118
    3.6 Singular case: With distribution dependent noise 126
        3.6.1 Some lemmas 128
        3.6.2 Proof of Theorem 3.6.1 134
    3.7 Singular density dependent SDEs 135
        3.7.1 Density free noise 137
        3.7.2 Density dependent noise 150
    3.8 Notes and further results 156
        3.8.1 DDSDEs with linear functional derivative of noise coefficients 156
        3.8.2 Singular DDSDEs with integral type, Kato class and critical drifts 157
        3.8.3 Singular distribution dependent semilinear SPDEs 158
        3.8.4 Path-distribution dependent SDEs 159
        3.8.5 Singular path-distribution dependent nonlinear SPDEs 159
        3.8.6 Singular degenerate DDSDEs 160

4. DDSDEs: Harnack Inequality and Derivative Estimates — 163

    4.1 Log-Harnack inequality 163

		4.1.1	Monotone and non-degenerate case	164
		4.1.2	Degenerate case	167
		4.1.3	Singular case	169
	4.2	Power Harnack inequality		172
	4.3	Chain rule for intrinsic/L-derivatives		174
	4.4	Bismut formula for singular DDSDEs		181
		4.4.1	Main results	182
		4.4.2	Some lemmas	184
		4.4.3	Proof of Theorem 4.4.1(1)	192
		4.4.4	Proof of Theorem 4.4.1(2)	196
		4.4.5	Proof of Theorem 4.4.3	201
	4.5	Notes and further results		208
		4.5.1	Bismut formula for degenerate DDSDEs	209
		4.5.2	L-derivative estimate for distribution dependent noise	211
		4.5.3	Derivative estimates for the transition density	211
		4.5.4	Log-Harnack inequality and Bismut formula for DDSDEs with drifts singular in distributions	213
5.	DDSDEs: Long Time Behaviors			219
	5.1	A general result with application to $\mathbb{W}_k$-exponential ergodicity		219
	5.2	Ergodicity in variation distance: Singular case		221
		5.2.1	Two lemmas	222
		5.2.2	Proof of Theorem 5.2.1	225
	5.3	Exponential ergodicity in relative entropy and $\mathbb{W}_2$: Dissipative case		227
		5.3.1	A criterion with application to Granular media type equations	227
		5.3.2	The non-degenerate case	234
		5.3.3	The degenerate case	239
	5.4	Exponential ergodicity: Non-dissipative case		245
		5.4.1	Proof of Theorem 5.4.1	250
	5.5	Exponential ergodicity in $\mathbb{W}_\psi$: Partially dissipative case		258
		5.5.1	Main results and example	259
	5.6	Donsker-Varadhan large deviations		264
		5.6.1	Main result and examples	264
		5.6.2	LDP for Markov processes	268
		5.6.3	Proof of Theorem 5.6.1	271

	5.7	Notes	275
6.	**DDSDEs with Reflecting Boundary**		**277**
	6.1	Reflected DDSDE for nonlinear Neumann problem	277
	6.2	Well-posedness: Singular case	280
	6.3	Well-posedness: Monotone case	290
	6.4	Log-Harnack inequality and applications	296
		6.4.1 Singular case	296
		6.4.2 Monotone case	299
	6.5	Exponential ergodicity	304
		6.5.1 Singular case	305
		6.5.2 Dissipative case	305
		6.5.3 Partially dissipative case	309
		6.5.4 Non-dissipative case	313
	6.6	Notes	316
7.	**Killed DDSDEs**		**319**
	7.1	Killed DDSDE for nonlinear Dirichlet problem	319
	7.2	Monotone case	323
		7.2.1 Monotonicity in $\hat{\mathbb{W}}_1$	323
		7.2.2 Monotonicity in $\mathbb{W}_1$	328
	7.3	Singular case with distribution dependent noise	333
		7.3.1 For D-distributions in $\mathcal{P}^D$	334
		7.3.2 For D-distributions in $\mathcal{P}_1^D$	337
	7.4	Singular case with distribution independent noise	340
		7.4.1 Main result	341
		7.4.2 An extension of Theorem 7.4.1	342

Bibliography 347

Index 359

Chapter 1

Singular Stochastic Differential Equations

Let Δ be the Laplace operator on the d-dimensional Euclidean space $\mathbb{R}^d$. It was observed by A. Einstein (1905) that the distribution density function $p_t(x,\cdot)$ of the d-dimensional Brownian motion starting at point x is the fundamental solution of the heat equation

$$\partial_t u_t = \frac{1}{2}\Delta u_t, \quad t \geq 0.$$

To characterize the Fokker-Planck equation where Δ is replaced by a second order differential operator, K. Itô (1944) developed his stochastic calculus and then established the chain rule (Itô's formula) of stochastic differentials in 1951, which settled the foundation of stochastic differential equations (SDEs). There are plentiful references concerning SDEs and applications, which include, among many others, the books [Ikeda and Watanabe (1977)], [Oksendal (2014)] and [Situ (2005)].

In this chapter, we summarize some recent progress made to singular SDEs with non-degenerate noise, where the drift only satisfies an integrability condition and thus can be unbounded on bounded domains. We first introduce the link between SDEs and Fokker-Planck equations, then study properties of singular SDEs: well-posedness, Harnack inequalities, Bismut formula, and the exponential ergodicity.

1.1 Itô's SDE and linear Fokker-Planck equation

Let W_t be the m-dimensional Brownian motion on a complete filtration probability space $(\Omega, \{\mathcal{F}_t\}_{t\geq 0}, \mathbb{P})$, and let $\mathbb{R}^{d\otimes m}$ be the space of $d\times m$-matrices. Consider the following Itô's SDE on $\mathbb{R}^d$ for a fixed time $T > 0$:

$$dX_t = b_t(X_t)dt + \sigma_t(X_t)dW_t, \quad t \in [0,T], \tag{1.1.1}$$

where
$$b: [0,T] \times \mathbb{R}^d \to \mathbb{R}^d, \quad \sigma: [0,T] \times \mathbb{R}^d \to \mathbb{R}^{d\otimes m}$$
are measurable.

To define solutions of (1.1.1), let $|\cdot|$ and $\langle \cdot, \cdot \rangle$ be the norm and inner product in $\mathbb{R}^d$, let $\|\cdot\|$ and $\|\cdot\|_{HS}$ be the operator norm and the Hilbert-Schmidt norm of matrices, and denote by $\mathcal{L}_\xi$ the distribution of a random variable ξ. When different probabilities are concerned, we denote $\mathcal{L}_\xi = \mathcal{L}_{\xi|\mathbb{P}}$ to emphasize the distribution under the probability measure $\mathbb{P}$.

Without loss of generality, we assume that $(\Omega, \mathcal{F}_0, \mathbb{P})$ is atomless, such that for any probability measure μ on a Polish space E, there exists an $\mathcal{F}_0$-measurable E-valued random variable ξ such that $\mathcal{L}_\xi = \mu$. In case $(\Omega, \mathcal{F}_0, \mathbb{P})$ has an atom, i.e. there exists $A \in \mathcal{F}_0$ with $\mathbb{P}(A) > 0$ such that $\mathcal{F}_0 \ni B \subset A$ implies $B = \emptyset$ or $B = A$, we may use $(\Omega \times \mathbb{R}, \mathcal{F}_t \times \mathcal{B}, \mathbb{P} \times \mu_0)$ to replace $(\Omega, \mathcal{F}_t, \mathbb{P})$, where $(\mathbb{R}, \mathcal{B}, \mu_0)$ is an atomless complete probability space such that $(\Omega \times \mathbb{R}, \mathcal{F}_t \times \mathcal{B}, \mathbb{P} \times \mu_0)$ is atomless and $W_t(\omega, x) := W_t(\omega)$ for $(\omega, x) \in \Omega \times \mathbb{R}$ is m-dimensional Brownian motion.

Definition 1.1.1.

(1) A stochastic process $(X_t)_{t \in [0,T]}$ on $\mathbb{R}^d$ is called a (strong) solution of (1.1.1), if it is continuous (i.e. $\mathbb{P}$-a.s. X_t is continuous in $t \in [0,T]$), adapted (i.e. X_t is $\mathcal{F}_t$-measurable for $t \in [0,T]$), and $\mathbb{P}$-a.s.
$$\int_0^t |b_s(X_s)| \mathrm{d}s + \int_0^t \|\sigma_s(X_s)\|^2 \mathrm{d}s < \infty,$$
$$X_t = X_0 + \int_0^t b_s(X_s) \mathrm{d}s + \int_0^t \sigma_s(X_s) \mathrm{d}W_s, \quad t \in [0,T].$$

(2) (1.1.1) is said to have pathwise (strong) uniqueness, if $X_t = Y_t (t \in [0,T])$ holds for any two solutions $(X_t)_{t \in [0,T]}$ and $(Y_t)_{t \in [0,T]}$ of the SDE with $X_0 = Y_0$.

(3) We call (1.1.1) (strongly) well-posed, if for any initial value it has a unique solution.

(4) A couple $(X_t, W_t)_{t \in [0,T]}$ is called a weak solution of (1.1.1), if there exists a complete filtration probability space such that W_t is an m-dimensional Brownian motion and X_t solves (1.1.1).

(5) (1.1.1) is said to have weak uniqueness (joint weak uniqueness), if for any two weak solutions $(X_t^i, W_t^i)_{i=1,2}$ under probabilities $(\mathbb{P}^i)_{i=1,2}$, $\mathcal{L}_{X_0^1|\mathbb{P}^1} = \mathcal{L}_{X_0^2|\mathbb{P}^2}$ implies $\mathcal{L}_{X_\cdot^1|\mathbb{P}^1} = \mathcal{L}_{X_\cdot^2|\mathbb{P}^2}$ ($\mathcal{L}_{(X^1,W^1)|\mathbb{P}^1} = \mathcal{L}_{(X^2,W^2)|\mathbb{P}^2}$). The SDE is called weakly well-posed, if for any initial distribution it has a unique weak solution.

By Yamada-Watanabe principle [Yamada and Watanabe (1971)] (see Lemma 1.3.2 below for a general version), the weak existence together with the pathwise uniqueness implies the strong and weak well-posedness. In particular, the strong well-posedness implies the weak one.

Let $\mathcal{P}$ be the space of all probability measures on $\mathbb{R}^d$ equipped with the weak topology. Let $(X_t)_{t \in [0,T]}$ be a (weak) solution of (1.1.1). By Itô's formula,

$$df(X_t) = L_t f(X_t) dt + \langle \nabla f(X_t), \sigma_t(X_t) dW_t \rangle, \quad t \in [0,T], f \in C^2(\mathbb{R}^d)$$

holds for

$$L_t := \frac{1}{2} \mathrm{tr}(\sigma_t \sigma_t^* \nabla^2) + \nabla_{b_t}, \quad t \in [0,T],$$

where $\nabla_{b_t} := b_t \cdot \nabla = \langle b_t, \nabla \rangle$ is the directional derivative along b_t, ∇ and ∇^2 are the gradient and Hessian operators on $\mathbb{R}^d$ respectively, and for any $k \in \mathbb{Z}_+$ (the set of nonnegative integers), $C^k(\mathbb{R}^d)$ denotes the space of real functions on $\mathbb{R}^d$ with continuous derivatives up to order k. When $k = 0$ we simply denote $C(\mathbb{R}^d) = C^0(\mathbb{R}^d)$. Then for any solution X_t to (1.1.1) with $\mu_t := \mathcal{L}_{X_t}$ satisfying

$$\int_0^t ds \int_{B(0,r)} (|b_s| + \|\sigma_s\|^2) d\mu_s < \infty, \quad r > 0, \qquad (1.1.2)$$

where $B(0,r) := \{x \in \mathbb{R}^d : |x| \leq r\}$, μ_t solves the linear Fokker-Planck equation

$$\partial_t \mu_t = L_t^* \mu_t, \quad t \in [0,T] \qquad (1.1.3)$$

in the sense that $\mu \in C([0,\infty); \mathcal{P})$ satisfying (1.1.2) and

$$\mu_t(f) := \int_{\mathbb{R}^d} f d\mu_t = \mu_0(f) + \int_0^t \mu_s(L_s f) ds, \quad t \in [0,T], f \in C_0^2(\mathbb{R}^d),$$

where for any $k \subset \mathbb{Z}_+ \cup \{\infty\}$, $C_0^k(\mathbb{R}^d)$ is the space of functions in $C^k(\mathbb{R}^d)$ with compact supports.

On the other hand, by the superposition principle presented in [Trevisan (2016)], a solution of (1.1.3) is the time-marginal of the distribution of a weak solution to (1.1.1). So, we have the following correspondence between SDEs and linear Fokker-Planck equations.

Theorem 1.1.1 ([Trevisan (2016)]). *For any weak solution (X_t, W_t) of (1.1.1) with $\mu_t := \mathcal{L}_{X_t}$ satisfying (1.1.2), μ_t solves (1.1.3). On the other hand, if μ is a solution of (1.1.3), then (1.1.1) has a weak solution such that $\mu_t = \mathcal{L}_{X_t}, t \in [0,T]$.*

1.2 Krylov's and Khasminskii's estimates

As a crucial tool in the study of singular SDEs, Krylov's estimate [Krylov (1980)] bounds the conditional expectation for time-integrals of (unbounded) singular functions of a solution to the SDEs. To understand this type estimate, let us simply consider the d-dimensional Brownian motion W_t (i.e. $m=d$). For any $p,q \geq 1, 0 \leq s < t$, and a measurable function f on $[s,t] \times \mathbb{R}^d$ with

$$\|f\|_{L_q^p(s,t)} := \left(\int_s^t \|f_r\|_{L^p(\mathbb{R}^d)}^q \mathrm{d}r\right)^{\frac{1}{q}} < \infty,$$

where $\|f_r\|_{L^p} := \left(\int_{\mathbb{R}^d} |f_r|^p(x)\mathrm{d}x\right)^{\frac{1}{p}}$, we have

$$\left|\mathbb{E}(f_r(W_r)|\mathcal{F}_s)\right| = \left|(2\pi(r-s))^{-\frac{d}{2}} \int_{\mathbb{R}^d} f_r(x) \mathrm{e}^{-\frac{|W_s-x|^2}{2(r-s)}} \mathrm{d}x\right|$$
$$\leq (2\pi(r-s))^{-\frac{d}{2p}} \|f_r\|_{L^p}, \quad r \in (s,t].$$

Consequently,

$$\left|\mathbb{E}\left(\int_s^t f_r(W_r)\mathrm{d}r \Big| \mathcal{F}_s\right)\right| \leq \int_s^t (2\pi(r-s))^{-\frac{d}{2p}} \|f_r\|_{L^p} \mathrm{d}r$$
$$\leq \left(\int_s^t (2\pi(r-s))^{-\frac{dq}{2p(q-1)}} \mathrm{d}r\right)^{\frac{q-1}{q}} \|f\|_{L_q^p(s,t)}.$$

When $\frac{d}{p} + \frac{2}{q} < 2$, we have $\frac{dq}{2p(q-1)} < 1$ so that

$$c_{p,q} := \sup_{0 \leq s \leq t \leq T} \left(\int_s^t (2\pi(r-s))^{-\frac{dq}{2p(q-1)}} \mathrm{d}r\right)^{\frac{q-1}{q}} < \infty.$$

Therefore,

$$\left|\mathbb{E}\left(\int_s^t f_r(W_r)\mathrm{d}r \Big| \mathcal{F}_s\right)\right| \leq c_{p,q} \|f\|_{L_q^p(s,t)}, \quad 0 \leq s \leq t \leq T.$$

Krylov's estimate extends this inequality to solutions of singular SDEs, which also implies an exponential estimate which is called Khasminskii's estimate. In the spirit of [Xia et al. (2020)], we will establish these estimates using the local norm $\tilde{L}_q^p$ replacing L_q^p, see (1.2.7) and (1.2.17) below for details.

For any $p \geq 1$, let $L^p(\mathbb{R}^d)$ be the class of measurable functions f on $\mathbb{R}^d$ such that $\|f\|_{L^p(\mathbb{R}^d)} < \infty$. For any $\epsilon > 0$ and $p \geq 1$, let

$$\|f\|_{H^{\epsilon,p}(\mathbb{R}^d)} := \|(1-\Delta)^{\frac{\epsilon}{2}}f\|_{L^p(\mathbb{R}^d)} < \infty, \quad f \in H^{\epsilon,p}(\mathbb{R}^d) := (1-\Delta)^{-\frac{\epsilon}{2}}L^p(\mathbb{R}^d),$$

where $(1-\Delta)^{-\frac{\epsilon}{2}}$ is defined by spectral representations of Δ. In general, letting
$$e^{t\Delta}f(x) := \int_{\mathbb{R}^d} (4\pi t)^{-\frac{d}{2}} e^{-\frac{|x-y|^2}{4t}} f(y) dy,$$
for any $\lambda \geq 0$ we have
$$\begin{aligned}(\lambda-\Delta)^{-\frac{\epsilon}{2}} &:= \frac{1}{\int_0^\infty t^{\frac{\epsilon}{2}-1} e^{-\lambda t} dt} \int_0^\infty t^{\frac{\epsilon}{2}-1} e^{-t} e^{t\Delta} dt, \\ (\lambda-\Delta)^{\frac{\epsilon}{2}} &:= (\lambda-\Delta)^n (\lambda-\Delta)^{\frac{\epsilon}{2}-n}, \quad n \in \mathbb{N}, n \geq \frac{\epsilon}{2}.\end{aligned} \quad (1.2.1)$$
For any $z \in \mathbb{R}^d$ and $r > 0$, let $B(z,r) := \{x \in \mathbb{R}^d : |x-z| \leq r\}$. For any $p, q > 1$ and $t_0 < t_1$, let $\tilde{L}_q^p(t_0, t_1)$ denote the class of measurable functions f on $[t_0, t_1] \times \mathbb{R}^d$ such that
$$\|f\|_{\tilde{L}_q^p(t_0,t_1)} := \sup_{z \in \mathbb{R}^d} \left(\int_{t_0}^{t_1} \|1_{B(z,1)} f_t\|_{L^p(\mathbb{R}^d)}^q dt \right)^{\frac{1}{q}} < \infty.$$

Let $g \in C_0^\infty(\mathbb{R}^d)$ satisfy $g|_{B(0,1)} = 1$, where $C_0^\infty(\mathbb{R}^d)$ is the class of C^∞ functions on $\mathbb{R}^d$ with compact support. For any $\epsilon > 0$, let $\tilde{H}_q^{\epsilon,p}(t_0, t_1)$ be the space of $f \in \tilde{L}_q^p(t_0, t_1)$ with
$$\|f\|_{\tilde{H}_q^{\epsilon,p}(t_0,t_1)} := \sup_{z \in \mathbb{R}^d} \left(\int_{t_0}^{t_1} \|g(z+\cdot) f_t\|_{H^{\epsilon,p}(\mathbb{R}^d)}^q dt \right)^{\frac{1}{q}} < \infty.$$
We remark that the space $\tilde{H}_q^{\epsilon,p}(t_0, t_1)$ does not depend on the choice of g. When $t_0 = 0$, we simply denote
$$\tilde{L}_q^p(t_1) := \tilde{L}_q^p(0, t_1), \quad \tilde{H}_q^{\epsilon,p}(t_1) := \tilde{H}_q^{\epsilon,p}(0, t_1), \quad t_1 > 0.$$

Finally, let $\tilde{L}^p$ (resp. $\tilde{H}^{\epsilon,p}$) denote the set of functions in $\tilde{L}_1^p(T)$ (resp. $\tilde{H}^{\epsilon,p}(T)$) which are independent of the time parameter t.

A vector or matrix valued function is said in one of the above introduced spaces, if so are its components.

We will take (p, q) from the class
$$\mathcal{K} := \left\{ (p,q) : p, q \in (1, \infty), \frac{d}{p} + \frac{2}{q} < 1 \right\},$$
and use the following assumptions on the coefficients b and σ, where $\|\cdot\|_\infty$ denotes the uniform norm for real (or vector/matrix) valued functions. Let ∇^i be the i-th order gradient in the spatial variable $x \in \mathbb{R}^d$, and for a Lipschitz continuous function f on $\mathbb{R}^d$,
$$\|\nabla f\|_\infty := \sup_{x \neq y} \frac{|f(x) - f(y)|}{|x-y|}$$
is its Lipschitz constant.

$(A^{1.1})$ Let $T > 0$. There exist a constant $K > 0$ and $(p_0, q_0) \in \mathcal{K}$ such that σ and b satisfy the following conditions on $[0, T] \times \mathbb{R}^d$.
(1) $a := \sigma\sigma^*$ is invertible with $\|a\|_\infty + \|a^{-1}\|_\infty \leq K$, where σ^* is the transposition of σ, and
$$\lim_{\varepsilon \to 0} \sup_{|x-y|\leq \varepsilon, t\in[0,T]} \|a_t(x) - a_t(y)\| = 0.$$
(2) $b = b^{(0)} + b^{(1)}$, $b^{(1)}$ is locally bounded and
$$\|b^{(0)}\|_{\tilde{L}^{p_0}_{q_0}(T)} \vee \|\nabla b^{(1)}\|_\infty \leq K. \tag{1.2.2}$$

To establish Krylov's estimate, we first introduce two lemmas, where the first is taken from [Chapter III, Theorem 2.4] in [Krylov (1985)], and the second follows from Theorem 2.1 in [Zhang and Yuan (2021)] which extends Theorem 3.2 in [Xia et al. (2020)] for $b^{(1)} = 0$, see also [Zhang (2011)] and references therein.

Lemma 1.2.1. For any $0 \leq f \in C^\infty([0,\infty) \times \mathbb{R}^d)$ and $\lambda > 0$, there exists $0 \leq u \in C^\infty([0,\infty) \times \mathbb{R}^d)$ satisfying
$$\beta \partial_t u_t + \mathrm{tr}\{a_t \nabla^2 u_t\} - \lambda(\beta + \mathrm{tr} a_t)u_t + (\beta \det a_t)^{\frac{1}{d+1}} f_t \leq 0, \quad \beta \geq 0, t \geq 0,$$
$$|\nabla u_t| \leq \sqrt{\lambda} u_t, \quad u_t \leq K_d \lambda^{-\frac{d}{2(d+1)}} \|f\|_{L^{d+1}([0,\infty)\times\mathbb{R}^d)},$$
for any measurable symmetric nonnegative definite matrix valued function a, and some constant $K_d > 0$ depending only on d.

Lemma 1.2.2. Assume $(A^{1.1})$ and let $p, q > 1$.
(1) For any $0 \leq t_0 < t_1 \leq T$ and $f \in \tilde{L}^p_q(t_0, t_1)$, the PDE
$$(\partial_t + L_t) u^\lambda_t = \lambda u^\lambda_t + f_t, \quad t \in [t_0, t_1], u_{t_1} = 0, \tag{1.2.3}$$
has a unique solution in $\tilde{H}^{2,p}_q(t_0, t_1)$ with $\|w u^\lambda\|_{\tilde{L}^p_q(t_0,t_1)} < \infty$, where $w_t(x) = w(x) = (1 + |x|)^{-1}$.
(2) If $(2p, 2q) \in \mathcal{K}$, then for any $\theta \in [0, 2), p' \in [p, \infty]$ and $q' \in [q, \infty]$ with
$$\frac{d}{p} + \frac{2}{q} < 2 - \theta + \frac{d}{p'} + \frac{2}{q'},$$
there exist constants $\lambda_0, c > 0$ depending only on $T, d, p, q, p', q', \theta, K$ and the continuity modulus of a, such that for any $0 \leq t_0 < t_1 \leq T$ and $f \in \tilde{L}^p_q(t_0, t_1)$,
$$\lambda^{\frac{1}{2}(2-\theta+\frac{d}{p'}+\frac{1}{q'}-\frac{d}{p}-\frac{2}{q})} \|u\|_{\tilde{H}^{\theta,p'}_{q'}(t_0,t_1)} + \|(\partial_t + \nabla_{b^{(1)}}) u^\lambda\|_{\tilde{L}^p_q(t_0,t_1)}$$
$$+ \|u^\lambda\|_{\tilde{H}^{2,p}_q(t_0,t_1)} \leq c \|f\|_{\tilde{L}^p_q(t_0,t_1)}, \quad \lambda \geq \lambda_0.$$

According to Theorem 2.1 in [Zhang and Yuan (2021)], (1.2.3) has a unique solution in $\tilde{W}_{1,q}^{2,p,w}(t_0,t_1)$, which consists of $f \in \tilde{H}_q^{2,p}(t_0,t_1)$ such that for $w(x) := (1+|x|)^{-1}$ (note that $w(x)$ therein should be $(1+|x|^2)^{-\frac{p}{2}}$ according to Lemma 2.3 in [Yang and Zhang (2023)]),

$$|wf| + |w\nabla f| + |w\nabla^2 f| + |w\partial_t f| \in \tilde{L}_q^p(t_0,t_1).$$

Since $w \leq 1$ and $f \in \tilde{H}_q^{2,p}(t_0,t_1)$, the condition $|wf| + |w\nabla f| + |w\nabla^2 f| \in \tilde{L}_q^p(t_0,t_1)$ can be dropped. When $(2p,2q) \in \mathcal{K}$, by Lemma 1.2.2(2) we have $\|(\partial_t + \nabla_{b^{(1)}})u^\lambda\|_{\tilde{L}_q^p(t_0,t_1)} < \infty$, since ∇u^λ is bounded and $(A^{1.1})$ implies $|b_t^{(1)}(x)| \leq k(1+|x|)$ for some constant $k > 0$, we derive $|w\partial_t f| \in \tilde{L}_q^p(t_0,t_1)$, so that in this case we have $\tilde{W}_{1,q}^{2,p,w}(t_0,t_1) = \tilde{H}_q^{2,p}(t_0,t_1)$.

We also need the following mollifying approximations of functions. Let $0 \leq \mathcal{S} \in C_0^\infty(\mathbb{R}^{d+1})$ with $\int_{\mathbb{R}^{d+1}} \mathcal{S}(s,x) \mathrm{d}s\mathrm{d}x = 1$. For a bounded measurable function g on $[t_0,t_1] \times \mathbb{R}^d$, its mollifying approximations $\{\mathcal{S}_n(g)\}_{n \geq 1} \subset C_b^\infty(\mathbb{R}^{d+1})$ are defined by

$$\{\mathcal{S}_n(g)\}_t(x) := n^{d+1} \int_{\mathbb{R}^{d+1}} g_{[(t+s) \vee t_0] \wedge t_1}(x+y) \mathcal{S}(ns,ny) \mathrm{d}s\mathrm{d}y. \quad (1.2.4)$$

Moreover, for a topological space E, let $C(E)$ (respectively $C_b(E)$) be the space of all continuous (respectively bounded continuous) functions on E.

Theorem 1.2.3 (Krylov's estimate). *Assume* $(A^{1.1})$. *Let* $(X_t)_{t \in [0,T]}$ *be a continuous adapted process on* $\mathbb{R}^d$ *satisfying*

$$X_t = X_0 + \int_0^t b_s(X_s)\mathrm{d}s + \int_0^t \sigma_s(X_s)\mathrm{d}W_s + \int_0^t \beta_s \mathrm{d}A_s, \quad t \in [0,T] \quad (1.2.5)$$

for some continuous adapted increasing process A_t and a bounded adapted process β_t.

(1) *For any $(p,q) \in \mathcal{K}$ and any $\varepsilon \in (0,1)$, there exists a constant $c > 0$ depending only on $T, d, p, q, \mathcal{K}$ and the continuity modulus of a, such that for any X_t satisfying (1.2.5),*

$$\mathbb{E}\left(\int_{t_0}^{t_1} |f_s(X_s)|\mathrm{d}s \bigg| \mathcal{F}_{t_0}\right) \leq \{c + \varepsilon \mathbb{E}(A_{t_1} - A_{t_0}|\mathcal{F}_{t_0})\}\|f\|_{\tilde{L}_q^p(t_0,t_1)} \quad (1.2.6)$$

holds for $0 \leq t_0 \leq t_1 \leq T$, and $f \in \tilde{L}_q^p(t_0,t_1)$.

(2) Let $\mathrm{d}A_t = B_t \mathrm{d}t$. For any $p, q > 1$ with $(2p, 2q) \in \mathcal{K}$ and any $\varepsilon \in (0,1)$, there exists a constant $c > 0$ such that for any X_t satisfying (1.2.5),

$$\mathbb{E}\bigg(\int_{t_0}^{t_1} |f_s(X_s)| \mathrm{d}s \bigg| \mathcal{F}_{t_0}\bigg) \\ \leq \bigg\{c + \varepsilon \bigg[\mathbb{E}\bigg(\int_{t_0}^{t_1} |B_s|^2 \mathrm{d}s \bigg| \mathcal{F}_{t_0}\bigg)\bigg]^{\frac{1}{2}}\bigg\} \|f\|_{\tilde{L}_q^p(t_0, t_1)} \qquad (1.2.7)$$

holds for all $0 \leq t_0 \leq t_1 \leq T$ and $f \in \tilde{L}_q^p(t_0, t_1)$.

(3) For any $u \in C([0, T] \times \mathbb{R}^d)$ with

$$\|u\|_\infty + \|\nabla u\|_\infty + \|\nabla^2 u\|_{\tilde{L}_q^p(T)} + \|(\partial_t + \nabla_{b^{(1)}})u\|_{\tilde{L}_q^p(T)} < \infty,$$

$\{u_t(X_t)\}_{t \in [0,T]}$ is a semimartingale satisfying

$$\mathrm{d}u_t(X_t) = (\partial_t + L_t)u_t(X_t)\mathrm{d}t + \langle \nabla u_t(X_t), \sigma_t(X_t)\mathrm{d}W_t + \mathrm{d}A_t\rangle. \quad (1.2.8)$$

Proof. By Jensen's inequality, for assertions (1) and (2) we only need to consider $f \geq 0$.

(a) By $(A^{1.1})$ and conditions on β_s and A_s, (1.2.5) implies that

$$[t_0, T] \ni t \mapsto |X_t| + \int_{t_0}^t |b_s(X_s)|\mathrm{d}s + A_t - A_{t_0}$$

is a continuous adapted process. For any $k > 0$, let

$$\tau_k := \inf\bigg\{t \in [t_0, T] : |X_t| + \int_{t_0}^t |b_s(X_s)|\mathrm{d}s + A_t - A_{t_0} \geq k\bigg\}.$$

Then $\tau_k \to T$ as $k \to \infty$. We claim that for some constant $c(k) > 0$

$$\mathbb{E}\bigg(\int_{t_0}^{t_1 \wedge \tau_k} f_s(X_s)\mathrm{d}s \bigg| \mathcal{F}_{t_0}\bigg) \leq c(k)\|f\|_{L^{d+1}_{d+1}(t_0, t_1)}, \qquad (1.2.9)$$

$f \in L^{d+1}([t_0, t_1] \times \mathbb{R}^d)$.

Let ν be the (random) finite measure on $[0, \infty) \times \mathbb{R}^d$ given by

$$\nu_k(A) := \mathbb{E}\bigg(\int_{t_0}^{t_1 \wedge \tau_k} 1_A(s, X_s)\mathrm{d}s \bigg| \mathcal{F}_{t_0}\bigg), \quad A \in \mathcal{B}([0,\infty) \times \mathbb{R}^d),$$

where $\mathcal{B}(E)$ is the Borel σ-algebra of a measurable space E. Since $C_0^\infty([0, \infty) \times \mathbb{R}^d)$ is dense in $L^{d+1}(\nu_k + \mathrm{d}s\mathrm{d}x)$, it suffices to prove (1.2.9) for $0 \leq f \in C_0^\infty([0, \infty) \times \mathbb{R}^d)$.

Let $u \geq 0$ be given in Lemma 1.2.1 for $a_t := \frac{1}{2}\sigma_t\sigma_t^*, \beta = \lambda = 1$. By Itô's formula, we find constants $c_1, c_2 > 0$ such that

$$\mathrm{d}u_t(X_t) = (\partial_t + L_t)u_t(X_t)\mathrm{d}t + \langle \beta_t, \nabla u_t(X_t)\rangle \mathrm{d}A_t + \mathrm{d}M_t \\ \leq \{(\nabla_{b_t} u_t)(X_t) + c_2 u_t(X_t)\}\mathrm{d}t + \nabla_{\beta_t} u_t(X_t)\mathrm{d}A_t - c_1 f_t(X_t)\mathrm{d}t + \mathrm{d}M_t$$

for some martingale M_t. Combining this with the definition of τ_k, the boundedness of β_t, and using estimates on $|u|$ and $|\nabla u|$ in Lemma 1.2.1, we derive (1.2.9) for some constant $c(k) > 0$.

(b) Proofs of (1.2.6) and (1.2.7). Let $\{x_n\}_{n\geq 1} \subset \mathbb{R}^d$ such that
$$\|f\|_{\tilde{L}_q^p(t_0,t_1)} \leq \|1_{B(x_n,1)}f\|_{L_q^p(t_0,t_1)} + \frac{1}{n}, \quad n \geq 1.$$
Let
$$\nu(A) := \mathbb{E}\left(\int_{t_0}^{t_1} 1_A(s, X_s)\mathrm{d}s \Big| \mathcal{F}_{t_0}\right), \quad A \in \mathcal{B}([0,\infty) \times \mathbb{R}^d).$$
Since $C_0^\infty([0,\infty) \times \mathbb{R}^d)$ is dense in $L^{p\vee q}(\nu + 1_{[t_0,t_1]\times B(x_n,1)}(s,x)\mathrm{d}s\mathrm{d}x)$ for each $n \geq 1$, it suffices to prove for $0 \leq f \in C_0^\infty([0,\infty) \times \mathbb{R}^d)$.

Let u solve (1.2.3). By Lemma 1.2.2 with $\theta = 1$ and $p', q' = \infty$, we find constants $c, \varepsilon_0 > 0$ such that
$$\lambda^{\varepsilon_0}\big(\|u\|_\infty + \|\nabla u\|_\infty\big) \leq c\|f\|_{\tilde{L}_q^p(t_0,t_1)}. \tag{1.2.10}$$
Since $\|f\|_\infty < \infty$, we have $|\nabla^2 u| + |(\partial_t + \nabla_{b^{(1)}})u| \in L_m^m(t_0,t_1)$ for any $m > 1$. To apply Itô's formula, we consider the mollifying approximation $u^{\lambda,n} := \mathcal{S}_n(u)$ of u, see (1.2.4). Then (1.2.10) and $|\nabla^2 u| + |(\partial_t + \nabla_{b^{(1)}})u| \in L_{d+1}^{d+1}(t_0,t_1)$ imply
$$\lambda^{\varepsilon_0}\big(\|u^{\lambda,n}\|_\infty + \|\nabla u^{\lambda,n}\|_\infty\big) \leq c\|f\|_{\tilde{L}_q^p(t_0,t_1)}, \tag{1.2.11}$$
$$\lim_{n\to\infty}\big\{\|u - u^{\lambda,n}\|_\infty + \|\nabla(u - u^{\lambda,n})\|_\infty\big\} \\ + \lim_{n\to\infty} \|(\partial_t + L_t - \nabla_{b^{(0)}})(u - u^{\lambda,n})\|_{\tilde{L}_{d+1}^{d+1}(t_0,t_1)} = 0. \tag{1.2.12}$$
Moreover, by Itô's formula and (1.2.3), we obtain
$$\mathrm{d}u_t^{\lambda,n}(X_t) = (\partial_t + L_t)u_t^{\lambda,n}(X_t)\mathrm{d}t + \{\nabla_{\beta_t}u_t^{\lambda,n}(X_t)\}\mathrm{d}A_t + \mathrm{d}M_t^{(n)}$$
$$= \big\{f_t + \lambda u_t^\lambda + (\partial_t + L_t - \nabla_{b_t^{(0)}})(u_t^{\lambda,n} - u_t^\lambda) + \nabla_{b_t^{(0)}}(u_t^{\lambda,n} - u_t^\lambda)\big\}(X_t)\mathrm{d}t$$
$$+ \{\nabla_{\beta_t}u_t^{\lambda,n}(X_t)\}\mathrm{d}A_t + \mathrm{d}M_t^{(n)}, \quad n \geq 1, t \in [t_0,t_1],$$
where $\mathrm{d}M_t^{(n)} := \langle\nabla u_t^{\lambda,n}(X_t), \sigma_t(X_t)\mathrm{d}W_t\rangle$ is a martingale. Let $h \in C_0^\infty(\mathbb{R}^d)$ with $h|_{B(0,1)} = 1$ and $0 \leq h \leq 1$, and take $h_m(x) := h(m^{-1}x)$ for $m \geq 1$. By the above formula for $\mathrm{d}u_t^{\lambda,n}(X_t)$ and applying $(A^{1.1})$, we find a constant $c_1 > 0$ such that up to a martingale,
$$\mathrm{d}(h_m u_t^{\lambda,n})(X_t) \geq (h_m f_t)(X_t)\mathrm{d}t$$
$$- \big\{h_m\big[|\lambda u_t^\lambda + (\partial_t + L_t - \nabla_{b_t^{(0)}})(u_t^{\lambda,n} - u_t^\lambda) + \nabla_{b_t^{(0)}}(u_t^{\lambda,n} - u_t^\lambda)|\big]\big\}(X_t)\mathrm{d}t$$
$$- c_1 m^{-1}\big\{(|u_t^{\lambda,n}| + |b_t^{(0)}|\cdot|\nabla u_t^{\lambda,n}|)(X_t) + |\nabla u_t^{\lambda,n}(X_t)|1_{\{|X_t|>m\}}\big\}\mathrm{d}t$$
$$- c_1\big\{|\nabla u_t^{\lambda,n}(X_t)| + c_1 m^{-1}|u_t^{\lambda,n}|(X_t)\big\}\mathrm{d}A_t.$$

Combining this with (1.2.9), (1.2.11), and using the definition of τ_k, we find constants $c_2, c(k), c(k,m) > 0$ such that for any $n, m, k \geq 1$,

$$\mathbb{E}\left(\int_{t_0}^{t_1 \wedge \tau_k} (h_m f_s)(X_s)\mathrm{d}s \bigg| \mathcal{F}_{t_0}\right)$$
$$\leq c_2\big\{1 + \lambda + \lambda^{-\varepsilon_0}\mathbb{E}(A_{t_1} - A_{t_0}|\mathcal{F}_{t_0}) + c(k)m^{-1}\big\}\|f\|_{\tilde{L}_q^p(t_0, t_1)}$$
$$+ c_2 \mathbb{E}\left(\int_{t_0}^{t_1 \wedge \tau_k} \big\{1_{\{|X_s|>m\}}\mathrm{d}s + m^{-1}\mathrm{d}A_s\big\}\bigg|\mathcal{F}_{t_0}\right)\|f\|_{\tilde{L}_q^p(t_0, t_1)}$$
$$+ c(k)\|\nabla(u^\lambda - u^{\lambda,n})\|_\infty + c(k,m)\|(\partial_t + L_t - \nabla_{b^{(0)}})(u^\lambda - u^{\lambda,n})\|_{\tilde{L}_{d+1}^{d+1}(t_0, t_1)}.$$

By (1.2.12), letting first $n \to \infty$, then $m \to \infty$ and finally $k \to \infty$, and taking large enough λ such that $c_2 \lambda^{-\varepsilon_0} \leq \varepsilon$, we derive (1.2.6).

In the situation of (2), we find a constant $c_3 > 0$ such that

$$\mathbb{E}\left(\int_{t_0}^{t_1} |\nabla_{\beta_s} u_s^{\lambda,n}(X_s)|\mathrm{d}A_s \bigg| \mathcal{F}_{t_0}\right)$$
$$\leq c_3 \mathbb{E}\left(\int_{t_0}^{t_1} |B_s| \cdot |\nabla u_s^{\lambda,n}(X_s)|\mathrm{d}s \bigg| \mathcal{F}_{t_0}\right)$$
$$\leq c_3 \left\{\mathbb{E}\left(\int_{t_0}^{t_1} |B_s|^2 \mathrm{d}s \bigg| \mathcal{F}_{t_0}\right)\right\}^{\frac{1}{2}} \left\{\mathbb{E}\left(\int_{t_0}^{t_1} |\nabla u_s^{\lambda,n}(X_s)|^2 \mathrm{d}s \bigg| \mathcal{F}_{t_0}\right)\right\}^{\frac{1}{2}}.$$

On the other hand, by Lemma 1.2.2 with $\theta = 0$ and $p', q' = \infty$, as well as $\theta = 1$ and $(p', q') = (2p, 2q)$, we find constants $c, \varepsilon_0 > 0$ such that

$$\lambda^{\varepsilon_0}\big\{\|u^\lambda\|_\infty + \|\nabla u^\lambda\|_{\tilde{L}_{2q}^{2p}(t_0, t_1)}\big\} \leq c\|f\|_{\tilde{L}_q^p(t_0, t_1)}. \tag{1.2.13}$$

Combining these with (1.2.6) and taking large enough λ, we obtain

$$\mathbb{E}\left(\int_{t_0}^{t_1} |\nabla_{\beta_s} u_s^{\lambda,n}(X_s)|\mathrm{d}A_s \bigg| \mathcal{F}_{t_0}\right) \leq \varepsilon \left[\mathbb{E}\left(\int_{t_0}^{t_1} |B_s|^2 \mathrm{d}s \bigg| \mathcal{F}_{t_0}\right)\right]^{\frac{1}{2}} \|f\|_{\tilde{L}_q^p(t_0, t_1)}.$$

Substituting into the above lower bound estimate on $\mathrm{d}(h_m u_t^{\lambda,n})(X_t)$ and applying (1.2.13), as in above we find a constant $c > 0$ such that

$$\mathbb{E}\left(\int_{t_0}^{t_1 \wedge \tau_k} (h_m f_s)(X_s)\mathrm{d}s \bigg| \mathcal{F}_{t_0}\right)$$
$$\leq \left\{c + \varepsilon\left[\mathbb{E}\left(\int_{t_0}^{t_1} |B_s|^2 \mathrm{d}s \bigg| \mathcal{F}_{t_0}\right)\right]^{\frac{1}{2}}\right\}\|f\|_{\tilde{L}_q^p(t_0, t_1)} + \xi_{k,m,n}$$

holds for random variables $\{\xi_{k,m,n}\}$ satisfying

$$\lim_{k \to \infty} \lim_{m \to \infty} \lim_{n \to \infty} \xi_{k,m,n} = 0.$$

Then (1.2.7) holds.

(c) Proof of (3). Let u satisfy

$$\|u\|_\infty + \|\nabla u\|_\infty + \|\nabla^2 u\|_{\tilde L_q^p(T)} + \|(\partial_t + \nabla_{b^{(1)}})u\|_{\tilde L_q^p(T)} < \infty.$$

Then the mollifying approximations $\{u^{(n)}\}_{n\geq 1}$ satisfy the same estimate and

$$\begin{aligned}\lim_{n\to\infty}\{\|u-u^{(n)}\|_\infty + \|\nabla(u-u^{(n)})\|_\infty\} \\ + \lim_{n\to\infty}\|(\partial_t + L_t)(u-u^{(n)})\|_{\tilde L_q^p(T)} = 0.\end{aligned} \quad (1.2.14)$$

By Itô's formula we have

$$\begin{aligned}u_t^{(n)}(X_t) = u_0(X_0) + \int_0^t (\partial_s + L_s)u_s^{(n)}(X_s)\mathrm{d}s + \int_0^t \langle \beta_s, \nabla u_s^{(n)}(X_s)\rangle \mathrm{d}A_s \\ + \int_0^t \langle \nabla u_s^{(n)}(X_s), \sigma_s(X_s)\mathrm{d}W_s\rangle, \quad t\in[0,T], n\geq 1.\end{aligned}$$

By (1.2.6) and (1.2.14), we may let $n\to\infty$ to derive (1.2.8). $\square$

Since Krylov's estimate is uniform in $0 \leq t_0 < t_1 \leq T$, we have the following exponential estimate (1.2.17) due to [Khasminskii (1959)].

Theorem 1.2.4 (Khasminskii's estimate). *Let $\tilde{\mathcal{K}}$ be a non-empty open subset of $\{(p,q): p,q > 1\}$. Let X_t be a continuous adapted process on $\mathbb{R}^d$ satisfying the Krylov estimate for some map $c: \tilde{\mathcal{K}} \to (0,\infty)$:*

$$\mathbb{E}\left(\int_{t_0}^{t_1} |f_s(X_s)|\mathrm{d}s \Big| \mathcal{F}_{t_0}\right) \leq c(p,q)\|f\|_{\tilde L_q^p(t_0,t_1)}, \quad (1.2.15)$$

$$0 \leq t_0 < t_1 < T, (p,q) \in \tilde{\mathcal{K}}, f \in \tilde L_q^p(t_0, t_1)$$

Then for any $(p,q) \in \tilde{\mathcal{K}}$ and $j \in \mathbb{N}$ (the set of natural numbers),

$$\mathbb{E}\left(\left[\int_{t_0}^{t_1} |f_s(X_s)|\mathrm{d}s\right]^j \Big| \mathcal{F}_{t_0}\right) \leq c(p,q)^j (j!)\|f\|_{\tilde L_q^p(t_0,t_1)}^j, \quad (1.2.16)$$

$$0 \leq t_0 \leq t_1 \leq T, (p,q) \in \tilde{\mathcal{K}}, f \in \tilde L_q^p(t_0, t_1),$$

and for any $(p,q) \in \tilde{\mathcal{K}}$, there exist constants $c', k \geq 1$ such that

$$\mathbb{E}\left(e^{\int_{t_0}^{t_1}|f_s(X_s)|\mathrm{d}s}\Big|\mathcal{F}_{t_0}\right) \leq e^{c'+c'\|f\|_{\tilde L_q^p(t_0,t_1)}^k}, \quad f \in \tilde L_q^p(t_0,t_1) \quad (1.2.17)$$

holds for all $0 \leq t_0 \leq t_1 \leq T$.

Proof. We first prove (1.2.16) by induction. By (1.2.15), (1.2.16) holds for $j = 1$. Assume that it holds for $m = k$ for some $k \in \mathbb{N}$, then

$$\mathbb{E}\left(\left[\int_{t_0}^{t_1} |f_s(X_s)| \mathrm{d}s\right]^{k+1} \bigg| \mathcal{F}_{t_0}\right)$$

$$= (k+1)\mathbb{E}\left\{\int_{t_0}^{t_1} |f_{s_1}(X_{s_1})| \mathrm{d}s_1 \mathbb{E}\left(\left[\int_{s_1}^{t_1} |f_s(X_s)| \mathrm{d}s\right]^{k} \bigg| \mathcal{F}_{s_1}\right) \bigg| \mathcal{F}_{t_0}\right\}$$

$$\leq c^k(k+1)!\mathbb{E}\left\{\int_{t_0}^{t_1} \|f\|_{\tilde{L}_q^p(s_1,t_1)}^k |f_{s_1}(X_{s_1})| \mathrm{d}s_1 \bigg| \mathcal{F}_{t_0}\right\}$$

$$\leq c^{k+1}(k+1)!\|f\|_{\tilde{L}_q^p(t_0,t_1)}^{k+1}.$$

Next, since $\tilde{\mathcal{K}}$ is open, there exists a map
$$k : \tilde{\mathcal{K}} \to (1,\infty], \quad k_{p,q} < q, \quad (p, k_{p,q}) \in \tilde{\mathcal{K}},$$
such that (1.2.15) with $q' := k_{p,q}$ replacing q yields

$$\mathbb{E}\left(\int_{t_0}^{t_1} |f_s(X_s)| \mathrm{d}s \bigg| \mathcal{F}_{t_0}\right) \leq c(p,q')\|f\|_{\tilde{L}_{q'}^p(t_0,t_1)} \leq c_1(t_1-t_0)^{\frac{q-q'}{qq'}} \|f\|_{\tilde{L}_q^p(t_0,t_1)}$$

for some constant $c_1 = c_1(p,q) > 0$. Let

$$n := \inf\left\{N \in \mathbb{N} : c_1(T/N)^{\frac{q-q'}{qq'}} \|f\|_{\tilde{L}_q^p(t_0,t_1)} \leq 2^{-1}\right\},$$

$$\delta_i := t_0 + \frac{i(t_1-t_0)}{n}, \quad 0 \leq i \leq n,$$

$$D_n^{i,j} := \{(s_1,\ldots,s_j) : \delta_i \leq s_1 < s_2 < \ldots < s_j \leq \delta_{i+1}\}, \quad j \geq 1.$$

We find constants $c_2 = c_2(p,q) > 0$ and $k := \frac{qq'}{q-q'} > 1$ such that

$$n \leq c_2 + c_2\|f\|_{\tilde{L}_q^p(t_0,t_1)}^k, \tag{1.2.18}$$

$$\mathbb{E}\left(\int_{\delta_i}^{\delta_{i+1}} f_s(X_s) \mathrm{d}s \bigg| \mathcal{F}_r\right) \leq \frac{1}{2}, \quad 0 \leq i \leq n-1. \tag{1.2.19}$$

By (1.2.19), for any $0 \leq i \leq n-1$ and $j \geq 1$, we have

$$\frac{1}{j!}\mathbb{E}\left[\left(\int_{\delta_i}^{\delta_{i+1}} f_s(X_s) \mathrm{d}s\right)^j \bigg| \mathcal{F}_{\delta_i}\right] = \mathbb{E}\left[\int_{D_n^{i,j}} \prod_{j=1}^j f_{s_j}(X_{s_j}) \mathrm{d}s_j \bigg| \mathcal{F}_{\delta_i}\right]$$

$$= \mathbb{E}\left[\int_{D_n^{i,j-1}} \mathbb{E}\left(\int_{s_{j-1}}^{\delta_{i+1}} f_{s_j}(X_{s_j}) \mathrm{d}s_j \bigg| \mathcal{F}_{s_{j-1}}\right) \prod_{j=1}^{j-1} f_{s_j}(X_{s_j}) \mathrm{d}s_j \bigg| \mathcal{F}_{\delta_i}\right]$$

$$\leq 2^{-1}\mathbb{E}\left[\int_{D_n^{i,j-1}} \prod_{j'=1}^{j-1} f_{s_{j'}}(X_{s_{j'}}) \mathrm{d}s_{j'} \bigg| \mathcal{F}_{\delta_i}\right]$$

$$= \frac{2^{-1}}{(j-1)!}\mathbb{E}\left[\left(\int_{\delta_i}^{\delta_{i+1}} f_s(X_s) \mathrm{d}s\right)^{j-1} \bigg| \mathcal{F}_{\delta_i}\right], \quad j \geq 1.$$

By induction we obtain
$$\frac{1}{j!}\mathbb{E}\left[\left(\int_{\delta_i}^{\delta_{i+1}} f_s(X_s)\mathrm{d}s\right)^j \Big| \mathcal{F}_{\delta_i}\right] \leq 2^{-j}, \quad j \geq 1,$$
so that
$$\mathbb{E}\left[e^{\int_{\delta_i}^{\delta_{i+1}} f_s(X_s)\mathrm{d}s} \Big| \mathcal{F}_{\delta_i}\right] \leq \sum_{j=0}^{\infty} 2^{-j} = 2, \quad 0 \leq i \leq n-1.$$
Combining with (1.2.18) we obtain
$$\mathbb{E}\left[e^{\int_{t_0}^{\delta_i} f_s(X_s)\mathrm{d}s} \Big| \mathcal{F}_{t_0}\right] = \mathbb{E}\left[e^{\int_{t_0}^{\delta_{i-1}} f_s(X_s)\mathrm{d}s} \cdot e^{\int_{\delta_{i-1}}^{\delta_i} f_s(X_s)\mathrm{d}s} \Big| \mathcal{F}_{t_0}\right]$$
$$= \mathbb{E}\left[e^{\int_{t_0}^{\delta_{i-1}} f_s(X_s)\mathrm{d}s} \mathbb{E}\left(e^{\int_{\delta_{i-1}}^{\delta_i} f_s(X_s)\mathrm{d}s} \Big| \mathcal{F}_{\delta_{i-1}}\right) \Big| \mathcal{F}_{t_0}\right]$$
$$\leq 2\,\mathbb{E}\left[e^{\int_{t_0}^{\delta_{i-1}} f_s(X_s)\mathrm{d}s} \Big| \mathcal{F}_{t_0}\right], \quad 1 \leq i \leq n.$$
By induction, $t_1 = \delta_n$ and the definition of n, we derive
$$\mathbb{E}\left[e^{\int_{t_0}^{t_1} f_s(X_s)\mathrm{d}s} \Big| \mathcal{F}_{t_0}\right] \leq 2^n \leq 2^{c_2 + c_2 \|f\|_{\tilde{L}_q^p(t_0,t_1)}^k}.$$
Taking $c' = c_2 \log 2$, we obtain
$$\mathbb{E}\left[e^{\int_{t_0}^{t_1} f_s(X_s)\mathrm{d}s} \Big| \mathcal{F}_{t_0}\right] \leq e^{c' + c' \|f\|_{\tilde{L}_q^p(t_0,t_1)}^k}, \quad 0 \leq t_0 < t_1 \leq T,$$
and hence finish the proof. □

1.3 Well-posedness

We first state the main result on the well-posedness of singular SDEs; then introduce some lemmas including a general version of the Yamada-Watanabe principle, the stochastic Gronwall's inequality, estimates on the maximal functionals, the BDG (Burkholder-Davis-Gundy) inequality and the Girsanov theorem; and finally prove the main result.

1.3.1 The main result

When (1.1.1) is well-posed, let X_t^x be the solution starting from x. We consider the associated semigroup P_t given by
$$P_t f(x) := \mathbb{E}[f(X_t^x)], \quad t \in [0,T], f \in \mathcal{B}_b(\mathbb{R}^d), x \in \mathbb{R}^d, \qquad (1.3.1)$$
where $\mathcal{B}_b(\mathbb{R}^d)$ is the set of all bounded measurable functions on $\mathbb{R}^d$. We call P_t strong Feller if $P_t \mathcal{B}_b(\mathbb{R}^d) \subset C_b(\mathbb{R}^d)$.

Let
$$\Phi := \left\{ \phi : [0, \infty) \to [1, \infty) \text{ is increasing}, \int_0^\infty \frac{\mathrm{d}s}{s + \phi(s)} < \infty \right\}.$$

$(A^{1.2})$ σ and $b = b^{(0)} + b^{(1)}$ satisfy the following conditions on $[0, T] \times \mathbb{R}^d$.
(1) σ satisfies $(A^{1.1})(1)$, and there exist $\{(p_i, q_i)\}_{0 \leq i \leq l} \subset \mathcal{K}$ with $p_i > 2$, and $0 \leq f_i \in \tilde{L}_{q_i}^{p_i}(T), 1 \leq i \leq l$, such that

$$|b^{(0)}| \in \tilde{L}_{q_0}^{p_0}(T), \quad \|\nabla \sigma\| \leq \sum_{i=1}^{l} f_i.$$

(2) $b^{(1)}$ is locally bounded, and there exist constants $K, \varepsilon > 0$, increasing $\phi \in C^1([0, \infty); [1, \infty))$ with $\int_1^\infty \frac{\mathrm{d}s}{s+\phi(s)} = \infty$, and $V \in C^2(\mathbb{R}^d; [1, \infty))$ with $\lim_{|x| \to \infty} V(x) = \infty$, such that

$$\langle b_t^{(1)}(x), \nabla V(x) \rangle + \varepsilon |b_t^{(1)}(x)| \sup_{B(x,\varepsilon)} \{|\nabla V| + \|\nabla^2 V\|\} \leq K\phi(V(x)),$$

$$\sup_{B(x,\varepsilon)} \{|\nabla V| + \|\nabla^2 V\|\} \leq KV(x), \quad x \in \mathbb{R}^d, t \in [0, T].$$

$(A^{1.3})$ $(A^{1.2})$ holds with $(A^{1.2})(2)$ strengthened as (1.2.2).

When $|b_t^{(1)}(x)| \leq c\{1 + |x|\log(1 + |x|)\}$ for some constant $c > 0$, then $(A^{1.2})(2)$ holds for $V(x) := 1 + |x|^2, \phi(r) := 1 + r\log(1 + r)$ and some constant $K > 0$. In particular, (1.2.2) is stronger than $(A^{1.2})(2)$.

For a signed measure φ, let $|\varphi|$ be its total variation, and define the variation norm as

$$\|\varphi\|_{var} := |\varphi|(\mathbb{R}^d).$$

In general, for a positive measurable function V on $\mathbb{R}^d$, let

$$\|\varphi\|_V := |\varphi(V)| = \int_{\mathbb{R}^d} V \mathrm{d}|\varphi|.$$

Theorem 1.3.1.

(1) Assume $(A^{1.1})$. Then (1.1.1) is weakly well-posed, and for any $k \in \mathbb{R}$ there exists a constant $c(k) > 0$ such that the (weak) solution X_t^x starting at x satisfies

$$\mathbb{E}\left[\sup_{t \in [0,T]} (1 + |X_t^x|)^k\right] \leq c(k)(1 + |x|)^k, \quad x \in \mathbb{R}^d. \tag{1.3.2}$$

(2) Assume $(A^{1.2})$. Then (1.1.1) is well-posed and

$$\lim_{\varepsilon\downarrow 0} \sup_{x,y\in B(0,k),|x-y|\leq\varepsilon} \mathbb{E}\left[\sup_{t\in[0,T]} |X_t^x - X_t^y|\wedge 1\right] = 0, \quad k\geq 1. \quad (1.3.3)$$

Moreover, for any $t \in (0,T]$,

$$\lim_{y\to x} \|P_t^*\delta_x - P_t^*\delta_y\|_{var} = 0, \quad t\in(0,T], x\in\mathbb{R}^d, \quad (1.3.4)$$

and P_t has a transition density $p_t(x,y)$ satisfying

$$\inf_{|x|\vee|y|\leq r} p_t(x,y) > 0, \quad t\in(0,T], r\in(0,\infty). \quad (1.3.5)$$

If $\phi(r) = r$, then for any $k \geq 1$ there exists a constant $c(k) > 0$ such that

$$\mathbb{E}\left[\sup_{t\in[0,T]} V(X_t^x)^k\right] \leq c(k)V(x)^k, \quad x\in\mathbb{R}^d. \quad (1.3.6)$$

(3) Assume $(A^{1.3})$. Then (1.1.1) is well-posed, and for any $k\in\mathbb{R}$ there exists a constant $c(k) > 0$ such that

$$\mathbb{E}\left[\sup_{t\in[0,T]} |X_t^x - X_t^y|^k\right] \leq c(k)|x-y|^k, \quad x\neq y\in\mathbb{R}^d. \quad (1.3.7)$$

Consequently, for any $p > 1$,

$$|\nabla P_t f|(x) := \limsup_{y\to x} \frac{|P_t f(x) - P_t f(y)|}{|x-y|} \leq c(p/(p-1))^{\frac{p-1}{p}} \left(P_t |\nabla f|^p(x)\right)^{\frac{1}{p}},$$
$$f\in C_b^1(\mathbb{R}^d), x\in\mathbb{R}^d, t\in[0,T]. \quad (1.3.8)$$

(4) Assume $(A^{1.3})$. Then for $\mathbb{P}$-a.e. $\omega\in\Omega$ and all $t\in[0,T]$, $x\mapsto X_t^x$ is a homeomorphism on $\mathbb{R}^d$.

Assertions in this result are taken from [Ren (2023)] and [Wang (2023c)], and will be proved by using Zvonkin's transform [Zvonkin (1974)] and the above introduced Krylov's and Khasminskii's estimates. The main idea of the proof goes back to [Veretennikov (1981)] where the well-posedness is proved for (1.1.1) with $d = m, \sigma = I_d$ (the $d\times d$ identity matrix) and bounded b, which is then improved in [Krylov and Röckner (2005)] for $\sigma = I_d$ and $|b| \in L_q^p(T) := L^q([0,T] \to L^p(\mathbb{R}^d); dt)$ for some $(p,q) \in \mathcal{K}$, in [Zhang (2011)] with $|\nabla\sigma| \in L_q^p(T)$ for some $(p,q)\in\mathcal{K}$, and in [Xia et al. (2020)] under some local integrability conditions. In recent years, this method has been applied to various different models. For readers' reference we summarize below some related studies.

Remark 1.3.1.

(1) Consider the critical case that $p, q \in [2, \infty)$ with $\frac{d}{p} + \frac{2}{q} = 1$. When $\sigma = I_d$, the existence and uniqueness of (1.1.1) for a.e. starting points have been proved in [Beck et al. (2019)] for $|b| \in L_q^p(T)$; the weak existence is proved in [Kinzebulatov and Semenov (2020)], and the well-posedness is proved in [Nam (2020)] for $|b|$ in the Lorentz space $L_{q,1}^p(T)$, i.e.
$$\int_0^\infty \left(\int_0^T 1_{\{\|b_t\|_{L^p(\mathbb{R}^d)} \geq r\}} dt \right)^{\frac{1}{q}} dr < \infty.$$
Moreover, the weak well-posedness has been proved in [Xia et al. (2020)] for the case that σ satisfies $(A^{1.1})(1)$ and $|b| \in \tilde{L}_\infty^{d;\text{uni}}(T)$ in the sense that $|b| \in \tilde{L}_\infty^d(T)$ and
$$\lim_{n \to \infty} \sup_{t \in [0,T]} \|\mathcal{S}_n(|b|)(t, \cdot) - |b_t|\|_{\tilde{L}^d(\mathbb{R}^d)} = 0,$$
where $\mathcal{S}_n(|b|)$ is the mollifying approximation of $|b|$, see (1.2.4).

(2) The well-posedness is proved in [Yang and Zhang (2020b)] for $\sigma = I_d$ and $|b|^2$ belonging to the Kato class $\mathbf{K}_{d,\alpha}$ for some $\alpha \in (0, 2)$, see Subsection 1.7.3 for details. Moreover, in the early paper [Yan (1988)] the weak existence is proved for $(\sigma_t, b_t) = (\sigma, b)$ independent of t, $\sigma\sigma^*$ is bounded, uniformly invertible and Hörlder continuous, and b satisfies
$$\sup_{x \in \mathbb{R}^d} \int_{B(x, \frac{1}{2})} \left\{ 1 + 1_{\{d=2\}} \log|x-y|^{-1} + 1_{\{d \geq 3\}} |x-y|^{2-d} \right\} |b(y)|^2 dy < \infty.$$

(3) Concerning singular degenerate SDEs, the stochastic Hamiltonian systems have been investigated in [Chaudru de Raynal (2017)], [Huang and Lv (2020)], [Wang and Zhang (2016)], [Zhang (2018)], the singular SDEs on Heisenberg groups are studied in [Huang and Wang (2018)], and the Gruchin type singular SDEs are considered in [Wang and Zhang (2018)]. See Subsections 1.7.1 and 1.7.2 for concrete results.

(4) Singular SDEs with jumps are studied in [Chen et al. (2021)] and [Xie and Zhang (2020)], the singular semilinear SPDEs are investigated in [Da Prato et al. (2013)], [Da Prato et al. (2015)] and [Wang (2016)], and the singular functional (i.e. path dependent) SPDEs are considered in [Huang and Wang (2017)] and [Lv and Huang (2021)].

1.3.2 Some lemmas

A. Yamada-Watanabe principle. This principle goes back to [Yamada and Watanabe (1971)] which says that the weak existence and pathwise uniqueness for an SDE imply the well-posedness. Furthermore, [Cherny (2003)] proved a stronger statement: the weak existence together with the pathwise uniqueness is equivalent to the strong existence together with joint weak uniqueness. In the following we state a general version of this principle due to [Kurtz (2014)].

Let S_1 and S_2 be two Polish spaces. For a measurable space E, let $\mathcal{P}(E)$ denote the set of all probability measures on E. Let $\nu \in \mathcal{P}(S_2)$ and Γ be a collection of constraints for random variables on $S_1 \times S_2$. We write $(X,Y) \in \Gamma$ if (X,Y) is a random variable on $S_1 \times S_2$ under a probability space satisfying all constraints in Γ. In the SDE set up, S_1 and S_2 stand for the path spaces of the solution and the noise respectively, and we denote $(X,Y) \in \Gamma$ if X solves the equation with noise Y.

Definition 1.3.1. Let $\nu \in \mathcal{P}(S_2)$ and Γ be a set of constraints for random variables on $S_1 \times S_2$.

(1) A weak solution for (Γ, ν) is a random variable (X,Y) on $S_1 \times S_2$ under a probability space such that $(X,Y) \in \Gamma$ and $\mathcal{L}_Y = \nu$. (Γ, ν) is said to have joint weak uniqueness if for any two weak solutions $(X^i, Y^i)_{i=1,2}$ under probabilities $(\mathbb{P}^i)_{i=1,2}$, we have $\mathcal{L}_{(X^1,Y^1)|\mathbb{P}^1} = \mathcal{L}_{(X^2,Y^2)|\mathbb{P}^2}$.
(2) A random variable (X,Y) on $S_1 \times S_2$ is called a strong solution for (Γ, ν) if it is a weak solution and there exists a measurable function $F : S_2 \to S_1$ such that $X = F(Y)$ a.s.
(3) (Γ, ν) is said to have pointwise (pathwise for stochastic processes) uniqueness, if for any random variables $\{X^1, X^2, Y\}$ under the same probability space such that $(X^1, Y), (X^2, Y) \in \Gamma$ and $\mathcal{L}_Y = \nu$, we have $X^1 = X^2$ a.s.

Lemma 1.3.2 (Yamada-Watanabe Principle [Kurtz (2014)]). Let $\nu \in \mathcal{P}(S_2)$ and Γ be a set of constraints for random variables on $S_1 \times S_2$. The following statements are equivalent:

(1) (Γ, ν) has a weak solution and the pointwise uniqueness.
(2) (Γ, ν) has a strong solution and the joint weak uniqueness.

B. Stochastic Gronwall inequality. This inequality was first found by [Scheutzow (2013)] for continuous martingales. The following version is taken from [Xie and Zhang (2020)].

Lemma 1.3.3 (Stochastic Gronwall Inequality). Let ξ_t and η_t be nonnegative progressively measurable processes under a complete filtration *probability space such that ξ_t is cádlág (i.e. right continuous with left limit), let A_t be a continuous adapted increasing process with $A_0 = 0$, and let M_t be a local martingale with $M_0 = 0$. If*

$$\xi_t \leq \xi_0 + \int_0^t \eta_s \mathrm{d}s + \int_0^t \xi_s \mathrm{d}A_s + M_t, \quad t \geq 0,$$

then for any $0 < q < p < 1$ and $t \geq 0$,

$$\left(\mathbb{E}\Big[\sup_{s\in[0,t]} \xi_s^q\Big]\right)^{\frac{1}{q}} \leq \left(\frac{p}{p-q}\right)^{\frac{1}{q}} \left(\mathbb{E}\big[e^{\frac{pA_t}{1-p}}\big]\right)^{\frac{1-p}{p}} \mathbb{E}\left(\xi_0 + \int_0^t \eta_s \mathrm{d}s\right).$$

C. Maximal functional. Consider the local Hardy-Littlewood maximal function for a nonnegative function f on $\mathbb{R}^d$:

$$\mathcal{M}f(x) := \sup_{r\in(0,1)} \frac{1}{|B(0,r)|} \int_{B(0,r)} f(x+y) \mathrm{d}y, \quad x \in \mathbb{R}^d.$$

The following result is taken from Lemma 2.1 in [Xia et al. (2020)], which goes back to [Stein (1970)].

Lemma 1.3.4 (Maximal function estimates). *There exists a constant $c > 0$ such that*

(1) *For any nonnegative function f on $\mathbb{R}^d$ with $|\nabla f| \in L^1_{loc}(\mathbb{R}^d)$,*

$$|f(x)-f(y)| \leq c|x-y|\big(\mathcal{M}|\nabla f|(x) + \mathcal{M}|\nabla f|(y) + \|f\|_\infty\big), \quad \text{a.e. } x, y \in \mathbb{R}^d.$$

(2) *For any nonnegative measurable function f on $[0,T] \times \mathbb{R}^d$,*

$$\|\mathcal{M}f\|_{\tilde{L}^p_q(T)} \leq c\|f\|_{\tilde{L}^p_q(T)}, \quad p, q \geq 1.$$

D. BDG (Burkholder-Davis-Gundy) inequality. This inequality goes back to [Burkholder and Gundy (1970)] and [Davis (1970)].

Lemma 1.3.5 (BDG inequality). *For any $p > 0$ there exist constants $C_p > c_p > 0$ such that for any continuous local martingale M_t with $M_0 = 0$, and any stopping time τ,*

$$c_p \mathbb{E}[\langle M \rangle_{t\wedge\tau}^p] \leq \mathbb{E}\Big[\sup_{s\in[0,t\wedge\tau]} |M_s|^{2p}\Big] \leq C_p \mathbb{E}[\langle M \rangle_{t\wedge\tau}^p], \quad t \geq 0,$$

where $\langle M \rangle_t$ is the quadratic variational of M_t, i.e. the unique continuous increasing process such that $M_t^2 - \langle M \rangle_t$ is a local martingale.

E. Girsanov theorem.
The following result is initiated in [Girsanov (1960)].

Lemma 1.3.6 (Girsanov theorem). *Let $(\xi_s)_{s\in[0,T]}$ be an adapted process on $\mathbb{R}^m$ such that $\mathbb{P}$-a.s. $\int_0^T |\xi_s|^2 ds < \infty$. If*

$$R_t := e^{-\int_0^t \langle \xi_s, dW_s\rangle - \frac{1}{2}\int_0^t |\xi_s|^2 ds}, \quad t \in [0,T]$$

is a martingale, then

$$\tilde{W}_t := \int_0^t \xi_s ds + W_t, \quad t \in [0,T]$$

is an m-dimensional Brownian motion under the probability $\mathbb{Q} := R_T\mathbb{P}$ (i.e. $d\mathbb{Q} = R_T d\mathbb{P}$).

The proof is straightforward by verifying that under $\mathbb{Q}$, $(\tilde{W}_t)_{t\in[0,T]}$ is a local martingale with $\langle\tilde{W}\rangle_t = t$. In general, $(R_t)_{t\in[0,T]}$ is only a local martingale, and it becomes a martingale if and only if the following uniform integrability condition hold:

$$\lim_{n\to\infty} \sup_{t\in[0,T]} \mathbb{E}[(R_t - n)^+] = 0.$$

It is in particular the case under Novikov's condition:

$$\mathbb{E}[e^{\frac{1}{2}\int_0^T |\xi_s|^2 ds}] < \infty.$$

1.3.3 Proof of Theorem 1.3.1

Proof of Theorem 1.3.1(1).

(a) The weak existence and (1.3.2). We first consider $b^{(0)} = 0$. In this case, $b = b^{(1)}$ is locally bounded and Lipschitz continuous in $x \in \mathbb{R}^d$ uniformly in $t \in [0,T]$. Since σ is bounded and $|b_t(x)|$ has linear growth in x uniformly in $t \in [0,T]$, it is easy to prove (1.3.2) by applying Itô's formula to $(1+|X_t|^2)^{\frac{k}{2}}$ and the BDG inequality in Lemma 1.3.5. Let

$$b^{(n)} = 1_{B(0,n)} b, \quad n \geq 1.$$

Since $b^{(n)}$ is bounded, (1.1.1) with $b^{(n)}$ replacing b is weakly well-posed, see for instance Theorem 1.4 in [Xia et al. (2020)]. Let $P^{(x,n)}$ be the distribution of the weak solution $(X_t^n, W_t^n)_{t\in[0,T]}$ starting from $(x,0) \in \mathbb{R}^d \times \mathbb{R}^m$, which is a probability measure on the path space $C([0,T];\mathbb{R}^d \times \mathbb{R}^m)$. By the boundedness of σ, the linear growth of $\sup_{t\in[0,T]} |b_t|$ and (1.3.2), we find a constant $K > 0$ such that for any $n \geq 1$,

$$\mathbb{E}\left[\sup_{0\leq s\leq t\leq T, |t-s|\leq \varepsilon} \{|X_t^n - X_s^n| + |W_t^n - W_s^n|\}\right] \leq K(1+|x|)\varepsilon^{\frac{1}{2}}, \quad \varepsilon \in (0,1),$$

so that by the Arzelá-Ascoli theorem, $\{P^{(x,n)}\}_{n\geq 1}$ is tight. By the continuity of the coefficients in the space variable, the weak limit for a convergent subsequence of $\{P^{(x,n)}\}_{n\geq 1}$ gives to the distribution of a weak solution $(X_t, W_t)_{t\in[0,T]}$ of (1.1.1) with $b^{(0)} = 0$; that is, W_t is an m-dimensional Brownian motion under a complete filtration probability space $(\Omega, \{\mathcal{F}_t\}_{t\geq 0}, \mathbb{P})$, and X_t solves the SDE

$$dX_t = b_t^{(1)}(X_t)dt + \sigma_t(X_t)dW_t, \quad X_0 = x, t \in [0,T]. \tag{1.3.9}$$

Next, when $b^{(0)} \neq 0$, we reformulate (1.3.9) as

$$dX_t = b_t(X_t)dt + \sigma_t(X_t)d\tilde{W}_t, \quad X_0 = x, t \in [0,T],$$

where $\tilde{W}_t := W_t - \int_0^t \{\sigma_s^*(\sigma_s\sigma_s^*)^{-1}b_s^{(0)}\}(X_s)ds$. By $(A^{1.1})$ and (1.2.17) for $(p,q) = (p_0/2, q_0/2)$ due to (1.2.7),

$$R_t := e^{\int_0^t \langle\{(\sigma_s\sigma_s^*)^{-1}b_s^{(0)}\}(X_s), \sigma_s(X_s)dW_s\rangle - \frac{1}{2}\int_0^t |\sigma_s^*(\sigma_s\sigma_s^*)^{-1}b_s^{(0)}|^2(X_s)ds}, \quad t \in [0,T]$$

is a martingale with $\mathbb{E}[|R_T|^2] \leq c_1$ for some constants $c_1 > 0$ independent of x. By Girsanov's theorem, $\tilde{W}_t$ is an m-dimensional Brownian motion under the probability $\mathbb{Q} := R_T\mathbb{P}$. Consequently, $(X_t, \tilde{W}_t)$ under $\mathbb{Q}$ is a weak solution of (1.1.1), and by (1.3.2) for the SDE (1.3.9) implied by $(A^{1.1})$, we find a constant $c_2 > 0$ such that

$$\mathbb{E}_{\mathbb{Q}}\Big[\sup_{t\in[0,T]}(1+|X_t|)^k\Big] \leq \Big(\mathbb{E}\Big[\sup_{t\in[0,T]}(1+|X_t|)^{2k}\Big]\Big)^{\frac{1}{2}}(\mathbb{E}R_T^2)^{\frac{1}{2}} \leq c_2(1+|x|)^k.$$

(b) To prove the weak uniqueness, let $(X_t^i, W_t^i)_{i=1,2}$ under probabilities $(\mathbb{P}^i)_{i=1,2}$ be two weak solutions of (1.1.1) starting from x, i.e.

$$dX_t^i = b_t(X_t^i)dt + \sigma_t(X_t^i)dW_t^i, \quad t \in [0,T], X_0^i = x. \tag{1.3.10}$$

It suffices to show

$$\mathcal{L}_{(X_t^1)_{t\in[0,T]}|\mathbb{P}^1} = \mathcal{L}_{(X_t^2)_{t\in[0,T]}|\mathbb{P}^2}. \tag{1.3.11}$$

To this end, let

$$\tau_n^i := \inf\{t \in [0,T] : |X_t^i| \geq N\}, \quad i = 1,2, \quad n \geq 1,$$

where we set $\inf \emptyset = T$ by convention. By $(A^{1.1})$, (1.2.17) and Girsanov's theorem, for each $i = 1, 2$,

$$R_{n,t}^i := e^{-\int_0^{t\wedge\tau_n^i}\langle\{(\sigma_s\sigma_s^*)^{-1}b_s\}(X_s^i), \sigma_s(X_s^i)dW_s^i\rangle - \frac{1}{2}\int_0^{t\wedge\tau_n^i}|\sigma_s^*(\sigma_s\sigma_s^*)^{-1}b_s|^2(X_s^i)ds}$$

for $t \in [0,T]$ is a $\mathbb{P}^i$-martingale, and

$$\tilde{W}_t^i := W_t^i + \int_0^{t\wedge\tau_n^i}\{\sigma_s^*(\sigma_s\sigma_s^*)^{-1}b_s\}(X_s^i)ds, \quad t \in [0,T]$$

is an m-dimensional Brownian motion under the probability $\mathbb{Q}_n^i := R_{n,T}^i \mathbb{P}^i$. Consequently, $(X_t^i, \tilde{W}_t^i)_{i=1,2}$ under $(\mathbb{Q}_n^i)_{i=1,2}$ are weak solutions of the reference SDE

$$dY_t = \sigma_t(Y_t)dW_t, \quad Y_0 = x, t \in [0, T], \tag{1.3.12}$$

which has weak uniqueness according to Theorem 1.4 in [Xia et al. (2020)]. Since (1.3.10) implies $\sigma_s(X_s)dW_s^i = dX_s^i - b_s(X_s^i)ds$, and $R_{n,T}^i = G_n(X^i), \tau_n^i = H_n(X^i)$ hold for some measurable functions G_n and H_n, the weak uniqueness of (1.3.12) implies that for any bounded measurable function F on $C([0,T]; \mathbb{R}^d)$,

$$\mathbb{E}_{\mathbb{P}^1}[1_{\{\tau_n^1 > T\}} F(X^1)] = \mathbb{E}_{\mathbb{Q}_n^1}[1_{\{H_n(X^1) > T\}} G_n(X^1)^{-1} F(X^1)]$$
$$= \mathbb{E}_{\mathbb{Q}_n^2}[1_{\{H_n(X^2) > T\}} G_n(X^2)^{-1} F(X^2)] = \mathbb{E}_{\mathbb{P}^2}[1_{\{\tau_n^2 > T\}} F(X^2)], \quad n \geq 1.$$

Since $\tau_n^i \to T$ as $n \to \infty$, by letting $n \to \infty$ we obtain (1.3.11) since F is an arbitrary bounded measurable function on $C([0,T]; \mathbb{R}^d)$. □

Proof of Theorem 1.3.1(2).

(a) For any $n \geq 1$, let

$$b^n := 1_{B(0,n)} b^{(1)} + b^{(0)}.$$

By Theorem 1.1 in [Xia et al. (2020)], for any $x \in \mathbb{R}^d$, the following SDE is well-posed:

$$dX_t^n = b^n(X_t^n)dt + \sigma(X_t^n)dW_t, \quad X_0^n = x,$$

and

$$\sup_{x \neq y} \mathbb{E}\left[\sup_{t \in [0,T]} \frac{|X_t^{x,n} - X_t^{y,n}|}{|x-y|}\right] < \infty, \tag{1.3.13}$$

where $X_t^{x,n}$ is the solution starting at x.

Let $\tau_n^x := \inf\{t \in [0,T] : |X_t^{x,n}| > n\}$. Then $X_t^{x,n}$ solves (1.1.1) up to time τ_n^x, and by the uniqueness we have

$$X_t^{x,n} = X_t^{x,m}, \quad t \leq \tau_n^x \wedge \tau_m^x, n, m \geq 1.$$

So, it suffices to prove that $\tau_n^x \to T$ as $n \to T$.

By Lemma 1.2.2 and $(A^{1.1})$, for any $\lambda \geq 0$, the PDE

$$(\partial_t + L_t - \nabla_{b_t^{(1)}})u_t = \lambda u_t - b_t^{(0)}, \quad t \in [0,T], u_T = 0 \tag{1.3.14}$$

has a unique solution $u \in \tilde{H}_{q_0}^{p_0}(T)$, and there exist constants $\lambda_0, c, \theta > 0$ such that

$$\lambda^\theta(\|u\|_\infty + \|\nabla u\|_\infty) + \|\partial_t u\|_{\tilde{L}_{q_0}^{p_0}(T)} + \|\nabla^2 u\|_{\tilde{L}_{q_0}^{p_0}(T)} \leq c, \quad \lambda \geq \lambda_0. \tag{1.3.15}$$

So, we may take $\lambda \geq \lambda_0$ such that
$$\|u\|_\infty + \|\nabla u\|_\infty \leq \varepsilon. \tag{1.3.16}$$
Let $\Theta_t(x) = x + u_t(x)$. By Itô's formula (1.2.8), $Y_t^n := \Theta_t(X_t^n)$ satisfies
$$\begin{aligned}\mathrm{d}Y_t^n &= \{1_{B(0,n)}b^{(1)} + \lambda u_t + 1_{B(0,n)}\nabla_{b^{(1)}}u_t\}(X_t^n)\mathrm{d}t \\ &\quad + \{(\nabla\Theta_t)\sigma\}(X_t^n)\mathrm{d}W_t.\end{aligned} \tag{1.3.17}$$
By (1.3.16) and $(A^{1.2})(2)$, there exist constants $c_0, c_1, c_1 > 0$ such that for some martingale M_t,
$$\begin{aligned}\mathrm{d}\{V(Y_t^n) + M_t\} &\leq \Big[\langle 1_{B(0,n)}\{b^{(1)} + \nabla_{b^{(1)}}u_t\}(X_t^n), \nabla V(Y_t^n)\rangle \\ &\quad + c_0(|\nabla V(Y_t^n)| + \|\nabla^2 V(Y_t^n)\|)\Big]\mathrm{d}t \\ &\leq \Big\{1_{B(0,n)}\big[\langle b^{(1)}, \nabla V\rangle + \varepsilon|b^{(1)}|\sup_{B(\cdot,\varepsilon)}(|\nabla V| + \|\nabla^2 V\|)\big](X_t^n) + c_0 KV(Y_t^n)\Big\}\mathrm{d}t \\ &\leq \{K\phi(V(X_t^n)) + c_0 KV(Y_t^n)\}\mathrm{d}t \\ &\leq K\{\phi((1+\varepsilon K)V(Y_t^n)) + c_0 V(Y_t^n)\}\mathrm{d}t, \quad t \leq \tau_n.\end{aligned}$$
Let $H(r) := \int_1^r \frac{\mathrm{d}s}{s+\phi((1+\varepsilon K)s)}$. Then $\int_1^\infty \frac{\mathrm{d}s}{s+\phi(s)} = \infty$ implies
$$H(\infty) := \lim_{r\to\infty} H(r) = \infty. \tag{1.3.18}$$
Since $\phi \in C^1([0,\infty); [1,\infty))$ is increasing, we have $H'' \leq 0$, so that by Itô's formula we obtain
$$\mathrm{d}H(V(Y_t^n)) \leq c_3 \mathrm{d}t + \mathrm{d}\tilde{M}_t, \quad t \in [0, \tau_n]$$
for some constant $c_3 > 0$ and some martingale $\tilde{M}_t$. Thus,
$$\mathbb{E}[(H \circ V)(Y_{t\wedge\tau_n}^n)] \leq V(x + u(x)) + c_3 t, \quad t \in [0, T], n \geq 1.$$
Since (1.3.16) and $|z| \geq n$ imply $|\Theta_t(z)| \geq |z| - |u(z)| \geq n - \varepsilon$, we derive
$$\mathbb{P}(\tau_n^x < t) \leq \frac{V(x + \Theta_0(x)) + c_3 t}{\inf_{|y|\geq n-\varepsilon} H(V(y))} =: \varepsilon_{t,n}(x), \quad t \in [0,T]. \tag{1.3.19}$$
By $\lim_{|x|\to\infty} H(V)(x) = \infty$, we have $\lim_{n\to\infty} \varepsilon_{t,n}(x) = 0$. Therefore, $\tau_n^x \to T$ when $n \to \infty$ as desired.

(b) Let X_t^x and X_t^y solve (1.1.1) with initial values x, y respectively. Then
$$X_t^{x,n} = X_t^x, \quad X_t^{y,n} = X_t^y, \quad t \in [0, T \wedge \tau_n^x \wedge \tau_n^y].$$

Combining this with (1.3.13) and (1.3.19), we obtain
$$\sup_{x,y\in B(0,k), |x-y|\le \varepsilon} \mathbb{E}\Big[\sup_{t\in[0,T]} |X_t^x - X_t^y| \wedge 1\Big]$$
$$\le \sup_{x,y\in B(0,k), |x-y|\le \varepsilon} \Big\{\mathbb{E}\Big[\sup_{t\in[0,T]} |X_t^{x,n} - X_t^{y,n}| \wedge 1\Big] + \mathbb{P}(\tau_n^x \wedge \tau_n^y < T)\Big\}$$
$$\le c(n)\varepsilon + \varepsilon_{T,n}(x) + \varepsilon_{T,n}(y), \quad n \ge k \ge 1.$$
By letting first $\varepsilon \downarrow 0$ then $n \to \infty$, we derive (1.3.3).

(c) Let P_t^n and $(P_t^n)^*$ be defined as P_t and P_t^* for X_t^n solving (1.1.1). By Theorem 1.4.2 below, (1.4.3) holds for P_t^n such that for some constant $c_n > 0$,
$$\|(P_t^n)^*\delta_x - (P_t^n)^*\delta_y\|_{var} \le \frac{c_n}{\sqrt{t}}|x-y|, \quad x,y \in \mathbb{R}^d, t \in (0,T].$$
Next, by (1.3.19) and $X_t = X_t^n$ for $t \le \tau_n$, we obtain
$$\sup_{|f|\le 1} |P_t f(x) - P_t^n f(x)| \le 2\mathbb{P}(\tau_n \le t) \le \varepsilon_{t,n}(x) \to 0 \text{ as } n \to \infty.$$
Then
$$\limsup_{y\to x} \|P_t^*\delta_x - P_t^*\delta_y\|_{var}$$
$$\le \limsup_{n\to\infty} \limsup_{y\to x} \sup_{|f|\le 1} \Big\{|P_t^n f(x) - P_t^n f(y)| + \varepsilon_{t,n}(x) + \varepsilon_{t,n}(y)\Big\}$$
$$= 0, \quad t \in (0,T].$$
So, (1.3.4) holds.

(d) By Theorem 6.2.7(ii)–(iii) in [Bogachev et al. (2015)], P_t has a heat kernel $p_t(x,y)$ continuous in y such that for some $c: (0,T] \times \mathbb{N} \to (0,\infty)$,
$$\inf_{|y|\le n} p_t(x,y) \ge c(t,n) \sup_{|y|\le n} p_{t/2}(x,\cdot) > 0, \quad t \in (0,T], x \in \mathbb{R}^d, n \ge 1.$$
Combining this with (1.3.4) implies (1.3.5).

(e) When $\phi(r) = r$, by $(A^{1.2})(2)$, (1.3.17) and Itô's formula, for any $k \ge 1$ we find a constant $c_1(k) > 0$ such that
$$\mathrm{d}\{V(Y_t^n)^k\} \le c_1(k)V(Y_t^n)^k \mathrm{d}t + \mathrm{d}M_t^k$$
holds for some martingale M_t^k with $\mathrm{d}\langle M^k\rangle_t \le \{c_1(k)V(Y_t^n)^k\}^2 \mathrm{d}t$. Combining this with $|X_t^n - Y_t^n| \le \|u\|_\infty \le \varepsilon$ due to (1.3.16) and $Y_t^n := X_t^n + u_t(X_t^n)$, BDG's inequality, (1.3.16) and $(A^{1.2})(2)$, we find constants $c_2(k), c_3(k), c_4(k) > 0$ such that
$$\mathbb{E}\Big[\sup_{t\in[0,T]} V(X_t^n)^k\Big] \le c_2(k)\mathbb{E}\Big[\sup_{t\in[0,T]} V(Y_t^n)^k\Big]$$
$$\le c_3(k)V(x+u_0(x))^k \le c_4(k)V(x)^k, \quad n \ge 1.$$
By Fatou's lemma with $n \to \infty$, we derive (1.3.6). □

Proof of Theorem 1.3.1(3). By Theorem 1.3.1(1) and the Yamada-Watanabe principle, it suffices to prove (1.3.7), which implies the pathwise uniqueness as well as (1.3.8).

To prove (1.3.7), we use Zvonkin's transform. By Lemma 1.2.2, there exists $\lambda_0 > 0$ such that for any $\lambda \geq \lambda_0$, the PDE

$$(\partial_t + L_t - \lambda)u_t = -b_t^{(0)}, \quad u_T = 0, \tag{1.3.20}$$

for $u_t : \mathbb{R}^d \to \mathbb{R}^d$ has a unique solution in $\tilde{H}_q^{2,p}(T)$, and there exist constants $\varepsilon, c > 0$ such that

$$\lambda^\varepsilon \big(\|u\|_\infty + \|\nabla u\|_\infty\big) + \|(\partial_t + \nabla_{b^{(1)}})u\|_{\tilde{L}_{q_0}^{p_0}(T)} + \|\nabla^2 u\|_{\tilde{L}_{q_0}^{p_0}(T)} \\ \leq c, \quad \lambda \geq \lambda_0. \tag{1.3.21}$$

Then for large enough $\lambda_0 > 0$, $\Theta_t := id + u_t$ satisfies

$$\frac{1}{2}|x-y|^2 \leq |\Theta_t(x) - \Theta_t(y)|^2 \leq 2|x-y|^2, \quad \lambda \geq \lambda_0, x, y \in \mathbb{R}^d. \tag{1.3.22}$$

For $(X_t^i)_{i=1,2}$ solving (1.1.1) starting at $(x^i)_{i=1,2}$ respectively, by (1.3.20) and Itô's formula in Theorem 1.2.3(3), we obtain

$$d\Theta_t(X_t^i) = \{b_t^{(1)} + \lambda u_t\}(X_t^i)dt + \{(\nabla \Theta_t)^* \sigma_t\}(X_t^i)dW_t, \quad t \in [0, T]. \tag{1.3.23}$$

So, by $(A^{1.1})$, (1.3.21), (1.3.22) and Lemma 1.3.4,

$$H_t := |\Theta_t(X_t^x) - \Theta_t(X_t^y)|^{2k}, \quad t \in [0, T]$$

satisfies

$$dH_t \leq A_t H_t dt + dM_t, \tag{1.3.24}$$

where M_t is a local martingale and for some constant $K > 0$

$$A_t := K\Big\{1 + \sum_{i=1}^{2} \mathcal{M}\Big(\sum_{j=1}^{l} |f_i(t,\cdot)|^2 + \|\nabla^2 u_t\|^2\Big)(X_t^i)\Big\}. \tag{1.3.25}$$

By (1.2.7) and (1.2.17) for $(p, q) = \frac{1}{2}(p_i, q_i), 0 \leq i \leq l$, and (1.3.21), we see that

$$\mathbb{E}\big[e^{\lambda A_T}\big] < \infty, \quad \lambda > 0.$$

So, applying the stochastic Gronwall inequality in Lemma 1.3.7 for $q = \frac{1}{2}$, $p = \frac{3}{4}$, we find a constant $c > 0$ such that

$$\Big(\mathbb{E}\Big[\sup_{t \in [0,T]} |X_t^1 - X_t^2|^{2k}\Big]\Big)^2 \leq \big(3\mathbb{E}e^{3A_T}\big)^{\frac{1}{3}}|x^1 - x^2|^{2k} \leq c|x^1 - x^2|^{2k}.$$

Therefore, (1.3.7) holds for some constant $c(k) > 0$, and for $p > 1$, (1.3.8) follows by noting that

$$|\nabla P_t f|(x) \le \limsup_{y \to x} \mathbb{E}\Big[|\nabla f(X_t^x)| \frac{|X_t^x - Y_t^y|}{|x-y|}\Big]$$
$$\le (P_t|\nabla f|^p)^{\frac{1}{p}}(x) \limsup_{y \to x} \frac{(\mathbb{E}[|X_t^x - Y_t^y|^{\frac{p}{p-1}}])^{\frac{p-1}{p}}}{|x-y|}$$
$$\le c(p/(p-1))^{\frac{p-1}{p}} (P_t|\nabla f|^p)^{\frac{1}{p}}(x).$$

Proof of Theorem 1.3.1(4). As in (1.3.23), we have

$$d\Theta_t(X_t^x) = \{b_t^{(1)} + \lambda u_t\}(X_t^x)dt + \{(\nabla \Theta_t)^* \sigma_t\}(X_t^x)dW_t, \quad X_0^x = x.$$

Since $|u| + \|(\nabla \Theta)^* \sigma\|$ is bounded and $\sup_{t \in [0,T]} |b_t^{(1)}(x)|$ has linear growth in $|x|$, for any $p \ge 1$ we find a constant $C_p > 0$ such that

$$\mathbb{E}\big[|\Theta_t(X_t^x) - \Theta_s(X_s^x)|^{2p}\big]$$
$$\le C_p(1+|x|)^{2p}|t-s|^p, \quad 0 \le s \le t \le T, x \in \mathbb{R}^d. \tag{1.3.26}$$

Next, by (1.3.2) and (1.3.22), we find a function $c : \mathbb{R} \to (0, \infty)$ such that

$$\sup_{t \in [0,T]} \mathbb{E}\big[|\Theta_t(X_t^x) - \Theta_t(X_t^y)|^{2p}\big] \le c(p)|x-y|^{2p}, \quad x,y \in \mathbb{R}^d, p \ge 1,$$

$$\sup_{t \in [0,T]} \mathbb{E}\big[(1+|\Theta_t(X_t^x)|)^k\big] \le c(k)(1+|x|)^k, \quad x \in \mathbb{R}^d, k \in \mathbb{R}.$$

By a standard argument with Kolmogorov's continuity theorem, this together with (1.6.14) yields that for $\mathbb{P}$-a.s. ω and all $t \in [0,T]$, $x \mapsto X_t^x(\omega)$ is a homeomorphism in $\mathbb{R}^d$, see the proof of Theorem 4.5.1 in [Kutani (1990)] or Theorem 3.4 in [Zhang (2011)]. □

1.4 Bismut formula

The following type of derivative formula

$$\nabla P_t^M f = \mathbb{E}\big[f(X_t)M_t\big], \quad t > 0, f \in \mathcal{B}_b(\mathbb{R}^d)$$

for some random variable M_t was first established by [Bismut (1984)] for the heat semigroup P_t^M on a Riemannian manifold M using Malliavin calculus, then by [Elworthy and Li (1994)] using martingale argument, and has been intensively developed for many different models including SDEs and SPDEs.

In the following we first recall the integration by parts formula in Malliavin calculus, then establish Bismut formula for P_t associated with (1.1.1).

1.4.1 Malliavin calculus

Malliavin calculus, also known as Stochastic Calculus of Variations, was developed by Malliavin [Malliavin (1978)] to study hypoelliptic operators using stochastic analysis. Roughly speaking, Malliavin calculus is analysis on the Wienner space for functionals of the Brownian motion. There are many articles and books on Malliavin calculus and applications, see for instance [Fang (2004)] and references therein.

For fixed $T > 0$, let Λ_T be the Wiener measure on the path space $\mathcal{C}_T := C([0,T]; \mathbb{R}^m)$, which is the distribution of the m-dimensional Brownian motion $(W_t)_{t \in [0,T]}$. To develop analysis on the Wiener space $(\mathcal{C}_T, \mathcal{B}(\mathcal{C}_T), \Lambda_T)$, we first define the directional derivative of a nice function $f \in L^2(\Lambda_T)$ along a direction $h \in \mathcal{C}_T$:

$$D_h f(\gamma) := \lim_{\varepsilon \downarrow 0} \frac{f(\gamma + \varepsilon h) - f(\gamma)}{\varepsilon}, \quad \gamma \in \mathcal{C}_T.$$

Noting that each $f \in L^2(\Lambda_T)$ is an equivalent class, i.e. $f = g$ in $L^2(\Lambda_T)$ if $f = g$ Λ_T-a.s., to ensure that $D_h f$ is well-defined in $L^2(\Lambda_T)$, one needs to show that the limit does not depend on the choice of f from its equivalent class, i.e. $f = g$ Λ_T-a.s. should imply $f(\cdot + \varepsilon h) = g(\cdot + \varepsilon h)$ Λ_T-a.s.. This property is called the quasi-invariance of Λ_T under shift by εh. According to the Carmon-Martin theorem, this property holds if and only if h belongs to the Carmon-Martin space

$$\mathbf{H} := \left\{ h \in \mathcal{C}_T : \|h\|_{\mathbf{H}} := \left(\int_0^T |\dot{h}_t|^2 \mathrm{d}t \right)^{\frac{1}{2}} < \infty \right\},$$

where $\dot{h}_t := \frac{\mathrm{d}}{\mathrm{d}t} h_t$ is the derivative of h_t in the weak (i.e. integration by parts) sense. Observe that $\mathbf{H}$ is a separable Hilbert space under the inner product

$$\langle h, \phi \rangle_{\mathbf{H}} := \int_0^T \langle \dot{h}_t, \dot{\phi}_t \rangle \mathrm{d}t, \quad h, \phi \in \mathbf{H}.$$

Definition 1.4.1. (1) A function $f \in L^2(\Lambda_T)$ is called Malliavin differentiable, denoted by $f \in D^{1,2}$, if

$$\mathbf{H} \ni h \mapsto D_h f := \lim_{\varepsilon \downarrow 0} \frac{f(\cdot + \varepsilon h) - f}{\varepsilon} \in L^2(\Lambda_T)$$

is a well-defined bounded linear functional. In this case, there exists a unique bounded linear operator Df from $\mathbf{H}$ to $L^2(\mathcal{C}_T \to \mathbf{H}, \Lambda_T)$ such that

$$\langle Df, h \rangle_{\mathbf{H}} = D_h f.$$

We call Df the Malliavin derivative (or gradient) of f.

Singular Stochastic Differential Equations 27

(2) A measurable map $h : \mathcal{C}_T \to \mathbf{H}$ is called a vector field. We denote $h \in \mathcal{D}(D^*)$ if there exists a unique $D^*h \in L^2(\Lambda_T)$ such that
$$\int_{\mathcal{C}_T} (D_h f) \mathrm{d}\Lambda_T = \int_{\mathcal{C}_T} (fD^*h) \mathrm{d}\Lambda_T, \quad f \in D^{1,2}.$$
In this case, D^*h is called the Malliavin divergence of h.

Theorem 1.4.1 (Integration by parts formula). *Let $W_{[0,T]} : \Omega \to \mathcal{C}_T$ be the m-dimensional Brownian motion. Then*
$$\mathbb{E}\big[(D_h f)(W_{[0,T]})\big] = \mathbb{E}\big[(fD^*h)(W_{[0,T]})\big], \quad f \in D^{1,2}, h \in \mathcal{D}(D^*).$$
In particular, if $h_t(W_{[0,T]})$ is adapted in the natural filtration of W_t, then
$$(D^*h)(W_{[0,T]}) = \int_0^T \langle \dot{h}_t(W_{[0,T]}), \mathrm{d}W_t \rangle.$$

In applications we may simply take $(\Omega, \mathbb{P}) = (\mathcal{C}_T, \Lambda_T)$ such that the Brownian motion becomes the coordinate process $W_t(\gamma) = \gamma_t, \gamma \in \Omega = \mathcal{C}_T$. In this case, a vector field h coincides with $h(W_{[0,T]})$, hence the composition of $W_{[0,T]}$ can be dropped from Theorem 1.4.1.

1.4.2 The main result

The following result is taken from [Wang (2023d)], which establishes such a formula for (1.1.1) under assumption $(A^{1.3})$. See Theorem 1.1(ii) in [Xia et al. (2020)] for the case with $b^{(1)} = 0$, $\beta_s = \frac{s}{t}$ and $f \in C_b^1(\mathbb{R}^d)$.

Theorem 1.4.2. *Assume $(A^{1.3})$, and let P_t be given in (1.3.1) for X_t^x solving (1.1.1) with $X_0 = x$. Then for any $v, x \in \mathbb{R}^d$,*
$$\nabla_v X_s^x := \lim_{\varepsilon \downarrow 0} \frac{X_s^{x+\varepsilon v} - X_s^x}{\varepsilon}, \quad s \in [0,T]$$
exists in $L^j(\Omega \to C([0,T]; \mathbb{R}^d), \mathbb{P})$ for any $j \geq 1$, and
$$\sup_{x \in \mathbb{R}^d} \mathbb{E}\Big[\sup_{t \in [0,T]} |\nabla_v X_t^x|^j\Big] \leq c(j)|v|^j, \quad x, v \in \mathbb{R}^d \qquad (1.4.1)$$
holds for some constant $c(j) > 0$. Moreover, for any $\beta \in C^1([0,t])$ with $\beta_0 = 0$ and $\beta_t = 1$, and any $f \in \mathcal{B}_b^1(\mathbb{R}^d)$,
$$\nabla_v P_t f(x) = \mathbb{E}\Big[f(X_t^x) \int_0^t \beta_s' \langle \{\sigma_s^*(\sigma_s\sigma_s^*)^{-1}\}(X_s^x) \nabla_v X_s^x, \mathrm{d}W_s \rangle\Big]. \qquad (1.4.2)$$
Consequently, for any $p > 1$ there exists a constant $c(p) > 0$ such that
$$|\nabla P_t f| \leq \frac{c(p)}{\sqrt{t}} \big(P_t |f|^p\big)^{\frac{1}{p}}, \quad t > 0, f \in \mathcal{B}_b(\mathbb{R}^d). \qquad (1.4.3)$$

Proof.
(a) Let $\lambda > 0$ be large enough such that the unique solution of (1.3.20) satisfies (1.3.21) and
$$\|u\|_\infty + \|\nabla u\|_\infty \le \frac{1}{2}.$$
Let $\Theta_t := id + u_t$ and
$$\tilde{b} := \{\lambda u + b^{(1)}\} \circ \Theta^{-1}, \quad \tilde{\sigma} := \left(\{I_d + \nabla u\}\sigma\right) \circ \Theta^{-1}.$$
Then $(A^{1.3})$ implies
$$\|\nabla \tilde{b}\|_\infty + \|\tilde{\sigma}\|_\infty + \|\nabla \tilde{\sigma}\|_{\tilde{L}_{q_0}^{p_0}(T)} + \|(\tilde{\sigma}\tilde{\sigma}^*)^{-1}\|_\infty < \infty. \tag{1.4.4}$$
By (1.3.20), (1.3.21) and Theorem 1.2.3(iii), $Y_t^x := \Theta_t(X_t^x)$ solves
$$\mathrm{d}Y_t^x = \tilde{b}_t(Y_t^x)\mathrm{d}t + \tilde{\sigma}_t(Y_t^x)\mathrm{d}W_t, \quad Y_0^x = \Theta_0(x). \tag{1.4.5}$$
By (1.4.4), (1.2.7) and (1.2.17), the increasing process
$$A_t := \int_0^t (\|\nabla \tilde{b}_s\| + \|\nabla \tilde{\sigma}_s\|^2)(Y_s^x)\mathrm{d}s, \quad t \in [0,T]$$
satisfies
$$\mathbb{E}[e^{\alpha A_T}] < \infty, \quad \alpha > 0.$$
So, for any $v, x \in \mathbb{R}^d$, the linear SDE
$$\mathrm{d}v_t = (\nabla_{v_t}\tilde{b}_t)(Y_t^x) + (\nabla_{v_t}\tilde{\sigma}_t)(Y_t^x)\mathrm{d}W_t, \quad v_0 = v + \nabla_v u_0(x) \tag{1.4.6}$$
has a unique solution, and by Itô's formula and the stochastic Gronwall inequality, for any $j \ge 1$ there exists a constant $c(j) > 0$ such that
$$\sup_{x \in \mathbb{R}^d} \mathbb{E}\left[\sup_{t \in [0,T]} |v_t|^j\right] \le c(j)|v|^j, \quad j \ge 1. \tag{1.4.7}$$

(b) Proof of assertion (1). Let $Y_t^{x+\varepsilon v} := \Theta_t(X_t^{x+\varepsilon v})$. By (1.4.7), for the first assertion it suffices to prove
$$\lim_{\varepsilon \to 0} \mathbb{E}\left[\sup_{t \in [0,T]} \left|\frac{Y_t^{x+\varepsilon v} - Y_t^x}{\varepsilon} - v_t^x\right|^j\right] = 0, \quad j \ge 1. \tag{1.4.8}$$
Indeed, by an approximation argument indicated in Remark 1.4.1 below, see also Remark 2.1 in [Zhang and Yuan (2021)], we may assume that $\nabla^2 b_t^{(1)}$ is bounded so that by Lemma 2.3(3) in [Zhang and Yuan (2021)],
$$|\nabla \Theta_t(x) - \nabla \Theta_t(y)| \le c|x-y|^\alpha, \quad t \in [0,T], x, y \in \mathbb{R}^d \tag{1.4.9}$$

holds for some constants $c > 0$ and $\alpha \in (0,1)$. Since $X_t^x = \Theta_t^{-1}(Y_t^x)$, (1.4.8) implies that $\nabla_v X_t^x$ exists in $L^j(\Omega \to C([0,T]; \mathbb{R}^d), \mathbb{P})$ with

$$\nabla_v X_t^x = (\nabla \Theta_t(X_t^x))^{-1} \nabla_v Y_t^x = (\nabla \Theta_t(X_t^x))^{-1} v_t^x, \quad t \in [0,T].$$

To prove (1.4.8), let

$$v_s^\varepsilon := \frac{Y_s^{x+\varepsilon v} - Y_s^x}{\varepsilon}, \quad s \in [0,T], \varepsilon \in (0,1].$$

By (1.4.4), (1.2.17), Lemma 1.3.4, and the stochastic Gronwall inequality in Lemma 1.3.3, we have

$$\sup_{\varepsilon \in (0,1]} \mathbb{E}\left[\sup_{t \in [0,T]} |v_t^\varepsilon|^j \right] < \infty, \quad j \geq 1. \tag{1.4.10}$$

Write

$$v_r^\varepsilon = \int_0^r (\nabla_{v_s^\varepsilon} \tilde{b}_s)(Y_s^x) ds + \int_0^r (\nabla_{v_s^\varepsilon} \tilde{\sigma}_s)(Y_s^x) dW_s + \alpha_r^\varepsilon, \quad r \in [0,t], \tag{1.4.11}$$

where

$$\alpha_r^\varepsilon := \int_0^r \xi_s^\varepsilon ds + \int_0^t \eta_s^\varepsilon dW_s \tag{1.4.12}$$

for

$$\xi_s^\varepsilon := \frac{\tilde{b}_s(Y_s^{x+\varepsilon v}) - \tilde{b}_s(Y_s^x)}{\varepsilon} - (\nabla_{v_s^\varepsilon} \tilde{b}_s)(Y_s^x),$$

$$\eta_s^\varepsilon := \frac{\tilde{\sigma}_s(Y_s^{x+\varepsilon v}) - \tilde{\sigma}_s(Y_s^x)}{\varepsilon} - (\nabla_{v_s^\varepsilon} \tilde{\sigma}_s)(Y_s^x).$$

We aim to prove

$$\lim_{\varepsilon \to 0} \mathbb{E}\left[\sup_{t \in [0,T]} |\alpha_t^\varepsilon|^n \right] = 0, \quad n \geq 1. \tag{1.4.13}$$

Firstly, since $\nabla \tilde{b}_s$ and $\nabla \tilde{\sigma}_s$ exists a.e., for a.e. $x \in \mathbb{R}^d$ we have

$$\lim_{\varepsilon \downarrow 0} \sup_{|v| \leq 1} \left\{ \left| \frac{\tilde{b}_s(x+\varepsilon v) - \tilde{b}_s(x)}{\varepsilon} - \nabla_v \tilde{b}_s(x) \right| \right. $$
$$\left. + \left\| \frac{\tilde{\sigma}_s(x+\varepsilon v) - \tilde{\sigma}_s(x)}{\varepsilon} - \nabla_v \tilde{\sigma}_s(x) \right\| \right\} = 0.$$

Combining this with (1.4.10) and noting that $\mathcal{L}_{Y_s^x}(s \in (0,T])$ is absolutely continuous with respect to the Lebesgue measure, see for instance Theorem 6.3.1 in [Bogachev et al. (2015)], we obtain

$$\lim_{\varepsilon \to 0} \{|\xi_s^\varepsilon| + \|\eta_s^\varepsilon\|\} = 0, \quad \mathbb{P}\text{-a.s.}, \quad s \in (0,T]. \tag{1.4.14}$$

Next, let $\theta > 1$ such that $(\theta^{-1}p_i, \theta^{-1}q_i) \in \mathcal{K}, 0 \le i \le l$. By $f_i \in \tilde{L}_{q_i}^{p_i}(T)$, Lemma 1.3.4 and (1.2.17) with $f = f_i^\theta$ and $(p,q) = (\theta^{-1}p_i, \theta^{-1}q_i)$, we obtain

$$\sup_{\varepsilon \in [0,1]} \mathbb{E}\left[\left(\int_0^T (\mathcal{M}f_i^{2\theta})(X_t^{x+\varepsilon v})\mathrm{d}t \Big| \mathcal{F}_0\right)^n\right] \le K_n, \quad 0 \le i \le l \quad (1.4.15)$$

for some constant $K_n > 0$. By $(A^{1.3})$ and Lemma 1.3.4, there exists a constant $c_1 > 0$ such that

$$|\xi_s^\varepsilon|^{2\theta} + \|\eta_s^\varepsilon\|^{2\theta} \le c_1 |\tilde{v}_t^\varepsilon|^2 \Big(1 + \sum_{i=0}^l \{(\mathcal{M}f_i^{2\theta}(s,\cdot))(X_s^x) + (\mathcal{M}f_i^{2\theta}(s,\cdot))(X_s^{x+\varepsilon v})\}\Big).$$

Combining this with (1.4.10) and (1.4.15), for any $n \ge 1$ we find constants $c_1(n), c_2(n) > 0$ such that

$$\mathbb{E}\left[\left(\int_0^T \{|\xi_s^\varepsilon|^{2\theta} + \|\eta_s^\varepsilon\|^{2\theta}\}\mathrm{d}s\right)^n\right]$$
$$\le c_1(n)\mathbb{E}\left[\left(\sup_{s\in[0,T]} |v_s^\varepsilon|^{2n}\right)\right.$$
$$\times \left(\int_0^T \Big\{1 + \sum_{i=0}^l (\mathcal{M}f_i^{2\theta}(s,\cdot))(X_s^x) + (\mathcal{M}f_i^{2\theta}(s,\cdot))(X_s^{x+\varepsilon v})\Big\}\mathrm{d}s\right)^n\right]$$
$$\le c_1(n)\left(\mathbb{E}\left[\left(\sup_{s\in[0,T]} |v_s^\varepsilon|^{4n}\right)\right]\right)^{\frac{1}{2}}$$
$$\times \left(\mathbb{E}\left[\left(\int_0^T \Big\{1 + \sum_{i=0}^l (\mathcal{M}f_i^{2\theta}(s,\cdot))(X_s^x) + (\mathcal{M}f_i^{2\theta}(s,\cdot))(X_s^{x+\varepsilon v})\Big\}\mathrm{d}s\right)^{2n}\right]\right)^{\frac{1}{2}}$$
$$\le c_2(n) < \infty, \quad \varepsilon \in (0,1].$$

Thus, by (1.4.14) and the dominated convergence theorem, we derive

$$\lim_{\varepsilon \to 0} \mathbb{E}\left[\left(\int_0^T \{|\xi_s^\varepsilon|^2 + \|\eta_s^\varepsilon\|^2\}\mathrm{d}s\right)^n\right] = 0, \quad n \ge 1.$$

Therefore, (3.7.17) and BDG's inequality in Lemma 1.3.5 imply (1.4.13).

Finally, by (1.4.4), (1.4.6), (1.4.11), and Lemma 1.3.4, for any $j \ge 1$, we find a constant $c(j) > 0$ such that

$$\mathrm{d}|v_s - v_s^\varepsilon|^{2j} \le c(j)\Big\{1 + \sum_{i=0}^l f_i^2(s, Y_s^x)\Big\}|v_s - v_s^\varepsilon|^{2j}\mathrm{d}s$$
$$+ c(j) \sup_{r \in [0,s]} |\alpha_r^\varepsilon|^{2j} + \mathrm{d}M_s, \quad s \in [0,t]$$

holds for some local martingale M_s. Since $\lim_{\varepsilon \to 0} |v_0 - v_0^\varepsilon| = 0$, by combining this with (1.2.17), (1.4.13), and the stochastic Gronwall inequality in Lemma 1.3.3, we derive (1.4.8).

(c) Proof of (1.4.2) for $f \in C_{Lip}(\mathbb{R}^d)$, the space of Lipschitz continuous functions on $\mathbb{R}^d$. Let $t \in (0,T]$ be fixed, and consider

$$h_s := \int_0^s \beta_r' [\tilde{\sigma}_r^* \{\sigma_r \sigma_r^*\}^{-1}](Y_r^x) v_r \mathrm{d}r, \quad s \in [0,t]. \tag{1.4.16}$$

By the same reason leading to (1.4.7), the SDE

$$\mathrm{d}w_s = \{\nabla_{w_s} \tilde{b}_s(Y_s^x) + \tilde{\sigma}_s(Y_s^x) h_s'\} \mathrm{d}s + (\nabla_{w_s} \tilde{\sigma}_s)(Y_s^x) \mathrm{d}W_s,$$
$$w_0 = 0, s \in [0,t] \tag{1.4.17}$$

has a unique solution satisfying

$$\sup_{x \in \mathbb{R}^d} \mathbb{E}\left[\sup_{t \in [0,T]} |w_s|^j\right] < \infty, \quad j \geq 1. \tag{1.4.18}$$

We aim to prove that the Malliavin derivative $D_h Y_t^x$ of Y_t^x along h exists and

$$D_h Y_t^x = w_t. \tag{1.4.19}$$

By Theorem 1.3.1, for any $\varepsilon > 0$ the following SDE is well-posed:

$$\mathrm{d}Y_s^{x,\varepsilon} = \{\tilde{b}_t(Y_s^{x,\varepsilon}) + \varepsilon \tilde{\sigma}_s(Y_s^{x,\varepsilon}) h_s'\} \mathrm{d}s + \tilde{\sigma}_s(Y_s^{x,\varepsilon}) \mathrm{d}W_s,$$
$$s \in [0,t], Y_0^{x,\varepsilon} = Y_0^x. \tag{1.4.20}$$

By (1.4.4), (1.4.16), Lemma 1.3.4 and Itô's formula, for any $j \geq 1$ we find a constant $c_1(j) > 0$ such that

$$\mathrm{d}|Y_s^{x,\varepsilon} - Y_s^x|^{2j} \leq c_1(j) |Y_s^{x,\varepsilon} - Y_s^x|^{2j} \sum_{i=0}^{l} \{1 + \mathcal{M}\{f_i(s,\cdot)\}^2(Y_s^x)$$
$$+ \{\mathcal{M}f_i(s,\cdot)\}^2(Y_s^{x,\varepsilon})\} \mathrm{d}s + c_1(j) \varepsilon^{2j} |v_s|^{2j} \mathrm{d}s + \mathrm{d}M_s, \quad s \in [0,t]$$

holds for some local martingale M_s. Noting that $Y_0^{x,\varepsilon} - Y_0^x = 0$, by combining this with Lemma 1.3.3 and Lemma 1.3.4, we obtain

$$\sup_{\varepsilon \in (0,1]} \mathbb{E}\left[\sup_{t \in [0,T]} \frac{|Y_s^{x,\varepsilon} - Y_s^x|^j}{\varepsilon^j}\right] < \infty, \quad j \geq 1. \tag{1.4.21}$$

Let $w_s^\varepsilon = \frac{Y_s^{x,\varepsilon} - Y_s^x}{\varepsilon}$. Then

$$w_r^\varepsilon = \int_0^r \{(\nabla_{w_s^\varepsilon} \tilde{b}_s)(Y_s^x) + \tilde{\sigma}_s(Y_s^x) h_s'\} \mathrm{d}s$$
$$+ \int_0^r (\nabla_{w_s^\varepsilon} \tilde{\sigma}_s)(Y_s^x) \mathrm{d}W_s + \tilde{\alpha}_r^\varepsilon, \quad r \in [0,t] \tag{1.4.22}$$

holds for
$$\tilde{\alpha}_r^\varepsilon := \int_0^r \left\{ \frac{\tilde{b}_s(Y_s^{x,\varepsilon}) - \tilde{b}_s(Y_s^x)}{\varepsilon} - (\nabla_{w_s^\varepsilon} \tilde{b}_s)(Y_s^x) \right\} ds$$
$$+ \int_0^r \left\{ \sigma_s(Y_s^{x,\varepsilon}) - \tilde{\sigma}_s(Y_s^x) \right\} h_s' \, ds$$
$$+ \int_0^r \left\{ \frac{\tilde{\sigma}_s(Y_s^{x,\varepsilon}) - \tilde{\sigma}_s(Y_s^x)}{\varepsilon} - (\nabla_{w_s^\varepsilon} \tilde{\sigma}_s)(Y_s^x) \right\} dW_s.$$

Combining this with (1.4.17) and using the same argument leading to (1.4.8), we derive (1.4.19).

By (1.4.16) and the SDE (1.4.6) for v_s, we see that $\beta_s v_s$ solves (1.4.17), so that by the uniqueness we obtain
$$\nabla_v Y_t^x = v_t = w_t = D_h Y_t^x.$$
For $f \in C_{Lip}(\mathbb{R}^d)$, ∇f exists a.e. and $\|\nabla f\|_\infty < \infty$. Since $\mathcal{L}_{X_t^x}$ is absolutely continuous, see for instance Theorem 6.3.1 in [Bogachev et al. (2015)], we conclude that $(\nabla f)(X_t^x)$ is a bounded random variable. By Theorem 1.4.1, $\nabla_v Y_t^x = D_h Y_t^x$ implies
$$\nabla_v P_t f(x) = \nabla_v \mathbb{E}[\{f \circ (\Theta_t)^{-1}\}(Y_t^x)] = \mathbb{E}\big[\langle \nabla (f \circ \Theta_t^{-1})(Y_t^x), \nabla_v Y_t^x \rangle\big]$$
$$= \mathbb{E}\big[D_h\{(f \circ \Theta_t^{-1})(Y_t^x)\}\big] = \mathbb{E}\left[f(X_t^x) \int_0^t \langle h_s', dW_s \rangle\right]$$
$$= \mathbb{E}\left[f(X_t^x) \int_0^t \beta_s' \langle \{\tilde{\sigma}_s^*(\tilde{\sigma}_s\tilde{\sigma}_s^*)^{-1}\}(Y_s^x) v_s, dW_s \rangle\right], \quad f \in C_{Lip}(\mathbb{R}^d).$$
By $v_t = \nabla Y_t^x$, $Y_t^x = \Theta_t(X_t^x)$ and $\tilde{\sigma}_t = \{(\nabla \Theta_t)\sigma_t\} \circ \Theta_t^{-1}$, we obtain
$$\{\tilde{\sigma}_s^*(\tilde{\sigma}_s\tilde{\sigma}_s^*)^{-1}\}(Y_s^x) v_s$$
$$= \left[\sigma_s^*(\sigma_s\sigma_s^*)^{-1}\left\{(\nabla\Theta_s)\sigma_s\sigma_s^*(\nabla\Theta_s)^*\right\}^{-1}\right](X_s^x)\{\nabla\Theta_s(X_s^x)\}\nabla X_s^x$$
$$= \{\sigma_s^*(\sigma_s\sigma_s^*)^{-1}\}(X_s^x) \nabla X_s^x, \quad s \in [0,T],$$
so that the previous formula implies
$$\nabla_v P_t f(x) = \mathbb{E}\left[f(X_t^x) \int_0^t \beta_s' \langle \{\sigma_s^*(\sigma_s\sigma_s^*)^{-1}\}(X_s^x) \nabla X_s^x, dW_s \rangle\right], \quad (1.4.23)$$
$$f \in C_{Lip}(\mathbb{R}^d).$$

(d) Proof of (1.4.3) and (1.4.2). Let $P_t^* \delta_x = \mathcal{L}_{X_t^x}$ and let ν_ε be the finite signed measure defined by
$$\nu_\varepsilon(A)$$
$$:= \int_0^\varepsilon \mathbb{E}\left[1_A(X_t^{x+rv}) \int_0^t \beta_s' \langle \{\sigma_s^*(\sigma_s(\sigma_s)^*)^{-1}\}(X_s^{x+rv}) \nabla X_s^{x+rv}, dW_s \rangle\right] dr$$

for $A \in \mathcal{B}(\mathbb{R}^d)$, the Borel σ-algebra on $\mathbb{R}^d$. Then (1.4.23) implies
$$(P_t^* \delta_{x+\varepsilon v} - P_t^* \delta_x)(f) = \nu_\varepsilon(f), \quad f \in C_{Lip}(\mathbb{R}^d),$$
where $\nu(f) := \int f d\nu$ for a (signed) measure ν and $f \in L^1(|\nu|)$. Since $C_{Lip}(\mathbb{R}^d)$ determines measures, we obtain
$$P_t^* \delta_{x+\varepsilon v} - P_t^* \delta_x = \nu_\varepsilon,$$
so that for any $f \in \mathcal{B}_b(\mathbb{R}^d)$,
$$P_t f(x+\varepsilon v) - P_t f(x)$$
$$= \int_0^\varepsilon \mathbb{E}\left[f(X_t^{x+rv}) \int_0^t \beta_s' \langle \{\sigma_s^*(\sigma_s(\sigma_s)^*)^{-1}\}(X_s^{x+rv})\nabla X_s^{x+rv} x, dW_s\rangle\right] dr.$$
Combining this with (1.4.1) and the boundedness of $\sigma^*(\sigma\sigma^*)^{-1}$, we derive (1.4.3).

Next, let $f \in \mathcal{B}_b(\mathbb{R}^d)$. For any $r \in (0, T)$, let $(X_{r,t}^x)_{t \in [r,T]}$ solve (1.1.1) from time r with $X_{r,r}^x = x$. Let
$$P_{r,t} f(x) := \mathbb{E}[f(X_{r,t}^x)], \quad f \subset \mathcal{B}_b(\mathbb{R}^d), x \in \mathbb{R}^d. \quad (1.4.24)$$
Then the well-posedness implies
$$P_t = P_r P_{r,t}, \quad 0 < r < t \leq T.$$
Moreover, considering the SDE from time r replacing 0, (1.4.3) implies
$$\|\nabla P_{r,t} f\|_\infty < \infty, \quad f \in \mathcal{B}_b(\mathbb{R}^d), 0 < r < t \leq T.$$
So, by (1.4.23) for $(P_r, \beta_s/\beta_r)$ replacing (P_t, β_s), we obtain
$$\nabla_v P_t f(x) = \nabla_v P_r(P_{r,t} f)(x)$$
$$= \frac{1}{\beta_r} \mathbb{E}\left[P_{r,t} f(X_r^x) \int_0^r \beta_s' \langle \{\sigma_s^*(\sigma_s(\sigma_s)^*)^{-1}\}(X_s^x)\nabla X_s^x, dW_s\rangle\right]$$
for all $f \in \mathcal{B}_b(\mathbb{R}^d)$ and $r \in (0, t)$ such that $\beta_r > 0$. Since the Markov property implies
$$\mathbb{E}[f(X_t^x)|\mathcal{F}_r] = P_{r,t} f(X_r^x),$$
we obtain
$$\nabla_v P_t f(x) = \frac{1}{\beta_r} \mathbb{E}\left[f(X_t^x) \int_0^r \beta_s' \langle \{\sigma_s^*(\sigma_s(\sigma_s)^*)^{-1}\}(X_s^x)\nabla X_s^x, dW_s\rangle\right],$$
so that letting $r \uparrow t$ gives (1.4.2). □

To conclude this section, we make the following remark which will be used to study distribution dependent SDEs in Chapter 3.

Remark 1.4.1. For fixed σ but may be variable b, the constants $c(\cdot)$ in Theorem 1.4.2 are uniformly in $b = b^{(0)} + b^{(1)}$ satisfying

$$\|b^{(0)}\|_{\tilde{L}_{q_0}^{p_0}} + \|\nabla b^{(1)}\|_\infty \leq N \qquad (1.4.25)$$

for a given constant $N > 0$. Indeed, letting γ be the standard Gaussian measure and taking

$$\tilde{b}_t^{(1)}(x) := \int_{\mathbb{R}^d} b_t^{(1)}(x+y)\gamma(\mathrm{d}y), \quad x \in \mathbb{R}^d, t \in [0,T],$$

we find constant $c > 0$ only depending on N such that (1.4.25) implies

$$\|\nabla \tilde{b}^{(1)}\|_\infty + \|\nabla^2 \tilde{b}^{(1)}\|_\infty + \|b_t^{(1)} - \tilde{b}_t^{(1)}\|_\infty \leq c.$$

Then $\tilde{b}^{(0)} := b^{(0)} + \tilde{b}^{(1)} - b^{(1)}$ satisfies

$$\|\tilde{b}^{(0)}\|_{\tilde{L}_{q_0}^{p_0}} \leq \|b^{(0)}\|_{\tilde{L}_{q_0}^{p_0}} + c\|1\|_{\tilde{L}_{q_0}^{p_0}} =: c'.$$

According to the proofs of Theorem 2.1 in [Zhang and Yuan (2021)] as well as Theorems 1.2.3 and 1.2.4 for $b = \tilde{b}^{(0)} + \tilde{b}^{(1)}$, the constant $\lambda_0 > 0$ before (1.3.20), upper bounds on $\|u\|_\infty + \|\nabla u\|_\infty \|\nabla^2 u\|_{\tilde{L}_{q_0}^{p_0}}$, and the constants in Krylov's and Khasminskii's estimates (1.2.7) and (1.2.17), are uniformly in b satisfying (1.4.25).

1.5 Dimension-free Harnack inequality

Let P be a Markov operator on $\mathcal{B}_b(\mathbb{R}^d)$, i.e. P is a bounded linear operator on $\mathcal{B}_b(\mathbb{R}^d)$ with $P1 = 1$ and $Pf \geq 0$ for $f \geq 0$. We consider the following type of Harnack inequality:

$$\Phi(Pf(x)) \leq (P\Phi(f)(y))e^{\Psi(x,y)}, \quad x, y \in E, f \in \mathcal{B}_b^+(\mathbb{R}^d), \qquad (1.5.1)$$

where Φ is a nonnegative convex function on $[0,\infty)$, Ψ is a nonnegative function on $\mathbb{R}^d \times \mathbb{R}^d$, and $\mathcal{B}_b^+(\mathbb{R}^d)$ is the set of bounded positive measurable functions on $\mathbb{R}^d$.

This type of inequality was first found in [Wang (1997)] for diffusion semigroups on Riemannian manifolds where $\Phi(r) := r^p$ for $p > 1$, $\Psi(x,y) = c\rho(x,y)^2$ for some constant $c > 0$ and the Riemannian distance ρ (we call it power Hanarck inequality), and was extended in [Wang (2010)] to $\Phi(r) = e^r$ for which (1.5.1) reduces to the log-Harnack inequality

$$P \log f(x) \leq \log Pf(y) + c\rho(x,y)^2.$$

For any $\mu \in \mathcal{P}$, the space of probability measures on $\mathbb{R}^d$, let $P^*\mu \in \mathcal{P}$ be defined as
$$(P^*\mu)(A) := \mu(P1_A) = \int_{\mathbb{R}^d} P1_A \mathrm{d}\mu, \quad A \in \mathcal{B}(\mathbb{R}^d).$$
Then the above log-Harnack inequality is equivalent to
$$\mathrm{Ent}(P^*\mu_1|P^*\mu_2) \leq c\mathbb{W}_2(\mu_1,\mu_2)^2, \quad \mu_1,\mu_2 \in \mathcal{P},$$
where
$$\mathrm{Ent}(\mu_1|\mu_2) := \sup_{f>0, \mu_2(f) \leq 1} \mu_1(\log f) = \begin{cases} \mu_2(f \log f), & \text{if } \mu_1 = f\mu_2, \\ \infty, & \text{otherwise} \end{cases}$$
is the relative entropy, and for any $p \geq 1$,
$$\mathbb{W}_p(\mu_1,\mu_2) := \inf_{\pi \in \mathcal{C}(\mu_1,\mu_2)} \left(\int_{\mathbb{R}^d \times \mathbb{R}^d} |x-y|^p \pi(\mathrm{d}x,\mathrm{d}y) \right)^{\frac{1}{p}}$$
for $\mathcal{C}(\mu_1,\mu_2)$ being the set of couplings for μ_1 and μ_2.

Comparing with classical Harnack inequalities, a crucial feature of (1.5.1) is dimension-free so that it applies to infinite-dimensional models. Due to this essential difference, in references this type of inequality is called Wang's Harnack inequality. The dimension-free Harnack inequality has been developed and applied to many different models of Markov processes and SDEs/SPDEs, see [Wang (2013)] for a general theory on dimension-free Harnack inequality and applications.

In this section, we establish the log-Harnack inequality and the power Harnack inequality for the singular SDE (1.1.1).

1.5.1 *Log-Harnack inequality*

The following result was presented in [Zhang and Yuan (2021)] using an approximation argument due to [Xia et al. (2020)] and the log-Harnack inequality proved in [Li et al. (2015)]. Below we give a simple proof using the idea of [Röckner and Wang (2010)].

Theorem 1.5.1. *Assume* $(A^{1.3})$. *Then there exists a constant $c > 0$ such that for any $f \in \mathcal{B}_b^+(\mathbb{R}^d)$, the class of nonnegative bounded measurable functions on $\mathbb{R}^d$,*
$$P_t \log f(x) \leq \log P_t f(y) + \frac{c|x-y|^2}{t}, \quad t \in (0,T], x,y \in \mathbb{R}^d. \quad (1.5.2)$$
Equivalently,
$$\mathrm{Ent}(P_t^*\mu_1|P_t^*\mu_2) \leq \frac{c\mathbb{W}_2(\mu_1,\mu_2)^2}{t}, \quad t \in (0,T], \mu_1,\mu_2 \in \mathcal{P}. \quad (1.5.3)$$

Proof. Let $P_{r,t}$ be given in (1.4.24). For any $f \in C_c^\infty(\mathbb{R}^d) := \mathbb{R} + C_0^\infty(\mathbb{R}^d)$, by Itô's formula we have
$$P_{s,t}f(x) = f(x) + \int_s^t P_{s,r}(L_r f)(x)\mathrm{d}r, \quad 0 \le s \le t \le T.$$
This implies the Kolmogorov forward equation
$$\partial_t P_{s,t}f = P_{s,t}(L_t f), \quad \text{a.e. } t \in [s,T]. \tag{1.5.4}$$
On the other hand, by Lemma 1.2.2, for any $t \in (0,T]$, the PDE
$$(\partial_s + L_s)u_s = -L_s f, \quad s \in [0,t], u_t = 0 \tag{1.5.5}$$
has a unique solution, such that by Itô's formula in Theorem 1.2.3(3),
$$\mathrm{d}u_s(X_s) = -L_s f(X_s) + \langle \nabla f(X_s), \sigma_s(X_s)\mathrm{d}W_s \rangle, \quad s \in [0,t].$$
This and (1.5.4) yield
$$0 = u_t(x) = u_s(x) - \int_s^t (P_{s,r} L_r f)\mathrm{d}r$$
$$= u_s(x) - \int_s^t \frac{\mathrm{d}}{\mathrm{d}r}(P_{s,r}f)\mathrm{d}r = u_s(x) - P_{s,t}f + f, \quad 0 \le s \le t \le T.$$
Combining this with (1.5.5) we derive the Kolmogorov backward equation
$$\partial_s P_{s,t}f = \partial_s u_s = -L_s(u_s + f) = -L_s P_{s,t}f, \quad 0 \le s \le t \le T. \tag{1.5.6}$$
Let $\gamma_s = x + s(y-x)/t$ for $s \in [0,t]$. By (1.5.6) and Itô's formula in Theorem 1.2.3(3), for any $0 < f \in C_c^\infty(\mathbb{R}^d)$, we have
$$\mathrm{d}\log P_{s,t}f(X_s^{\gamma_s}) = \left\{ L_s(\log P_{s,t}f) - \frac{L_s P_{s,t}f}{P_{s,t}f} \right\}(X_s^{\gamma_s})\mathrm{d}s$$
$$+ \left\langle \nabla_{\gamma_s'} X_r^{\gamma_s}, \nabla \log P_{s,t}f \right\rangle(X^{\gamma_s})\mathrm{d}s + \mathrm{d}M_s, \quad s \in [0,t]$$
for some martingale M_s. Since $\sigma\sigma^* \ge \lambda I_d$ for some constant $\lambda > 0$, this implies
$$\mathrm{d}\log P_{s,t}f(X_s^{\gamma_s})$$
$$\le \left\{ -\lambda |\nabla \log P_{s,t}f|^2 + \frac{|y-x|}{t}|\nabla \log P_{s,t}f| \right\}(X_s^{\gamma_s})\mathrm{d}s + \mathrm{d}M_s$$
$$\le \frac{|x-y|^2}{4\lambda t^2}|\nabla X_s^{\gamma_s}|^2 \mathrm{d}s + \mathrm{d}M_s, \quad s \in [0,t].$$
Combining this with (1.4.1), we find a constant $c > 0$ such that
$$P_t \log f(y) - \log P_t f(x) = \mathbb{E}\left[\log f(X_t^y) - \log P_{0,t}f(X_0^x) \right]$$
$$\le \frac{|x-y|^2}{4\lambda t^2} \int_0^t \mathbb{E}\big[|\nabla X_s^{\gamma_s}|^2\big]\mathrm{d}s \le \frac{c|x-y|^2}{t}, \quad t \in (0,T].$$
By an approximation argument, this implies (1.5.2) for $f \in \mathcal{B}_b^+(\mathbb{R}^d)$. $\square$

1.5.2 Power Harnack inequality

By constructing a new coupling to force two marginal processes to meet before a fixed time, under the monotone condition

$$2\langle b_t(x) - b_t(y), x - y\rangle + \|\sigma_t(x) - \sigma_t(y)\|_{HS}^2 \leq K|x - y|^2,$$
$$|\{\sigma_t(x) - \sigma_t(y)\}^*(x - y)| \leq K|x - y|, \quad x, y \in \mathbb{R}^d, t \in [0, T]$$

for some constant $K \geq 0$, the power Harnack inequality was established in [Wang (2011)]: there exist constants $c, p^* > 1$ such that for any $p > p^*$,

$$|P_t f(y)|^p \leq e^{\frac{c|x-y|^2}{t}} P_t |f|^p(x), \quad x, y \in \mathbb{R}^d, t \in (0, T], f \in \mathcal{B}_b(\mathbb{R}^d). \quad (1.5.7)$$

This result has been extended to singular SDEs.

Under $(A^{1.3})$ and that σ_t is Lipshitzi continuous uniformly in $t \in [0, T]$, the following inequality was established in [Shao (2013)] for $p > p^*$:

$$|P_t f(y)|^p \leq e^{c + \frac{c|x-y|^2}{t}} P_t |f|^p(x), \quad x, y \in \mathbb{R}^d, t \in (0, T], f \in \mathcal{B}_b(\mathbb{R}^d). \quad (1.5.8)$$

This inequality is less sharp for small $|x - y|$. To make the exponential term sharp for $x = y$, when $\frac{d}{p_0} + \frac{2}{q_0} < \frac{1}{2}$ and σ_t is α-Hölder continuous with $\alpha \in (\frac{1}{2}, 1 - \frac{d}{p_0} - \frac{2}{q_0})$, Theorem 4.3(2) in [Zhang and Yuan (2021)] gives

$$|P_t f(y)|^p \leq e^{ct^{-1}(|x-y|^2 \vee |x-y|^{2\alpha})} P_t |f|^p(x), \quad x, y \in \mathbb{R}^d, t \in (0, T].$$

The following result is due to [Ren (2023)], which establishes the sharp inequality (1.5.7) under $(A^{1.3})$ without any additional conditions.

Theorem 1.5.2. *Assume* $(A^{1.3})$ *and let*

$$\kappa_0 := \sup_{t \in [0, T], x, y \in \mathbb{R}^d} \|\sigma_t(x) - \sigma_t(y)\|^2, \quad \kappa_1 := \|\sigma^*(\sigma\sigma^*)^{-1}\|_\infty^2.$$

Then for any

$$p > p^* := \frac{3 + \sqrt{1 + (8\kappa_0 \kappa_1)^{-1}}}{\sqrt{1 + (8\kappa_0 \kappa_1)^{-1}} - 1},$$

there exists a constant $c > 0$ such that (1.5.7) *holds.*

To prove this result, we present the following lemma which gives the less sharp Harnack inequality (1.5.8).

Lemma 1.5.3. *Assume* $(A^{1.3})$. *Then for any $p > p^*$, there exists a constant $c > 0$ such that* (1.5.8) *holds.*

Proof. (a) We first observe that it suffices to prove for $b^{(0)} = 0$. Indeed, let $\hat{P}_t$ be the semigroup associated with the SDE

$$dX_t^x = b_t^{(1)}(X_t^x)dt + \sigma_t(X_t^x)dW_t, \quad t \in [0, T].$$

Let

$$R^x := e^{\int_0^T \langle \{\sigma_t^*(\sigma_t\sigma_t^*)^{-1} b_t^{(0)}\}(X_t^x), dW_t \rangle - \frac{1}{2}\int_0^T |\{\sigma_t^*(\sigma_t\sigma_t^*)^{-1} b_t^{(0)}\}(X_t^x)|^2 dt}.$$

By $(A^{1.3})$, (1.2.7) with $(p, q) = (p_0/2, q_0/2)$ and Khasminskii's estimate in Theorem 1.2.4, we obtain

$$\sup_{x \in \mathbb{R}^d} \mathbb{E}[|R^x|^q] < \infty, \quad q > 1.$$

Then by Girsanov's theorem, for any $p > 1$ there exists $c(p) > 0$ such that

$$|P_t f|^p(x) = \left|\mathbb{E}[R^x f(X_t^x)]\right|^p$$
$$\leq \left(\mathbb{E}[|R^x|^{\frac{p}{p-1}}]\right)^{p-1} \mathbb{E}[|f|^p(X_t^x)] \leq c(p)\hat{P}_t|f|^p(x), \quad p > 1.$$

Similarly, the same inequality holds by exchanging positions of P_t and $\hat{P}_t$, so that

$$|P_t f|^p \leq c(p)\hat{P}_t|f|^p, \quad |\hat{P}_t f|^p \leq c(p)P_t|f|^p, \quad p > 1, t \in [0, T]. \quad (1.5.9)$$

Thus, if the desired assertion holds for $\hat{P}_t$, it also holds for P_t. Indeed, assuming

$$\{\hat{P}_t f(x)\}^p \leq (\hat{P}_t f^p) e^{c_1(p) + c_1(p) t^{-1}|x-y|^2}, \quad x, y \in \mathbb{R}^d, t \in (0, T], p > p^*$$

for some $c_1 : (p^*, \infty) \to (0, \infty)$, then for any $p > p^*$ we have

$$p_1 := \left(\frac{2p}{p+p^*}\right)^{\frac{1}{2}} > 1, \quad p_2 := \frac{p+p^*}{2} > p^*, \quad p_1^2 p_2 = p,$$

so that this inequality and (1.5.9) yield

$$(P_t f(x))^p \leq c(p_1)^{\frac{p}{p_1}} \{\hat{P}_t f^{p_1}(x)\}^{\frac{p}{p_1}}$$
$$\leq c(p_1)^{\frac{p}{p_1}} \left\{e^{c_1(p_2)+c_1(p_2)t^{-1}|x-y|^2} \hat{P}_t f^{p_1 p_2}(y)\right\}^{p_1}$$
$$\leq c(p_1)^{\frac{p}{p_1}+1} e^{p_1 c_1(p_2)+p_1 c_1(p_2)t^{-1}|x-y|^2} P_t f^p(y) \leq e^{c+ct^{-1}|x-y|^2} P_t f^p(y)$$

for some constant $c > 0$.

(b) Now, we consider the regular case that $b = b^{(1)}$. In this case, by $(A^{1.3})$ and Lemma 1.3.4, there exists a constant $c_0 > 0$ such that

$$\|\sigma_t(x) - \sigma_t(y)\|_{HS}^2 \leq d(1+\kappa_0)\|\sigma_t(x) - \sigma_t(y)\|_{HS}$$
$$\leq c_0 |x-y| \sum_{i=1}^{l} \left(1 + \mathcal{M} f_i(t, x) + \mathcal{M} f_i(t, y)\right).$$

Combining this with $b = b^{(1)}$, we find a constant $c_1 > 0$ such that
$$2\langle x - y, b_t(x) - b_t(y)\rangle + \|\sigma_t(x) - \sigma_t(y)\|_{HS}^2$$
$$\leq c_1 |x-y|^2 + c_1 |x-y| \sum_{i=1}^{l} \big(1 + \mathcal{M}f_i(t,x) + \mathcal{M}f_i(t,y)\big). \quad (1.5.10)$$
For fixed $t \in (0, T]$, let
$$\gamma_s = \frac{1 - e^{c_1(s-t)}}{c_1}, \quad s \in [0, t],$$
so that for some constant $K_0 > 0$ such that
$$c_1 \gamma_s - 2 - \gamma_s' = -1, \quad \gamma_s \geq K_0(t-s), \quad s \in [0,t]. \quad (1.5.11)$$
Since the coefficients of the following SDE are continuous and of linear growth in x locally uniformly in $s \in [0, t)$, it has a weak solution:
$$\begin{cases} dX_s = b_s(X_s)ds + \sigma_s(X_s)dW_s, & X_0 = x, \\ dY_s = \{b_s(Y_s) + \sigma_s(Y_s)\xi_s\}ds + \sigma_s(Y_s)dW_s, & Y_0 = y, \end{cases} \quad (1.5.12)$$
where
$$\xi_s := \frac{\{\sigma_s^*(\sigma_s\sigma_s^*)^{-1}\}(X_s)(X_s - Y_s)}{\gamma_s}, \quad s \in [0, t]. \quad (1.5.13)$$
This construction of coupling is due to [Wang (2011)].

For any $n \geq 1$, let
$$\tau_n = \frac{nt}{n+1} \wedge \inf\{s \geq 0 : |X_s| \vee |Y_s| \geq n\},$$
$$R_r := e^{-\int_0^r \langle \xi_s, dW_s\rangle - \frac{1}{2}\int_0^{\tau_n} |\xi_s|^2 ds}, \quad r \in [0, t]. \quad (1.5.14)$$
By Girsanov's theorem,
$$\tilde{W}_s := W_s + \int_0^{s \wedge \tau_n} \xi_r dr, \quad s \in [0, t]$$
is an m-dimensional Brownian motion under the probability $\mathbb{Q}_n := R_{\tau_n}\mathbb{P}$. So, before time τ_n, (1.5.12) is reformulated as
$$\begin{cases} dX_s = \{b_s(X_s) - \frac{X_s - Y_s}{\gamma_s}\}ds + \sigma_s(X_s)d\tilde{W}_s, & X_0 = x, \\ dY_s = b_s(Y_s)ds + \sigma_s(Y_s)d\tilde{W}_s, & Y_0 = y, s \in [0, \tau_n]. \end{cases}$$
By (1.5.10) and Itô's formula, we obtain
$$d|X_s - Y_s|^2 - dM_s$$
$$\leq \bigg\{c_1|X_s - Y_s|^2 + c_1|X_s - Y_s|\sum_{i=1}^{l}\big[1 + \mathcal{M}f_i(s, X_s) + \mathcal{M}f_i(s, Y_s)\big]\bigg\}ds,$$

where M_s is a $\mathbb{Q}_n$-martingale with
$$\mathrm{d}\langle M\rangle_s \leq 4\kappa_0 |X_s - Y_s|^2. \tag{1.5.15}$$
Combining this with (1.5.11) and applying Itô's formula, we obtain
$$\mathrm{d}\Big\{\frac{|X_s - Y_s|^2}{\gamma_s}\Big\} - \frac{\mathrm{d}M_s}{\gamma_s}$$
$$\leq \frac{c_1|X_s - Y_s|}{\gamma_s} \sum_{i=1}^{l} \big(1 + \mathcal{M}f_i(s, X_s) + \mathcal{M}f_i(s, Y_s)\big)\mathrm{d}s$$
$$+ \frac{(c_1\gamma_s - 2 - \gamma'_s)|X_s - Y_s|^2}{\gamma_s^2}\mathrm{d}s$$
$$= \frac{c_1|X_s - Y_s|}{\gamma_s} \sum_{i=1}^{l} \big(1 + \mathcal{M}f_i(s, X_s) + \mathcal{M}f_i(s, Y_s)\big)\mathrm{d}s \tag{1.5.16}$$
$$- \frac{|X_s - Y_s|^2}{\gamma_s^2}\mathrm{d}s$$
$$\leq \frac{c_1^2}{2}\Big[\sum_{i=1}^{l}\big(1 + \mathcal{M}f_i(s, X_s) + \mathcal{M}f_i(s, Y_s)\big)\Big]^2\mathrm{d}s$$
$$- \frac{|X_s - Y_s|^2}{2\gamma_s^2}\mathrm{d}s, \quad s \in [0, \tau_n].$$
Thus, by $\gamma_0 \geq K_0 t$ in (1.5.11), we derive
$$\mathbb{E}_{\mathbb{Q}_n}\Big[\mathrm{e}^{\lambda \int_0^{\tau_n} \frac{|X_s - Y_s|^2}{\gamma_s^2}\mathrm{d}s}\Big] - \frac{2\lambda|x - y|^2}{K_0 t}$$
$$\leq \mathbb{E}_{\mathbb{Q}_n}\Big[\mathrm{e}^{\lambda c_1^2 \int_0^{\tau_n}[\sum_{i=1}^{l}(1+\mathcal{M}f_i(s,X_s)+\mathcal{M}f_i(s,Y_s))]^2\mathrm{d}s + 2\lambda \int_0^{\tau_n}\frac{\mathrm{d}M_s}{\gamma_s}}\Big]$$
$$\leq \Big(\mathbb{E}_{\mathbb{Q}_n}\Big[\mathrm{e}^{\frac{\lambda c_1^2 r}{r-1}\int_0^{\tau_n}[\sum_{i=1}^{l}(1+\mathcal{M}f_i(s,X_s)+\mathcal{M}f_i(s,Y_s))]^2\mathrm{d}s}\Big]\Big)^{\frac{r-1}{r}} \tag{1.5.17}$$
$$\times \Big(\mathbb{E}_{\mathbb{Q}_n}\Big[\mathrm{e}^{2\lambda r \int_0^{\tau_n}\frac{\mathrm{d}M_s}{\gamma_s}}\Big]\Big)^{\frac{1}{r}}, \quad \lambda > 0, r > 1.$$
By $(A^{1.3})$ and Lemma 1.3.8, there exists a constant $c_2 > 0$ such that
$$\sum_{i=1}^{l} \|\mathcal{M}f_i\|_{\tilde{L}_{q_i}^{p_i}} \leq c_2.$$
Noting that X_s solves (1.1.1) under probability $\mathbb{P}$ while Y_s solves the same equation under $\mathbb{Q}_n$, combining this with (1.2.7) for $(p,q) = (p_i/2, q_i/2)$ and Khasminskii's inequality in Theorem 1.2.4, we find an increasing map $\delta : (0, \infty) \to (0, \infty)$ such that
$$\mathbb{E}\Big[\mathrm{e}^{\lambda \int_0^{\tau_n}\sum_{i=1}^{l}|\mathcal{M}f_i(s,X_s)|^2\mathrm{d}s}\Big] + \mathbb{E}_{\mathbb{Q}_n}\Big[\mathrm{e}^{\lambda \int_0^{\tau_n}\sum_{i=1}^{l}|\mathcal{M}f_i(s,Y_s)|^2\mathrm{d}s}\Big]$$
$$\leq \delta(\lambda), \quad \lambda > 0, n \geq 1.$$

Consequently,

$$\mathbb{E}_{\mathbb{Q}_n}\left[e^{\lambda \int_0^{\tau_n} \sum_{i=1}^l |\mathcal{M}f_i(s,X_s)|^2 ds}\right] = \mathbb{E}\left[R_{\tau_n} e^{\lambda \int_0^{\tau_n} \sum_{i=1}^l |\mathcal{M}f_i(s,X_s)|^2 ds}\right]$$

$$\leq \left(\mathbb{E}[R_{\tau_n}^q]\right)^{\frac{1}{q}} \left(\mathbb{E}\left[e^{\frac{\lambda q}{q-1} \int_0^{\tau_n} \sum_{i=1}^l |\mathcal{M}f_i(s,X_s)|^2 ds}\right]\right)^{\frac{q-1}{q}}$$

$$\leq \delta(\lambda q/(q-1))^{\frac{q-1}{q}} \left(\mathbb{E}[R_{\tau_n}^q]\right)^{\frac{1}{q}}, \quad q > 1.$$

Therefore, for any $\lambda > 0, r, q > 1$, there exists a constant $c(\lambda, r, q) > 0$ such that

$$\sup_{n \geq 1} \left(\mathbb{E}_{\mathbb{Q}_n}\left[e^{\frac{\lambda c_1^2 r}{r-1} \int_0^{\tau_n} \{\sum_{i=1}^l (1+|\mathcal{M}f_i(s,X_s)|^2 + |\mathcal{M}f_i(s,Y_s)|^2)\} ds}\right]\right)^{\frac{q-1}{q}}$$

$$\leq c(\lambda, r, q) \left(\mathbb{E}[R_{\tau_n}^q]\right)^{\frac{r-1}{rq}}, \quad q, r > 1, \lambda > 0.$$

Thus, by (1.5.17), (1.5.15) and $\mathbb{E}_{\mathbb{Q}_n}[e^{N_t}] \leq (\mathbb{E}_{\mathbb{Q}_n}[e^{2\langle N \rangle_t}])^{\frac{1}{2}}$ for a continuous $\mathbb{Q}_n$-martingale N_t, we arrive at

$$\mathbb{E}_{\mathbb{Q}_n}\left[e^{\lambda \int_0^{\tau_n} \frac{|X_s - Y_s|^2}{\gamma_s^2} ds}\right] - \frac{2\lambda |x-y|^2}{K_0 t}$$

$$\leq c(\lambda, r, q) \left(\mathbb{E}[R_{\tau_n}^q]\right)^{\frac{r-1}{rq}} \left(\mathbb{E}_{\mathbb{Q}_n}\left[e^{8\lambda^2 r^2 \int_0^{\tau_n} \frac{d\langle M \rangle_s}{\gamma_s^2} ds}\right]\right)^{\frac{1}{2q}} \quad (1.5.18)$$

$$\leq c(\lambda, r, q) \left(\mathbb{E}[R_{\tau_n}^q]\right)^{\frac{r-1}{rq}} \left(\mathbb{E}_{\mathbb{Q}_n}\left[e^{32\lambda^2 r^2 \kappa_0 \int_0^{\tau_n} \frac{|X_s - Y_s|^2}{\gamma_s^2} ds}\right]\right)^{\frac{1}{2r}}$$

for any $\lambda > 0, q, r > 1$. For any $q > 1$ and $\lambda \in [0, \frac{1}{32\kappa_0})$, we take $r = \frac{1}{32\lambda\kappa_0} > 1$ such that for some constant $\beta(\lambda, q) > 0$,

$$\sup_{n \geq 1} \mathbb{E}_{\mathbb{Q}_n}\left[e^{\lambda \int_0^{\tau_n} \frac{|X_s - Y_s|^2}{\gamma_s^2} ds}\right]$$

$$\leq \left(\mathbb{E}[R_{\tau_n}^q]\right)^{\frac{2-64\lambda\kappa_0}{q(2-64\lambda\kappa_0)}} e^{\beta(\lambda,q) + \beta(\lambda,q)t^{-1}|x-y|^2} \quad (1.5.19)$$

$$\leq \left(\mathbb{E}[R_{\tau_n}^q]\right)^{\frac{1}{q}} e^{\beta(\lambda,q) + \beta(\lambda,q)t^{-1}|x-y|^2}, \quad q > 1, \lambda \in \left[0, \frac{1}{32\kappa_0}\right).$$

Noting that for any

$$1 \leq q < q^* := \frac{p^*}{p^* - 1} = \frac{3 + \sqrt{1 + (8\kappa_0\kappa_1)^{-1}}}{4},$$

we have $0 \leq (2q^2 - 3q + 1)\kappa_1 < \frac{1}{32\kappa_0}$, by (1.5.19) and

$$|\xi_s|^2 \leq \frac{\kappa_1 |X_s - Y_s|^2}{\gamma_s^2},$$

we find an increasing function $k : (1, q^*) \to (0, \infty)$ such that

$$\left(\mathbb{E}\big[|R_{\tau_n}|^q\big]\right)^{\frac{1}{q}} = \left(\mathbb{E}_{\mathbb{Q}_n} e^{-(q-1)\int_0^{\tau_n} \langle \xi_s, d\tilde{W}_s\rangle + \frac{q-1}{2}\int_0^{\tau_n} |\xi_s|^s ds}\right)^{\frac{1}{q}}$$

$$\leq \left(\mathbb{E}_{\mathbb{Q}_n} e^{-2(q-1)\int_0^{\tau_n} \langle \xi_s, d\tilde{W}_s\rangle - 2(q-1)^2 \int_0^t |\xi_s|^s ds}\right)^{\frac{1}{2q}}$$

$$\times \left(\mathbb{E}_{\mathbb{Q}_n} e^{(2q^2-3q+1)\int_0^{\tau_n} |\xi_s|^2 ds}\right)^{\frac{1}{2q}}$$

$$\leq \left(\mathbb{E}_{\mathbb{Q}_n} e^{(2q^2-3q+1)\kappa_1 \int_0^{\tau_n} \frac{|X_s-Y_s|^2}{\gamma_s^2} ds}\right)^{\frac{1}{2q}}$$

$$\leq \left(\mathbb{E}\big[|R_{\tau_n}|^q\big]\right)^{\frac{1}{2q^2}} e^{k_q + k_q t^{-1}|x-y|^2}, \quad q \in (1, q^*).$$

Consequently,

$$\sup_{n \geq 1} \left(\mathbb{E}\big[|R_{\tau_n}|^q\big]\right)^{\frac{1}{q}} \leq e^{2k_q + 2k_q t^{-1}|x-y|^2}, \quad q \in (1, q^*).$$

By the martingale convergence theorem with $n \to \infty$, this implies that $(R_s)_{s \in [0,t]}$ is a martingale with

$$\left(\mathbb{E}[R_t^q]\right)^{\frac{1}{q}} \leq e^{2k_q + 2k_q t^{-1}|x-y|^2}, \quad q \in (1, q^*), \tag{1.5.20}$$

such that Girsanov's theorem implies that $(\tilde{W}_s)_{s \in [0,t]}$ is an m-dimensional Brownian motion under $\mathbb{Q} := R_t \mathbb{P}$, and Y_s solves the SDE

$$dY_s = b_s(Y_s) ds + \sigma_s(Y_s) d\tilde{W}_s, \quad Y_0 = y, s \in [0, t].$$

By the weak uniqueness we obtain

$$P_t f(y) = \mathbb{E}_{\mathbb{Q}}[f(Y_t)]. \tag{1.5.21}$$

Moreover, (1.5.19) ensures

$$\mathbb{E}_{\mathbb{Q}}\left[e^{\lambda \int_0^t \frac{|X_s-Y_s|^2}{\gamma_s^2} ds}\right] < \infty$$

for some constant $\lambda > 0$, together with the continuity of $|X_s - Y_s|$ in $s \in [0, t]$ and $\int_0^t \frac{1}{\gamma_s^2} ds = \infty$ imply $\mathbb{Q}(X_t = Y_t) = 1$. Therefore, by (1.5.20), (1.5.21) and Hölder's inequality, for any $p > p^* = \frac{q^*}{q^*-1}$ so that $q := \frac{p}{p-1} \in [1, q^*)$, we find a constant $c > 0$ such that for any $t \in (0, T]$

$$|P_t f(y)|^p = |\mathbb{E}[R_t f(Y_t)]|^p = |\mathbb{E}[R_t f(X_t)]|^p$$

$$\leq (\mathbb{E} R_t^{\frac{p}{p-1}})^{\frac{p-1}{p}} \mathbb{E}[|f|^p(X_t)] \leq (P_t |f|^p)(x) e^{c + ct^{-1}|x-y|^2}, \quad f \in \mathcal{B}_b(\mathbb{R}^d). \quad \square$$

Proof of Theorem 1.5.2. By an approximation argument, it suffices to prove for $0 \leq f \in C_c^\infty(\mathbb{R}^d) := \mathbb{R} + C_0^\infty(\mathbb{R}^d)$.
By (1.4.1), for any $q > 1$ there exists a constant $k(q) > 0$ such that

$$|\nabla P_t g| \leq k(q)(P_t|\nabla g|^q)^{\frac{1}{q}}, \quad g \in C_b^1(\mathbb{R}^d), t \in (0, T]. \tag{1.5.22}$$

Below we show that this and the Harnack inequality in Lemma 1.6.5 imply the desired inequality.

For $p > p^*$, we have

$$p_1 := \frac{\sqrt{5p^2 + 4pp^*} - p}{p + p^*} \in (1, 2), \quad p_2 := \frac{p + p^*}{2} > p^*, \quad \frac{p_1^2 p_2}{2 - p_1} = p. \tag{1.5.23}$$

By Kolmogorov equations (1.5.4) and (1.5.6), for $0 \leq f \in C_c^2(\mathbb{R}^d)$,

$$\begin{aligned}\partial_s\{P_{st}(P_{st,t}f)^{p_1}\} &= tP_{st}\{|\sigma_{st}^* \nabla P_{st,t}f|^2 (P_{st,t}f)^{p_1-2}\} \\ &\geq c_1 t P_{st}\{|\nabla P_{st,t}f|^2 (P_{st,t}f)^{p_1-2}\}, \quad s \in [0,1].\end{aligned} \tag{1.5.24}$$

Next, by (1.5.22) and Hölder's inequality,

$$\begin{aligned}|\nabla P_{st}(P_{st,t}f)^{p_1}| &\leq k(p_1)\big(P_{st}|\nabla(P_{st,t}f)^{p_1}|^{p_1}\big)^{\frac{1}{p_1}} \\ &= p_1 k(p_1)\big(P_{st}\{|\nabla P_{st,t}f|^{p_1}(P_{st,t}f)^{(p_1-1)p_1}\}\big)^{\frac{1}{p_1}} \\ &\leq p_1 k(p_1)\big(P_{st}\{|\nabla P_{st,t}f|^2 (P_{st,t}f)^{p_1-2}\}\big)^{\frac{1}{2}} \big(P_{st}\{P_{st,t}f\}^{\frac{p_1^2}{2-p_1}}\big)^{\frac{2-p_1}{2p_1}}, \quad s \in [0,1].\end{aligned}$$

Combining this with (1.5.24), and letting $z_s := sx + (1-s)y, s \in [0,1]$, we find a constant $c_1 = c_1(p) > 0$ such that

$$\begin{aligned}&\frac{\mathrm{d}}{\mathrm{d}s}\big\{P_{st}(P_{st,t}f)^{p_1}(z_s)\big\} \\ &= \Big\{\frac{\mathrm{d}}{\mathrm{d}s}P_{st}(P_{st,t}f)^{p_1}\Big\}(z_s) + \big\langle x - y, \nabla P_{st}(P_{st,t}f)^{p_1}(z_s)\big\rangle \\ &\geq c_1 t P_{st}\{|\nabla P_{st,t}f|^2 (P_{st,t}f)^{p_1-2}\}(z_s) \\ &\quad - p_1|x - y|\Big\{P_{st}\big(|\nabla P_{st,t}f|^2(P_{st,t}f)^{p_1-2}\big)\Big\}^{\frac{1}{2}}(z_s) \\ &\qquad \times \Big\{P_{st}\big(P_{st,t}f\big)^{\frac{p_1^2}{2-p_1}}\Big\}^{\frac{2-p_1}{2p_1}}(z_s) \\ &\geq -\frac{p_1^2|x-y|^2}{4c_1 t}\Big\{P_{st}(P_{st,t}f)^{\frac{p_1^2}{2-p_1}}\Big\}^{\frac{2-p_1}{p_1}}(z_s), \quad s \in [0,1], x, y \in \mathbb{R}^d.\end{aligned} \tag{1.5.25}$$

By Jensen's inequality and Lemma 1.6.5, we find a constant $c_2 = c_2(p) > 0$ such that

$$\left\{P_{st}(P_{st,t}f)^{\frac{p_1^2}{2-p_1}}\right\}^{\frac{2-p_1}{p_1}}(z_s) \le \left\{P_{st}P_{st,t}f^{\frac{p_1^2}{2-p_1}}\right\}^{\frac{2-p_1}{p_1}}(z_s)$$
$$= \left(P_t f^{\frac{p_1^2}{2-p_1}}\right)^{\frac{2-p_1}{p_1}}(z_s) \le \left(P_t f^{\frac{p_1^2 p_2}{2-p_1}}\right)^{\frac{2-p_1}{p_1 p_2}}(x) e^{c_2 + c_2 t^{-1}|x-y|^2}$$
$$= (P_t f^p(x))^{\frac{p_1}{p}} e^{c_2 + c_2 t^{-1}|x-y|^2}, \quad x, y \in \mathbb{R}^d.$$

Combining this with (1.5.25) we derive

$$(P_t f)^{p_1}(y) \le P_t f^{p_1}(x) + \frac{p_1^2 |x-y|^2}{4c_1 t}(P_t f^p)^{\frac{p_1}{p}}(x) e^{c_2 + c_2 t^{-1}|x-y|^2}$$
$$\le (P_t f^p(x))^{\frac{p_1}{p}}\left(1 + \frac{p_1^2 e^{c_2}|x-y|^2}{4c_1 t}\right) e^{c_2 t^{-1}|x-y|^2}$$
$$\le (P_t f^p)^{\frac{p_1}{p}}(x) e^{ct^{-1}|x-y|^2}, \quad x, y \in \mathbb{R}^d, t \in (0, T],$$

where $c := c_2 + \frac{p_1^2 e^{c_2}}{4c_1}$. Then the proof is finished. $\square$

1.6 Exponential ergodicity

There are many results on the ergodicity of diffusion processes under dissipative or Lyapunov conditions. In this section we investigate the ergodicity of the following time-homogeneous singular SDE:

$$dX_t = b(X_t)dt + \sigma(X_t)dW_t, \quad t \ge 0, \tag{1.6.1}$$

for which dissipative or Lyapunov conditions are not available.

We first state the main results in this part, then recall two ergodic theorems and present a lemma on elliptic equations, and finally prove the main results.

1.6.1 *Main results*

$(A^{1.4})$ σ is weakly differentiable, $\sigma\sigma^*$ is invertible, and $b = b^{(0)} + b^{(1)}$ such that the following conditions hold.
(1) There exists $p > d \vee 2$ such that

$$\|\sigma\|_\infty + \|(\sigma\sigma^*)^{-1}\|_\infty + \|b^{(0)}\|_{\tilde{L}^p} + \|\nabla\sigma\|_{\tilde{L}^p} < \infty.$$

(2) $b^{(1)}$ is locally bounded, there exist constants $K > 0, \varepsilon \in (0,1)$, some compact function $V \in C^2(\mathbb{R}^d; [1,\infty))$, and a continuous increasing function $\Phi : [1,\infty) \to [1,\infty)$ with $\Phi(n) \to \infty$ as $n \to \infty$, such that

$$\langle b^{(1)}, \nabla V \rangle(x) + \varepsilon |b^{(1)}(x)| \sup_{B(x,\varepsilon)} \{|\nabla V| + \|\nabla^2 V\|\}$$
$$\leq K - \varepsilon(\Phi \circ V)(x), \qquad (1.6.2)$$
$$\lim_{|x| \to \infty} \sup_{B(x,\varepsilon)} \frac{\|\nabla^2 V\| + |\nabla V|}{V(x) \wedge (\Phi \circ V)(x)} = 0.$$

Theorem 1.6.1. *Assume* $(A^{1.4})$. *Then* (1.1.1) *is well-posed, and the associated Markov semigroup* P_t *has a unique invariant probability measure* $\bar{\mu}$ *such that* $\bar{\mu}(\Phi(\varepsilon_0 V)) < \infty$ *for some* $\varepsilon_0 \in (0,1)$, *and*

$$\lim_{t \to \infty} (P_t^* \nu)(f) = \bar{\mu}(f), \quad \nu \in \mathcal{P}, f \in \mathcal{B}_b(\mathbb{R}^d). \qquad (1.6.3)$$

Moreover:

(1) *If* $\Phi(r) \geq \delta r$ *for some constant* $\delta > 0$ *and all* $r \geq 0$, *then there exist constants* $c > 1, \lambda > 0$ *such that*

$$\|P_t^* \mu_1 - P_t^* \mu_2\|_V \leq c e^{-\lambda t} \|\mu_1 - \mu_2\|_V, \quad \mu_1, \mu_2 \in \mathcal{P}, t \geq 0. \qquad (1.6.4)$$

In particular,

$$\|P_t^* \nu - \bar{\mu}\|_V \leq c e^{-\lambda t} \|\nu - \bar{\mu}\|_V, \quad \nu \in \mathcal{P}, t \geq 0.$$

(2) *Let* $H(r) := \int_0^r \frac{ds}{\Phi(s)}$ *for* $r \geq 0$. *If* Φ *is convex, then there exist constants* $k > 1, \lambda > 0$ *such that for any* $x \in \mathbb{R}^d, t \geq 0$,

$$\|P_t^* \delta_x - \bar{\mu}\|_V \leq k\{1 + H^{-1}(H(V(x)) - k^{-1}t)\} e^{-\lambda t}, \qquad (1.6.5)$$

where H^{-1} *is the inverse of* H *with* $H^{-1}(r) := 0$ *for* $r \leq 0$. *Consequently, if* $H(\infty) < \infty$ *then there exist constants* $c, \lambda, t^* > 0$ *such that*

$$\|P_t^* \mu_1 - \mu_2\|_V \leq c e^{-\lambda t} \|\mu_1 - \mu_2\|_{var}, \quad t \geq t^*, \mu_1, \mu_2 \in \mathcal{P}. \qquad (1.6.6)$$

The above result is taken from [Wang (2023c)]. To illustrate this result, we present below a consequence which covers the situation that

$$\langle b^{(1)}(x), x \rangle \leq c_1 - c_2 |x|^{1+p}, \quad |b^{(1)}(x)| \leq c_1(1 + |x|)^{q+1}$$

for some constants $c_1, c_2 > 0$ and $q > \frac{1}{2}$, since (1.6.7) and (1.6.8) hold for $\phi(r) := (1+r)^{\frac{1+q}{2}}$, and (1.6.9) holds for $\psi(r) := (1+r^2)^k$ for any $k > 0$ when $q \geq 1$. See also [Xie and Zhang (2020)] for the case with jumps.

Corollary 1.6.2. *Assume* $(A^{1.4})(1)$ *and let* $b^{(1)}$ *satisfy*
$$\langle b^{(1)}(x), x\rangle \leq c_1 - c_2\phi(|x|^2), \quad |b^{(1)}(x)| \leq c_1\phi(|x|^2), \quad x \in \mathbb{R}^d \quad (1.6.7)$$
for some constants $c_1, c_2 > 0$ *and increasing function* $\phi : [0, \infty) \to [1, \infty)$ *with*
$$\alpha := \liminf_{r \to \infty} \frac{\log \phi(r)}{\log r} > \frac{1}{2}. \quad (1.6.8)$$
Then

(1) (1.6.1) *is well-posed, P_t has a unique invariant probability measure $\bar{\mu}$ such that $\bar{\mu}(V) < \infty$ and (1.6.4) hold for $V := \mathrm{e}^{(1+|\cdot|^2)^\theta}$ with $\theta \in ((1-\alpha)^+, \frac{1}{2})$. In general, for any increasing function $1 \leq \psi \in C^2([1, \infty))$ satisfying*
$$\liminf_{r \to \infty} \frac{\psi'(r)\phi(r)}{\psi(r)} > 0, \quad \lim_{r \to \infty} \frac{\psi''(r)r}{\psi(r)} = 0, \quad (1.6.9)$$
$\bar{\mu}(V) < \infty$ *and (1.6.4) hold for* $V := \psi(|\cdot|^2)$.

(2) *If* $\int_0^\infty \frac{\mathrm{d}s}{\phi(s)} < \infty$, *then (1.6.6) holds for* $V := (1+|\cdot|^2)^q (q > 0)$ *and some constants* $c, \lambda, t^* > 0$.

Remark 1.6.1. We have the following assertions on the invariant probability measure $\bar{\mu}$ and the ergodicity in Wasserstein distance and relative entropy.

(1) According to Corollary 1.6.7 and Theorem 3.4.2 in [Bogachev et al. (2015)], $(A^{1.4})$ implies that $\bar{\mu}$ has a strictly positive density function $\rho \in H_{loc}^{1,p}$, the space of functions f such that $fg \in H^{1,p}(\mathbb{R}^d)$ for all $g \in C_0^\infty(\mathbb{R}^d)$. Moreover, by Theorem 3.1.2 in [Bogachev et al. (2015)], when σ is Lipschitz continuous and $\bar{\mu}(|b|^2) < \infty$, we have $\sqrt{\rho} \in H^{1,2}(\mathbb{R}^d)$. So, when (1.6.7) holds for $\phi(r) \sim r^q$ for some $q > \frac{1}{2}$ and large $r > 0$, Corollary 1.6.2(1) implies that $\bar{\mu}$ has density with $\sqrt{\rho} \in H^{1,2}(\mathbb{R}^d)$. See also [Wang (2017)] and [Wang (2018b)] for different type of global regularity estimates on ρ under integrability conditions.

(2) Let $V := (1+|\cdot|^2)^{\frac{k}{2}}$ for some $k \geq 1$. By Theorem 6.15 in [Villani (2009)], there exists a constant $c(k) > 0$ such that
$$\mathbb{W}_k(\mu, \nu)^k \leq c(k)\|\mu - \nu\|_V.$$
So, by Corollary 1.6.2, if $(A^{1.4})$ holds with $\Phi(r) \geq \delta r$ for some $\delta > 0$, then there exist constants $c, \lambda > 0$ such that
$$\mathbb{W}_k(P_t^* \nu, \bar{\mu})^q \leq c(1+\nu(|\cdot|^k))\mathrm{e}^{-\lambda t}, \quad t \geq 0, \nu \in \mathcal{P};$$
and if moreover Φ is convex with $\int_0^\infty \frac{\mathrm{d}s}{\Phi(s)} < \infty$, then there exist constants $c, \lambda, t^* > 0$ such that
$$\mathbb{W}_k(P_t^* \nu, \bar{\mu})^k \leq c\mathrm{e}^{-\lambda t}, \quad t \geq t^*, \nu \in \mathcal{P}_k.$$

(3) When $b^{(1)}$ is Lipschitz continuous, the log-Harnack inequality in Theorem 1.5.1 implies
$$\text{Ent}(P_t^*\nu|\bar\mu) \le \frac{c'}{1\wedge t}W_2(\nu,\bar\mu)^2, \quad \nu\in\mathcal{P}, t>0$$
for some constant $c'>0$, where $\text{Ent}(\nu|\bar\mu)$ is the relative entropy. Thus, by Corollary 1.6.2, if $(A^{1.4})$ holds for $V(x):=1+|x|^2$ and $\Phi(r)\ge\delta r$ for some constant $\delta>0$, then there exist constants $c,\lambda>0$ such that
$$\text{Ent}(P_t^*\nu|\bar\mu) \le c(1+\nu(|\cdot|^2))\mathrm{e}^{-\lambda t}, \quad t\ge 1, \nu\in\mathcal{P};$$
and if moreover Φ is convex with $\int_0^\infty \frac{\mathrm{d}s}{\Phi(s)}<\infty$, then there exist $c,\lambda,t^*>0$ such that
$$\text{Ent}(P_t^*\nu|\bar\mu) \le c(1+\nu(|\cdot|^2))\mathrm{e}^{-\lambda t}, \quad t\ge t^*, \nu\in\mathcal{P}.$$

1.6.2 Ergodic theorems

In this part we recall two ergodic theorems. To this end, we first introduce the following notions. For a topological space E, let $\mathcal{B}_b(E)$ (respectively $C_b(E)$) be the classes of bounded measurable (respectively continuous) functions on E.

A family $\{P_t(x,\cdot)\}_{t\ge 0, x\in\mathbb{R}^d}$ is called a Markov transition kernel, if $P_t(x,A)$ is measurable in x for any $t\ge 0$ and measurable $A\subset\mathbb{R}^d$, $P_t(x,\cdot)$ is a probability measure on $\mathbb{R}^d$ for any $(t,x)\in[0,\infty)\times\mathbb{R}^d$, and the Chapman-Kolmogorov equation
$$P_{t+s}(x,\cdot) = \int_{\mathbb{R}^d} P_s(y,\cdot)P_t(x,\mathrm{d}z), \quad t,s\ge 0, x\in\mathbb{E}^d$$
holds. In this case,
$$P_t f(x) := \int_{\mathbb{R}^d} f(y)P_t(x,\mathrm{d}y), \quad t\ge 0, f\in\mathcal{B}_b(\mathbb{R}^d)$$
gives rise to a Markov semigroup P_t on $\mathcal{B}_b(\mathbb{R}^d)$. We call the transition kernel stochastically continuous, if for any $x\in\mathbb{R}^d$, $P_t(x,\cdot)$ is weakly continuous in t.

Definition 1.6.1. Let P_t be a Markov semigroup with stochastically continuous transition kernel $P_t(x,\cdot)$ on a Polish space E.

(1) P_t is called t_0-regular for some $t_0>0$ if the transition probabilities $\{P_{t_0}(x,\cdot)\}_{x\in E}$ are mutually equivalent.
(2) P_t is called strong Feller if $P_t\mathcal{B}_b(E)\subset C_b(E)$.

(3) P_t is called irreducible if for any non-empty open set $G \subset E$,
$$P_t(x,G) > 0, \quad x \in E.$$
(4) A set $K \subset E$ is called petite (or small) if there exists nontrivial measure ν such that for some $t > 0$
$$\inf_{x \in K} P_t(x, \cdot) \geq \nu.$$

Theorem 1.6.3. *Let P_t be a Markov semigroup with stochastically continuous transition kernel $P_t(x, \cdot)$ on a Polish space E.*

(1) *If P_{t_1} is strong Feller and P_{t_2} is irreducible, then P_t is (t_1+t_2)-regular.*
(2) *If P_t has an invariant probability measure $\bar{\mu}$ and is t_0-regular for some $t_0 > 0$, then $\bar{\mu}$ is equivalent to $P_t(x, \cdot)$ for all $t > t_0$ and $x \in E$, and*
$$\lim_{t \to \infty} P_t f(x) = \bar{\mu}(f), \quad x \in E, f \in \mathcal{B}_b(E).$$
(3) *Let $V \geq 1$ be measurable such that the level sets $\{V \leq r\}_{r>0}$ of V are petite and*
$$P_t V \leq c_1 + e^{-c_2 t} V, \quad t \geq 0 \tag{1.6.10}$$
holds for some constant $c_1, c_2 > 0$. Then P_t has a unique invariant probability measure such that for some $c, \lambda > 0$
$$\|P_t \delta_x - \bar{\mu}\|_V \leq c e^{-\lambda t} V(x), \quad t \geq 0, x \in \mathbb{R}^d.$$

The first result is due to Khasminskii [Khasminskii (1980)], the second is due to Doob [Doob (1948)], and the third is called Harris theorem (see Theorem 4.2.1 in [Da Prato and Zabczyk (1996)]), see also [Hairer *et al.* (2011)] for a weaker version of Harris theorem.

The condition (1.6.10) holds if V is in the weak domain of L, i.e.
$$V(X_t) - \int_0^t LV(X_s)\mathrm{d}s$$
is a locally martingale for the associated Markov process X_t, such that
$$LV \leq c_1 c_2 - c_2 V.$$

Finally, we present a result on the exponential ergodicity in a probability distance. Let $(E, \mathcal{B})$ be a measurable space and let $\hat{\mathcal{P}}$ be a non-empty convex set of probability measures on E equipped with a complete metric $\mathbb{W}$. A family $(P_t^*)_{t \geq 0}$ is called a semigroup on $\hat{\mathcal{P}}$, if P_0^* is identity, $P_{t+s}^* = P_t^* P_s^*$ for $s, t \geq 0$, and
$$[0, \infty) \times \hat{\mathcal{P}} \ni (t, \mu) \mapsto P_t^* \mu \in \hat{\mathcal{P}}$$
is measurable.

Theorem 1.6.4. Let $(P_t^*)_{t\geq 0}$ be a semigroup on $\hat{\mathcal{P}}$, a convex subspace of $\mathcal{P}$ equipped with a complete metric $\mathbb{W}$. If for any $\mu \in \hat{\mathcal{P}}$, the family $\{P_t^*\mu : t \geq 0\}$ is locally bounded in t with respect to $\mathbb{W}$, and there exist constants $t_0 > 0$ and $\varepsilon \in (0,1)$ such that

$$\mathbb{W}(P_{t_0}^*\mu, P_{t_0}^*\nu) \leq \varepsilon \mathbb{W}(\mu, \nu), \quad \mu, \nu \in \hat{\mathcal{P}}, \tag{1.6.11}$$

then P_t^* has a unique invariant probability measure $\bar{\mu} \in \hat{\mathcal{P}}$, and

$$\mathbb{W}(P_t^*\mu, \bar{\mu}) \leq \varepsilon^{(t/t_0-1)^+} \sup_{s\in[0,t_0\wedge t]} \mathbb{W}(P_s\mu, \bar{\mu}), \quad t \geq 0, \mu \in \hat{\mathcal{P}}. \tag{1.6.12}$$

Proof. By (1.6.11), $P_{t_0}^*$ is contractive in the complete metric space $(\hat{\mathcal{P}}, \mathbb{W})$, so it has a unique fixed point $\bar{\mu} \in \hat{\mathcal{P}}$. To prove that $\bar{\mu}$ is the unique invariant probability measure, let

$$\mu^* := \frac{1}{t_0}\int_0^{t_0} P_s^*\bar{\mu}\,\mathrm{d}s,$$

which is in $\hat{\mathcal{P}}$ by the boundedness of $\{P_s^*\bar{\mu}\}_{s\in[0,t_0]}$ as well as the convexity and completeness of $(\mathcal{P}, \mathbb{W})$. Then for any $t \in [0,t_0]$, $P_{t_0}^*\bar{\mu} = \bar{\mu}$ and the semigroup property imply

$$P_t^*\mu^* = \frac{1}{t_0}\int_t^{t+t_0} P_s^*\bar{\mu}\,\mathrm{d}s - \frac{1}{t_0}\int_t^{t_0} P_s^*\bar{\mu}\,\mathrm{d}s + \frac{1}{t_0}\int_0^t P_{s|t_0}^*\bar{\mu}\,\mathrm{d}s$$
$$-\frac{1}{t_0}\int_t^{t_0} P_s^*\bar{\mu}\,\mathrm{d}s + \frac{1}{t_0}\int_0^t P_s^*\mu\,\mathrm{d}s = \frac{1}{t_0}\int_0^{t_0} P_s^*\bar{\mu}\,\mathrm{d}s = \mu^*.$$

Thus, μ^* is an invariant probability measure of P_t^*. In particular, $\mu^* \in \hat{\mathcal{P}}$ is a fixed point of $P_{t_0}^*$. By the uniqueness of the fixed point we conclude that $\mu^* = \bar{\mu}$. It remains to prove estimate (1.6.12).

(1.6.12) is obvious for $t \subset [0, t_0]$. For any $t > t_0$, there exist $n \in \mathbb{N}$ and $s \in [0, t_0)$ such that $t = nt_0 + s$. By (1.6.11) and the semigroup property,

$$\mathbb{W}(P_t^*\mu, \bar{\mu}) = \mathbb{W}(P_{nt_0}^*(P_s^*\mu), P_{nt_0}^*\bar{\mu}) \leq \varepsilon^n \sup_{s\in[0,t_0]} \mathbb{W}(P_s^*\mu, \bar{\mu}),$$

so that (1.6.12) holds for $t > t_0$. $\square$

1.6.3 Elliptic equations

$(A^{1.5})$ $a := \sigma\sigma^*$ and $b = b^{(0)} + b^{(1)}$ satisfy the following conditions.
(1) a is invertible, uniformly continuous, and $\|a\|_\infty + \|a^{-1}\|_\infty < \infty$.
(2) $|b^{(0)}| \in \tilde{L}^{p_0}$ for some $p_0 > d$, and $b^{(1)}$ is Lipschitz continuous.

When $b^{(1)} = 0$, the following lemma follows from [Xie and Zhang (2020)] where a jump term is also considered.

Lemma 1.6.5. *Assume* $(A^{1.5})$ *and let* $p \in (1, \infty)$. *There exist constants* $\lambda_0 > 0$ *increasing in* $\|b^{(0)}\|_{\tilde{L}^{p_0}}$ *such that for any* $\lambda \geq \lambda_0$ *and any* $f \in \tilde{L}^p$, *the elliptic equation*

$$(L - \lambda)u = f \tag{1.6.13}$$

has a unique solution $u \in \tilde{H}^{2,p}$. *Moreover, for any* $p' \in [p, \infty]$ *and* $\theta \in [0, 2 - \frac{d}{p} + \frac{d}{p'})$, *there exists a constant* $c > 0$ *increasing in* $\|b^{(0)}\|_{\tilde{L}^{p_0}}$ *such that*

$$\lambda^{\frac{1}{2}(2-\theta+\frac{d}{p'}-\frac{d}{p})}\|u\|_{\tilde{H}^{\theta,p'}} + \|u\|_{\tilde{H}^{2,p}} \leq c\|f\|_{\tilde{L}^p}, \quad f \in \tilde{L}^p. \tag{1.6.14}$$

Proof. (a) Let us verify the priori estimate (1.6.14) for a solution u to (1.6.13), which in particular implies the uniqueness, since the difference of two solutions solves the equation with $f = 0$.

For u solving (1.6.13), let

$$\bar{u}_t = u(1 - t), \quad t \in [0, 1].$$

By (1.6.13) we have

$$(\partial_t + L - \lambda)\bar{u}_t = f(1 - t) - u, \quad t \in [0, 1], \bar{u}_1 = 0.$$

By Lemma 1.2.2, there exist constants $\lambda_1, c_1 > 1$ increasing in $\|b^{(0)}\|_{\tilde{L}^{p_0}}$ and sufficient large $q > 2$ such that

$$\begin{aligned}&\lambda^{\frac{1}{2}(2-\theta+\frac{d}{p'}-\frac{d}{p})}\|\bar{u}\|_{\tilde{H}^{\theta,p'}_q} + \|\bar{u}\|_{\tilde{H}^{2,p}_q} \\ &\leq c_1\|f(1-t) - u\|_{\tilde{L}^p_q} \leq c_1\|f\|_{\tilde{L}^p} + c_1\|u\|_{\tilde{L}^p}.\end{aligned} \tag{1.6.15}$$

Taking $\theta = 0, p = p'$ and $c_2 = \|1 - \cdot\|_{L^q([0,1])}$, we obtain

$$\lambda\|u\|_{\tilde{L}^p} \leq \frac{c_1}{c_2}\big(\|f\|_{\tilde{L}^p} + \|u\|_{\tilde{L}^p}\big), \quad \lambda \geq \lambda_1.$$

Letting $\lambda_0 > \lambda_1$ such that

$$\lambda_0 \geq 2\frac{c_1}{c_2},$$

we obtain

$$\|u\|_{\tilde{L}^p} \leq \|f\|_{\tilde{L}^p}, \quad \lambda \geq \lambda_0.$$

Combining this with (1.6.15) implies (1.6.14) for some constant $c > 0$.

(b) Existence of solution for $f \in \tilde{L}^p$. Let $\{f_n\}_{n\geq 1} \subset C_b^\infty(\mathbb{R}^d)$ such that $\|f_n - f\|_{\tilde{L}^p} \to 0$ as $n \to \infty$. Let

$$u_n = \int_0^\infty e^{-\lambda t} P_t f_n \mathrm{d}t.$$

By Kolmogorov equations, see (1.5.4) and (1.5.6) with $P_{s,t} = P_{t-s}$ for the present setting, we have

$$\partial_t P_t f_n = L P_t f_n = P_t L f_n$$

so that

$$L u_n = \int_0^\infty e^{-\lambda t} L P_t f_n \mathrm{d}t = \int_0^\infty e^{-\lambda t} \partial_t P_t f_n \mathrm{d}t = \lambda u_n - f_n.$$

Then

$$(L - \lambda)(u_n - u_m) = f_n - f_m, \quad n, m \geq 1.$$

By (1.6.14),

$$\lim_{n,m \to \infty} \left\{ \|u_n - u_m\|_{\tilde{H}^{\theta,r'}} + \|\nabla^2(u_n - u_m)\|_{\tilde{L}^p} \right\} = 0,$$

so that $u := \lim_{n \to \infty} u_n$ exists in $\tilde{H}^{\theta,p'} \cap \tilde{H}^{2,p}$, which solves (1.6.13). □

1.6.4 Proofs of Theorem 1.6.1 and Corollary 1.6.2

Proof of Theorem 1.6.1. By ($\Lambda^{1.4}$), conditions in Theorem 1.3.1 hold for $\phi(r) = 1$, so that we have the well-posedness, strong Feller property and irreducibility of (1.6.1). According to Theorem 1.6.3(1)–(2), it remains to prove the existence of the invariant probability measure $\bar{\mu}$ and the claimed assertions on the ergodicity.

(a) Let u solve (1.6.13) for $b^{(1)} = 0$ and $f = -b^{(0)}$ for large enough $\lambda > 0$, i.e.

$$(L - \nabla_{b^{(1)}} - \lambda)u = f, \tag{1.6.16}$$

such that (1.6.14) implies (1.3.16). Moreover, for $\Theta(x) := x + u(x)$, let $\hat{P}_t$ be the Markov semigroup associated with $Y_t := \Theta(X_t)$, so that

$$\hat{P}_t f(x) = \{P_t(f \circ \Theta)\}(\Theta^{-1}(x)), \quad t \geq 0, x \in \mathbb{R}^d, f \in \mathcal{B}_b(\mathbb{R}^d). \tag{1.6.17}$$

Since $\lim_{|x| \to \infty} \sup_{|y-x| \leq \varepsilon} \frac{|\nabla V(y)|}{V(x)} = 0$, by (1.3.16) and $V \geq 1$, we find a constant $\theta \in (0,1)$ such that

$$\begin{aligned} &\|\nabla u(x)\| \vee |x - \Theta(x)| \leq \varepsilon, \\ &\theta V(\Theta(x)) \leq V(x) \leq \theta^{-1} V(\Theta(x)), \quad x \in \mathbb{R}^d. \end{aligned} \tag{1.6.18}$$

Thus, it suffices to prove the desired assertions for $\hat{P}_t$ replacing P_t, where the unique invariant probability measure $\hat{\mu}$ of $\hat{P}_t$ and that $\bar{\mu}$ of P_t satisfies

$$\hat{\mu} = \bar{\mu} \circ \Theta^{-1}. \tag{1.6.19}$$

(b) Let X_t^n, Y_t^n and τ_n be given in the proof of Theorem 1.3.1(2) for the present time-homogeneous setting. Since $Y_t^n = Y_t$ and $1_{B(0,n)}(X_t^n) = 1$ for $t \leq \tau_n$, and since $\tau_n \to \infty$ as $n \to \infty$, (1.6.16) implies that $Y_t := \Theta(X_t)$ solves the SDE

$$dY_t = \{b^{(1)} + \lambda u + \nabla_{b^{(1)}} u\}(X_t)dt + \{(\nabla\Theta)\sigma\}(X_t)dW_t,$$

so that for any $\varepsilon \in (0, 1 \wedge r_0)$, by Itô's formula and (2.4.8), we find a constant $c_\varepsilon > 0$ such that

$$d\{V(Y_t) + M_t\}$$
$$\leq \Big[\langle \{b^{(1)} + \nabla_{b^{(1)}} u\}(X_t), \nabla V(Y_t)\rangle + c_\varepsilon(|\nabla V(Y_t)| + \|\nabla^2 V(Y_t)\|)\Big] dt$$
$$\leq \Big\{\langle b^{(1)}(X_t), \nabla V(X_t)\rangle$$
$$+ \varepsilon |b^{(1)}(X_t)| \sup_{B(X_t,\varepsilon)} \{|\nabla V| + \|\nabla^2 V\|\} + c_\varepsilon \sup_{B(X_t,\varepsilon)} (|\nabla V| + \|\nabla^2 V\|)\Big\} dt.$$

Combining this with (1.6.2), when $\varepsilon > 0$ is small enough we find constants $c_1, c_2 > 0$ such that

$$d\{V(Y_t) + M_t\} \leq \{c_1 - c_2\Phi(V(X_t))\}dt.$$

By (1.6.18), this implies that for some constant $c_4 > 0$,

$$dV(Y_t) \leq \{c_4 - c_2\Phi(\theta V(Y_t))\}dt - dM_t. \tag{1.6.20}$$

Thus,

$$\int_0^t \mathbb{E}\Phi(\theta V(Y_s))ds \leq \frac{c_4 + V(x)}{c_2} < \infty, \quad t > 0, Y_0 = x \in \mathbb{R}^d.$$

Since $\Phi(\theta V)$ is a compact function, this implies the existence of invariant probability $\hat{\mu}$ according to the standard Bogoliubov-Krylov's tightness argument. Moreover, (1.6.20) implies $\hat{\mu}(\Phi(\theta V)) < \infty$, so that by (1.6.18) and (1.6.19), $\bar{\mu}(\Phi(\varepsilon_0 V)) < \infty$ holds for $\varepsilon_0 = \theta^2$.

(c) By (1.3.5) and (1.6.17), $\hat{P}_t$ is t-regular for any $t > 0$, and any compact set $\mathbf{K} \subset \mathbb{R}^d$ is a petite set of $\hat{P}_t$, i.e. there exist $t > 0$ and a nontrivial measure ν such that

$$\inf_{x \in \mathbf{K}} P_t^* \delta_x \geq \nu.$$

By Theorem 1.6.3(2), (1.6.3) holds. Let $\hat{L}$ be the generator of $\hat{P}_t$. When $\Phi(r) \geq kr$ for some constant $k > 0$, (1.6.20) implies

$$\hat{L}V(x) \leq k_1 - k_2 V(x), \quad t \geq 0, x \in \mathbb{R}^d \qquad (1.6.21)$$

for some constants $k_1, k_2 > 0$. Since $\lim_{|x| \to \infty} V(x) = \infty$ and as observed above that any compact set is a petite set for $\hat{P}_t$, by Theorem 1.6.3(3), we obtain

$$\|\hat{P}_t^* \delta_x - \hat{\mu}\|_V \leq c e^{-\lambda t} V(x), \quad x \in \mathbb{R}^d, t \geq 0$$

for some constants $c, \lambda > 0$. Thus,

$$\|\hat{P}_t^* \delta_x - \hat{P}_t^* \delta_y\|_V \leq c e^{-\lambda t}(V(x) + V(y)), \quad t \geq 0, x, y \in \mathbb{R}^d.$$

Therefore, for any $\mu_1, \mu_2 \in \mathcal{P}$,

$$\|\hat{P}_t^* \mu_1 - \hat{P}_t^* \mu_2\|_V = \|\hat{P}_t^*(\mu_1 - \mu_2)^+ - \hat{P}_t^*(\mu_1 - \mu_2)^-\|_V$$

$$= \frac{1}{2}\|\mu_1 - \mu_2\|_{var} \left\| \hat{P}_t^* \frac{2(\mu_1 - \mu_2)^+}{\|\mu_1 - \mu_2\|_{var}} - \hat{P}_t^* \frac{2(\mu_1 - \mu_2)^-}{\|\mu_1 - \mu_2\|_{var}} \right\|_V$$

$$\leq \frac{c}{2} e^{-\lambda t} \|\mu_1 - \mu_2\|_{var} \left(\frac{2(\mu_1 - \mu_2)^+}{\|\mu_1 - \mu_2\|_{var}} + \frac{2(\mu_1 - \mu_2)^-}{\|\mu_1 - \mu_2\|_{var}} \right)(V)$$

$$\leq c e^{-\lambda t} \|\mu_1 - \mu_2\|_V.$$

This together with (1.6.17) and (1.6.18) implies (1.6.4) for some constants $c, \lambda > 0$.

(d) Let Φ be convex. By Jensen's inequality and (1.6.20), $\gamma_t := \theta \mathbb{E}[V(Y_t)]$ satisfies

$$\frac{\mathrm{d}}{\mathrm{d}t} \gamma_t \leq \theta c_4 - \theta c_2 \Phi(\gamma_t), \quad t \geq 0. \qquad (1.6.22)$$

Let

$$H(r) := \int_0^r \frac{\mathrm{d}s}{\Phi(s)}, \quad r \geq 0.$$

We aim to prove that for some constant $k > 1$

$$\gamma_t \leq k + H^{-1}(H(\gamma_0) - tk^{-1}), \quad t \geq 0, \qquad (1.6.23)$$

where $H^{-1}(r) := 0$ for $r \leq 0$. We prove this estimate by considering three situations.

(1) Let $\Phi(\gamma_0) \leq \frac{c_4}{c_2}$. Since (1.6.22) implies $\gamma_t' \leq 0$ for $\gamma_t \geq \Phi^{-1}(\frac{c_4}{c_2})$, so

$$\gamma_t \leq \Phi^{-1}(c_4/c_2), \quad t \geq 0. \qquad (1.6.24)$$

(2) Let $\frac{c_4}{c_2} < \Phi(\gamma_0) \leq \frac{2c_4}{c_2}$. Then (1.6.22) implies $\gamma'_t \leq 0$ for all $t \geq 0$ so that
$$\gamma_t \leq \Phi^{-1}(2c_4/c_2), \quad t \geq 0. \tag{1.6.25}$$

(3) Let $\Phi(\gamma_0) > \frac{2c_4}{c_2}$. If
$$t \leq t_0 := \inf\left\{t \geq 0 : \Phi(\gamma_t) \leq \frac{2c_4}{c_2}\right\},$$
then (1.6.22) implies
$$\frac{dH(\gamma_t)}{dt} = \frac{\gamma'_t}{\Phi(\gamma_t)} \leq -\frac{\theta c_2}{2},$$
so that
$$H(\gamma_t) \leq H(\gamma_0) - \frac{\theta c_2}{2}t, \quad t \in [0, t_0], \tag{1.6.26}$$
which implies
$$\gamma_t \leq H^{-1}(H(\gamma_0) - \theta c_2 t/2), \quad t \in [0, t_0].$$

Note that when $t > t_0$, $(\gamma_t)_{t \geq t_0}$ satisfies (1.6.22) with γ_{t_0} satisfying $\frac{c_4}{c_2} < \Phi(\gamma_{t_0}) \leq \frac{2c_4}{c_2}$, so that (1.6.24) holds, i.e.
$$\gamma_t \leq \Phi^{-1}(2c_4/c_2).$$

In conclusion, we obtain
$$\gamma_t \leq \Phi^{-1}(2c_4/c_2) + H^{-1}(H(\gamma_0) - \theta c_2 t/2), \quad t \geq 0.$$
Combining this with (1) and (2), we derive (1.6.23) for some constant $k > 1$.

(e) Since $1 \leq \Phi(r) \to \infty$ as $r \to \infty$, we find a constant $\delta > 0$ such that $\Phi(r) \geq \delta r, r \geq 0$. So, by step (b), (1.6.4) holds. Combining this with (1.6.23) we derive
$$\|\hat{P}^*_{t+s}\delta_x - \hat{\mu}\|_V = \sup_{|f| \leq V} |\hat{P}_s(\hat{P}_t f - \hat{\mu}(f))(x)|$$
$$\leq ce^{-\lambda t}\hat{P}_s V(x) \leq c\{k + H^{-1}(H(\theta V(x)) - k^{-1}s)\}e^{-\lambda t}.$$
Combining this with (1.6.17), (1.6.18) and (1.6.19), we derive (1.6.5) for some constants $k, \lambda > 0$.

Finally, if $H(\infty) < \infty$, we take $t^* = kH(\infty)$ in (1.6.5) to derive
$$\sup_{x \in \bar{D}} \|P_t \delta_x - \bar{\mu}\|_V \leq ce^{-\lambda t}, \quad t \geq t^*$$
for some constants $c, \lambda > 0$, which implies (1.6.6) by the argument leading to (1.6.4) in step (c). $\square$

Proof of Corollary 1.6.2. By (1.6.8), for any $\theta \in ((1-\alpha)^+, \frac{1}{2})$ there exists a constant $c_3 > 0$ such that

$$\phi(r) \geq c_3(1+r)^{1-\theta}, \quad r \geq 0.$$

Then (1.6.2) holds for $V := e^{(1+|\cdot|^2)^\theta}$ and $\Phi(r) = r$. So the first assertion in (1) follows from Theorem 1.6.1(1).

Next, (1.6.7) and (1.6.9) imply (1.6.2) for $V := \psi(|\cdot|^2)$ and $\Phi(r) = r$, so that the second assertion in (1) holds by Theorem 1.6.1(1).

Finally, if $\int_0^\infty \frac{ds}{\phi(s)} < \infty$, then for any $q > 0$, (1.6.2) holds for $V := (1+|\cdot|^2)^q$ and $\Phi(r) = (1+r)^{1-\frac{1}{q}}\phi(r^{\frac{1}{q}})$, so that $\int_0^\infty \frac{ds}{\Phi(s)} < \infty$. Then the proof is finished by Theorem 1.6.1(2). □

1.7 Notes and further results

In the previous sections we studied singular SDEs using Zvonkin's transform such that the non-degenerate noise kills the singular drift.

Intuitively, a degenerate noise should be able to kill a singular drift taking values in the image of the noise coefficients. In this spirit, the following SDE where $\sigma\sigma^*$ may be non-invertible has been studied in [Huang and Wang (2018)]:

$$dX_t = \{b_t^{(1)} + \sigma_t b_t^{(0)}\}(X_t)dt + \sigma_t(X_t)dW_t, \quad t \in [0, T],$$

where $\|\nabla b^{(1)}\|_\infty < \infty$ and $|\sigma_t b_t^{(0)}| \geq \lambda |b_t^{(0)}|$ holds for some constant $\lambda > 0$. This model contains two typical degenerate singular SDEs:

(1) singular SDEs on Heisenberg groups;
(2) singular stochastic Hamiltonian systems.

In the following we introduce some results on these two typical degenerate models. We will also introduce a result of [Yang and Zhang (2020b)] for the well-posedness of SDEs with a Kato class drift which is not included in assumption $(A^{1.3})$. See also [Wang (2016)] and references therein for the study of singular SDEs on Hilbert space as well as [Huang and Wang (2018)] for path-dependent singular SDEs.

1.7.1 Singular SDEs on Heisenberg groups

Consider the following vector fields on $\mathbb{R}^{m+d}$, where $m \geq 2, d \geq 1$:

$$U_i(x,y) = \sum_{k=1}^{m} \theta_{ki}\partial_{x_k} + \sum_{l=1}^{d}(A_l x)_i \partial_{y_l}, \quad 1 \leq i \leq m, \tag{1.7.1}$$

where $(x,y) = (x_1,\ldots,x_m,y_1,\ldots,y_d) \in \mathbb{R}^{m+d}$, $\Theta := (\theta_{ij})$ and $A_l(1 \leq l \leq d)$ are $m \times m$-matrices satisfying the following assumption:

$(A^{1.6})$ Θ is invertible, $G_l := A_l\Theta - \Theta^* A_l^* \neq 0 (1 \leq l \leq d)$, and there exists $\varepsilon \in [0,1)$ such that

$$\varepsilon \sum_{l=1}^{d} a_l^2 |G_l u|^2 \geq \sum_{1 \leq l \neq k \leq d} |a_l a_k \langle G_l u, G_k u \rangle|, \quad a \in \mathbb{R}^d, u \in \mathbb{R}^m.$$

Under this condition, the operator

$$\mathcal{L} := \frac{1}{2}\sum_{i=1}^{m} U_i^2$$

is hypoelliptic and symmetric in $L^2(\mathbb{R}^{m+d})$, and the associated diffusion process solves the SDE for $(X_t, Y_t) \in \mathbb{R}^{m+d}$:

$$d(X_t, Y_t) = \sum_{i=1}^{m} U_i(X_t) \circ dW_t^i = Z dt + \sigma(X_t)dW_t, \quad t \in [0,T], \tag{1.7.2}$$

where $W_t := (W_t^i)_{1 \leq i \leq m}$ is the m-dimensional Brownian motion, and

$$\sigma(x) := (\Theta, A_1 x,\ldots, A_d x), \quad Z \cdot \nabla := \sum_{i=1}^{m} \nabla_{U_i} U_i = \sum_{l=1}^{d} \text{tr}(\Theta A_l)\partial_{y_l}.$$

We now consider the following SDE with a singular drift $b : [0,T] \times \mathbb{R}^{m+d} \to \mathbb{R}^m$:

$$d(\tilde{X}_t, \tilde{Y}_t) = \{\sigma(\tilde{X}_t)b_t(\tilde{X}_t, \tilde{Y}_t) + Z\}dt + \sigma(\tilde{X}_t)dW_t. \tag{1.7.3}$$

A typical example is $d = m-1, \Theta = I_{m \times m}$ and for some constants $\alpha_l \neq \beta_l$,

$$(A_l)_{ij} = \begin{cases} \alpha_l, & \text{if } i=1, j=l+1, \\ \beta_l, & \text{if } i=l+1, j=1, \\ 0, & \text{otherwise.} \end{cases}$$

Then $G_l^* G_k = 0$ for $l \neq k$, so that $(A^{1.6})$ holds with $\varepsilon = 0$. In particular, for $a_l = -\beta_l = \frac{1}{2}$, $\mathcal{L}$ is the Kohn-Laplacian operator on the $(2m-1)$-dimensional Heisenberg group. In general, $\mathbb{R}^{m+d}$ is a group under the action

$$(x,y) \bullet (x',y') := (x+x', y+y' + \langle (\Theta^*)^{-1} \Lambda.x, x' \rangle),$$
$$(x,y),(x',y') \in \mathbb{R}^{m+d}, \qquad (1.7.4)$$

and $U_i, 1 \leq i \leq m$ are left-invariant vector fields. So, we call (1.7.3) a singular SDE on the generalized Heisenberg group.

Let $\Delta_y = \sum_{l=1}^d \partial_{y_l}^2$. Then $(\Delta_y, W^{2,2}(\mathbb{R}^d))$ is a negative definite operator in $L^2(\mathbb{R}^d)$. For any $\alpha > 0$ and $\lambda \geq 0$, we consider the operator $(\lambda - \Delta_y)^\alpha$ defined on domain $\mathcal{D}((-\Delta_y)^\alpha) := H^{2\alpha,2}(\mathbb{R}^d)$, see (1.2.1). This operator extends naturally to a measurable function f on the product space $\mathbb{R}^{m+d}$ such that $f(x,\cdot) \in \mathcal{D}((-\Delta_y)^\alpha)$ for $x \in \mathbb{R}^m$:

$$(\lambda - \Delta_y)^\alpha f(x,y) := (\lambda - \Delta)^\alpha f(x,\cdot)(y).$$

For any $\beta > 0, p \geq 1$, let $H_y^{\alpha,p}$ be the space of measurable functions on $\mathbb{R}^{m+d}$ such that

$$\|f\|_{H_y^{\beta,p}} := \|(1-\Delta_y)^{\frac{\beta}{2}} f\|_p \asymp \|f\|_p + \|(-\Delta_y)^{\frac{\beta}{2}} f\|_p < \infty,$$

where $f \asymp g$ for two positive functions means that $c^{-1} f \leq g \leq cf$ holds for some constant $c > 1$. Recall that for $\beta \in (0,2)$ and $z \in \mathbb{R}^{m+d}$,

$$-(-\Delta_y)^{\frac{\beta}{2}} f(z) = c(\beta) \int_{\mathbb{R}^d} (f(z+(0,y')) - f(z)) |y'|^{-(m+\beta)} dy'$$

holds for some constant $c(\beta) > 0$.

For any $\beta > 0, p, q \geq 1$, let $H_y^{\beta,p,q}$ be the completion of $C_0^\infty([0,T] \times \mathbb{R}^{m+d})$ with respect to the norm

$$\|f\|_{H_y^{\beta,p,q}} := \|(1-\Delta_y)^{\frac{\beta}{2}} f\|_{L_p^q} \asymp \|f\|_{L_p^q} + \|(-\Delta_y)^{\frac{\beta}{2}} f\|_{L_p^q}.$$

The following result is taken from Theorem 3.1 in [Huang and Wang (2018)].

Theorem 1.7.1. *Assume* $(A^{1.6})$ *and let* $p, q \geq 1$ *satisfy* $\frac{2}{q} + \frac{m+2d}{p} < 1$.

(1) *If* $|b| \in L_p^q([0,T] \times \mathbb{R}^{m+d})$, *then for any initial value* $x \in \mathbb{R}^{m+d}$, *the SDE* (1.7.3) *has a weak solution* $(X_t)_{t \in [0,T]}$ *starting at* x *with* $\mathbb{E}[e^{\lambda \int_0^T |b_t(X_t)|^2 dt}] < \infty$ *for all* $\lambda > 0$.

(2) *If* $(hb) \in H_y^{\frac{1}{2},p,q}$ *holds for any* $h \in C_0^\infty(\mathbb{R}^{m+d})$, *then for any initial value* $x \in \mathbb{R}^{m+d}$, *the SDE* (1.7.3) *has a unique strong solution up to life time*.

1.7.2 Singular stochastic Hamiltonian systems

As a probability model characterizing Langevin kinetic equations, the following degenerate SDE for $(X_t, Y_t) \in \mathbb{R}^{2d}$ is known as stochastic Hamiltonian equation system:

$$\begin{cases} \mathrm{d}X_t = Z_t(X_t, Y_t)\mathrm{d}t, \\ \mathrm{d}Y_t = b_t(X_t, Y_t)\mathrm{d}t + \sigma_t(X_t, Y_t)\mathrm{d}W_t, \quad t \in [0, T], \end{cases} \quad (1.7.5)$$

where W_t is the d-dimensional Brownian motion and

$$\sigma : [0, T] \times \mathbb{R}^{2d} \to \mathbb{R}^d \otimes \mathbb{R}^d, \quad Z, b : [0, T] \times \mathbb{R}^{2d} \to \mathbb{R}^d$$

are measurable. A typical model is that $Z_t(x, y) = y$, for which the SDE becomes

$$\begin{cases} \mathrm{d}X_t = Y_t\mathrm{d}t, \\ \mathrm{d}Y_t = b_t(X_t, Y_t)\mathrm{d}t + \sigma_t(X_t, Y_t)\mathrm{d}W_t, \quad t \in [0, T], \end{cases} \quad (1.7.6)$$

where X_t stands for the position at time t of a moving random particle while Y_t is the speed of X_t, so that the noise perturbs the speed variable.

Let $\nabla^{(i)}$ be the gradient in the i-th component of $(x, y) \in \mathbb{R}^d \times \mathbb{R}^d$, $i = 1, 2$. The following well-posedness result of (1.7.5) is taken from [Wang and Zhang (2016)] where moments and continuity estimates are also presented.

Theorem 1.7.2. *The SDE* (1.7.5) *is well-posed if there exist a closed convex subspace* $\mathbb{M}_{iv}$ *of invertible* $d \times d$-*matrices, an increasing function* $\phi : [0, \infty) \to [0, \infty)$ *with*

$$\int_0^1 \frac{\phi(t)}{t}\mathrm{d}t < \infty, \quad \lim_{t \downarrow 0} \frac{\phi(\lambda t)}{\phi(t)} = 1, \quad \lambda > 0,$$

and a function $\gamma \in C^1([0, \infty); [0, \infty))$ *with*

$$\int_0^1 \frac{\mathrm{d}t}{t\gamma(t)} = \infty, \quad \liminf_{t \downarrow 0} \left\{ \frac{\gamma(t)}{4} + t\gamma'(t) \right\} > 0,$$

such that

(1) $\nabla^{(2)}Z_t(x, y) \in \mathbb{M}_{iv}$ *and* $\sigma_t(x)$ *is invertible such that*

$$\|\nabla^{(2)}Z\|_\infty + \|\sigma\|_\infty + \|\sigma^{-1}\|_\infty < \infty.$$

(2) *For any* $t \in [0, T]$ *and* $x, y, x', y' \in \mathbb{R}^d$,

$$|Z_t(x, y) - Z_t(x', y)| \leq |x - x'|^{\frac{2}{3}}\phi(|x - x'|),$$
$$\|\nabla^{(2)}Z_t(x, y) - \nabla^{(2)}Z_t(x, y')\| \leq \phi(|y - y'|).$$

(3) *Either*

$$|b_t(x,y) - b_t(x',y')| \le \{|x-x'|^{\frac{2}{3}}\phi(|x-x'|) + \phi^{\frac{7}{2}}(|y-y'|)\},$$
$$\|\sigma_t(x) - \sigma_t(y)\| \le |x-y|\sqrt{\gamma(|x-y|)}, \quad t \in [0,T], x, x', y, y' \in \mathbb{R}^d;$$

or $\|\nabla^{(2)}\sigma\|_\infty < \infty$ *and*

$$|b_t(x,y) - b_t(x',y')| \le \{|x-x'|^{\frac{2}{3}}\phi(|x-x'|) + \phi(|y-y'|)\},$$
$$\|\nabla^{(2)}\sigma_t(x,y) - \nabla^{(2)}\sigma_t(x',y)\| \le |x-x'|\sqrt{\gamma(|x-x'|)},$$
$$\|\sigma_t(x,y) - \sigma_t(x',y)\| \le |x-x'|\sqrt{\gamma(|x-x'|)}, t \in [0,T], x, x', y, y' \in \mathbb{R}^d.$$

By a standard truncation argument, if conditions in Theorem 1.7.2 hold for x, x', y, y' in any compact set of $\mathbb{R}^d$, then the SDE is locally well-posed; i.e. it is well-posed up to life time. The weak well-posedness of (1.7.5) is also derived in [Wang and Zhang (2016)] under slightly weaker conditions.

The following result for (1.7.6) is modified from [Zhang (2018)], where by a localization argument, we have replaced L^p by $\tilde{L}^p$, see [Zhang (2021)] for the weak existence of (1.7.6) where the drift b only satisfies a locally integrable condition and may depend on the distribution density of the solution (see Subsection 3.6.3).

Theorem 1.7.3. *The SDE* (1.7.6) *is well-posed if* $\|\sigma\|_\infty + \|\sigma^{-1}\|_\infty < \infty$ *and there exists* $p > 2(2d+1)$ *such that*

$$\sup_{t \in [0,T]} \left\{ \|\nabla\sigma\|_{\tilde{L}^p} + \int_0^T \|(1-\Delta_x)^{\frac{1}{3}} b_s\|_{\tilde{L}^p}^p \mathrm{d}s \right\} < \infty,$$

where Δ_x *is the Laplacian for the first variable* x *and* $\tilde{L}^p$ *is with respect to the Lebesgue measure on* $\mathbb{R}^{2d}$.

For other references on the SDE (1.7.6), see [Chaudru de Raynal (2017)] and [Wang and Zhang (2018)] for a stronger situation where the drift is Hölder continuous, [Wang and Zhang (2013)] for Bismut formula for this type of degenerate SDEs, [Wang (2017b)] and [Wang and Zhang (2014)] for the hypercontractivity and dimension-free Harnack inequalities for finite and infinite-dimensional stochastic Hamiltonian systems, and [Grothause and Stigenbauer (2014)], [Grothause and Wang (2019)], [Baudoin *et al.* (2021)] and references therein for the ergodicity of stochastic Hamiltonian systems.

1.7.3 Singular SDEs with Kato and critical drifts

Consider the following SDE on $\mathbb{R}^d$ with additive noise:

$$dX_t = b_t(X_t)dt + dW_t, \quad t \in [0,T], \tag{1.7.7}$$

i.e. in (1.1.1) we set $m = d$ and $\sigma = I_d$. A measurable function f on $[0,T] \times \mathbb{R}^d$ is said in the Kato class $\mathbf{K}_{d,\alpha}$ for $\alpha > 0$, if

$$\lim_{\varepsilon \downarrow 0} \sup_{t \in [0, T+\varepsilon], x \in \mathbb{R}^d} \int_{-\varepsilon}^{\varepsilon} 1_{[-t, T-t]}(s)|s|^{-\frac{d+2-\alpha}{2}} ds \int_{\mathbb{R}^d} e^{-\frac{\lambda |x-y|^2}{2|s|}} |f_{t+s}(y)| dy = 0.$$

The following result is due to Theorems 2.2 and 5.1 in [Yang and Zhang (2020b)].

Theorem 1.7.4. (1) If $|b|^2 \in \mathbf{K}_{d,\alpha}$ for some $\alpha > 0$, then (1.7.7) is well-posed.

(2) If $|b| \in \mathbf{K}_{d,1}$ then (1.7.7) is weakly well-posed, and the solution has a transition density satisfying the following Gaussian upper bounded estimate for some constants $c_1, c_2 > 0$:

$$p_{s,t}(x,y) \le \frac{c_1}{(t-s)^{\frac{d}{2}}} e^{-\frac{c_2|x-y|^2}{t-s}}, \quad 0 \le s < t \le T, x, y \in \mathbb{R}^d.$$

Finally, the following well-posedness result has been proved in [Röckner and Zhao (2023)] and [Röckner and Zhao (2020b)] for b satisfying the critical integrability condition $|b| \in L_{q_0}^{p_0}$ with $\frac{d}{p_0} + \frac{2}{q_0} = 1$.

Theorem 1.7.5. If either $b \in C([0,T]; L^d(\mathbb{R}^d))$ or $|b| \in L_{q_0}^{p_0}$ for some $(p_0, q_0) \in (2, \infty)$ with $\frac{d}{p_0} + \frac{2}{q_0} = 1$, then (1.7.7) is well-posed. If $|b| \in L_\infty^d$ then (1.7.7) is weakly well-posed.

Chapter 2

Singular Reflected SDEs

When the SDE (1.1.1) is restricted to a domain $D \subset \mathbb{R}^d$, a natural model is the following reflected SDE:

$$\mathrm{d}X_t = b_t(X_t)\mathrm{d}t + \sigma_t(X_t)\mathrm{d}W_t + \mathbf{n}(X_t)\mathrm{d}l_t, \quad t \in [0, T], \qquad (2.0.1)$$

where $\mathbf{n}$ is the inward normal vector field of the boundary ∂D (see Definition 2.1.1), and l_t is a continuous adapted increasing process with $\mathrm{d}l_t$ supported on $\{t \in [0, T] : X_t \in \partial D\}$.

The problem of confining a stochastic process to a domain goes back to [Skorohod (1961, 1962)], and has been well developed under monotone (or locally semi-Lipschitz) conditions, see the recent work of [Hino et al. (2021)] and references therein. In this chapter, we study (2.0.1) with singular coefficients based on [Wang (2023b)].

2.1 Reflected SDE and Neumann problem

Let $D \subset \mathbb{R}^d$ be a connected open domain with boundary ∂D. For any $x \in \partial D$ and $r > 0$, let

$$\mathcal{N}_{x,r} := \{\mathbf{n} \in \mathbb{R}^d : |\mathbf{n}| = 1, B(x - r\mathbf{n}, r) \cap D = \emptyset\},$$

where $B(x, r) := \{y \in \mathbb{R}^d : |x - y| < r\}$. Since $\mathcal{N}_{x,r}$ is decreasing in $r > 0$, we have

$$\mathcal{N}_x := \cup_{r>0} \mathcal{N}_{x,r} = \lim_{r \downarrow 0} \mathcal{N}_{x,r}, \quad x \in \partial D.$$

We call $\mathcal{N}_x$ the set of inward unit normal vectors of ∂D at point x. When ∂D is differentiable at x, $\mathcal{N}_x$ is a singleton set. Otherwise $\mathcal{N}_x$ may be empty or contain more than one vector. For instance, letting D be the interior of a triangle in $\mathbb{R}^2$, at each vertex x, the set $\mathcal{N}_x$ contains infinite many vectors, whereas for D being the exterior of the triangle, $\mathcal{N}_x$ is empty at each vertex point x.

Definition 2.1.1. A measurable map $\mathbf{n} : \partial D \to \mathbb{R}^d$ with $\mathbf{n}(x) \in \mathcal{N}_x(x \in \partial D)$ is called an inward normal vector field of ∂D.

The following assumption on D goes back to [Lions and Sznitman (1984); Saisho (1987)]. Recall that D is called convex, if $rx + (1-r)y \in D$ for $x, y \in D$ and $r \in [0, 1]$.

(D) *Either D is convex, or there exists a constant $r_0 > 0$ such that $\mathcal{N}_x = \mathcal{N}_{x,r_0} \neq \emptyset$ for $x \in \partial M$, and*

$$\sup_{v \in \mathbb{R}^d, |v|=1} \inf \big\{ \langle v, \mathbf{n}(y) \rangle : y \in B(x, r_0) \cap \partial D, \mathbf{n}(y) \in \mathcal{N}_y \big\} \tag{2.1.1}$$
$$\geq r_0, \quad x \in \partial D.$$

Remark 2.1.1. We present below some facts on assumption **(D)**.

(1) According to Remark 1.1 in [Saisho (1987)], for any $x \in \partial D$ and $r > 0$, $\mathbf{n} \in \mathcal{N}_{x,r}$ if and only if

$$\langle y - x, \mathbf{n} \rangle \geq -\frac{|y-x|^2}{2r}, \quad y \in \bar{D},$$

so that the condition $\mathcal{N}_x = \mathcal{N}_{x,r_0}$ in **(D)** implies

$$\langle y - x, \mathbf{n}(x) \rangle \geq -\frac{|y-x|^2}{2r_0}, \quad y \in \bar{D}, x \in \partial D, \mathbf{n}(x) \in \mathcal{N}_x. \tag{2.1.2}$$

When D is convex, $\mathcal{N}_x = \mathcal{N}_{x,r_0}$ holds for all $x \in \partial D$ and $r_0 > 0$,

$$\langle y - x, \mathbf{n}(x) \rangle \geq 0, \quad y \in \bar{D}, x \in \partial D, \mathbf{n}(x) \in \mathcal{N}_x, \tag{2.1.3}$$

and (2.1.1) holds if $d = 2$ or D is bounded, see [Tanaka (1979)].
(2) When ∂D is C^1-smooth, for each $x \in \partial D$ the set $\mathcal{N}_x$ is singleton. If $\mathbf{n}(x) \in \mathcal{N}_x$ is uniformly continuous in $x \in \partial D$, then (2.1.1) holds for small $r_0 > 0$. In particular, **(D)** holds when $\partial D \in C_b^2$ in the following sense.

Definition 2.1.2. For any $r > 0$, let

$$\partial_r D := \big\{ x \in \bar{D} : \text{dist}(x, \partial D) \leq r \big\}, \quad \partial_{-r} D := \big\{ x \in D^c : \text{dist}(x, \partial D) \leq r \big\},$$
$$D_r := D \cup (\partial_{-r} D) = \big\{ x \in \mathbb{R}^d : \text{dist}(x, D) \leq r \big\}.$$

For any $k \in \mathbb{N}$, we write $\partial D \in C^k$ (respectively, $\partial D \in C_b^k$) if there exists a constant $r_0 > 0$ such that the polar coordinate map

$$I : \partial D \times [-r_0, r_0] \ni (\theta, \rho_\partial) \mapsto \theta + \rho_\partial \mathbf{n}(\theta) \in (\partial_{r_0} D) \cup \partial_{-r_0} D$$

is a C^k-diffeomorphism, such that $(\theta(x), \rho_\partial(x)) := I^{-1}(x)$ having continuous (respectively, bounded and continuous) derivatives in $x \in (\partial_{r_0} D) \cup \partial_{-r_0} D$ up to the k-th order, where $\theta(x)$ is the projection of x to ∂D and

$$\rho_\partial(x) = \mathrm{dist}(x, \partial D) 1_{\{\partial_{r_0} D\}}(x) - \mathrm{dist}(x, \partial D) 1_{\{\partial_{-r_0} D\}}(x), \quad x \in (\partial_{r_0} D) \cup \partial_{-r_0} D. \tag{2.1.4}$$

Moreover, for $\varepsilon \in (0, 1)$, we denote $\partial D \in C_b^{k+\varepsilon}$ if it is in C_b^k with $\nabla^k \rho_\partial$ and $\nabla^k \theta$ being ε-Hölder continuous on $\partial_{r_0} D$. Finally, we write $\partial D \in C_b^{k,L}$ if it is C_b^k with $\nabla^k \rho_\partial$ being Lipschitz continuous on $\partial_{r_0} D$.

Note that $\partial D \in C_b^k$ does not imply the boundedness of D or ∂D, but any bounded C^k domain satisfies $\partial D \in C_b^k$.

$(A^{2.1})$ **(D)** holds, $a := \sigma \sigma^*$ and b are measurable functions on $[0, T] \times \mathbb{R}^d$, b has decomposition $b = b^{(0)} + b^{(1)}$ with $b_t^{(0)}|_{\bar{D}^c} = 0$, such that the following conditions hold:
(1) a_t is invertible with $\|a\|_\infty + \|a^{-1}\|_\infty < \infty$, and

$$\lim_{\varepsilon \to 0} \sup_{|x-y| \le \varepsilon, t \in [0,T]} \|a_t(x) - a_t(y)\| = 0. \tag{2.1.5}$$

(2) There exists $(p_0, q_0) \in \mathcal{K}$ such that $|b^{(0)}| \in \tilde{L}_{q_0}^{p_0}(T)$. Moreover, $b^{(1)}$ is locally bounded on $[0, T] \times \mathbb{R}^d$, and there exist a constant $L > 1$ and a function $\tilde{\rho}_\partial \in C_b^2(\bar{D})$ such that

$$\|\nabla b^{(1)}\|_\infty := \sup_{t \in [0,T], x \neq y} \frac{|b_t^{(1)}(x) - b_t^{(1)}(y)|}{|x - y|} \le L, \tag{2.1.6}$$

$$\langle b_t^{(1)}, \nabla \tilde{\rho}_\partial \rangle|_D \ge -L, \quad \langle \nabla \tilde{\rho}_\partial, \mathbf{n} \rangle|_{\partial D} \ge 1, \quad t \in [0, T]. \tag{2.1.7}$$

$(A^{2.2})$ $(A^{2.1})$ holds, and there exist $l \in \mathbb{N}$, $\{(p_i, q_i)\}_{0 \le i \le l} \subset \mathcal{K}$ and $\{f_i \in \tilde{L}_{q_i}^{p_i}(T)\}_{1 \le i \le l}$ such that

$$|b^{(0)}|^2 \in \tilde{L}_{q_0}^{p_0}(T), \quad \|\nabla \sigma\|^2 \le \sum_{i=1}^{l} f_i.$$

Remark 2.1.2. Each of the following two conditions implies the existence of $\tilde{\rho}_\partial$ in (2.1.7):

(a) $\partial D \in C_b^2$ and there exists a constant $K > 0$ such that $\langle b_t^{(1)}, \mathbf{n}\rangle|_{\partial D} \geq -K$ for $t \in [0, T]$;

(b) D is bounded and there exist $\varepsilon \in (0, 1)$ and $x_0 \in D$ such that

$$\langle x_0 - x, \mathbf{n}(x)\rangle \geq \varepsilon|x - x_0|, \quad x \in \partial D. \tag{2.1.8}$$

Indeed, if (a) holds then there exists $r_0 > 0$ such that $\rho_\partial \in C_b^2(\partial_{r_0} D)$. Let $h \in C^\infty([0, \infty))$ with $h(r) = r$ for $r \in [0, r_0/4]$ and $h(r) = r_0/2$ for $r \geq r_0/2$. By taking $\tilde{\rho}_\partial = h \circ \rho_\partial$ we have $\tilde{\rho}_\partial \in C_b^2(\bar{D})$, $\langle \nabla \tilde{\rho}_\partial, \mathbf{n}\rangle|_{\partial D} = 1$, and for any $x \in D$ letting $\bar{x} \in \partial D$ such that $|x - \bar{x}| = \rho_\partial(x)$, we deduce from (2.1.6) that

$$\langle b_t^{(1)}(x), \nabla \tilde{\rho}_\partial(x)\rangle = h'(\rho_\partial(x))\{\langle b_t^{(1)}(\bar{x}), \mathbf{n}(\bar{x})\rangle + \langle b_t^{(1)}(x) - b_t^{(1)}(\bar{x}), \mathbf{n}(\bar{x})\rangle\}$$
$$\geq -(1 + r_0)L\|h'\|_\infty.$$

Therefore, (2.1.7) holds for some (different) constant L. Next, if (b) holds, by (2.1.8) we may take $\tilde{\rho}_\partial(x) = N\sqrt{1 + |x - x_0|^2}$ for large enough $N \geq 1$ such that $\langle \nabla \tilde{\rho}_\partial, \mathbf{n}\rangle|_{\partial D} \geq 1$. So, by the boundedness of D and $b^{(1)} \in C([0, T] \times \mathbb{R}^d)$, (2.1.7) holds for some constant $L > 0$.

Assumption $(A^{2.1})$ will be used to establish Krylov's estimate for functions $f \in \cap_{(p,q) \in \mathcal{K}} \tilde{L}_q^p(T)$, see Lemma 2.2.1 below. To improve this estimate for (p, q) satisfying $\frac{d}{p} + \frac{2}{q} < 2$ as in Theorem 1.2.3(2), we introduce one more assumption.

Consider the following differential operators on $\bar{D}$:

$$L_t^{\sigma, b^{(1)}} := \frac{1}{2}\mathrm{tr}\left(\sigma_t \sigma_t^* \nabla^2\right) + \nabla_{b_t^{(1)}}, \quad t \in [0, T]. \tag{2.1.9}$$

Let $\{P_{s,t}^{\sigma, b^{(1)}}\}_{T \geq t_1 \geq t \geq s \geq 0}$ be the Neumann semigroup on $\bar{D}$ generated by $L_t^{\sigma, b^{(1)}}$, that is, for any $\phi \in C_b^2(\bar{D})$, and any $t \in (0, T]$, $(P_{s,t}^{\sigma, b^{(1)}}\phi)_{s \in [0,t]}$ is the unique solution of the PDE

$$\partial_s u_s = -L_s^{\sigma, b^{(1)}} u_s, \quad \nabla_\mathbf{n} u_s|_{\partial D} = 0 \text{ for } s \in [0, t), u_t = \phi. \tag{2.1.10}$$

For any $t > 0$, let $C^{1,2}([0, t) \times \bar{D})$ be the set of functions $f \in C_b([0, t) \times \bar{D})$ with continuous derivatives $\partial_t f, \nabla f$ and $\nabla^2 f$.

For any $p, q \geq 1$, any $0 \leq s < t \leq T$, let $\tilde{L}_q^p(s, t, D)$ be the class of measurable functions on $[s, t] \times \bar{D}$ such that $1_{\bar{D}} f \in \tilde{L}_q^p(s, t)$. Moreover, we denote $\tilde{L}_q^p(t, D) := \tilde{L}_q^p(0, t, D), t \in (0, T]$.

$(A^{2.3})$ $\partial D \in C_b^{2,L}$ and the following conditions hold for σ and b on $[0,T] \times \bar{D}$:
(1) $a_t := \sigma_t \sigma_t^*$ is invertible, (2.1.5) holds for $x, y \in \bar{D}$, and for some $\{f_i \in \tilde{L}_{q_i}^{p_i}(T, D), (p_i, q_i) \in \mathcal{K}\}_{1 \le i \le l}$,

$$\|a\|_\infty + \|a^{-1}\|_\infty < \infty, \quad \|\nabla \sigma\| \le \sum_{i=1}^l f_i$$

holds on $[0,T] \times \bar{D}$.
(2) $b = b^{(1)} + b^{(0)}$ with $\nabla_\mathbf{n} b_t^{(1)}|_{\partial D} = 0$, $\|\nabla b^{(1)}\|_\infty + \|1_{\partial D}\langle b^{(1)}, \mathbf{n}\rangle\|_\infty < \infty$ and $|b^{(0)}| \in \tilde{L}_{q_0}^{p_0}(T, D)$ for some $(p_0, q_0) \in \mathcal{K}$ with $p_0 > 2$.
(3) For any $\phi \in C_b^2(\bar{D})$ and $t \in (0, T]$, the PDE (2.1.10) has a unique solution $P_{\cdot,t}^{\sigma,b^{(1)}} \phi \in C^{1,2}([0,t) \times \bar{D})$, such that for $\nabla^0 \phi := \phi$ and some constant $c > 0$, we have

$$\|\nabla^i P_{s,t}^{\sigma,b^{(1)}} \phi\|_\infty \le c(t-s)^{-\frac{1}{2}} \|\nabla^{i-1}\phi\|_\infty,$$
$$\|\partial_s P_{s,t}^{\sigma,b^{(1)}} \phi\|_\infty \le c(t-s)^{-\frac{1}{2}} \|\nabla \phi\|_\infty, \qquad (2.1.11)$$
$$0 \le s < t \le T, \ i = 1, 2, \phi \in C_b^2(\bar{D}).$$

Remark 2.1.3. (1) Let $\rho_\partial \in C_b^2(\partial_{r_0} D)$ for some $r_0 > 0$. Since $\nabla \rho_\partial|_{\partial D} = \mathbf{n}$, $\|\nabla b^{(1)}\|_\infty + \|1_{\partial D}\langle b^{(1)}, \mathbf{n}\rangle\|_\infty < \infty$ implies $\|1_{\partial_{r_0} D}\langle b^{(1)}, \nabla \rho_\partial\rangle\|_\infty < \infty$, which will be used in the proof of Lemma 2.2.2 below.
(2) $(A^{2.3})(3)$ holds if D is bounded with $\partial D \subset C^{2+\alpha}$ for some $\alpha \in (0,1)$, and there exists $c > 0$ such that

$$\{|b_t^{(1)}(x) - b_s^{(1)}(y)| + \|a_t(x) - a_s(y)\|\} \le c(|t-s|^\alpha + |x-y|^{\frac{\alpha}{2}}),$$
$$s, t \in [0,T], x, y \in \bar{D}. \qquad (2.1.12)$$

Indeed, $\partial D \in C^{2+\alpha}$ implies $\mathbf{n} \in C^{1+\alpha}(\partial D)$, so that (2.1.12) implies estimates (3.4) and (3.6) in Theorem VI.3.1 in [Carroni and Menaldi (1992)] with $\varrho = \infty$ for the Neumann heat kernel $p_{s,t}^{\sigma,b^{(1)}}(x,y)$ of $P_{s,t}^{\sigma,b^{(1)}}$. We note that according to its proof, the condition (3.3) therein is assumed for some $\alpha \in (0,1)$ rather than all $\alpha \in (0,1)$. In particular, $\nabla^2 p_{s,t}^{\sigma,b^{(1)}}(\cdot,y)(x)$ and $\partial_s p_{s,t}^{\sigma,b^{(1)}}(x,y)$ are continuous in $(s,x) \in [0,t] \times \bar{D}$, and there exists a constant $c > 1$ such that

$$|\partial_s p_{s,t}^{\sigma,b^{(1)}}(x,y)| = |L_s^{\sigma,b^{(1)}} p_{s,t}^{\sigma,b^{(1)}}(\cdot,y)(x)| \le c|t-s|^{-\frac{d+2}{2}} e^{-\frac{|x-y|^2}{c(t-s)}},$$
$$|\nabla^i p_{s,t}^{\sigma,b^{(1)}}(\cdot,y)(x)| \le c|t-s|^{-\frac{d+i}{2}} e^{-\frac{|x-y|^2}{c(t-s)}},$$
$$0 \le s < t \le T, \ x, y \in \bar{D}, \ i = 0, 1, 2.$$

These properties imply (2.1.11). For instance, by $\int_D p_{s,t}(x,y)dy = 1$, the second estimate implies that for some constant $c' > 0$,

$$|\partial_s P_{s,t}^{\sigma,b^{(1)}}\phi(x)| = \left|\partial_s \int_D p_{s,t}^{\sigma,b^{(1)}}(x,y)\phi(y)dy\right|$$

$$= \left|\partial_s \int_D p_{s,t}^{\sigma,b^{(1)}}(x,y)\{\phi(y) - \phi(x)\}dy\right|$$

$$\leq c\|\nabla\phi\|_\infty \int_D |x-y|\cdot|t-s|^{-\frac{d+2}{2}} e^{-\frac{|x-y|^2}{c(t-s)}} dy$$

$$\leq c'(t-s)^{-\frac{1}{2}}, \quad 0 \leq s < t \leq T, x \in \bar{D}.$$

When $D = \mathbb{R}^d$, these estimates (hence (2.1.11)) hold for more general σ and $b^{(1)}$, see [Menozzi et al. (2021)].

Definition 2.1.3. (1) A pair $(X_t, l_t)_{t\in[0,T]}$ is called a solution of (2.0.1), if X_t is an adapted continuous process on $\bar{D}$, l_t is an adapted continuous increasing process with dl_t supported on $\{t \in [0,T] : X_t \in \partial D\}$, such that $\mathbb{P}$-a.s.

$$\int_0^t \{|b_r(X_r)| + \|\sigma_r(X_r)\|^2\}dr < \infty, \quad t \in [0,T],$$

and for an inward normal vector field $\mathbf{n}$ of ∂D, $\mathbb{P}$-a.s.

$$X_t = X_0 + \int_0^t b_r(X_r, \mathcal{L}_{X_r})dr + \int_0^t \sigma_r(X_r, \mathcal{L}_{X_r})dW_r + \int_0^t \mathbf{n}(X_r)dl_r, \quad t \in [0,T].$$

In this case, l_t is called the local time of X_t on ∂D. We call (2.0.1) strongly well-posed if for any $X_0 \in \bar{D}$, the equation has a unique solution.

(2) A triple $(X_t, l_t, W_t)_{t\in[0,T]}$ is called a weak solution of (2.0.1), if W_t is an m-dimensional Brownian motion under a probability space and $(X_t, l_t)_{t\in[0,T]}$ solves (2.0.1). (2.0.1) is said to have weak uniqueness (resp. jointly weak uniqueness), if for any two weak solutions $(X_t, l_t, W_t)_{t\in[0,T]}$ under probability $\mathbb{P}$ and $(\tilde{X}_t, \tilde{l}_t, \tilde{W}_t)_{t\in[0,T]}$ under probability $\tilde{\mathbb{P}}$, $\mathcal{L}_{X_0|\mathbb{P}} = \mathcal{L}_{\tilde{X}_0|\tilde{\mathbb{P}}}$ implies $\mathcal{L}_{(X_t,l_t)_{t\in[0,T]}|\mathbb{P}} = \mathcal{L}_{(\tilde{X}_t,\tilde{l}_t)_{t\in[0,T]}|\tilde{\mathbb{P}}}$ (resp. $\mathcal{L}_{(X_t,l_t,W_t)_{t\in[0,T]}|\mathbb{P}} = \mathcal{L}_{(\tilde{X}_t,\tilde{l}_t,\tilde{W}_t)_{t\in[0,T]}|\tilde{\mathbb{P}}}$). We call (2.0.1) weakly well-posed, if it has a unique weak solution for any initial value.

To characterize the linear Fokker-Planck equation associated with (2.0.1), consider the time-distribution dependent second order differential operator on D:

$$L_t := \frac{1}{2}\text{tr}\{\sigma_t\sigma_t^*\nabla^2\} + \nabla_{b_t}, \quad t \in [0,T]. \tag{2.1.13}$$

Let $C_N^2(\bar{D})$ be the class of C_0^2-functions on $\bar{D}$ satisfying the Neumann boundary condition $\nabla_{\mathbf{n}} f|_{\partial D} = 0$. By Itô's formula, for any (weak) solution X_t to (2.0.1), $\mu_t := \mathcal{L}_{X_t}$ solves the nonlinear Fokker-Planck equation

$$\partial_t \mu_t = L_t^* \mu_t \text{ with respect to } C_N^2(\bar{D}), \ t \in [0, T] \tag{2.1.14}$$

for probability measures on $\bar{D}$, in the sense that $\mu_. \in C([0, \infty); \mathcal{P}(\bar{D}))$ and

$$\mu_t(f) := \int_{\bar{D}} f \mathrm{d}\mu_t = \mu_0(f) + \int_0^t \mu_s(L_s f) \mathrm{d}s, \tag{2.1.15}$$
$$t \in [0, T], f \in C_N^2(\bar{D}).$$

To understand (2.1.14) as a linear Neumann problem on D, let L_t^* be the adjoint operator of L_t: for any $g \in L_{loc}^1(D, (\|\sigma_t(x)\|^2 + |b_t(x)|)\mathrm{d}x)$, $L_t^* g$ is the linear functional on $C_0^2(D)$ given by

$$C_0^2(D) \ni f \mapsto \int_D \{f L_t^* g\}(x) \mathrm{d}x := \int_D \{g L_t f\}(x) \mathrm{d}x. \tag{2.1.16}$$

Assume that $\mathcal{L}_{X_t}$ has a density function ρ_t, i.e. $\mu_t := \mathcal{L}_{X_t} = \rho_t(x) \mathrm{d}x$. It is the case under a general non-degenerate or Hörmander condition (see for instance [Bogachev et al. (2015)]). When $\partial D \in C^2$, (2.1.14) implies that ρ_t solves the following linear Neumann problem on $\bar{D}$:

$$\partial_t \rho_t = L_t^* \rho_t, \ \nabla_{t,\mathbf{n}} \rho_t|_{\partial D} = 0, \ t \in [0, T] \tag{2.1.17}$$

in the weak sense, where for a function g on ∂D

$$\nabla_{t,\mathbf{n}} g := \nabla_{\sigma_t \sigma_t^* \mathbf{n}} g + \mathrm{div}_{\partial D}(g \pi \sigma_t \sigma_t^* \mathbf{n})$$

for the divergence $\mathrm{div}_{\partial D}$ on ∂D and the projection π to the tangent space of ∂D:

$$\pi_x v := v - \langle v, \mathbf{n}(x) \rangle \mathbf{n}(x), \ v \in \mathbb{R}^d, x \in \partial D.$$

If in particular $\sigma \sigma^* \mathbf{n} = \lambda \mathbf{n}$ holds on $[0, \infty) \times \partial D$ for a function $\lambda \neq 0$ a.e., $\nabla_{t,\mathbf{n}} \rho_t|_{\partial D} = 0$ is equivalent to the standard Neumann boundary condition $\nabla_{\mathbf{n}} \rho_t|_{\partial D} = 0$.

We now deduce (2.1.17) from (2.1.15). Firstly, by (2.1.16), (2.1.15) implies

$$\int_D (f \rho_t)(x) \mathrm{d}x = \int_D (f \rho_0)(x) \mathrm{d}x + \int_0^t \mathrm{d}s \int_D (f L_s^* \rho_s)(x) \mathrm{d}x,$$
$$f \in C_0^2(D), t \in [0, T],$$

so that $\partial_t \rho_t = L_t^* \rho_t$. Next, by the integration by parts formula, (2.1.15) implies

$$\int_D (f\rho_t)(x)\mathrm{d}x = \int_D (f\rho_0)(x)\mathrm{d}x + \int_0^t \mathrm{d}s \int_D (\rho_s L_s f)(x)\mathrm{d}x$$

$$= \int_D (f\rho_0)(x)\mathrm{d}x + \int_0^t \mathrm{d}s \int_D (fL_s^* \rho_s)(x)\mathrm{d}x$$

$$+ \int_0^t \mathrm{d}s \int_{\partial D} \{f \nabla_{\sigma_s \sigma_s^* \mathbf{n}} \rho_s - \rho_s \nabla_{\sigma_s \sigma_s^* \mathbf{n}} f\}(x)\mathrm{d}x$$

$$= \int_D (f\rho_0)(x)\mathrm{d}x + \int_0^t \mathrm{d}s \int_D (f\partial_s \rho_s)(x)\mathrm{d}x$$

$$+ \int_0^t \mathrm{d}s \int_{\partial D} \{f \nabla_{\sigma_s \sigma_s^* \mathbf{n}} \rho_s + f \mathrm{div}_{\partial D}(\rho_s \pi \sigma_s \sigma_s^* \mathbf{n})\}(x)\mathrm{d}x$$

$$= \int_D (f\rho_t)(x)\mathrm{d}x + \int_0^t \mathrm{d}s \int_{\partial D} \{f(\nabla_{t,\mathbf{n}} \rho_t)\}(x)\mathrm{d}x, \quad f \in C_N^2(\bar{D}), t \in [0,T].$$

Thus, $\nabla_{t,\mathbf{n}} \rho_t|_{\partial D} = 0$.

2.2 Krylov's and Khasminskii's estimates

Let us first explain the main difficulty in the study of singular reflected SDEs using Zvonkin's transform. Consider the following simple reflected SDE on $\bar{D}$:

$$\mathrm{d}X_t = b_t(X_t)\mathrm{d}t + \sqrt{2}\mathrm{d}W_t + \mathbf{n}(X_t)\mathrm{d}l_t, \quad t \in [0,T], \tag{2.2.1}$$

where W_t is the d-dimensional Brownian motion and $\int_0^T \|b_t\|_{L^p(\mathbb{R}^d)}^q \mathrm{d}t < \infty$ for some $p, q > 2$ with $\frac{d}{p} + \frac{2}{q} < 1$. By Lemma 1.2.2, when $\lambda > 0$ is large enough, the unique solution of the PDE

$$(\partial_t + \Delta + \nabla_{b_t})u_t = \lambda u_t - b_t, \quad t \in [0,T], u_T = 0$$

satisfies

$$\|u\|_\infty + \|\nabla u\|_\infty \leq \frac{1}{2}, \quad \|\nabla^2 u\|_{L_q^p} := \left(\int_0^T \|\nabla^2 u_t\|_{L^p(\mathbb{R}^d)}^q \mathrm{d}t\right)^{\frac{1}{q}} < \infty.$$

Thus, for any $t \in [0,T]$, $\Theta_t := id + u_t$ is a homeomorphism on $\mathbb{R}^d$, and by Itô's formula, $Y_t := \Theta_t(X_t)$ solves

$$\mathrm{d}Y_t = \lambda\{u_t \circ \Theta_t^{-1}\}(Y_t)\mathrm{d}t + \mathrm{d}W_t$$
$$+ \{(\nabla u_t) \circ \Theta_t^{-1}\}(Y_t)\mathrm{d}W_t + \{\mathbf{n}(X_t) + \nabla_\mathbf{n} u_t(X_t)\}\mathrm{d}l_t.$$

To prove the pathwise uniqueness of Y_t by applying Itô's formula to $|Y_t - \tilde{Y}_t|^2$, where $\tilde{Y}_t := \Theta_t(\tilde{X}_t)$ for another solution $\tilde{X}_t$ of (2.2.1) with local time $\tilde{l}_t$, one needs to find a constant $c > 0$ such that

$$\langle \Theta_t(X_t) - \Theta_t(\tilde{X}_t), (\mathbf{n} + \nabla_\mathbf{n} u_t)(X_t) \rangle \mathrm{d}l_t$$
$$+ \langle \Theta_t(\tilde{X}_t) - \Theta_t(X_t), (\mathbf{n} + \nabla_\mathbf{n} u_t)(\tilde{X}_t) \rangle \mathrm{d}\tilde{l}_t \qquad (2.2.2)$$
$$\leq c|X_t - \tilde{X}_t|^2 (\mathrm{d}l_t + \mathrm{d}\tilde{l}_t).$$

This is not implied by (2.1.2) except for $d = 1$, since only in this case the vectors $\Theta_t(x) - \Theta_t(y)$ and $(\mathbf{n} + \nabla_\mathbf{n} u_t)(x)$ are in the same directions of $x - y$ and $\mathbf{n}(x)$ respectively for large $\lambda > 0$.

To overcome this difficulty, we will construct a Zvokin's transform by solving the associated Neumann problem on $\bar{D}$, for which $\nabla_\mathbf{n} u_t|_{\partial D} = 0$. Even in this case, Θ_t may also map a point from $\bar{D}$ to $\bar{D}^c$ such that (2.1.2) does not apply. To this end, we will construct a modified process of $|X_t - \tilde{X}_t|^2$ by using a function from [Dupuis and Ishii (1990)], see [Yang and Zhang (2023)] for the study of bounded b and bounded C^3 domain D.

In the following we first deduce Krylov's estimate and Khasminskii's estimate by using Lemma 1.2.2, then make improvements by solving a Neumann problem on D.

Lemma 2.2.1. *Assume $(A^{2.1})$. Let $(p,q) \in \mathcal{K}$.*

(1) *There exist a constant $i \geq 1$ depending only on (p,q), and a constant $c \geq 1$ increasing in $\|b^{(0)}\|_{\tilde{L}_{q_0}^{p_0}(T)}$, such that for any solution X_t of (2.0.1), any $0 \leq t_0 \leq t_1 \leq T$, and any $f \in \tilde{L}_q^p(t_0, t_1)$,*

$$\mathbb{E}\left[\left(\int_{t_0}^{t_1} |f_s(X_s)|\right)^j \mathrm{d}s \Big| \mathcal{F}_{t_0}\right] \leq c^j (j!) \|f\|_{\tilde{L}_q^p(t_0,t_1)}^j, \quad j \geq 1, \qquad (2.2.3)$$

$$\mathbb{E}\left(\mathrm{e}^{\int_{t_0}^{t_1} |f_t(X_t)| \mathrm{d}t} \Big| \mathcal{F}_{t_0}\right) \leq \exp\left[c + c\|f\|_{\tilde{L}_q^p(t_0,t_1)}^i\right], \qquad (2.2.4)$$

$$\sup_{t_0 \in [0,T]} \mathbb{E}\left(\mathrm{e}^{\lambda(l_T - l_{t_0})} \Big| \mathcal{F}_{t_0}\right) < \mathrm{e}^{c(1+\lambda^2)}, \quad \lambda > 0. \qquad (2.2.5)$$

(2) *For any $u \in C([0,T] \times \mathbb{R}^d)$ with continuous ∇u and*

$$\|u\|_\infty + \|\nabla u\|_\infty + \|(\partial_t + \nabla_{b^{(1)}})u\|_{\tilde{L}_q^p(T)} + \|\nabla^2 u\|_{\tilde{L}_q^p(T)} < \infty, \qquad (2.2.6)$$

we have the following Itô's formula for a solution X_t to (2.0.1):

$$\mathrm{d}u_t(X_t) = (\partial_t + L_t)u_t(X_t)\mathrm{d}t + \langle \nabla u_t(X_t), \sigma_t(X_t)\mathrm{d}W_t \rangle$$
$$+ (\nabla_\mathbf{n} u_t)(X_t)\mathrm{d}l_t, \quad t \in [0,T]. \qquad (2.2.7)$$

Proof. (1) We first prove (2.2.3) for $j = 1$. By Theorem 1.2.3(1), for any $\varepsilon \in (0, 1)$ there exists a constant $c > 0$ such that

$$\mathbb{E}\left(\int_{t_0}^{t_1} f_s(X_s)\mathrm{d}s \Big| \mathcal{F}_{t_0}\right) \leq \{c + \varepsilon\mathbb{E}(l_{t_1} - l_{t_0}|\mathcal{F}_{t_0})\}\|f\|_{\tilde{L}_q^p(t_0,t_1)}. \quad (2.2.8)$$

On the other hand, by (2.1.7) and the boundedness of σ, we find a constant $c_1 > 0$ such that

$$\mathrm{d}\tilde{\rho}_\theta(X_t) \geq -c_1\mathrm{d}t - c_1|b_t^{(0)}(X_t)|\mathrm{d}t + \mathrm{d}l_t + \langle \nabla\tilde{\rho}_\theta(X_t), \sigma_t(X_t)\mathrm{d}W_t\rangle. \quad (2.2.9)$$

So, (2.2.8) with $(p, q) = (p_0, q_0)$ implies that

$$\mathbb{E}(l_{t_1} - l_{t_0}|\mathcal{F}_{t_0}) \leq c_1(t_1 - t_0) + c_1\mathbb{E}\left(\int_{t_0}^{t_1} |b_s^{(0)}(X_s)|\mathrm{d}s \Big| \mathcal{F}_{t_0}\right) + \|\tilde{\rho}_\theta\|_\infty$$
$$\leq c_2 + c_1\varepsilon\mathbb{E}(l_{t_1} - l_{t_0}|\mathcal{F}_{t_0}), \quad t \in [t_0, T]$$

holds for some constant $c_2 > 0$ increasing in $\|b^{(0)}\|_{\tilde{L}_q^p(T)}$. By an approximation argument we may assume that $\mathbb{E}l_T < \infty$, so that by taking $\varepsilon > 0$ small enough such that $c_1\varepsilon \leq \frac{1}{2}$, we arrive at

$$\mathbb{E}(l_{t_1} - l_{t_0}|\mathcal{F}_{t_0}) \leq c_3, \quad t_0 \leq t_1 \leq T \quad (2.2.10)$$

for some constant $c_3 > 0$ increasing in $\|b^{(0)}\|_{\tilde{L}_q^p(T)}$. This and (2.2.8) imply (2.2.3) for $j = 1$, which further yields the inequality for any $j \geq 1$ as well as (2.2.4) according to the proof of Theorem 1.2.4. Finally, combining (2.2.4) with (2.2.9), $b^{(0)} \in \tilde{L}_{q_0}^{p_0}(T)$ and $\|\sigma^*\nabla\tilde{\rho}_\theta\|_\infty < \infty$, we derive (2.2.5).

(2) We first extend u to $\mathbb{R}^{d+1}$ by letting $u_t = u_{t^+\wedge T}$ for $t \in \mathbb{R}$, and consider its mollifying approximation $u^{\{n\}} := \mathcal{S}_n(u)$ in (1.2.4). Then $\|\sigma\|_\infty < \infty$ and (2.2.6) imply

$$\lim_{n\to\infty}\big\{\|u - u^{\{n\}}\|_\infty + \|\nabla(u - u^{\{n\}})\|_\infty$$
$$+ \|(\partial_t + L_t)(u - u^{\{n\}})\|_{\tilde{L}_q^p(T)}\big\} = 0. \quad (2.2.11)$$

Combining this with $\|\sigma\|_\infty < \infty$ and (2.2.3), we obtain

$$\lim_{n\to\infty}\sup_{t\in[0,T]}|u_t^{\{n\}}(X_t) - u_t(X_t)| = 0, \quad \mathbb{P}\text{-a.s.}$$

$$\lim_{n\to\infty}\int_0^t \nabla_n u_s^{\{n\}}(X_s)\mathrm{d}l_s = \int_0^t \nabla_n u_s(X_s)\mathrm{d}l_s, \quad \mathbb{P}\text{-a.s.} \quad (2.2.12)$$

$$\lim_{n\to\infty}\mathbb{E}\int_0^T |(\partial_s + L_s)(u_s^{\{n\}} - u_s)|(X_s)\mathrm{d}s = 0, \quad \mathbb{P}\text{-a.s.}$$

$$\lim_{n\to\infty}\mathbb{E}\sup_{t\in[0,T]}\left|\int_0^t \langle\nabla(u_s^{\{n\}} - u_s)(X_s), \sigma_s(X_s)\mathrm{d}W_s\rangle\right| = 0.$$

Therefore, we derive (2.2.7) by letting $n \to \infty$ in the following Itô's formula:
$$u_t^{\{n\}}(X_t) = u_0^{\{n\}}(X_0) + \int_0^t (\partial_s + L_s)(u_s^{\{n\}})(X_s)ds$$
$$+ \int_0^t \langle \nabla u_s^{\{n\}}(X_s), \sigma_s(X_s) dW_s \rangle + \int_0^t (\nabla_{\mathbf{n}} u_s^{\{n\}})(X_s) dl_s, \quad t \in [0, T]. \qquad \square$$

To improve Lemma 2.2.1 for $(p,q) \in \mathcal{K}$ with $\frac{d}{p} + \frac{2}{q} < 2$, we extend Lemma 1.2.2 to the Neumann boundary case. For any $k \in \mathbb{N}$, let $C_b^{0,k}([t_0, t_1] \times \bar{D}; \mathbb{R}^d)$ be the space of $f \in C_b([t_0, t_1] \times \bar{D}; \mathbb{R}^d)$ with bounded and continuous derivatives in $x \in \bar{D}$ up to order k. Let $C_b^{1,2}([t_0, t_1] \times \bar{D}; \mathbb{R}^d)$ denote the space of $f \in C_b^{0,2}([t_0, t_1] \times \bar{D}; \mathbb{R}^d)$ with bounded and continuous $\partial_t f$.

Lemma 2.2.2. *Assume $(A^{2.3})$ but without the condition on $\|\nabla \sigma\|$. Then assumption $(A^{2.1})$ and the following assertions hold.*

(1) *For any $\lambda \geq 0$, $0 \leq t_0 < t_1 \leq T$ and $\tilde{b}, f \in C_b^{0,2}([t_0, t_1] \times \bar{D}; \mathbb{R}^d)$, the PDE*
$$(\partial_t + L_t^{\sigma, b^{(1)}} + \nabla_{\tilde{b}_t} - \lambda)\tilde{u}_t^\lambda = f_t, \quad \tilde{u}_{t_1}^\lambda = \nabla_{\mathbf{n}} \tilde{u}_t^\lambda|_{\partial D} = 0, t \in [t_0, t_1] \quad (2.2.13)$$
has a unique solution $\tilde{u}^\lambda \in C_b^{1,2}([t_0, t_1] \times \bar{D}; \mathbb{R}^d)$.

(2) *For any $(p,q), (p', q') \in \mathcal{K}$ and $\tilde{b} \in C_b^{0,2}([0, T] \times \bar{D}; \mathbb{R}^d)$, there exist a constant $\varepsilon > 0$ depending only on (p, q) and (p', q'), and constants $\lambda_0, c > 0$ increasing in $\|\tilde{b}\|_{\tilde{L}_{q'}^{p'}(T, D)}$, such that for any $0 \leq t_0 < t_1 \leq T$ and $f \in C_b^{0,2}([t_0, t_1] \times \bar{D}; \mathbb{R}^d)$,*
$$\lambda^\varepsilon(\|\tilde{u}^\lambda\|_\infty + \|\nabla \tilde{u}^\lambda\|_{\tilde{L}_q^p(t_0, t_1, D)})$$
$$\leq c\|f\|_{\tilde{L}_{q/2}^{p/2}(t_0, t_1, D)} \quad (\text{when } p > 2), \quad \lambda \geq \lambda_0, \qquad (2.2.14)$$
$$\lambda^\varepsilon \|\nabla \tilde{u}^\lambda\|_\infty \leq c\|f\|_{\tilde{L}_q^p(t_0, t_1, D)}, \quad \lambda \geq \lambda_0, \qquad (2.2.15)$$
and there exists decomposition $\tilde{u}^\lambda = \tilde{u}^{\lambda, 1} + \tilde{u}^{\lambda, 2}$ such that
$$\|\nabla^2 \tilde{u}^{\lambda, 1}\|_{\tilde{L}_q^p(t_0, t_1, D)} + \|(\partial_t + \nabla_{b^{(1)}})\tilde{u}^{\lambda, 1}\|_{\tilde{L}_q^p(t_0, t_1, D)}$$
$$+ \|\nabla^2 \tilde{u}^{\lambda, 2}\|_{\tilde{L}_{q'}^{p'}(t_0, t_1, D)} + \|(\partial_t + \nabla_{b^{(1)}})\tilde{u}^{\lambda, 2}\|_{\tilde{L}_{q'}^{p'}(t_0, t_1, D)} \qquad (2.2.16)$$
$$\leq c\|f\|_{\tilde{L}_q^p(t_0, t_1, D)}, \quad \lambda \geq \lambda_0.$$

Proof. (1) Let $\mathbb{V} := C_b^{0,2}([t_0, t_1] \times \bar{D}; \mathbb{R}^d)$, which is a Banach space under the norm
$$\|u\|_{\mathbb{V}, N} := \sup_{t \in [t_0, t_1]} e^{-N(t_1 - t)} \{\|u_t\|_\infty + \|\nabla u_t\|_\infty + \|\nabla^2 u_t\|_\infty\}, \quad u \in \mathbb{V}$$

for $N > 0$. To solve (2.2.13), for any $\lambda \geq 0$ and $u \in \mathbb{V}$, let

$$\Phi_s^\lambda(u) := \int_s^{t_1} e^{-\lambda(t-s)} P_{s,t}^{\sigma,b^{(1)}} \{\nabla_{\tilde{b}_t} u_t - f_t\} dt, \quad s \in [t_0, t_1].$$

Then $(A^{2.3})$ implies $\Phi^\lambda(u) \in C_b^{1,2}([t_0, t_1] \times \bar{D})$ with

$$(\partial_s + L_s^{\sigma,b^{(1)}} - \lambda)\Phi_s^\lambda(u) = f_s - \nabla_{\tilde{b}_s} u_s, \qquad (2.2.17)$$
$$s \in [t_0, t_1], \nabla_{\mathbf{n}} \Phi_t^\lambda(u)|_{\partial D} = 0, \Phi_{t_1}^\lambda(u) = 0.$$

So, it suffices to prove that Φ^λ has a unique fixed point $\tilde{u}^\lambda \in \mathbb{V}$:

$$\tilde{u}_s^\lambda = \int_s^{t_1} e^{-\lambda(t-s)} P_{s,t}^{\sigma,b^{(1)}} \{\nabla_{\tilde{b}_t} \tilde{u}_t^\lambda - f_t\} dt, \quad s \in [t_0, t_1], \qquad (2.2.18)$$

which, according to (2.2.17), is the unique solution of (2.2.13) in $C_b^{1,2}([t_0, t_1] \times \bar{D}; \mathbb{R}^d)$.

For any $u, \bar{u} \in \mathbb{V}$, by $\|\tilde{b}\|_\infty < \infty$, we find a constant $c_1 > 0$ such that

$$\|\Phi_s^\lambda(u) - \Phi_s^\lambda(\bar{u})\|_\infty \leq \int_s^{t_1} \|\tilde{b}_t\|_\infty \|\nabla(u_t - \bar{u}_t)\|_\infty dt$$
$$\leq c_1 \int_s^{t_1} \|\nabla(u_t - \bar{u}_t)\|_\infty dt, \quad s \in [t_0, t_1].$$

Similarly, (2.1.11) with $i = 1$ implies

$$\|\nabla\{\Phi^\lambda(u)_s - \Phi^\lambda(\bar{u})_s\}\|_\infty \leq c \int_s^{t_1} (t-s)^{-\frac{1}{2}} \|\tilde{b}_t\|_\infty \|\nabla(u_t - \bar{u}_t)\|_\infty dt$$
$$\leq c_1 \int_s^t (t-s)^{-\frac{1}{2}} \|\nabla(u_t - \bar{u}_t)\|_\infty dt,$$

while (2.1.11) with $i = 2$ and $\|\tilde{b}\|_\infty + \|\nabla \tilde{b}_t\|_\infty < \infty$ yield

$$\|\nabla^2\{\Phi_s^\lambda(u) - \Phi_s^\lambda(\bar{u})\}\|_\infty \leq c \int_s^{t_1} (t-s)^{-\frac{1}{2}} \|\nabla\{\nabla_{\tilde{b}_t}(u_t - \bar{u}_t)\}\|_\infty dt$$
$$\leq c_1 \int_s^{t_1} (t-s)^{-\frac{1}{2}} \{\|\nabla(u_t - \bar{u}_t)\|_\infty + \|\nabla^2(u_t - \bar{u}_t)\|_\infty\} dt.$$

Combining these with (2.2.17) and the boundedness of a and $\tilde{b} \in C_b^{0,1}([t_0, t_1] \times \bar{D}; \mathbb{R}^d)$, we find a constant $c_2 > 0$ such that

$$\|\Phi^\lambda(u) - \Phi^\lambda(\bar{u})\|_{\mathbb{V},N}$$
$$\leq c_2 \sup_{s \in [t_0, t_1]} \int_s^{t_1} \frac{e^{-N(t_1-s)}}{\sqrt{t-s}} \Big\{\sum_{i=0}^2 \|\nabla^i(u_t - \bar{u}_t)\|_\infty\Big\} dt$$
$$\leq c_2 \|u - \bar{u}\|_{\mathbb{V},N} \sup_{s \in [t_0, t_1]} \int_s^{t_1} e^{-N(t-s)} (t-s)^{-\frac{1}{2}} dt.$$

So, Φ^λ is contractive under the norm $\|\cdot\|_{\mathbb{V},N}$ for large enough $N > 0$, and hence has a unique fixed point $\tilde{u}^\lambda$ in $\mathbb{V}$.

(2) To prove (2.2.14) and (2.2.16), we extend the PDE (2.2.13) to a global one such that estimates in Lemma 1.2.2 apply. By $(A^{2.3})$, there exists $r_0 > 0$ such that

$$\varphi: \partial_{-r_0}D \to \partial_{r_0}D;\quad \theta - r\mathbf{n}(\theta) \mapsto \theta + r\mathbf{n}(\theta),\quad r \in [0, r_0], \theta \in \partial D$$

is a $C_b^{1,L}$-diffeomorphism (i.e. it is a homeomorphism with $\nabla \varphi$ bounded and Lipschitz continuous) and $\rho_D := \mathrm{dist}(\cdot, D) \in C_b^2(D_{r_0} \setminus \partial D)$, recall that $D_{r_0} = \{\rho_D \le r_0\}$. For any vector field v on $\partial_{r_0}D$, $v^\star := (\varphi^{-1})^\star v$ is the vector field on $\partial^0_{-r_0}D := \partial_{-r_0}D \setminus \partial D$ given by

$$\langle v^\star, \nabla g\rangle(x) := \langle v, \nabla(g \circ \varphi^{-1})\rangle(\varphi(x)),\quad x \in \partial^0_{-r_0}D,\ g \in C^1(\partial^0_{-r_0}D).$$

We then extend $b_t^{(1)}$ and $\tilde{b}_t$ to $\mathbb{R}^d$ by taking

$$b_t^{(1)} := 1_{\bar{D}} b_t^{(1)} + h(\rho_D/2) 1_{\partial^0_{-r_0}D}(b_t^{(1)})^\star,\quad \tilde{b}_t := 1_{\bar{D}} \tilde{b}_t + 1_{\partial^0_{-r_0}D}(\tilde{b}_t)^\star,\quad (2.2.19)$$

where $h \in C^\infty(\mathbb{R})$ such that

$$0 \le h \le 1,\quad h|_{(-\infty, r_0/4]} = 1,\quad h|_{[r_0/2, \infty)} = 0.$$

Since $(A^{2.3})$ implies $\|1_{\bar{D}} \nabla b^{(1)}\|_\infty < \infty$ and $\nabla_\mathbf{n} b^{(1)}|_{\partial D} = 0$, we have $\|\nabla b^{(1)}\|_\infty < \infty$. Let

$$\tilde{\varphi}(x) := x 1_{\bar{D}}(x) + \varphi(x) 1_{\partial^0_{-r_0}D}(x),\quad x \in D_{r_0}. \qquad (2.2.20)$$

We extend $\tilde{u}^\lambda$ to $[t_0, t_1] \times \mathbb{R}^d$ by setting

$$u_t^\lambda = h(\rho_D)(\tilde{u}_t^\lambda \circ \tilde{\varphi}),\quad t \in [t_0, t_1]. \qquad (2.2.21)$$

We claim that

$$u_t^\lambda \in C_b^{1,L}(\mathbb{R}^d),\quad t \in [t_0, t_1], \qquad (2.2.22)$$

where $C_b^{1,L}(D_{r_0})$ is the class of C_b^1-functions f on D_{r_0} with Lipschitz continuous ∇f. Indeed, since φ is a $C_b^{1,L}$-diffeomorphism from $\partial_{-r_0}D$ to $\partial_{r_0}D$, $\tilde{\varphi} \in C_b^{1,L}(D_{r_0} \setminus \partial D)$ with bounded and continuous first and second order derivatives, which together with $\tilde{u}_t^\lambda \in C_b^2(\bar{D})$ yields $u_t^\lambda \in C_b^{1,L}(\mathbb{R}^d \setminus \partial D)$. So, we only need to verify that $\tilde{u}_t^\lambda \circ \tilde{\varphi} \in C_b^{1,L}(D_{r_0})$. To this end, for any $x \in \partial_{-r_0}D$ and $v \in \mathbb{R}^d$, let

$$\pi_x v := v - \langle v, \mathbf{n}(\theta(x))\rangle \mathbf{n}(\theta(x))$$

be the projection of $v \in T_x \mathbb{R}^d$ to the tangent space of ∂D, recall that $\theta(x)$ is the projection of x to ∂D, i.e. $x = \theta(x) - \rho_D(x)\mathbf{n}(\theta(x))$ for $\rho_D(x) := \mathrm{dist}(x, D)$. We have

$$\begin{aligned}\nabla_v \tilde{\varphi}(x) &= \nabla_{\langle v, \mathbf{n}(\theta(x))\rangle \mathbf{n}(\theta(x))} \tilde{\varphi}(x) + \nabla_{\pi_x v} \tilde{\varphi}(x) \\ &= 1_{\partial D}(x) |\langle v, \mathbf{n}(\theta(x))\rangle| \mathbf{n}(\theta(x)) \\ &\quad + \{1_D - 1_{\partial^0_{-r_0} D}\}(x)\langle v, \mathbf{n}(\theta(x))\rangle \mathbf{n}(\theta(x)) \\ &\quad + \pi_x v + \rho_D(x)(\nabla_{\pi_x v}\mathbf{n})(\theta(x)).\end{aligned} \quad (2.2.23)$$

Since $\tilde{u}_t^\lambda \in C_b^2(\bar{D})$ with $\nabla_\mathbf{n} \tilde{u}_t^\lambda|_{\partial D} = 0$, (2.2.23) yields

$$\begin{aligned}\nabla_v(\tilde{u}_t^\lambda \circ \tilde{\varphi})(x) &= (\nabla_v \tilde{u}_t^\lambda) \circ \tilde{\varphi}(x) \\ &\quad - 2 1_{\partial^0_{-r_0} D}(x) \langle v, \mathbf{n}(\theta(x))\rangle \cdot \langle \mathbf{n}(\theta(x)), (\nabla \tilde{u}_t^\lambda) \circ \tilde{\varphi}(x)\rangle \\ &\quad + \rho_D(x)\big(\nabla_{(\nabla_{\pi_x v} \mathbf{n})(\theta(x))} \tilde{u}_t^\lambda\big) \circ \tilde{\varphi}(x), \quad x \in D_{r_0}.\end{aligned} \quad (2.2.24)$$

Combining this with $\nabla \tilde{u}_t^\lambda \in C_b^1(\bar{D}), \nabla_\mathbf{n} \tilde{u}_t^\lambda|_{\partial D} = 0$ and that $\mathbf{n}, \nabla \mathbf{n}$ are Lipschitz continuous on $\partial_{-r_0} D$ due to $\partial D \in C_b^{2,L}$, we conclude that $\nabla(\tilde{u}_t^\lambda \circ \tilde{\varphi})$ is Lipschitz continuous on D_{r_0}.

Next, we construct the PDE satisfied by u^λ. By (2.2.23), we see that

$$(\nabla \tilde{\varphi})(\nabla \tilde{\varphi})^* = Q \text{ holds on } D_{r_0} \setminus \partial D, \quad (2.2.25)$$

where Q is a $d \times d$ symmetric matrix valued function given by

$$\begin{aligned}\langle Q(x) v_1, v_2\rangle &:= \langle v_1, v_2\rangle + \rho_D(x)^2 \langle (\nabla_{\pi_x v_1} \mathbf{n})(\theta(x)), (\nabla_{\pi_x v_2} \mathbf{n})(\theta(x))\rangle \\ &\quad + \rho_D(x)\Big\{\langle v_1 - 2 1_{\partial_{-r_0} D}(x) \langle v_1, \mathbf{n}(\theta(x))\rangle \mathbf{n}(\theta(x)), (\nabla_{\pi_x v_2} \mathbf{n})(\theta(x))\rangle \\ &\quad + \langle v_2 - 2 1_{\partial_{-r_0} D}(x)\langle v_2, \mathbf{n}(\theta(x))\rangle \mathbf{n}(\theta(x)), (\nabla_{\pi_x v_1} \mathbf{n})(\theta(x))\rangle\Big\}, \\ &\quad x \in D_{r_0}, v_1, v_2 \in \mathbb{R}^d.\end{aligned}$$

Then by taking $r_0 > 0$ small enough, on D_{r_0} the matrix-valued functional Q is bounded, invertible, Lipchitz continuous, and symmetric with

$$Q^{-1}(x) \geq \frac{1}{2} \mathbf{I}_d, \quad x \in D_{r_0}. \quad (2.2.26)$$

We extend $a_t := \frac{1}{2} \sigma_t \sigma_t^*$ from $\bar{D}$ to $\mathbb{R}^d$ by letting

$$a_t := h(\rho_D/2)(a_t \circ \tilde{\varphi}) Q^{-1} + (1 - h(\rho_D/2)) \mathbf{I}_d. \quad (2.2.27)$$

Since (2.1.5) holds for $x, y \in \bar{D}$, with this extension of a, it holds for all $x, y \in \mathbb{R}^d$. Combining this with (2.2.19), Remark 2.1(a) for the existence of $\tilde{\rho}_\partial$, and noting that $b_t = b_t^{(1)} + 1_{\bar{D}} b_t^{(0)}$ extends b from $\bar{D}$ to $\mathbb{R}^d$, we see that ($A^{2.1}$) holds.

Since $h(\rho_D/2), h(\rho_D) \in C_b^2(\mathbb{R}^d)$ with $h(\rho_D/2) = 1$ on $\{h(\rho_D) \neq 0\}$, and since $(\nabla\tilde\varphi)^2 = Q$ on $D_{r_0} \setminus \partial D$, by (2.2.13), (2.2.19), (2.2.27) and (2.2.22), we see that u_t^λ in (2.2.21) solves the PDE

$$(\partial_t + \mathrm{tr}\{a_t \nabla^2\} + \nabla_{b_t^{(1)} + \tilde b_t}) u_t^\lambda = \lambda u_t^\lambda + f_t^{(1)} + f_t^{(2)},$$
$$t \in [t_0, t_1], \quad u_{t_1}^\lambda = 0, \qquad (2.2.28)$$

where outside the null set ∂D,

$$f_t^{(1)} := (h \circ \rho_D) f_t \circ \tilde\varphi + 2\langle a_t \nabla(h \circ \rho_D), \nabla\{\tilde u_t^\lambda \circ \tilde\varphi\}\rangle,$$
$$f_t^{(2)} := (\tilde u_t^\lambda \circ \tilde\varphi)(L_t^{\sigma, b^{(1)}} + \nabla_{\tilde b_t})(h \circ \rho_D).$$

By (2.2.23), $h \in C^\infty([0,\infty))$ with support $\mathrm{supp}\, h \subset [0, r_0/2]$, $\|a\|_\infty + \|1_{\partial_{r_0} D} \nabla_{b^{(1)}} \rho_\partial\|_\infty < \infty$ according to $(A^{2.3})$ and Remark 2.2(1), we find a constant $c > 0$ such that

$$|f_t^{(1)}| \le 1_{\{\rho_D \le \frac{r_0}{2}\}}(|f_t| + |\nabla \tilde u_t^\lambda|) \circ \tilde\varphi,$$
$$|f_t^{(2)}| \le c 1_{\{\rho_D \le \frac{r_0}{2}\}}\{(1 + |\tilde h_t|)|\tilde u_t^\lambda|\} \circ \tilde\varphi. \qquad (2.2.29)$$

Since $|f| + |\tilde b| + |\tilde u^\lambda|$ is bounded on $[0,T] \times \bar D$, so is $|f^{(1)}| + |f^{(2)}|$ on $[0,T] \times \mathbb{R}^d$. Hence, by Lemma 1.2.2, the PDE (2.2.28) has a unique solution in $\tilde H_q^{2,p}(t_0, t_1)$. Moreover, for each $i = 1, 2$ and $\lambda \ge 0$, the PDE

$$(\partial_t + \mathrm{tr}\{a_t \nabla^2\} + \nabla_{b_t^{(1)} + \tilde b_t}) u_t^{\lambda, i} = \lambda u_t^{\lambda, i} + f_t^{(i)}, \quad t \in [t_0, t_1], \quad u_{t_1}^{\lambda, i} = 0 \quad (2.2.30)$$

has a unique solution in $\tilde H_q^{2,p}(t_0, t_1)$, and there exist constants $c_1, c_2 > 0$ increasing in $\|\tilde b\|_{\tilde L_{q'}^{p'}(T, D)}$ such that

$$\lambda^{1 - \frac{d}{p} - \frac{2}{q}} \|u^{\lambda, 1}\|_\infty + \lambda^{\frac{1}{2}(1 - \frac{d}{p} - \frac{2}{q})} \|\nabla u^{\lambda, 1}\|_{\tilde L_q^p(t_0, t_1)} \le c_1 \|f^{(1)}\|_{\tilde L_q^p(t_0, t_1)}$$
$$\le c_2 \big(\|f\|_{\tilde L_{q/2}^{p/2}(t_0, t_1, D)} + \|\tilde u_t^\lambda\|_{\tilde L_q^p(t_0, t_1, D)}\big), \quad p > 2, \qquad (2.2.31)$$

$$\lambda^{\frac{1}{2}(1 - \frac{d}{p} - \frac{2}{q})} \|\nabla u^{\lambda, 1}\|_\infty + \|\nabla^2 u^{\lambda, 1}\|_{\tilde L_q^p(t_0, t_1)}$$
$$+ \|(\partial_t + \nabla_{b^{(1)}}) u^{\lambda, 1}\|_{\tilde L_q^p(t_0, t_1)} \le c_1 \|f^{(1)}\|_{\tilde L_q^p(t_0, t_1)} \qquad (2.2.32)$$
$$\le c_2 (\|f\|_{\tilde L_q^p(t_0, t_1, D)} + \|\tilde u^\lambda\|_{\tilde L_q^p(t_0, t_1, D)}),$$

and

$$\lambda^{\frac{1}{2}(1 - \frac{d}{p'} - \frac{2}{q'})}(\|u^{\lambda, 2}\|_\infty + \|\nabla u^{\lambda, 2}\|_\infty) + \|\nabla^2 u^{\lambda, 2}\|_{\tilde L_{q'}^{p'}(t_0, t_1)}$$
$$+ \|(\partial_t + \nabla_{b^{(1)}}) u^{\lambda, 2}\|_{\tilde L_{q'}^{p'}(t_0, t_1)} \le c_1 \|f^{(2)}\|_{\tilde L_{q'}^{p'}(t_0, t_1)} \qquad (2.2.33)$$
$$\le c_2 (1 + \|\tilde b\|_{\tilde L_{q'}^{p'}(t_0, t_1, D)}) \|\tilde u^\lambda\|_\infty,$$

where the last step in these estimates follows from (2.2.29) and the integral transform
$$\tilde{\varphi} : D_{r_0} \setminus \bar{D} \to D$$
with $\|(\nabla\tilde{\varphi})^{-1}\|_\infty < \infty$ due to (2.2.25) and (2.2.26). By taking large enough $\lambda_0 > 0$ increasing in $\|\tilde{b}\|_{\tilde{L}_{q'}^{p'}(T,D)}$, we derive from (2.2.31) and (2.2.33) that

$$\|u^{\lambda,1}\|_\infty + \|\nabla u^{\lambda,1}\|_{\tilde{L}_q^p(t_0,t_1)} \le \frac{1}{2}\big(\|f\|_{\tilde{L}_{q/2}^{p/2}(t_0,t_1,D)} + \|\tilde{u}_t^\lambda\|_{\tilde{L}_q^p(t_0,t_1,D)}\big),$$

$$\|u^{\lambda,2}\|_\infty + \|\nabla u^{\lambda,2}\|_\infty \le \frac{1}{2}\|\tilde{u}^\lambda\|_\infty, \quad \lambda \ge \lambda_0.$$

Noting that the uniqueness of (2.2.28) and (2.2.30) implies $u_t^\lambda = u_t^{\lambda,1} + u_t^{\lambda,2}$, this and the definition of u_t^λ yield

$$\|\tilde{u}^\lambda\|_\infty + \|\nabla\tilde{u}^\lambda\|_{\tilde{L}_q^p(t_0,t_1,D)} \le \sum_{i=1}^{2}(\|u_t^{\lambda,i}\|_\infty + \|\nabla u^{\lambda,i}\|_{\tilde{L}_q^p(t_0,t_1)})$$

$$\le \frac{1}{2}\big\{\|\tilde{u}^\lambda\|_\infty + \|f\|_{\tilde{L}_{q/2}^{p/2}(t_0,t_1,D)} + \|\tilde{u}_t^\lambda\|_{\tilde{L}_q^p(t_0,t_1,D)}\big\},$$

so that

$$\|\tilde{u}^\lambda\|_\infty + \|\nabla\tilde{u}^\lambda\|_{\tilde{L}_q^p(t_0,t_1,D)} \le \|f\|_{\tilde{L}_{q/2}^{p/2}(t_0,t_1,D)}, \quad \lambda \ge \lambda_0.$$

This together with (2.2.31)–(2.2.33) imply (2.2.14), (2.2.15) and (2.2.16) for some $c,\varepsilon > 0$. □

Lemma 2.2.3. *Assume* $(A^{2.3})$ *but without the condition on* $\|\nabla\sigma\|$. *For any* $(p,q) \in \mathcal{K}$ *with* $p > 2$, *there exist a constant* $i \ge 1$ *depending only on* (p,q), *and a constant* $c \ge 1$ *increasing in* $\|b^{(0)}\|_{\tilde{L}_{q_0}^{p_0}(T,D)}$, *such that for any solution* $(X_t)_{t\in[0,T]}$ *of* (2.0.1), *any* $0 \le t_0 \le t_1 \le T$, *and any* $f \in \tilde{L}_{q/2}^{p/2}(t_0,t_1)$,

$$\mathbb{E}\left[\left(\int_{t_0}^{t_1}|f_s(X_s)|\mathrm{d}s\right)^j\Big|\mathcal{F}_{t_0}\right] \le c^j j! \|f\|_{\tilde{L}_{q/2}^{p/2}(t_0,t_1)}^j, \quad j \ge 1, \qquad (2.2.34)$$

$$\mathbb{E}\big(e^{\int_{t_0}^T |f_t(X_t)|\mathrm{d}t}\big|\mathcal{F}_{t_0}\big) \le \exp\Big[c + c\|f\|_{\tilde{L}_{q/2}^{p/2}(t_0,T)}^i\Big], \quad t_0 \in [0,T]. \qquad (2.2.35)$$

Proof. According to the proofs of Theorems 1.2.3 and 1.2.4, for (2.2.34), it suffices to consider $j = 1$ and $f \in C_0^\infty([t_0,t_1]\times\mathbb{R}^d)$. In the following, all constants are increasing in $\|b^{(0)}\|_{\tilde{L}_{q_0}^{p_0}(T)}$ when $b^{(0)}$ varies.

Let $(b^{0,n})_{n\ge 1}$ be the mollifying approximations of $b^{(0)} = 1_{\bar{D}} b^{(0)}$. We have

$$\|b^{0,n}\|_{\tilde{L}_{q_0}^{p_0}(T)} \le \|b^{(0)}\|_{\tilde{L}_{q_0}^{p_0}(T)}, \quad \lim_{n\to\infty}\|b^{0,n} - b^{(0)}\|_{\tilde{L}_{q_0}^{p_0}(T)} = 0. \qquad (2.2.36)$$

By Lemma 2.2.2 for $(f, 0, \ldots, 0)$ replacing f for $f \in C_0^\infty([t_0, t_1] \times \mathbb{R}^d)$, there exist constants $c, \lambda_0 > 0$ such that for any $\lambda \geq \lambda_0$, the following PDE on $\bar{D}$

$$(\partial_t + L_t^{\sigma, b^{(1)}} + \nabla_{b_t^{0,n}} - \lambda) u_t^{\lambda, n} = f_t,$$
$$t \in [t_0, t_1), \quad \nabla_{\mathbf{n}} u_t^{\lambda, n}|_{\partial D} = 0, \quad u_{t_1}^{\lambda, n} = 0 \tag{2.2.37}$$

has a unique solution in $C^{1,2}([t_0, t_1] \times \bar{D})$, and for some constant $c_1 > 0$, we have

$$\|u^{\lambda, n}\|_\infty \leq c_1 \|f\|_{\tilde{L}_{q/2}^{p/2}(t_0, t_1, D)},$$
$$\|\nabla u^{\lambda, n}\|_\infty \leq c_1 \|f\|_\infty, \quad \lambda \geq \lambda_0, n \geq 1. \tag{2.2.38}$$

Moreover, since $(A^{2.3})$ implies $(A^{2.1})$ due to Lemma 2.2.2, by (2.2.3) for $f = |b^{(0)} - b^{0,n}|$, we find a constant $c_2 > 0$ such that

$$\mathbb{E}\left(\int_{t_0}^{t_1} |b^{(0)} - b^{0,n}|(X_s) \mathrm{d}s \bigg| \mathcal{F}_{t_0} \right) \leq c_2 \|b^{(0)} - b^{0,n}\|_{\tilde{L}_{q_0}^{p_0}(t_0, t_1)}, \quad n \geq 1. \tag{2.2.39}$$

By (2.2.37) and $u^{\lambda, n} \in C_b^{1,2}([t_0, t_1] \times \bar{D})$, we have the following Itô's formula

$$\mathrm{d} u_t^{\lambda, n}(X_t) = (\partial_t + L_t) u_t^{\lambda, n}(X_t) \mathrm{d}t + \mathrm{d}M_t$$
$$= \{f_t + \nabla_{b_t^{(0)} - b_t^{0,n}} u_t^{\lambda, n}\}(X_t) \mathrm{d}t + \mathrm{d}M_t$$

for some martingale M_t. Combining this with (2.2.38) and (2.2.39), we obtain

$$\mathbb{E}\left(\int_{t_0}^{t_1} f_t(X_t) \mathrm{d}t \bigg| \mathcal{F}_{t_0} \right) \leq c_1 \|f\|_{\tilde{L}_{q/2}^{p/2}(t_0, t_1)} + c_1 c_2 \|f\|_\infty \|b_t^{(0)} - b_t^{0,n}\|_{\tilde{L}_{q_0}^{p_0}(t_0, t_1)}.$$

Therefore, by (2.2.36), we may let $n \to \infty$ to derive (2.2.34) for $j = 1$. □

2.3 Weak well-posedness

The following is the main result of this section.

Theorem 2.3.1 (Weak well-posedness). *If either* $(A^{2.2})$ *or* $(A^{2.3})$ *holds, then* (2.0.1) *is weakly well-posed. Moreover, for any* $k \geq 1$ *there exists a constant* $c > 0$ *such that*

$$\mathbb{E}\left[\sup_{t \in [0, T]} |X_t^x|^k \right] \leq c(1 + |x|^k), \quad \mathbb{E} e^{k l_T^x} \leq c, \quad x \in \bar{D}, \tag{2.3.1}$$

where (X_t^x, l_t^x) *is the (weak) solution of* (2.0.1) *with* $X_0^x = x$.

Below we first introduce some results for the reflected SDE with random coefficients, then present two lemmas which will be used in the proof of Theorem 2.3.1.

2.3.1 Preparations

Consider the following reflected SDE with random coefficients:

$$dX_t = J_t(X_t)dt + S_t(X_t)dW_t + \mathbf{n}(X_t)dl_t, \quad t \in [0,T], \tag{2.3.2}$$

where $(W_t)_{t\in[0,T]}$ is an m-dimensional Brownian motion on a complete filtration probability space $(\Omega, \{\mathcal{F}_t\}_{t\in[0,T]}, \mathbb{P})$,

$$J: [0,T] \times \Omega \times \mathbb{R}^d \to \mathbb{R}^d, \quad S: [0,T] \times \Omega \times \mathbb{R}^d \to \mathbb{R}^d \otimes \mathbb{R}^m$$

are progressively measurable, and l_t is the local time of X_t on ∂D. Let Λ be the set of increasing functions $h:(0,1] \to (0,\infty)$ such that $\int_0^1 \frac{ds}{h(s)} = \infty$, and let Γ be the class of increasing functions $\gamma: [0,\infty) \to [1,\infty)$ such that $\int_0^\infty \frac{ds}{\gamma(s)} = \infty$.

A continuous adapted process $(X_t, l_t)_{t\in[0,\tau)}$ is called a solution of (2.3.2) with life time τ, if τ is a stopping time, $\lim_{t\uparrow\tau} \sup_{s\in[0,t]} |X_s| = \infty$ holds on $\{\tau \leq T\}$, l_t is an increasing process with dl_t supported on $\{t \in [0,\tau) : X_t \in \partial D\}$, and $\mathbb{P}$-a.s.

$$X_t = X_0 + \int_0^t J_s(X_s)ds + \int_0^t S_s(X_s)dW_s + \int_0^t \mathbf{n}(X_s)dl_s, \quad t \in [0,T], t < \tau.$$

When $\mathbb{P}(\tau > T) = 1$, we call the solution non-explosive. A weak solution (X_t, l_t, W_t) is defined in the same spirit where W_t is an m-dimensional Brownian motion under a (not given) complete filtration probability space.

We have the following result.

Theorem 2.3.2. *Assume* **(D)**.

(1) *For any two solutions X_t and Y_t of (2.3.2) with $X_0 = Y_0 \in \bar{D}$, if there exist $h \in \Lambda$ and a positive $L^1([0,T])$-valued random variable g such that $\mathbb{P}$-a.s.*

$$\|S_t(X_t) - S_t(Y_t)\|_{HS}^2 + 2\langle X_t - Y_t, J_t(X_t) - J_t(Y_t)\rangle$$
$$\leq g_t h(|X_t - Y_t|^2), \quad t \in [0,T],$$

then $X_t = Y_t$ up to life time.

(2) *If $\mathbb{P}$-a.s. S and J are continuous and locally bounded on $[0,\infty) \times \bar{D}$, then for any initial value in $\bar{D}$, (2.3.2) has a weak solution up to life time. If S and J are bounded and deterministic S and J on $[0,T] \times \bar{D}$, (2.3.2) has a non-explosive weak solution.*

(3) *If either D is bounded, or there exist $1 \leq V \in C^{1,2}([0,T] \times \bar{D})$ with*

$$\lim_{x\in\bar{D}, |x|\to\infty} \inf_{t\in[0,T]} V_t(x) = \infty, \quad \nabla_\mathbf{n} V_t|_{\partial D} \leq 0,$$

and a positive $L^1([0,T])$-valued random variable g such that $\mathbb{P}$-a.s.
$$\operatorname{tr}\{S_t S_t^* \nabla^2 V_t\} + 2\langle \nabla V(x), J_t(x)\rangle + 2\partial_t V_t(x)$$
$$\leq g_t \gamma(V(x)), \quad t \in [0,T], x \in \bar{D}$$
holds for some $\gamma \in \Gamma$, then any solution to (2.3.2) is non-explosive.

Remark 2.3.1. When D is convex, this result goes back to [Tanaka (1979)], and in general it is mainly summarized from Theorem 1, Corollary 1 and Theorem 2 in [Hino et al. (2021)].

The condition in the first assertion is modified from that in [Hino et al. (2021)]:
$$\|S_t(x) - S_t(y)\|_{HS}^2 + 2\langle x-y, J_t(x) - J_t(y)\rangle \leq g_t h(|x-y|^2), \quad t \in [0,T], x, y \in \bar{D},$$
since in the proof of this assertion, one only uses the upper bound of
$$\|S_t(X_t) - S_t(Y_t)\|_{HS}^2 + 2\langle X_t - Y_t, J_t(X_t) - J_t(Y_t)\rangle,$$
so that the present condition is enough for the pathwise uniqueness. The present version of the condition is weaker when $\mathcal{L}_{(X_t, Y_t)}$ does not have full support $\mathbb{R}^d \times \mathbb{R}^d$.

In assertion (3), the term $\operatorname{tr}\{S_t S_t^* \nabla^2 V_t\}$ was formulated in Theorem 1.1 in [Hino et al. (2021)] as $\|S_t(x)\|^2 \Delta V_t(x)$, which should be changed into the present one according to Itô's formula of $V_t(X_t)$.

Moreover, when S and J are bounded and deterministic, the weak existence is given in Theorem 2.1 in [Rozkosz and Slominski (1997)].

Next, we apply Theorem 2.3.2 to (2.0.1) with coefficients satisfying the following assumption, where (1_b) is known as monotone or semi-Lipschitz condition, which comparing with (1_a) allows σ to be unbounded.

(H) b and σ are locally bounded and satisfy the following conditions.
(1) One of the following conditions hold:
(1_a) $(A^{2.1})$ holds with $\|\nabla \sigma\|^2 \leq \sum_{i=1}^l f_i$ for some $\{f_i \in \tilde{I}_{q_i}^{p_i}(T), (p_i, q_i) \subset \mathcal{K}\}_{1 \leq i \leq l}$, or $(A^{2.3})$ holds. Moreover, there exists a constant $K > 0$ such that
$$\langle x-y, b_t(x) - b_t(y)\rangle \leq K|x-y|^2, \quad t \in [0,T], x, y \in \bar{D}. \tag{2.3.3}$$
(1_b) There exists an increasing function $h : [0, \infty) \to [0, \infty)$ with $\int_0^1 \frac{dr}{r+h(r)} = \infty$, such that
$$2\langle x-y, b_t(x) - b_t(y)\rangle^+ + \|\sigma_t(x) - \sigma_t(y)\|_{HS}^2$$
$$\leq h(|x-y|^2), \quad t \in [0,T], x, y \in \bar{D}. \tag{2.3.4}$$

(2) $\|\sigma\| \leq c(1+|\cdot|^2)$ holds for some constant $c > 0$, there exist $x_0 \in D$ and $\tilde{\partial} D \subset \partial D$ such that
$$\langle x - x_0, \mathbf{n}(x) \rangle \leq 0, \quad x \in \partial D \setminus \tilde{\partial} D, \ \mathbf{n}(x) \in \mathcal{N}_x; \tag{2.3.5}$$
and when $\tilde{\partial} D \neq \emptyset$ there exists a function $\tilde{\rho}_\partial \in C_b^2(\bar{D})$ such that
$$\langle \nabla \tilde{\rho}_\partial, \mathbf{n} \rangle|_{\partial D} \geq 1_{\tilde{\partial} D},$$
$$\sup_{[0,T] \times \bar{D}} \left\{ \|\sigma^* \nabla \tilde{\rho}_\partial\| + \|\mathrm{tr}\{\sigma \sigma^* \nabla^2 \tilde{\rho}_\partial\}\| + \langle b, \nabla \tilde{\rho}_\partial \rangle^- \right\} \leq K. \tag{2.3.6}$$

According to (2.1.3) and Remark 2.1.2, $(H)(2)$ holds with $\tilde{\rho}_\partial = 0$ if D is convex, and it holds with $\tilde{\rho}_\partial = \rho_\partial$ in $\partial_{r_0/2} D$ for some $r_0 > 0$ when $\partial D \in C_b^2$ and $\|\sigma\| + \langle b, \nabla \rho_\partial \rangle^-$ is bounded on $[0,T] \times \partial_{r_0} D$.

To estimate $|X_t|$ and l_t, we need the following lemma on the maximal functional for nonnegative functions f on $\bar{D}$:
$$\mathcal{M}_D f(x) := \sup_{r \in (0,1)} \frac{1}{|B(0,r)|} \int_{B(0,r)} (1_D f)(x+y) \mathrm{d}y, \quad x \in \bar{D}.$$

Lemma 2.3.3. *Let $\partial D \in C_b^2$.*

(1) *For any real function f on $\bar{D}$ with $|\nabla f| \in L^1_{loc}(\bar{D})$,*
$$|f(x) - f(y)| \leq c|x-y|\big(\mathcal{M}_D|\nabla f|(x) + \mathcal{M}_D|\nabla f|(y) + \|f\|_\infty\big), \text{ a.e. } x, y \in \bar{D}.$$
(2) *There exists a constant $c > 0$ such that for any nonnegative measurable function f on $[0,T] \times \bar{D}$,*
$$\|\mathcal{M}_D f\|_{\tilde{L}_q^p(T,\bar{D})} \leq c\|f\|_{\tilde{L}_q^p(T,\bar{D})}, \quad p, q \geq 1.$$

Proof. We only prove (1), since (2) follows from Lemma 1.3.4(2) with $1_{\bar{D}} f$ replacing f. Let $\tilde{\varphi}$ be in (2.2.20). Take $0 \leq h \in C_b^\infty(\mathbb{R})$ with $h(r) = 1$ for $r \leq r_0/4$ and $h(r) = 0$ for $r \geq r_0/2$. We then extend a function f on $\bar{D}$ to $\tilde{f}$ on $\mathbb{R}^d$ by letting
$$\tilde{f}(x) := \{h \circ \rho_D\} f \circ \tilde{\varphi},$$
where ρ_D is the distance function to D. Then there exists a constant $c > 0$ such that
$$|\nabla \tilde{f}| \leq 1_{\bar{D}} |\nabla f| + c 1_{\partial_{-r_0/2} D}(|f \circ \tilde{\varphi}| + |\nabla f| \circ \tilde{\varphi}).$$
By Lemma 1.3.4(1) and the integral transform $x \mapsto \tilde{\varphi}(x)$ with $\|(\nabla \tilde{\varphi})^{-1}\|$ bounded on $\partial_{-r_0} D$, we find constants $c_1, c_2 > 0$ such that for any $x, y \in \bar{D}$,
$$|f(x) - f(y)| = |\tilde{f}(x) - \tilde{f}(y)|$$
$$\leq c_1 |x-y| \{\mathcal{M}|\nabla \tilde{f}|(x) + \mathcal{M}|\nabla \tilde{f}|(y) + \|f\|_\infty\}$$
$$\leq c_2 |x-y| \{\mathcal{M}_D |\nabla f|(x) + \mathcal{M}_D |\nabla f|(y) + \|f\|_\infty\},$$
where $\mathcal{M} := \mathcal{M}_D$ for $D = \mathbb{R}^d$. $\square$

We are now able to prove the following result.

Lemma 2.3.4. *Assume* **(D)** *and* (H)(1). *Then the reflected SDE* (2.0.1) *is well-posed up to life time. If* (H)(2) *holds, then the solution is non-explosive, and for any $k > 0$ there exists a constant $c > 0$ such that*

$$\mathbb{E}\Big[\sup_{t \in [0,T]} |X_t^x|^k\Big] \leq c(1 + |x|^k), \quad x \in \bar{D}, t \in [0,T], \tag{2.3.7}$$

$$\sup_{x \in \bar{D}} \mathbb{E}\big(e^{k(\tilde{l}_{t_1}^x - \tilde{l}_{t_0}^x)} | \mathcal{F}_{t_0}\big) \leq c, \quad 0 \leq t_0 \leq t_1 \leq T, \tag{2.3.8}$$

where (X_t^x, l_t^x) is the solution with $X_0^x = x$, and $\tilde{l}_t^x := \int_0^t 1_{\tilde{\partial}(D)}(X_s^x) \mathrm{d} l_s^x$.

Proof. (1) We first prove the existence and uniqueness up to life time. Since σ and b are locally bounded, by a truncation argument we may and do assume that σ and b are bounded. Indeed, let for any $n \geq 1$, we take

$$\sigma_t^{\{n\}}(x) := \sigma_t\big(\{1 \wedge (n/|x|)\}x\big), \quad b_t^{\{n\}}(x) := h(|x|/n)b_t(x), \quad t \in [0,T], x \in \bar{D},$$

where $h \in C_0^\infty([0,\infty))$ with $0 \leq h \leq 1$ and $h|_{[0,1]} = 1$. Then $\sigma^{\{n\}}$ and $b^{\{n\}}$ are bounded on $[0,T] \times \bar{D}$, and for some constant $K_n > 0$,

$$\langle b_t^{\{n\}}(x) - b_t^{\{n\}}(y), x - y \rangle^+$$
$$\leq h(|x|/n)\langle b_t(x) - b_t(y), x - y \rangle^+ + \big|h(|x|/n) - h(|y|/n)\big|\langle b_t(y), x - y \rangle^+$$
$$\leq \langle b_t(x) - b_t(y), x - y \rangle^+ + K_n|x - y|^2, \quad t \in [0,T], x, y \in \bar{D}, |y| \leq |x|.$$

So, by the symmetry of $\langle b_t^{\{n\}}(x) - b_t^{\{n\}}(y), x - y \rangle^+$ in (x, y), (1_a) implies that σ and $b^{\{n\}}$ are bounded on $[0,T] \times \bar{D}$ and satisfy (2.3.3) with $K + K_n$ replacing K; while (1_b) and

$$|\{1 \wedge (n/|x|)\}x - \{1 \wedge (n/|y|)\}y| \leq |x - y|$$

yield that $\sigma^{\{n\}}$ and $b^{\{n\}}$ are bounded and satisfy (2.3.4) for $2h(r) + K_n r$ replacing $h(r)$. Therefore, if the well-posedness is proved under (H) for bounded b and σ, then the SDE is well-posed up to the hitting time of $\partial B(0, n)$ for any $n \geq 1$, i.e. it is well-posed up to life time.

When σ and b are bounded, the weak existence is implied by Theorem 2.3.2(2). By the Yamada-Watanabe principle, it suffices to verify the pathwise uniqueness. Let X_t and Y_t be two solutions starting from $x \in \bar{D}$. By Lemma 2.3.3(1) and $(H)(1)$,

$$\|\sigma_t(X_t) - \sigma_t(Y_t)\|_{HS}^2 + 2\langle X_t - Y_t, b_t(X_t) - b_t(Y_t) \rangle$$
$$\leq \begin{cases} g_t|X_t - Y_t|^2, & \text{under } (1_a), \\ h(|X_t - Y_t|^2), & \text{under } (1_b), \end{cases}$$

where for some constant $c > 0$,
$$g_t := c\{1 + \mathcal{M}_D \|\nabla \sigma_t\|^2(X_t) + \mathcal{M}_D \|\nabla \sigma_t\|^2(Y_t)\}.$$
So, by Theorem 2.3.2(1), it suffices to prove $\int_0^T g_t \mathrm{d}t < \infty$ under (1_a). By Lemma 2.3.3, this follows from (2.2.3) under condition $(A^{2.1})$ with $\|\nabla \sigma\|^2 \leq \sum_{i=1}^l f_i$, or (2.2.34) under condition $(A^{2.3})$.

(2) To prove the non-explosion, we simply denote $(X_t, l_t) = (X_t^x, l_t^x)$ and let
$$\tau_n := \inf\{t \in [0, T] : |X_t| \geq n\}, \quad n \geq 1.$$
By $(H)(2)$, we find a constant $c_1 > 0$ such that
$$\mathrm{d}\tilde{\rho}_\partial(X_t) \geq -K\mathrm{d}t + \mathrm{d}M_t + \mathrm{d}\tilde{l}_t, \quad t \in [0, T] \qquad (2.3.9)$$
holds for $\mathrm{d}M_t := \langle \sigma_t(X_t)^* \nabla \tilde{\rho}_\partial(X_t), \mathrm{d}W_t \rangle$ satisfying $\mathrm{d}\langle M \rangle_t \leq K^2 \mathrm{d}t$. This implies (2.3.8). Next, by (H), we find a constant $c_1 > 0$ such that
$$2\langle b_t(x), x - x_0 \rangle + \|\sigma_t(x)\|_{HS}^2$$
$$= 2\langle b_t(x) - b_t(x_0), x - x_0 \rangle + \|\sigma_t(x) - \sigma_t(x_0)\|_{HS}^2$$
$$+ 2\langle b_t(x_0), x - x_0 \rangle + \|\sigma_t(x_0)\|_{HS}^2 + 2\langle \sigma_t(x_0), \sigma_t(x) \rangle_{HS}$$
$$\leq c_1(1 + |x - x_0|^2), \quad x \in \bar{D}.$$
Then by $(H)(2)$ and Itô's formula, for any $k \geq 2$ we find a constant $c_2 > 0$ such that
$$\mathrm{d}|X_t - x_0|^k \leq c_2(1 + |X_t - x_0|^k)\mathrm{d}t + \mathrm{d}\tilde{M}_t + k|X_t - x_0|^{k-1}\mathrm{d}\tilde{l}_t,$$
where $\tilde{M}_t$ is a local martingale with $\mathrm{d}\langle \tilde{M} \rangle_t \leq c_2(1 + |X_t - x_0|^k)^2 \mathrm{d}t$. By BDG's inequality and (2.3.8), we find constants $c_3, c_4 > 0$ such that
$$\eta_t^{\{n\}} := \sup_{s \in [0, t \wedge \tau_n]} (1 + |X_s - x_0|^k), \quad n \geq 1, t \in [0, T]$$
satisfies
$$\mathbb{E}\eta_t^{\{n\}} \leq 1 + |x - x_0|^k + c_3 \mathbb{E} \int_0^t \eta_s^{\{n\}} \mathrm{d}s$$
$$+ 2c_3 \mathbb{E}^x \left(\int_0^t |\eta_s^{\{n\}}|^2 \mathrm{d}s \right)^{\frac{1}{2}} + k\mathbb{E}\left[|\eta_t^{\{n\}}|^{\frac{k-1}{k}} \tilde{l}_t \right]$$
$$\leq \frac{1}{2} \mathbb{E}\eta_t^{\{n\}} + c_4(1 + |x|^k) + c_4 \int_0^t \mathbb{E}\eta_s^{\{n\}} \mathrm{d}s, \quad t \in [0, T].$$
By Gronwall's lemma, we obtain
$$\mathbb{E}[\eta_t^{\{n\}}] \leq 2c_4(1 + |x|^k)\mathrm{e}^{2c_4 t}, \quad t \in [0, T], x \in \bar{D}, n \geq 1,$$
which implies that X_t is non-explosive and (2.3.7) holds for some constant $c > 0$. □

2.3.2 Proof of Theorem 2.3.1

Let $X_0 = x \in \bar{D}$. We consider the following two cases respectively.

(a) Let $(A^{2.2})$ hold. Then (H) holds for $b^{(1)}$ replacing b. By Lemma 2.3.4, the reflected SDE

$$\mathrm{d}X_t = b_t^{(1)}(X_t)\mathrm{d}t + \sigma_t(X_t)\mathrm{d}W_t + \mathbf{n}(X_t)\mathrm{d}l_t \tag{2.3.10}$$

is well-posed with (2.3.7) holding for all $k \geq 1$ and some constant $c > 0$ depending on k. By Lemmas 2.2.1–2.2.3, (2.3.8) and $(A^{2.1})$ with $|b^{(0)}|^2 \in \tilde{L}_{q_0}^{p_0}(T)$, we see that (2.2.4) holds for $f := |b^{(0)}|^2$, so that for some map $c : [1, \infty) \to (0, \infty)$ independent of the initial value x,

$$\sup_{x \in \bar{D}} \mathbb{E}^x |R_T|^k \leq c(k), \quad k \geq 1 \tag{2.3.11}$$

holds for

$$R_t := \mathrm{e}^{\int_0^t \langle \{\sigma_s^*(\sigma_s\sigma_s^*)^{-1}b_s^{(0)}\}(X_s), \mathrm{d}W_s \rangle - \frac{1}{2}\int_0^t |\sigma_s^*(\sigma_s\sigma_s^*)^{-1}b_s^{(0)}|^2(X_s)\mathrm{d}s}, \quad t \in [0, T].$$

By Girsanov's theorem,

$$\tilde{W}_t := W_t - \int_0^t \{\sigma_s^*(\sigma_s\sigma_s^*)^{-1}b_s^{(0)}\}(X_s)\mathrm{d}s, \quad t \in [0, T]$$

is an m-dimensional Brownian motion under the probability measure $\mathbb{Q} := R_T \mathbb{P}$. Rewriting (2.3.10) as

$$\mathrm{d}X_t = b_t(X_t)\mathrm{d}t + \sigma_t(X_t)\mathrm{d}\tilde{W}_t + \mathbf{n}(X_t)\mathrm{d}l_t,$$

we see that $(X_t, l_t, \tilde{W}_t)_{t \in [0,T]}$ under probability $\mathbb{Q}$ is a weak solution of (2.0.1). Moreover, letting $\mathbb{E}_\mathbb{Q}$ be the expectation under $\mathbb{Q}$, by (2.3.7) and (2.3.11), for any $k \geq 1$ we find a constant $\tilde{c}(k) > 0$ independent of x such that

$$\mathbb{E}_\mathbb{Q}\left[\sup_{t \in [0,T]} |X_t|^k\right] = \mathbb{E}\left[R_T \sup_{t \in [0,T]} |X_t|^k\right]$$

$$\leq \left(\mathbb{E}[R_T^2]\right)^{\frac{1}{2}} \left(\left[\mathbb{E} \sup_{t \in [0,T]} |X_t|^{2k}\right]\right)^{\frac{1}{2}} \leq \tilde{c}(k)(1 + |x|^k), \quad x \in \bar{D}$$

for some constant $c > 0$. Similarly, (2.3.8) and (2.3.11) imply

$$\mathbb{E}_\mathbb{Q}[\mathrm{e}^{kl_T}] \leq C(k), \quad k \geq 1$$

for constants $C(k) > 0$ independent of x. So, (2.3.1) holds for this weak solution.

To prove the weak uniqueness, let $(\bar{X}_t, \bar{l}_t, \bar{W}_t)_{t \in [0,T]}$ under probability $\bar{\mathbb{P}}$ be another weak solution of (2.0.1) with $\bar{X}_0 = x$, i.e.

$$\mathrm{d}\bar{X}_t = b_t(\bar{X}_t)\mathrm{d}t + \sigma_t(\bar{X}_t)\mathrm{d}\bar{W}_t + \mathbf{n}(\bar{X}_t)\mathrm{d}\bar{l}_t, \quad t \in [0, T], \bar{X}_0 = x. \tag{2.3.12}$$

It suffices to show

$$\mathcal{L}_{(\bar{X}_t,\bar{l}_t)_{t\in[0,T]}|\bar{\mathbb{P}}} = \mathcal{L}_{(X_t,l_t)_{t\in[0,T]}|\mathbb{Q}}. \tag{2.3.13}$$

By Lemma 2.2.1 the estimate (2.2.4) holds for $\bar{X}_t$ and $f = |b^{(0)}|^2$, so that

$$\mathbb{E}_{\bar{\mathbb{P}}}\left[e^{\lambda \int_0^T |b_t^{(0)}(\bar{X}_t)|^2 dt}\right] < \infty, \quad \lambda > 0. \tag{2.3.14}$$

By Girsanov's theorem, this and $\|\sigma^*(\sigma\sigma^*)^{-1}\|_\infty < \infty$ imply that

$$G_t(\bar{X}, \bar{W}) := \bar{W}_t + \int_0^t \{\sigma_s^*(\sigma_s\sigma_s^*)^{-1} b_s^{(0)}\}(\bar{X}_s) ds, \quad t \in [0, T]$$

is an m-dimensional Brownian motion under the probability $\bar{\mathbb{Q}} := R(\bar{X}, \bar{W})\bar{\mathbb{P}}$, where

$$R(\bar{X}, \bar{W}) := e^{-\int_0^T \langle \{\sigma_s^*(\sigma_s\sigma_s^*)^{-1} b_s^{(0)}\}(\bar{X}_s), d\bar{W}_s\rangle - \frac{1}{2}\int_0^T |\{\sigma_s^*(\sigma_s\sigma_s^*)^{-1} b_s^{(0)}\}(\bar{X}_s)|^2 ds}.$$

Reformulating (2.3.12) as

$$d\bar{X}_t = b_t^{(1)}(\bar{X}_t) dt + \sigma_t(\bar{X}_t) dG_t(\bar{X}, \bar{W}) + \mathbf{n}(\bar{X}_t) d\bar{l}_t, \quad t \in [0, T],$$

and applying the well-posedness of (2.3.10) which implies the joint weak uniqueness, we conclude that

$$\mathcal{L}_{(\bar{X}_t,\bar{l}_t,G_t(\bar{X},\bar{W}))_{t\in[0,T]}|\bar{\mathbb{Q}}} = \mathcal{L}_{(X_t,l_t,W_t)_{t\in[0,T]}|\mathbb{P}}.$$

Noting that

$$R(\bar{X}, \bar{W})^{-1} = e^{-\int_0^T |\{\sigma_s^*(\sigma_s\sigma_s^*)^{-1} b_s^{(0)}\}(\bar{X}_s)|^2 ds} R(\bar{X}, G(\bar{X}, \bar{W}))^{-1},$$

this implies that for any bounded continuous function F on $C([0,T]; \mathbb{R}^d \times [0,\infty))$,

$$\mathbb{E}_{\bar{\mathbb{P}}}[F(\bar{X}, \bar{l})] = \mathbb{E}_{\bar{\mathbb{Q}}}[R(\bar{X}, \bar{W})^{-1} F(\bar{X}, \bar{l})]$$
$$= \mathbb{E}_{\bar{\mathbb{Q}}}[R(\bar{X}, G(\bar{X}, \bar{W}))^{-1} e^{-\int_0^T |\{\sigma_s^*(\sigma_s\sigma_s^*)^{-1} b_s^{(0)}\}(\bar{X}_s)|^2 ds} F(\bar{X}, \bar{l})]$$
$$= \mathbb{E}_{\mathbb{P}}[R(X, W)^{-1} e^{-\int_0^T |\{\sigma_s^*(\sigma_s\sigma_s^*)^{-1} b_s^{(0)}\}(X_s)|^2 ds} F(X, l)]$$
$$= \mathbb{E}_{\mathbb{P}}[R_T F(X, l)] = \mathbb{E}_{\mathbb{Q}}[F(X, l)].$$

Therefore, (2.3.13) holds.

(b) Let $(A^{2.3})$ hold. By Lemma 2.2.3, (2.3.11) and (2.3.14) hold, so that the desired assertions follow from Girsanov's transforms as in step (a).

2.4 Strong well-posedness and gradient estimates

Let $\mathcal{B}_b^+(\bar{D})$ be the space of bounded strictly positive measurable functions on $\bar{D}$. The first result in this section is the following.

Theorem 2.4.1. *Assume that one of the following conditions holds:*

(i) $d = 1$ and $(A^{2.2})$ holds;
(ii) $(A^{2.3})$ holds with $p_i > 2, 1 \le i \le l$.

Then (2.0.1) *is well-posed, and for any $k \ge 1$, there exists a constant $c > 0$ such that*

$$\mathbb{E}\left[\sup_{t \in [0,T]} |X_t^x - X_t^y|^k\right] \le c|x-y|^k, \quad x, y \in \bar{D}. \tag{2.4.1}$$

Consequently,

(1) *For any $p > 1$ there exists a constant $c(p) > 0$ such that*

$$P_t f(x) := \mathbb{E}[f(X_t^x)], \quad x \in \bar{D}, t \in [0,T], f \in \mathcal{B}_b(\bar{D})$$

satisfies

$$|\nabla P_t f| \le c(p)(P_t|\nabla f|^p)^{\frac{1}{p}}, \quad f \in C_b^1(\bar{D}), \quad t \in [0,T]. \tag{2.4.2}$$

(2) *There exist a constant $C > 0$ and a map $c : (1, \infty) \to (0, \infty)$ such that*

$$|\nabla P_t f| \le \frac{c(p)}{\sqrt{t}}(P_t|f|^p)^{\frac{1}{p}}, \quad t \in (0,T], f \in \mathcal{B}_b(\bar{D}), \, p > 1, \tag{2.4.3}$$

$$P_t f^2 - (P_t f)^2 \le tCP_t|\nabla f|^2, \quad f \in C_b^1(\bar{D}), \quad t \in [0,T], \tag{2.4.4}$$

$$P_t \log f(x) \le \log P_t f(y) + \frac{C|x-y|^2}{t},$$
$$t \in (0,T], x, y \in \bar{D}, f \in \mathcal{B}_b^+(\bar{D}). \tag{2.4.5}$$

To relax the condition on $b^{(1)}$ as in $(A^{1.2})(2)$, we consider the following time dependent differential operator on $\bar{D}$:

$$L_t^\sigma := \frac{1}{2}\mathrm{tr}(\sigma_t \sigma_t^* \nabla^2), \quad t \in [0,T]. \tag{2.4.6}$$

Let $\{P_{s,t}^\sigma\}_{T \ge t_1 \ge t \ge s \ge 0}$ be the Neumann semigroup on $\bar{D}$ generated by L_t^σ; that is, for any $\varphi \in C_b^2(\bar{D})$, and any $t \in (0,T]$, $(P_{s,t}^\sigma \varphi)_{s \in [0,t]}$ is the unique solution of the PDE

$$\partial_s u_s = -L_s^\sigma u_s, \quad \nabla_\mathbf{n} u_s|_{\partial D} = 0 \text{ for } s \in [0,t), u_t = \varphi. \tag{2.4.7}$$

$(A^{2.4})$ $(A^{2.3})$ holds for $b^{(1)} = 0$. Moreover, there exist constants $K, \varepsilon > 0$, increasing $\phi \in C^1([0,\infty);[1,\infty))$ with $\int_0^\infty \frac{\mathrm{d}s}{r+\phi(s)} = \infty$, and a compact function $V \in C^2(\mathbb{R}^d;[1,\infty))$ satisfying

$$\nabla_{\mathbf{n}(x)} V(y) \leq 0, \quad x \in \partial D, |y-x| \leq r_0 \tag{2.4.8}$$

for some constant $r_0 > 0$, such that

$$\sup_{B(x,\varepsilon)} \{|\nabla V| + \|\nabla^2 V\|\} \leq KV(x),$$

$$\langle b_t^{(1)}(x), \nabla V(x) \rangle + \varepsilon |b_t^{(1)}(x)| \sup_{B(x,\varepsilon)} \{|\nabla V| + \|\nabla^2 V\|\} \tag{2.4.9}$$

$$\leq K\phi(V(x)), \quad (t,x) \in [0,T] \times \mathbb{R}^d.$$

(2.4.8) can be dropped when ∂D is bounded. Indeed, for V satisfying (2.4.9), when ∂D is bounded, we may take $1 \leq \tilde{V} \in C^2(\mathbb{R}^d)$ such that $\tilde{V} = 1$ on $\partial_{r_0}(\partial D)$ and $\tilde{V} = V$ outside a compact set, so that (2.4.8) and (2.4.9) hold for $\tilde{V}$ replacing V with a different constant K. Similarly, (2.4.8) holds for $V(x_1, x_2) := V_1(x_1) + V_2(x_2)$ and $D = D_1 \times \mathbb{R}^l$ where $l \in \mathbb{N}$ is less than d, $\partial D_1 \subset \mathbb{R}^{d-l}$ is bounded, and $V_1 = 1$ in a neighborhood of ∂D_1.

Theorem 2.4.2. *Assume* $(A^{2.4})$. *Then* (2.0.1) *is well-posed up to time* T. *Moreover, for any* $t \in (0,T]$,

$$\lim_{\bar{D} \ni y \to x} \|P_t^* \delta_x - P_t^* \delta_y\|_{var} = 0, \quad t \in (0,T], x \in \bar{D}, \tag{2.4.10}$$

and P_t *has probability density (i.e. heat kernel)* $p_t(x,y)$ *such that*

$$\inf_{x,y \in \bar{D} \cap B_N, \ \rho_\partial(y) \geq N^{-1}} p_t(x,y) > 0, \quad N > 1, t \in (0,T], \tag{2.4.11}$$

where $\inf \emptyset := \infty$.

2.4.1 Proof of Theorem 2.4.1

The weak existence is implied by Theorem 2.3.1. By the Yamada-Watanabe principle in Lemma 1.3.2, it suffices to prove estimate (2.4.1) which in particular implies the pathwise uniqueness as well as estimate (2.4.2), since

$$|\nabla P_t f|(x) := \limsup_{\bar{D} \ni y \to x} \frac{|P_t f(x) - P_t f(y)|}{|x-y|} \leq \limsup_{\bar{D} \ni y \to x} \mathbb{E}\left[\frac{|f(X_t^x) - f(X_t^y)|}{|x-y|}\right]$$

$$\leq \limsup_{\bar{D} \ni y \to x} \left(\mathbb{E}\frac{|f(X_t^x) - f(X_t^y)|^p}{|X_t^x - X_t^y|^p}\right)^{\frac{1}{p}} \left(\frac{\mathbb{E}[|X_t^x - X_t^y|^{\frac{p}{p-1}}]}{|x-y|^{\frac{p}{p-1}}}\right)^{\frac{p-1}{p}}$$

$$\leq c(p) \left(P_t |\nabla f|^p\right)^{\frac{1}{p}}(x), \quad x \in \bar{D}, t \in [0,T], f \in C_b^1(\bar{D}).$$

Let $(X_t^{(i)}, l_t^{(i)})$ be two solutions of (2.0.1) with $X_0^{(i)} = x^{(i)} \in \bar{D}, i = 1, 2$. Below we prove (2.4.1) in situations (i) and (ii) respectively, and prove inequalities in Theorem 2.4.1(2).

Proof of Theorem 2.4.1 under (i). In this case, D is an interval or a half-line. For any $\lambda > 0$, let u_t^λ be the unique solution to (1.2.3) with $t_0 = 0, t_1 = T$ and $f = -b^{(0)}$, that is,

$$(\partial_t + L_t)u_t^\lambda = \lambda u_t^\lambda - b_t^{(0)}, \quad t \in [0, T], u_T^\lambda = 0. \tag{2.4.12}$$

By Lemma 1.2.2 with $f = -b^{(0)} \in \tilde{L}_{2q_0}^{2p_0}(T)$, we take large enough $\lambda > 0$ such that

$$\|u^\lambda\|_\infty + \|\nabla u^\lambda\|_\infty \leq \frac{1}{2}, \quad \|u^\lambda\|_{\tilde{H}_{2q_0}^{2,2p_0}(T)} < \infty. \tag{2.4.13}$$

Then

$$\Theta_t^\lambda(x) := x + u_t^\lambda(x), \quad x \in \mathbb{R}$$

is a diffeomorphism and there exists a constant $C > 0$ such that

$$\frac{1}{2}|x - y| \leq |\Theta_t^\lambda(x) - \Theta_t^\lambda(y)| \leq 2|x - y|, \quad x, y \in \mathbb{R}, t \in [0, T]. \tag{2.4.14}$$

Let $(X_t^{(i)}, l_t^{(i)})$ solve (2.0.1) for $X_0^{(i)} = x^{(i)} \in \bar{D}, i = 1, 2$, and let

$$Y_t^{(i)} := \Theta_t^\lambda(X_t^{(i)}) = X_t^{(i)} + u_t^\lambda(X_t^{(i)}), \quad i = 1, 2.$$

By Itô's formula in Lemma 2.2.1(2),

$$dY_t^{(i)} = B_t(Y_t^{(i)})dt + \Sigma_t(Y_t^{(i)})dW_t + \{1 + \nabla u_t^\lambda(X_t^{(i)})\}\mathbf{n}(X_t^{(i)})dl_t^{(i)} \tag{2.4.15}$$

holds for $i = 1, 2$ and

$$\begin{aligned}B_t(x) &:= \{b_t^{(1)} + \lambda u_t^\lambda\}(\{\Theta_t^\lambda\}^{-1}(x)), \\ \Sigma_t(x) &:= \{(1 + \nabla u_t^\lambda)\sigma_t\}(\{\Theta_t^\lambda\}^{-1}(x)).\end{aligned} \tag{2.4.16}$$

By (2.4.13), (2.4.16) and $\|\nabla b^{(1)}\|_\infty \prec 1$ due to $(A^{2.1})$, we find nonnegative functions $F_i \in \tilde{L}_{q_i}^{p_i}(T), 0 \leq i \leq l$ such that

$$\|\nabla B\|_\infty, \ \|\nabla \Sigma\|^2 \leq \sum_{i=0}^l F_i. \tag{2.4.17}$$

Since $d = 1$, for any $x \in \partial D$ and $y \in D$ we have $y - x = |y - x|\mathbf{n}(x)$, so that (2.4.13) implies

$$\langle \Theta_t^\lambda(y) - \Theta_t^\lambda(x), \{1 + \nabla u_t^\lambda(x)\}\mathbf{n}(x)\rangle \geq |y - x|(1 - \|\nabla u^\lambda\|_\infty)^2 \geq 0. \tag{2.4.18}$$

Combining this with (2.4.15) and Itô's formula, up to a local martingale we have

$$d|Y_t^{(1)} - Y_t^{(2)}|^{2k} \leq 2k|Y_t^{(1)} - Y_t^{(2)}|^{2k}\left\{\frac{|B_t(Y_t^{(1)}) - B_t(Y_t^{(2)})|}{|Y_t^{(1)} - Y_t^{(2)}|} + \frac{k\|\Sigma_t(Y_t^{(1)}) - \Sigma_t(Y_t^{(2)})\|_{HS}^2}{|Y_t^{(1)} - Y_t^{(2)}|^2}\right\}dt.$$

So, by Lemma 2.3.3, we find a constant $c_1 > 0$ and a local martingale M_t such that

$$|Y_t^{(1)} - Y_t^{(2)}|^{2k} \leq |Y_0^{(1)} - Y_0^{2}|^{2k} + c_1\int_0^t |Y_s^{(1)} - Y_s^{(2)}|^{2k}d\mathcal{L}_s + dM_t,$$

where

$$\mathcal{L}_t := \int_0^t \left\{1 + \mathcal{M}_D\big(\|\nabla B_s\| + \|\nabla \Sigma_s\|^2\big)(Y_s^{(1)}) \right. \\ \left. + \mathcal{M}_D\big(\|\nabla B_s\| + \|\nabla \Sigma_s\|^2\big)(Y_s^{(2)})\right\}ds. \tag{2.4.19}$$

Combining this with (2.2.4), (2.4.17), Lemma 2.3.3 and the stochastic Gronwall inequality in Lemma 1.3.3, we find constants $c_2, c_3 > 0$ such that

$$\left(\mathbb{E}\Big[\sup_{s\in[0,t]}\Theta_s^\lambda(X_s^{(1)}) - \Theta_s^\lambda(X_s^{(2)})|^k\Big]\right)^2 = \left(\mathbb{E}\sup_{s\in[0,t]}|Y_s^{(1)} - Y_s^{(2)}|^k\right)^2$$
$$\leq c_2|Y_0^{(1)} - Y_0^{(2)}|^{2k}\big(\mathbb{E}e^{\frac{c_1 p}{p-1}\mathcal{L}_t}\big)^{\frac{p-1}{p}} \leq c_3|\Theta_0^\lambda(x^{(1)}) - \Theta_0^\lambda(x^{(2)})|^{2k}.$$

This together with (2.4.14) implies (2.4.1) for some constant $c > 0$. □

To prove (2.4.1) under $(A^{2.3})$, we need the following lemma due to Lemma 5.2 in [Yang and Zhang (2023)], which is contained in the proof of Lemma 4.4 in [Dupuis and Ishii (1990)]. Let $\nabla^{(1)}$ and $\nabla^{(2)}$ be the gradient operators in the first and second variables on $\mathbb{R}^d \times \mathbb{R}^d$.

Lemma 2.4.3. *There exists a function* $g \in C^1(\mathbb{R}^d \times \mathbb{R}^d) \cap C^2((\mathbb{R}^d \setminus \{0\}) \times \mathbb{R}^d)$ *having the following properties for some constants* $k_2 > 1$ *and* $k_1 \in (0,1)$:

(1) $k_1|x|^2 \leq g(x,y) \leq k_2|x|^2$, $x, y \in \mathbb{R}^d$;
(2) $\langle \nabla^{(1)}g(x,y), y\rangle \leq 0$, $|y| = 1, \langle x, y\rangle \leq k_1|x|$;
(3) $\big|(\nabla^{(1)})^i(\nabla^{(2)})^j g(x,y)\big| \leq k_2|x|^{2-i}$, $i, j \in \{0,1,2\}, i+j \leq 2, x, y \in \mathbb{R}^d$.

Proof of Theorem 2.4.1 under (ii). Let $b^{0,n}$ be the mollifying approximation of $b^{(0)} = 1_{\bar{D}} b^{(0)}$. By Lemma 2.2.2, there exists $\lambda_0 > 0$ such that for any $\lambda \geq \lambda_0$ and $n \geq 1$, the PDE

$$(\partial_t + L_t + \nabla_{b_t^{0,n} - b_t^{(0)}} - \lambda) u_t^{\lambda,n} = -b_t^{0,n}, \quad u_T^{\lambda,n} = \nabla_{\mathbf{n}} u_t^{\lambda,n}|_{\partial D} = 0, \quad (2.4.20)$$

has a unique solution in $C_b^{1,2}([0,T] \times \bar{D})$, and there exist constants $\varepsilon, c > 0$ such that

$$\lambda^\varepsilon \big(\|u^{\lambda,n}\|_\infty + \|\nabla u^{\lambda,n}\|_\infty \big) + \|(\partial_t + \nabla_{b^{(1)}}) u^{\lambda,n}\|_{\tilde{L}_{q_0}^{p_0}(T,D)} \\ + \|\nabla^2 u^{\lambda,n}\|_{\tilde{L}_{q_0}^{p_0}(T,D)} \leq c\|b^{(0)}\|_{\tilde{L}_{q_0}^{p_0}(T,D)}, \quad \lambda \geq \lambda_0, n \geq 1. \quad (2.4.21)$$

Then for large enough $\lambda_0 > 0$, $\Theta_t^{\lambda,n} := id + u_t^{\lambda,n}$ satisfies

$$\frac{1}{2}|x-y|^2 \leq |\Theta_t^{\lambda,n}(x) - \Theta_t^{\lambda,n}(y)|^2 \leq 2|x-y|^2, \quad \lambda \geq \lambda_0, x, y \in \bar{D}. \quad (2.4.22)$$

Since $\partial D \in C_b^{2,L}$, there exists a constant $r_0 > 0$ such that $\rho_\partial \in C_b^2(\partial_{r_0} D)$ with $\nabla^2 \rho_\partial$ Lipschitz continuous on $\partial_{r_0} D$. Take $h \in C^\infty([0,\infty);[0,\infty))$ such that $h' \geq 0$, $h(r) = r$ for $r \leq r_0/2$ and $h(r) = r_0$ for $r \geq r_0$.

Let $(X_t^{(i)}, l_t^{(i)})$ solve (2.0.1) starting at $x^{(i)} \in \bar{D}$ for $i = 1, 2$. Alternatively to $|X_t^{(1)} - X_t^{(2)}|^2$, we consider the process

$$H_t := g\big(\Theta_t^{\lambda,n}(X_t^{(1)}) - \Theta_t^{\lambda,n}(X_t^{(2)}), \nabla(h \circ \rho_\partial)(X_t^{(1)})\big), \quad t \in [0,T],$$

where g is in Lemma 2.4.3. By Lemma 2.4.3(1) and (2.4.22), we have

$$\frac{k_1}{2}|X_t^{(1)} - X_t^{(2)}|^2 \leq H_t \leq 2k_2|X_t^{(1)} - X_t^{(2)}|^2, \quad t \in [0,T]. \quad (2.4.23)$$

Simply denote

$$\xi_t := \Theta_t^{\lambda,n}(X_t^{(1)}) - \Theta_t^{\lambda,n}(X_t^{(2)}), \quad \eta_t := \nabla(h \circ \rho_\partial)(X_t^{(1)}).$$

By Itô's formula, (2.4.20) and $\nabla_{\mathbf{n}} \Theta_t^{\lambda,n}|_{\partial D} = \mathbf{n}$ due to $\nabla_{\mathbf{n}} u_t^{\lambda,n}|_{\partial D} = 0$, we have

$$\begin{aligned}
d\xi_t &= \big\{ \lambda u_t^{\lambda,n}(X_t^{(1)}) - \lambda u_t^{\lambda,n}(X_t^{(2)}) \\
&\quad + (b_t^{(0)} - b_t^{0,n})(X_t^{(1)}) - (b_t^{(0)} - b_t^{0,n})(X_t^{(2)}) \big\} dt \\
&\quad + \big\{ [(\nabla \Theta_t^{\lambda,n}) \sigma_t](X_t^{(1)}) - [(\nabla \Theta_t^{\lambda,n}) \sigma_t](X_t^{(2)}) \big\} dW_t \\
&\quad + \mathbf{n}(X_t^{(1)}) dl_t^{(1)} - \mathbf{n}(X_t^{(2)}) dl_t^{(2)}, \\
d\eta_t &= L_t \nabla(h \circ \rho_\partial)(X_t^{(1)}) dt + \big\{ [\nabla^2 (h \circ \rho_\partial)] \sigma_t \big\}(X_t^{(1)}) dW_t \\
&\quad + \big\{ \nabla_{\mathbf{n}} \nabla(h \circ \rho_\partial) \big\}(X_t^{(1)}) dl_t^{(1)}.
\end{aligned} \quad (2.4.24)$$

Hence, Itô's formula for H_t reads

$$dH_t = A_t dt + B_t^{(1)} dl_t^{(1)} - B_t^{(2)} dl_t^{(2)} + dM_t, \qquad (2.4.25)$$

where for $N_t := \{(\nabla \Theta_t^{\lambda,n})\sigma_t\}(X_t^{(1)}) - \{(\nabla \Theta_t^{\lambda,n})\sigma_t\}(X_t^{(2)})$,

$$\begin{aligned}
A_t := & \left\langle \nabla^{(1)} g(\xi_t, \eta_t), \lambda u_t^{\lambda,n}(X_t^{(1)}) - \lambda u_t^{\lambda,n}(X_t^{(2)}) \right\rangle \\
& + \left\langle \nabla^{(1)} g(\xi_t, \eta_t), \nabla_{b_t^{(0)} - b_t^{0,n}} \Theta_t^{\lambda,n}(X_t^{(1)}) - \nabla_{b_t^{(0)} - b_t^{0,n}} \Theta_t^{\lambda,n}(X_t^{(2)}) \right\rangle \\
& + \left\langle \nabla^{(2)} g(\xi_t, \eta_t), L_t \nabla(h \circ \rho_\partial)(X_t^{(1)}) \right\rangle + \left\langle (\nabla^{(1)})^2 g(\xi_t, \eta_t), N_t N_t^* \right\rangle_{HS} \\
& + \left\langle \nabla^{(1)} \nabla^{(2)} g(\xi_t, \eta_t), N_t \sigma_t(X_t^{(1)})^* \nabla^2(h \circ \rho_\partial)(X_t^{(1)}) \right\rangle_{HS} \\
& + \left\langle (\nabla^{(2)})^2 g(\xi_t, \eta_t), \{[\nabla^2(h \circ \rho_\partial)]\sigma_t \sigma_t^* \nabla^2(h \circ \rho_\partial)\}(X_t^{(1)}) \right\rangle_{HS},
\end{aligned} \qquad (2.4.26)$$

$$\begin{aligned}
B_t^{(1)} := & \left\langle \nabla^{(1)} g(\xi_t, \eta_t), \mathbf{n}(X_t^{(1)}) \right\rangle \\
& + \left\langle \nabla^{(2)} g(\xi_t, \eta_t), \nabla_{\mathbf{n}}\{\nabla(h \circ \rho_\partial)\}(X_t^{(1)}) \right\rangle, \qquad (2.4.27) \\
B_t^{(2)} := & \left\langle \nabla^{(1)} g(\xi_t, \eta_t), \mathbf{n}(X_t^{(2)}) \right\rangle,
\end{aligned}$$

$$dM_t := \left\langle \nabla^{(2)} g(\xi_t, \eta_t), [\{\nabla^2(h \circ \rho_\partial)\}\sigma_t](X_t^{(1)}) dW_t \right\rangle + \left\langle \nabla^{(1)} g(\xi_t, \eta_t), \right. \\
\left. [\{(\nabla \Theta_t^{\lambda,n})\sigma_t\}(X_t^{(1)}) - \{(\nabla \Theta_t^{\lambda,n})\sigma_t\}(X_t^{(2)})] dW_t \right\rangle. \qquad (2.4.28)$$

In the following we estimate $A_t, B_t^{(1)}$ and $B_t^{(2)}$ respectively.

Firstly, (2.1.2) implies

$$\langle \Theta_t^{\lambda,n}(x) - \Theta_t^{\lambda,n}(y), \mathbf{n}(x) \rangle \leq \frac{|x-y|^2}{2r_0} + \|\nabla u_t^{\lambda,n}\|_\infty |x-y|, \quad x \in \partial D, y \in \bar{D}.$$

Combining this with (2.4.21), we find constants $\varepsilon_0, \lambda_1 > 0$ such that for any $\lambda \geq \lambda_1$,

$$\langle \Theta_t^{\lambda,n}(x) - \Theta_t^{\lambda,n}(y), \mathbf{n}(x) \rangle \leq k_1 |\Theta_t^{\lambda,n}(x) - \Theta_t^{\lambda,n}(y)|,$$
$$x \in \partial D, y \in \bar{D}, |x-y| \leq \varepsilon_0, n \geq 1, t \in [0,T].$$

So, Lemma 2.4.3 yields

$$\begin{aligned}
& \langle \nabla^{(1)} g(\Theta_t^{\lambda,n}(x) - \Theta_t^{\lambda,n}(y), \mathbf{n}(x)), \mathbf{n}(x) \rangle \\
& \leq k_2 \mathbf{1}_{\{|x-y| > \varepsilon_0\}} |\Theta_t^{\lambda,n}(x) - \Theta_t^{\lambda,n}(y)| \\
& \leq k_2 \varepsilon_0^{-1} |\Theta_t^{\lambda,n}(x) - \Theta_t^{\lambda,n}(y)|^2, \quad x \in \partial D, y \in \bar{D}, n \geq 1, t \in [0,T].
\end{aligned} \qquad (2.4.29)$$

Next, by the same reason leading to (2.4.29), we find a constant $c_1 > 0$ such that

$$\langle \nabla^{(1)} g(\Theta_t^{\lambda,n}(x) - \Theta_t^{\lambda,n}(y), \nabla(h \circ \rho_\partial)(x)), \mathbf{n}(y) \rangle$$
$$\geq \langle \nabla^{(1)} g(\Theta_t^{\lambda,n}(x) - \Theta_t^{\lambda,n}(y), \mathbf{n}(y)), \mathbf{n}(y) \rangle$$
$$- |\nabla^{(1)} g(\Theta_t^{\lambda,n}(x) - \Theta_t^{\lambda,n}(y), \nabla(h \circ \rho_\partial)(y))$$
$$- \nabla^{(1)} g(\Theta_t^{\lambda,n}(x) - \Theta_t^{\lambda,n}(y), \nabla(h \circ \rho_\partial)(x))|$$
$$\geq -\mathbf{1}_{\{|x-y|>\varepsilon_0\}} k_2 \varepsilon_0^{-1} |\Theta_t^{\lambda,n}(x) - \Theta_t^{\lambda,n}(y)|^2 \qquad (2.4.30)$$
$$- \|h'\|_\infty \|\nabla^{(1)}\nabla^{(2)} g(\Theta_t^{\lambda,n}(x) - \Theta_t^{\lambda,n}(y), \cdot)\|_\infty |\Theta_t^{\lambda,n}(x) - \Theta_t^{\lambda,n}(y)|^2$$
$$\geq -c_1 |\Theta_t^{\lambda,n}(x) - \Theta_t^{\lambda,n}(y)|^2, \quad x \in \bar{D}, y \in \partial D, n \geq 1, t \in [0,T].$$

Moreover, by $(A^{2.3})$ and $h \circ \rho_\partial \in C_b^{2,L}(\bar{D})$, there exists a constant $C > 0$ such that

$$|L_t\{\nabla(h \circ \rho_\partial)\}| \leq C(1 + |b_t^{(0)}|), \quad t \in [0,T].$$

Combining this with Lemma 2.4.3, Lemma 2.3.3, (2.4.23), and (2.4.26)–(2.4.30), we find a constant $K > 0$ such that

$$|A_t| \leq K\{|b_t^{(0)} - b_t^{0,n}|^2(X_t^{(1)}) + |b_t^{(0)} - b_t^{0,n}|^2(X_t^{(2)})\}$$
$$+ K|X_t^{(1)} - X_t^{(2)}|^2 \Big\{1 + |b_t^{(0)}|(X_t^{(1)}) + \sum_{i=1}^{2} \mathcal{M}_D \|\nabla\{(\nabla\Theta_t^{\lambda,n})\sigma_t\}\|^2(X_t^{(i)})\Big\},$$
$$d\langle M \rangle_t \leq K|X_t^{(1)} - X_t^{(2)}|^4 \Big\{1 + \sum_{i=1}^{2} \mathcal{M}_D \|\nabla\{(\nabla\Theta_t^{\lambda,n})\sigma_t\}\|^2(X_t^{(i)})\Big\},$$
$$B_t^{(1)} \leq K|X_t^{(1)} - X_t^{(2)}|^2, \quad -B_t^{(2)} \leq K|X_t^{(1)} - X_t^{(2)}|^2.$$

Combining these with (2.4.23) and (2.4.25), for any $k \geq 1$, we find a constant $c_1 > 0$ such that

$$dH_t^k \leq c_1 |X_t^{(1)} - X_t^{(2)}|^{2(k-1)}$$
$$\times \{|b_t^{(0)} - b_t^{0,n}|^2(X_t^{(1)}) + |b_t^{(0)} - b_t^{0,n}|^2(X_t^{(2)})\}dt \qquad (2.4.31)$$
$$+ c_1 |X_t^{(1)} - X_t^{(2)}|^{2k} d\mathcal{L}_t + k H_t^{k-1} dM_t,$$

where

$$\mathcal{L}_t := l_t^{(1)} + l_t^{(2)}$$
$$+ \int_0^t \Big\{1 + |b_s^{(0)}|(X_s^{(1)}) + \sum_{i=1}^{2} \mathcal{M}_D \|\nabla\{(\nabla\Theta_s^{\lambda,n})\sigma_s\}\|^2(X_s^{(i)})\Big\}ds. \qquad (2.4.32)$$

For any $m \geq 1$, let
$$\tau_m := \inf\{t \in [0,T] : |X_t^{(1)} - X_t^{(2)}| \geq m\}.$$
By (2.4.23) and (2.4.31), we find a constant $c_2 > 0$ such that
$$|X_{t\wedge\tau_m}^{(1)} - X_{t\wedge\tau_m}^{(2)}|^{2k} \leq G_m(t) + c_2 \int_0^{t\wedge\tau_m} |X_s^{(1)} - X_s^{(2)}|^{2k} \mathrm{d}\mathcal{L}_s + \tilde{M}_t \quad (2.4.33)$$
holds for some local martingale $\tilde{M}_t$ and
$$G_m(t) := c_2|x^{(1)} - x^{(2)}|^{2k}$$
$$+ c_2 m^{2(k-1)} \int_0^{t\wedge\tau_m} \{|b_s^{(0)} - b_s^{0,n}|^2(X_s^{(1)}) + |b_s^{(0)} - b_s^{0,n}|^2(X_s^{(2)})\}\mathrm{d}s.$$
Since $(A^{2.3})$ and (2.4.21) imply
$$\sup_{n\geq 1} \|\nabla\{(\nabla\Theta^{\lambda,n})\sigma\}\| \leq \sum_{i=0}^{l} F_i$$
for some $0 \leq F_i \in \tilde{L}_{q_i}^{p_1}(T), 0 \leq i \leq l$, by (2.2.34), (2.2.35), the stochastic Gronwall lemma, and Lemma 2.3.3, for any $p \in (\frac{1}{2}, 1)$, there exist constants $c_3, c_4 > 0$ such that
$$\left(\mathbb{E}\Big[\sup_{s\in[0,t\wedge\tau_m]} |X_s^{(1)} - X_s^{(2)}|^k\Big]\right)^2 \leq c_3(\mathbb{E}\mathrm{e}^{\frac{c_2 p}{1-p}\mathcal{L}_t})^{\frac{1-p}{p}} \mathbb{E}G_m(t)$$
$$\leq c_4\big(|x^{(1)} - x^{(2)}|^{2k} + m^{2(k-1)}\|b^{(0)} - b^{0,n}\|_{\tilde{L}_{q_0}^{p_0}(T)}\big), \quad n,m \geq 1.$$
By first letting $n \to \infty$ then $m \to \infty$ and applying (2.2.36), we derive (2.4.1) for some constant $c > 0$. $\square$

Proof of Theorem 2.4.1(2). Let $\{P_{s,t}\}_{t\geq s\geq 0}$ be the Markov semigroup associated with (2.0.1), i.e.
$$P_{s,t}f(x) := \mathbb{E}[f(X_{s,t}^x)], \quad t \geq s, f \in \mathcal{B}_b(\bar{D}),$$
where $(X_{s,t}^x)_{t\geq s}$ is the unique solution of (2.0.1) starting from x at time s. We have
$$P_t f(x) = \mathbb{E}[(P_{s,t}f)(X_s^x)], \quad s \in [0,t], f \in C_b^1(\bar{D}), \quad (2.4.34)$$
where $X_s^x := X_{0,s}^x$. By (2.4.2) for (2.0.1) from time s, for any $p > 1$, we have
$$|\nabla P_{s,t}f| \leq c(p)(P_{s,t}|\nabla f|^p)^{\frac{1}{p}}, \quad 0 \leq s \leq t \leq T, f \in C_b^1(\bar{D}). \quad (2.4.35)$$
If $P_{\cdot,t}f \in C^{1,2}([0,t]\times\bar{D})$ for $f \in C_N^2(\bar{D})$ such that
$$(\partial_s + L_s)P_{s,t}f = 0, \quad f \in C_N^2(\bar{D}), \nabla_\mathbf{n} P_{s,t}f|_{\partial D} = 0, \quad (2.4.36)$$

then the desired inequalities follow from (2.4.35) by taking derivative in s to the following reference functions respectively:

$$P_s\{P_{s,t}(\varepsilon+f)\}^p, P_s\{P_{s,t}(\varepsilon+f)\}^2, P_s\{\log P_{s,t}(\varepsilon+f)\}(x+s(y-s)/t), s \in [0,t],$$

see for instance the proof of Theorem 3.1 in [Wang and Zhang (2014)]. However, in the present singular setting it is not clear whether (2.4.36) holds or not. So, below we make an approximation argument.

(a) Proof of (2.4.3). Let $\{b^{0,n}\}_{n\geq 1}$ be the mollifying approximations of $b^{(0)}$. By $(A^{2.3})$, for any $f \in C_N^2(\bar{D})$ and $t \in (0,T]$, the equation

$$u_{s,t}^n = P_{s,t}^{\sigma,b^{(1)}} f + \int_s^t P_{s,r}^{\sigma,b^{(1)}}(\nabla_{b_r^{0,n}} u_{s,t}^n) \mathrm{d}r, \quad s \in [0,t]$$

has a unique solution in $C^{1,2}([0,t] \times \bar{D})$, and $P_{s,t}^n f := u_{s,t}^n$ satisfies

$$(\partial_s + L_s^{\sigma,b^{(1)}} + \nabla_{b_s^{0,n}})P_{s,t}^n f = 0, \quad s \in [0,t], f \in C_N^2(\bar{D}). \tag{2.4.37}$$

By this and Itô's formula for the reflected SDE

$$\mathrm{d}X_{s,t}^{x,n} = (b_t^{(1)} + b_t^{0,n})(X_{s,t}^{x,n})\mathrm{d}t + \sigma_t(X_{s,t}^{x,n})\mathrm{d}W_t + \mathbf{n}(X_t^{x,n})\mathrm{d}l_t, \quad t \geq s, X_{s,s}^{x,n} = x,$$

we obtain

$$P_{s,t}^n f(x) = \mathbb{E}f(X_{s,t}^{x,n}), \quad 0 \leq s \leq t.$$

Let X_t solve (2.0.1) from time s with $X_s = x$, and define

$$\xi_s^n := \{\sigma_s^*(\sigma_s\sigma_s^*)^{-1}(b_s^{(0)} - b_t^{0,n})\}(X_s),$$
$$R_s := e^{\int_0^s \langle \xi_r, \mathrm{d}W_r\rangle - \frac{1}{2}\int_0^s |\xi_r|^2 \mathrm{d}r}, \quad s \in [0,t].$$

By Girsanov's theorem, we obtain

$$|P_{s,t}f - P_{s,t}^n f|(x) = |\mathbb{E}[f(X_t) - R_t f(X_t)]|$$
$$\leq \|f\|_\infty (\mathbb{E}e^{c\int_0^t |b_s^{(0)} - b_s^{0,n}|^2(X_s)} - 1) =: \|f\|_\infty \varepsilon_n, \quad 0 \leq s \leq t \leq T,$$

where $c > 0$ is a constant, and due to (2.2.35), $\varepsilon_n \to 0$ as $n \to \infty$. Consequently,

$$\|P_{s,t}f - P_{s,t}^n f\|_\infty \leq \varepsilon_n \|f\|_\infty, \quad n \geq 1, 0 \leq s \leq t \leq T. \tag{2.4.38}$$

Moreover, the proof of (2.4.35) implies that it holds for $P_{s,t}^n$ replacing $P_{s,t}$ uniformly in $n \geq 1$, since the constant is increasing in $\|b^{(0)}\|_{\tilde{L}_{q_0}^{p_0}(T)}$, which is not less than $\|b^{0,n}\|_{\tilde{L}_{q_0}^{p_0}}(T)$. Thus,

$$|\nabla P_{s,t}^n f| \leq c(p)(P_{s,t}^n |\nabla f|^p)^{\frac{1}{p}}, \quad 0 \leq s \leq t \leq T, f \in C_b^1(\bar{D}), n \geq 1. \tag{2.4.39}$$

Now, let $0 \leq f \in C_N^2(\bar{D})$ and $t \in (0,T]$. For any $\varepsilon > 0$ and $p \in (1,2]$, by (2.4.39), (2.4.37), (2.4.38), $(A^{2.3})$ and Itô's formula, we find constants $c_1, c_2 > 0$ such that

$$\mathrm{d}(\varepsilon + P_{s,t}^n f)^p(X_s) = \{p(\varepsilon + P_{s,t}^n f)^{p-1}\langle b_t^{(0)} - b_t^{0,n}, \nabla P_{s,t}^n f\rangle$$
$$+ p(p-1)(\varepsilon + P_{s,t}^n f)^{p-2}|\sigma_s^* \nabla P_{s,t}^n f|^2\}(X_s)\mathrm{d}s + \mathrm{d}M_s$$
$$\geq \{c_2(\varepsilon + P_{s,t}^n f)^{p-2}|\nabla P_{s,t}^n f|^2 - c_1\|\nabla f\|_\infty |b_t^{(0)} - b_t^{0,n}|\}(X_s)\mathrm{d}s + \mathrm{d}M_s$$

holds for $s \in [0,t], \varepsilon > 0$ and some martingale M_s. By (2.2.3), Hölder's inequality, and $\|b^{(0)} - b^{0,n}\|_{\tilde{L}_{q_0}^{p_0}(T)} \to 0$ as $n \to \infty$, we find a constant $c_3 > 0$ and sequence $\varepsilon_n \to 0$ as $n \to \infty$ such that

$$\varepsilon_n + P_t(\varepsilon + f)^p - (P_t^n f + \varepsilon)^p$$
$$\geq c_2 \int_0^t P_s\{(\varepsilon + P_{s,t}^n f)^{p-2}|\nabla P_{s,t}^n f|^2\}\mathrm{d}s$$
$$\geq c_2 \int_0^t \frac{(P_s|\nabla P_{s,t}^n f|^p)^{\frac{2}{p}}}{\{P_s(\varepsilon + P_{s,t}^n f)^p\}^{\frac{2-p}{p}}}\mathrm{d}s$$
$$\geq c_3 \int_0^t \frac{|\nabla P_s P_{s,t}^n f|^2}{\{P_s(\varepsilon + P_{s,t}^n f)^p\}^{\frac{2-p}{p}}}\mathrm{d}s, \quad \varepsilon \in (0,1).$$

Thus, for any $x \in D$ and $x \neq y \in B(x,\delta) \subset D$ for small $\delta > 0$ such that

$$x_r := x + r(y-x) \in D, \quad r \in [0,1],$$

this implies

$$\frac{|\int_0^t (P_s P_{s,t}^n f(x) - P_s P_{s,t}^n f(y))\mathrm{d}s|}{|x-y|} \leq \int_0^1 \mathrm{d}r \int_0^t |\nabla P_s P_{s,t}^n f|(x_r)\mathrm{d}s$$
$$\leq \int_0^1 \left(\int_0^t \frac{|\nabla P_s P_{s,t}^n f|^2}{\{P_s(\varepsilon + P_{s,t}^n f)^p\}^{\frac{2-p}{p}}}(x_r)\mathrm{d}s\right)^{\frac{1}{2}}$$
$$\times \left(\int_0^t \{P_s(\varepsilon + P_{s,t}^n f)^p)\}^{\frac{2-p}{p}}(x_r)\mathrm{d}s\right)^{\frac{1}{2}} \mathrm{d}r$$
$$\leq \int_0^1 c_3^{-1/2}\{\varepsilon_n + P_t(\varepsilon + f)^p\}^{\frac{1}{2}}(x + r(y-x))$$
$$\times \left(\int_0^t (\varepsilon + P_s P_{s,t}^n f^p)^{\frac{2-p}{p}}(x_r)\mathrm{d}s\right)^{\frac{1}{2}} \mathrm{d}r.$$

Combining this with (2.4.38) and letting $n \to \infty, \varepsilon \to 0$, we obtain

$$\frac{|P_t f(x) - P_t f(y)|}{|x-y|} \leq \frac{1}{t}\int_0^1 (c_3^{-1} P_t f^p)^{\frac{1}{2}}(x_r)\left(\int_0^t (P_t f^p)^{\frac{2-p}{p}}(x_r)\mathrm{d}s\right)^{\frac{1}{2}} \mathrm{d}r.$$

Letting $y \to x$, we derive (2.4.3) for some constant c depending on p for $p \in (1,2]$ and all $f \in C_N^2(\bar{D})$. By Jensen's inequality the estimate also holds for $p > 2$, and by approximation argument, it holds for all $f \in \mathcal{B}_b(\bar{D})$.

(b) Proof of (2.4.4). By (2.4.39), Itô's formula and $(A^{2.3})$, we find a constant $c_4 > 0$ and a martingale M_s such that

$$d(P_{s,t}^n f)^2(X_s) = 2\{\langle \nabla P_{s,t}^n f, b_s^{(0)} - b_s^{0,n}\rangle + |\sigma_s^* \nabla P_{s,t}^n f|^2\}(X_s)\mathrm{d}s + \mathrm{d}M_s$$
$$\leq c_4\{\|\nabla f\|_\infty |b_s^{(0)} - b_s^{0,n}| + P_{s,t}^n |\nabla f|^2\}(X_s)\mathrm{d}s + \mathrm{d}M_s, \quad s \in [0,t].$$

Integrating both sides over $s \in [0,t]$, taking expectations and letting $n \to \infty$, and combining with (2.2.3) and (2.4.38), we derive (2.4.4).

(c) Proof of (2.4.5). Let $0 < f \in C_N^2(\bar{D})$. By taking Itô's formula to $P_{s,t}^n(\varepsilon + f)(X_s)$ for $\varepsilon > 0$ and taking expectation, we derive

$$\frac{\mathrm{d}}{\mathrm{d}s} P_s \log P_{s,t}^n\{\varepsilon + f\} = -P_s|\sigma_s^* \nabla \log P_{s,t}^n f|^2 + P_s\langle b_s^{(0)} - b_s^{0,n}, \nabla \log P_{s,t}^n(\varepsilon + f)\rangle.$$

For any $x, y \in \bar{D}$, let $\gamma : [0,1] \to \bar{D}$ be a curve linking x and y such that $|\dot\gamma_r| \leq c|x - y|$ for some constant $c > 0$ independent of x, y. Combining these with $(A^{2.3})$ and (2.4.2), for $p = 2$ we find a constant $c_5 > 0$ such that

$$P_t \log\{\varepsilon + f\}(x) - \log P_t^n\{\varepsilon + f\}(y) = \int_0^t \frac{\mathrm{d}}{\mathrm{d}s} P_s \log P_{s,t}^n f(\gamma_{s/t})\mathrm{d}s$$
$$\leq \int_0^t \{ct^{-1}|x-y||\nabla P_s \log P_{s,t}^n f(\gamma_{s/t})| - P_s|\sigma_s^* \nabla \log P_{s,t}^n f|^2\}(\gamma_{s/t})\mathrm{d}s$$
$$\leq c_5 \int_0^t \frac{|x-y|^2}{t^2}\mathrm{d}s = \frac{c_5|x-y|^2}{t}, \quad t \in (0,T].$$

Therefore, (2.4.5) holds. □

2.4.2 Proof of Theorem 2.4.2

(a) The well-posedness. The proof is similar to that of Theorem 1.3.1(2). For any $n \geq 1$, let

$$b^n := 1_{B_n} b^{(1)} + b^{(0)}.$$

By Theorem 2.4.1, the following SDE is well-posed:

$$\mathrm{d}X_t^{x,n} = b^n(X_t^{x,n})\mathrm{d}t + \sigma(X_t^{x,n})\mathrm{d}W_t + \mathbf{n}(X_t^{x,n})\mathrm{d}l_t^{x,n}, \quad X_0^{x,n} = x.$$

Let $\tau_n^x := \inf\{t \in [0,T] : |X_t^{x,n}| \geq n\}$. Then $X_t^{x,n}$ solves (2.0.1) up to time τ_n^x, and by the uniqueness we have

$$X_t^{x,n} = X_t^{x,m}, \quad t \leq \tau_n^x \wedge \tau_m^x, n, m \geq 1.$$

So, it suffices to prove that $\tau_n^x \to \infty$ as $n \to \infty$.

Let $L_t^0 := L_t^\sigma + \nabla_{b_t^{(0)}}$. By Lemma 2.2.2, $(A^{2.5})$ implies that for any $\lambda \geq 0$, the PDE

$$(\partial_t + L_t^0)u_t = \lambda u_t - b_t^{(0)}, \quad t \in [0,T], u_T = 0, \nabla_{\mathbf{n}} u_t|_{\partial D} = 0 \quad (2.4.40)$$

has a unique solution $u \in \tilde{H}_{q_0}^{p_0}(T)$, and there exist constants $\lambda_0, c, \theta > 0$ such that

$$\lambda^\theta(\|u\|_\infty + \|\nabla u\|_\infty) + \|\partial_t u\|_{\tilde{L}_{q_0}^{p_0}(T)} + \|\nabla^2 u\|_{\tilde{L}_{q_0}^{p_0}(T)} \leq c, \quad \lambda \geq \lambda_0. \quad (2.4.41)$$

So, we may take $\lambda \geq \lambda_0$ such that

$$\|u\|_\infty + \|\nabla u\|_\infty \leq \varepsilon, \quad (2.4.42)$$

where we take $\varepsilon \leq r_0$ when ∂D exists. Let $\Theta_t(x) = x + u_t(x)$. By (2.4.8) and (2.4.42) for $\varepsilon \leq r_0$ when ∂D exists, we have

$$\langle \nabla V(Y_t^{x,n}), \mathbf{n}(X_t^{x,n}) \rangle \mathrm{d} l_t^{x,n} \leq 0.$$

So, by Itô's formula, $Y_t^{x,n} := \Theta_t(X_t^{x,n})$ satisfies

$$\begin{aligned}\mathrm{d} Y_t^{x,n} &= \{1_{B_n} b_t^{(1)} + \lambda u_t + 1_{B_n} \nabla_{b_t^{(1)}} u_t\}(X_t^{x,n}) \mathrm{d} t \\ &\quad + \{(\nabla \Theta_t)\sigma_t\}(X_t^{x,n}) \mathrm{d} W_t + \mathbf{n}(X_t^n)\mathrm{d} l_t^n.\end{aligned} \quad (2.4.43)$$

By (2.4.42) and $(A^{2.4})$, there exists a constant $c_0 > 0$ such that for some martingale M_t,

$$\begin{aligned}\mathrm{d}&\{V(Y_t^{x,n}) + M_t\} \\ &\leq \Big[\langle \{b^{(1)} + \nabla_{b^{(1)}} u_t\}(X_t^{x,n}), \nabla V(Y_t^{x,n})\rangle \\ &\quad + c_0(|\nabla V(Y_t^{x,n})| + \|\nabla^2 V(Y_t^{x,n})\|)\Big]\mathrm{d} t \\ &\leq \Big\{\langle b^{(1)}(X_t^{x,n}), \nabla V(X_t^{x,n})\rangle + c_0 K V(Y_t^{x,n}) \\ &\quad + \varepsilon |b^{(1)}(X_t^{x,n})| \sup_{B(X_t^{x,n}, \varepsilon)}(|\nabla V| + \|\nabla^2 V\|)\Big\}\mathrm{d} t \\ &\leq \{K\phi(V(X_t^{x,n})) + c_0 K V(Y_t^{x,n})\}\mathrm{d} t \\ &\leq K\{\phi((1+\varepsilon K)V(Y_t^{x,n})) + c_0 V(Y_t^{x,n})\}\mathrm{d} t, \quad t \leq \tau_n^x.\end{aligned}$$

Letting $H(r) := \int_0^r \frac{\mathrm{d} s}{r + \phi((1+\varepsilon K)s)}$, by Itô's formula and noting that $\phi' \geq 0$, we find a constant $c_1 > 0$ such that

$$\mathrm{d} H(V(Y_t^{x,n})) \leq c_1 \mathrm{d} t + \mathrm{d} \tilde{M}_t, \quad t \in [0, \tau_n^x]$$

holds for some martingale $\tilde{M}_t$. Thus,

$$\mathbb{E}[(H \circ V)(Y_{t \wedge \tau_n^x}^{x,n})] \leq V(x + u(x)) + c_1 t, \quad t \in [0,T], n \geq 1.$$

Since (2.4.42) and $|z| \geq n$ imply $|\Theta_t(z)| \geq |z| - |u(z)| \geq n - \varepsilon$, we derive

$$\mathbb{P}(\tau_n^x \leq t) \leq \frac{V(x + \Theta_0(x)) + c_1 t}{\inf_{|y| \geq n-\varepsilon} H(V(y))} =: \varepsilon_{t,n}(x), \quad t > 0. \tag{2.4.44}$$

Since $\lim_{|x| \to \infty} H(V)(x) = \int_0^\infty \frac{ds}{s + \phi((1+\varepsilon K)s)} = \infty$, we obtain $\tau_n^x \to \infty (n \to \infty)$ as desired.

(b) Proof of (2.4.10). By Proposition 1.3.8 in [Wang (2013)], the log-Harnack inequality

$$P_t \log f(y) \leq \log P_t f(x) + c|x - y|^2, \quad x, y \in \bar{D}, 0 < f \in \mathcal{B}_b(\bar{D})$$

for some constant $c > 0$ implies the gradient estimate

$$|\nabla P_t f|^2 \leq 2c P_t |f|^2, \quad f \in \mathcal{B}_b(\bar{D}),$$

and hence

$$\lim_{y \to x} \|P_t^* \delta_x - P_t^* \delta_y\|_{var} = 0, \quad x \in \bar{D}.$$

Let P_t^n be the Markov semigroup associated with X_t^n. Thus, by the log-Harnack inequality in Theorem 2.4.1(2), we have

$$\lim_{y \to x} \|(P_t^n)^* \delta_x - (P_t^n)^* \delta_y\|_{var} = 0, \quad t \in (0, T]. \tag{2.4.45}$$

On the other hand, by (2.4.44) and $X_t = X_t^n$ for $t \leq \tau_n$, we obtain

$$\lim_{n \to \infty} \sup_{y \in \bar{D} \cap D(x,1)} \|P_t^* \delta_y - (P_t^n)^* \delta_y\|_{var}$$

$$= \lim_{n \to \infty} \sup_{|f| \leq 1, y \in \bar{D} \cap B(x,1)} |P_t f(y) - P_t^n f(y)|$$

$$\leq 2 \lim_{n \to \infty} \sup_{y \in \bar{D} \cap B(x,1)} \mathbb{P}(\tau_n^y \leq t) = 0.$$

Combining this with (2.4.45) and the triangle inequality, we derive (2.4.10).

(c) Finally, let $L_t := L_t^\sigma + \nabla_{b_t}$. For any $f \in C_0^2((0,T) \times D)$, by Itô's formula,

$$df_t(X_t) = (\partial_t + L_t) f_t(X_t) dt + dM_t$$

holds for some martingale M_t, so that $f_0 = f_T = 0$ yields

$$\int_0^T P_t \{(\partial_t + L) f_t\} dt = 0, \quad f \in C_0^\infty((0,T) \times D).$$

By the Harnack inequality as in Theorem 3 in [Aronson and Serrin (1967)] (see also [Trudinger (1968)]), for any $0 < s < t \leq T$ and $N > 1$ with

$$\tilde{B}_N := \{x \in \bar{D} \cap B_N : \rho_\partial(x) \geq N^{-1}\}$$

having positive volume, there exists a constant $c(s,t,N) > 0$ such that the heat kernel $p_t(x,y)$ of P_t satisfies

$$\sup_{\tilde{B}_N} p_s(x,\cdot) \leq c(s,t,N) \inf_{\tilde{B}_N} p_t(x,\cdot), \quad x \in \bar{D}. \tag{2.4.46}$$

Since $\int_{\tilde{B}_N} p_s(x,y)\mathrm{d}y \to 1$ as $N \to \infty$, this implies $p_t(x,y) > 0$ for any $(t,x,y) \in (0,T] \times \bar{D} \times D$. In particular, $P_t 1_{\tilde{B}_N} > 0$. On the other hand, (2.4.10) implies that $P_t 1_{\tilde{B}_N}$ is continuous, so that

$$\inf_{x \in \bar{D} \cap B_N} P_t 1_{\tilde{B}_N}(x) > 0, \quad t \in (0,T].$$

This together with (2.4.46) gives

$$\inf_{(\bar{D} \cap B_N) \times \tilde{B}_N} p_t \geq \frac{1}{c(s,t,N)} \inf_{x \in \bar{D} \cap \tilde{B}_N} P_s 1_{\tilde{B}_N}(x) > 0, \quad 0 < s < t \leq T.$$

Therefore, (2.4.11) holds.

2.5 Power Harnack inequality

By repeating the proof of Theorem 1.5.2 with coupling in (1.5.12) for the reflected SDE, and noting that for convex D we have

$$\langle X_s - Y_s, \mathbf{n}(X_s) \rangle \mathrm{d}l_s^X \leq 0, \quad -\langle X_s - Y_s, \mathbf{n}(Y_s) \rangle \mathrm{d}l_s^Y \leq 0,$$

we obtain the following result.

Theorem 2.5.1. *Assume that D is convex and $(A^{2.3})$ holds with $p_i > 2$. Let*

$$\kappa_0 := \sup_{t \in [0,T], x,y \in \mathbb{R}^d} \|\sigma_t(x) - \sigma_t(y)\|^2, \quad \kappa_1 := \|\sigma^*(\sigma\sigma^*)^{-1}\|_\infty^2.$$

Then for any

$$p > p^* := \frac{3 + \sqrt{1 + (8\kappa_0\kappa)^{-1}}}{\sqrt{1 + (8\kappa_0\kappa_1)^{-1}} - 1},$$

there exists a constant $c > 0$ such that (1.5.7) holds for P_t associated with (2.0.1).

2.6 Exponential ergodicity

Consider the following time dependent differential operator on $\bar{D}$:

$$L_t^\sigma := \frac{1}{2}\text{tr}\big(\sigma_t \sigma_t^* \nabla^2\big), \quad t \in [0,T]. \tag{2.6.1}$$

Let $\{P_{s,t}^\sigma\}_{T \geq t_1 \geq t \geq s \geq 0}$ be the Neumann semigroup on $\bar{D}$ generated by L_t^σ; that is, for any $\varphi \in C_b^2(\bar{D})$, and any $t \in (0,T]$, $(P_{s,t}^\sigma \varphi)_{s \in [0,t]}$ is the unique solution of the PDE

$$\partial_s u_s = -L_s^\sigma u_s, \quad \nabla_{\mathbf{n}} u_s|_{\partial D} = 0 \text{ for } s \in [0,t), u_t = \varphi. \tag{2.6.2}$$

For any $t > 0$, let $C_b^{1,2}([0,t] \times \bar{D})$ be the set of functions $f \in C_b([0,t] \times \bar{D})$ with bounded and continuous derivatives $\partial_t f, \nabla f$ and $\nabla^2 f$.

$(A^{2.5})$ $\partial D \in C_b^{2,L}$ and the following conditions hold.
  (1) $(A^{1.4})$ holds for $\bar{D}$ replacing $\mathbb{R}^d$, and there exists $r_0 > 0$ such that (2.4.8) holds.
  (2) For any $\varphi \in C_b^2(\bar{D})$, the PDE (2.6.2) has a unique solution $P_t^\sigma \varphi \in C_b^{1,2}(\bar{D})$, such that for some constant $c > 0$ we have

$$\|\nabla^i P_t^\sigma \varphi\|_\infty \leq c(1 \wedge t)^{-\frac{1}{2}} \|\nabla^{i-1} \varphi\|_\infty, \quad t > 0, \ i = 1, 2, \varphi \in C_b^2(\bar{D}),$$

where $\nabla^0 \varphi := \varphi$.

By repeating the proof of Theorem 1.6.1 using Theorem 2.4.2 in place of Theorem 1.3.1, we derive the following result.

Theorem 2.6.1. *Assume* $(A^{2.5})$. *Then all assertions in Theorem 1.6.1 hold for the reflected SDE* (2.0.1).

Chapter 3

DDSDEs: Well-posedness

To characterize nonlinear PDEs in Vlasov's kinetic theory, the "propagation of chaos" using mean field particle systems was proposed by Kac [Kac (1954, 1959)]. To realize this proposal, McKean [McKean (1966)] introduced a stochastic differential equation with expectation dependent drift, which describes the evolution of a single particle in the mean field particle systems as the number of particles goes to infinity. So, in references, distribution dependent SDEs (DDSDEs) are called McKean-Valasov or mean field SDEs, see [Sznitman (1991)] and [Carmona and Delarue (2019)].

In this chapter, we first describe the correspondence between DDSDEs and nonlinear Fokker Planck equations, then introduce a general result to solve DDSDEs by using SDEs with fixed distribution parameters, and finally present results on the well-posedness for monotone and singular coefficients respectively. Most results in this part are organized from [Huang et al. (2021)], [Wang (2023b)], [Wang (2023e)] and [Ren (2023)]. Some additional results are introduced in the last section.

3.1 DDSDE and nonlinear Fokker Planck equation

For fixed $T > 0$, we consider the following DDSDE on $\mathbb{R}^d$:

$$dX_t = b_t(X_t, \mathcal{L}_{X_t})dt + \sigma_t(X_t, \mathcal{L}_{X_t})dW_t, \quad t \in [0, T], \tag{3.1.1}$$

where $(W_t)_{t \in [0,T]}$ is an m-dimensional Brownian motion on a complete filtration probability space $(\Omega, \{\mathcal{F}_t\}_{t \in [0,T]}, \mathbb{P})$, $\mathcal{L}_\xi := \mathbb{P} \circ \xi^{-1}$ is the distribution (i.e. the law) of a random variable ξ, and

$$b : [0, T] \times \mathbb{R}^d \times \mathcal{P} \to \mathbb{R}^d,$$
$$\sigma : [0, T] \times \mathbb{R}^d \times \mathcal{P} \to \mathbb{R}^d \otimes \mathbb{R}^m$$

are measurable. When different probability measures are concerned, we denote $\mathcal{L}_\xi$ by $\mathcal{L}_{\xi|\mathbb{P}}$ to emphasize the distribution of ξ under $\mathbb{P}$. Recall that $\mathcal{P}$ is the space of probability measures on $\mathbb{R}^d$ equipped with the weak topology.

We will solve (3.1.1) for distributions in a sub-space $\hat{\mathcal{P}}$ of $\mathcal{P}$ equipped with a complete metric $\hat{d}$ whose topology may be different from the weak topology, such that $\mathcal{L}_X$ belongs to the class

$$C^w([0,T]; \hat{\mathcal{P}}) := \{\mu : [0,T] \to \hat{\mathcal{P}} \text{ is weakly continuous}\},$$
$$C^w_b([0,T]; \hat{\mathcal{P}}) := \Big\{\mu \in C^w([0,T]; \hat{\mathcal{P}}) : \sup_{t \in [0,T]} \hat{d}(\mu_t, \mu_0) < \infty\Big\}. \quad (3.1.2)$$

Without specification, the complete metric on $\hat{\mathcal{P}} = \mathcal{P}$ defaults to the total variation distance which is bounded, so that $C^w_b([0,T]; \mathcal{P}) = C^w([0,T]; \mathcal{P})$.

Definition 3.1.1.

(1) A continuous adapted process $(X_t)_{t \in [0,T]}$ is called a solution of (3.1.1), if $\mathbb{P}$-a.s.

$$\int_0^T \big[|b_r(X_r, \mathcal{L}_{X_r})| + \|\sigma_r(X_r, \mathcal{L}_{X_r})\|^2\big] \mathrm{d}r < \infty,$$

$$X_t = X_0 + \int_0^t b_r(X_r, \mathcal{L}_{X_r}) \mathrm{d}r + \int_0^t \sigma_r(X_r, \mathcal{L}_{X_r}) \mathrm{d}W_r, \quad t \in [0,T].$$

(2) A couple $(\tilde{X}_t, \tilde{W}_t)_{t \in [0,T]}$ is called a weak solution of (3.1.1), if $(\tilde{W}_t)_{t \in [0,T]}$ is the m-dimensional Brownian motion on a complete filtration probability space $(\tilde{\Omega}, \{\tilde{\mathcal{F}}_t\}_{t \in [0,T]}, \tilde{\mathbb{P}})$ such that $(\tilde{X}_t)_{t \in [0,T]}$ is a solution of (3.1.1) for $(\tilde{W}_t, \tilde{\mathbb{P}})$ replacing $(W_t, \mathbb{P})$. (3.1.1) is called weakly unique for an initial distribution $\nu \in \mathcal{P}$, if for any two weak solutions (X^i_t, W^i_t) with $\mathcal{L}_{X^i_0 | \mathbb{P}^i} = \nu$, we have $\mathcal{L}_{X^1 | \mathbb{P}^1} = \mathcal{L}_{X^2 | \mathbb{P}^2}$.

(3) Let $\hat{\mathcal{P}}$ be a subspace of $\mathcal{P}$ equipped with a complete metric $\hat{d}$. (3.1.1) is called strongly (respectively, weakly) well-posed for distributions in $\hat{\mathcal{P}}$, if for any $\mathcal{F}_0$-measurable X_0 with $\mathcal{L}_{X_0} \in \hat{\mathcal{P}}$ (respectively, any initial distribution $\nu \in \hat{\mathcal{P}}$), it has a unique strong (respectively, weak) solution with $\mathcal{L}_X \in C^w_b([0,T]; \hat{\mathcal{P}})$. When $\hat{\mathcal{P}} = \mathcal{P}$, we drop "for distributions in $\mathcal{P}$" and simply call the equation strongly (respectively, weakly) well-posed.

We call the equation well-posed (for distributions in $\hat{\mathcal{P}}$) if it is both strongly and weakly well-posed (for distributions in $\hat{\mathcal{P}}$).

Remark 3.1.1. Assume that for any $s \in [0,T)$, (3.1.1) with $t \in [s,T]$ is well-posed for distributions in $\hat{\mathcal{P}}$. Let $\hat{\mathcal{P}}_s$ be the class of $\mathcal{F}_s$-measurable ξ with $\mathcal{L}_\xi \in \hat{\mathcal{P}}$. For any $s \in [0,T)$ and $\xi \in \hat{\mathcal{P}}_s$, let $(X_{s,t}^\xi)_{t \in [s,T]}$ be the unique solution of (3.1.1) for $t \in [s,T]$ and $X_{s,s}^\xi = \xi$. Then

$$P_{s,t}^* \mu := \mathcal{L}_{X_{s,t}^\xi}, \quad \mu = \mathcal{L}_\xi \in \hat{\mathcal{P}}$$

gives rise to a family of maps

$$P_{s,T}^* : \hat{\mathcal{P}} \to \hat{\mathcal{P}}, \quad 0 \le s \le t \le T,$$

which satisfy the semigroup property

$$P_{s,t}^* = P_{r,t}^* P_{s,r}^*, \quad 0 \le s \le t \le T. \tag{3.1.3}$$

Moreover, $(X_{s,t}^\xi)_{0 \le s \le t \le T, \xi \in \hat{\mathcal{P}}_s}$ is a Markov process satisfying the flow property

$$X_{s,t}^\xi = X_{r,t}^{X_{s,r}^\xi}, \quad 0 \le s \le r \le t \le T,\ \xi \in \hat{\mathcal{P}}_s. \tag{3.1.4}$$

Due to the distribution dependence of the SDE, this Markov process is nonlinear in spatial variable, i.e. the crucial property for linear Markov process

$$P_{s,t}^* \mu = \int_{\mathbb{R}^d} (P_{s,t}^* \delta_x) \mu(\mathrm{d}x), \quad 0 \le s < t \le T, \mu \in \hat{\mathcal{P}}$$

is no longer available. The study of nonlinear Markov process goes back to McKean [McKean (1966)], see [Ren et al. (2022)], [Rehmeier and Röckner (2022)] and references therein.

We also consider the density dependent SDE (also denote by DDSDE), which is known as Nemytskii-type McKean-Vlasov SDE:

$$\mathrm{d}X_t = b_t(X_t, \ell_{X_t}(X_t), \ell_{X_t})\mathrm{d}t + \sigma_t(X_t, \ell_{X_t}(X_t), \ell_{X_t})\mathrm{d}W_t, \tag{3.1.5}$$

where ℓ_{X_t} is the distribution density function of X_t, and for $\mathcal{D}$ being the class of probability density functions,

$$b : [0,T] \times \mathbb{R}^d \times [0,\infty) \times \mathcal{D} \to \mathbb{R}^d,$$
$$\sigma : [0,T] \times \mathbb{R}^d \times [0,\infty) \times \mathcal{D} \to \mathbb{R}^d \otimes \mathbb{R}^m$$

are measurable. When $b_t(x,r,\rho) = b_t(x,\rho)$ and $\sigma_t(x,r,\rho) = \sigma_t(x,\rho)$ do not depend on r, this SDE reduces to (3.1.1).

Next, consider the following nonlinear Fokker-Planck equation on $\mathcal{P}$:

$$\partial_t \mu_t = L_{t,\mu_t}^* \mu_t, \quad t \in [0,T], \tag{3.1.6}$$

where for any $(t,\mu) \in [0,T] \times \mathcal{P}$, the Kolmogorov operator $L_{t,\mu}$ on $\mathbb{R}^d$ is given by

$$L_{t,\mu} := \frac{1}{2}\mathrm{tr}\big\{(\sigma_t\sigma_t^*)(\cdot,\mu)\nabla^2\big\} + \nabla_{b_t(\cdot,\mu)}.$$

Definition 3.1.2. $\mu_\cdot \in C_b^w([0,T];\mathcal{P})$ is called a solution of (3.1.6), if

$$\int_0^T \mathrm{d}r \int_{\mathbb{R}^d} \big\{\|\sigma_r(x,\mu_r)\|^2 + |b_r(x,\mu_r)|\big\}\mu_r(\mathrm{d}x) < \infty,$$

and for any $f \in C_0^\infty(\mathbb{R}^d)$,

$$\mu_t(f) := \int_{\mathbb{R}^d} f\mathrm{d}\mu_t = \mu_0(f) + \int_0^t \mu_r(L_{r,\mu_r}f)\mathrm{d}r, \quad t \in [0,T]. \tag{3.1.7}$$

Assume that $(\tilde{X}_t, \tilde{W}_t)_{t\in[0,T]}$ is a weak solution of (3.1.1) under a complete filtration probability space $(\tilde{\Omega}, \{\tilde{\mathcal{F}}_t\}_{t\in[0,T]}, \tilde{\mathbb{P}})$ such that

$$\int_{[0,T]\times\tilde{\Omega}} \big[|b_r(\tilde{X}_r, \mathcal{L}_{\tilde{X}_r|\tilde{\mathbb{P}}})| + \|\sigma_r(\tilde{X}_r, \mathcal{L}_{\tilde{X}_r|\tilde{\mathbb{P}}})\|^2\big]\mathrm{d}r\mathrm{d}\tilde{\mathbb{P}} < \infty. \tag{3.1.8}$$

By Itô's formula we have

$$\mathrm{d}f(\tilde{X}_t) = \big\{L_{t,\mu_t}f(\tilde{X}_t)\big\}\mathrm{d}t + \langle\nabla f(\tilde{X}_t), \sigma(t,\tilde{X}_t,\mu_t)\mathrm{d}\tilde{W}_t\rangle.$$

Integrating both sides over $[0,t]$ and taking expectations, we obtain (3.1.7) for $\mu_t := \mathcal{L}_{\tilde{X}_t|\tilde{\mathbb{P}}}$, hence μ_t solves (3.1.6). On the other hand, by the superposition theorem, a solution of (3.1.6) also provides a weak solution of (3.1.1), see [Barbu and Röckner (2020)] and [Barbu and Röckner (2018)]. So, we have the following correspondence between (3.1.1) and (3.1.6).

Theorem 3.1.1. *Let $\nu \in \mathcal{P}$. Then the DDSDE (3.1.1) has a weak solution $(\tilde{X}_t, \tilde{W}_t)_{t\in[0,T]}$ with $\mathcal{L}_{\tilde{X}_0|\tilde{\mathbb{P}}} = \nu$ satisfying (3.1.8), if and only if (3.1.6) has a solution $(\mu_t)_{t\in[0,T]}$ with $\mu_0 = \nu$. In this case, $\mu_t = \mathcal{L}_{\tilde{X}_t|\tilde{\mathbb{P}}}$, $t \in [0,T]$.*

Similarly, we may formulate the nonlinear PDE for the density function $f_t := \ell_{X_t}$ associated with (3.1.5):

$$\partial_t f_t = L_{t,f_t}^* f_t,$$

where $L_{t,f} := \frac{1}{2}\mathrm{tr}\{(\sigma_t\sigma_t^*)(\cdot,f(\cdot),f)\nabla^2\} + \nabla_{b_t(\cdot,f(\cdot),f)}$.

To conclude this section, we introduce some typical nonlinear PDEs and state their corresponding DDSDEs.

Example 3.1.1 (Landau type equations). Consider the following nonlinear PDE for probability density functions $(f_t)_{t \in [0,T]}$ on $\mathbb{R}^d$:

$$\partial_t f_t = \frac{1}{2} \operatorname{div}\left\{ \int_{\mathbb{R}^d} a(\cdot - z)\big(f_t(z)\nabla f_t - f_t \nabla f_t(z)\big) \mathrm{d}z \right\}, \quad (3.1.9)$$

where $a : \mathbb{R}^d \to \mathbb{R}^d \otimes \mathbb{R}^d$ has weak derivatives. For the real-world model of homogeneous Landau equation, we have $d = 3$ and

$$a(x) = |x|^{2+r}\left(I - \frac{x \otimes x}{|x|^2}\right), \quad x \in \mathbb{R}^3$$

for some constant $r \in [-3, 1]$. In this case (3.1.9) is a limit version of Boltzmann equation (for thermodynamic system) when all collisions become grazing. To characterize this equation using SDE, let $m = d$, $b = \frac{1}{2}\operatorname{div} a$ and $\sigma = \sqrt{a}$. Consider the DDSDE

$$\mathrm{d}X_t = (b * \mathcal{L}_{X_t})(X_t)\mathrm{d}t + (\sigma * \mathcal{L}_{X_t})(X_t)\mathrm{d}W_t, \quad (3.1.10)$$

where

$$(f * \mu)(x) := \int_{\mathbb{R}^d} f(x - z)\mu(\mathrm{d}z).$$

Then the distribution density $f_t(x) := \frac{\ell_{X_t}(\mathrm{d}x)}{\mathrm{d}x}$ solves the Landau type equation (3.1.9).

There are many references studying Landau type equations, see [Desvillettes and Villani (2000a)], [Desvillettes and Villani (2000b)], [Carrapatoso (2015)], [Fournier and Guillin (2017)], [Funaki (1985)], [Guérin (2002)] and references within.

Example 3.1.2 (Porous media equation). Consider the following nonlinear PDE for probability density functions on $\mathbb{R}^d$:

$$\partial_t f_t = \Delta f_t^3. \quad (3.1.11)$$

Then for any solution to the (3.1.5) with coefficients

$$b = 0, \quad \sigma(x, r) = \sqrt{2r} I_d,$$

the probability density function solves the porous media equation (3.1.11).

Example 3.1.3 (Granular media equation). Consider the following nonlinear PDE for probability density functions on $\mathbb{R}^d$:

$$\partial_t f_t = \Delta f_t + \operatorname{div}\{f_t \nabla V + f_t \nabla(W * f_t)\}. \quad (3.1.12)$$

Then the associated DDSDE (3.1.1) has coefficients

$$b(x, \mu) = -\nabla V(x) - \nabla(W * \mu)(x), \quad \sigma(x, \mu) = \sqrt{2} I_d,$$

where I_d is the $d \times d$ identity matrix, and

$$(W * \mu)(x) := \int_{\mathbb{R}^d} W(x - y)\mu(\mathrm{d}y).$$

3.2 Fixed point in distribution and Yamada-Watanabe principle

To solve (3.1.1), we will fix a subspace $\hat{\mathcal{P}} \subset \mathcal{P}$. Typical examples of $(\hat{\mathcal{P}}, \hat{d})$ include the following $\mathcal{P}_k$ for a constant $k \in (0, \infty)$ and $\mathcal{P}_V$ for a measurable function V.

(1) L^k-Wasserstein space for $k > 0$:
$$\mathcal{P}_k := \{\mu \in \mathcal{P} : \mu(|\cdot|^k) < \infty\}.$$
It is a Polish space under the L^k-Wasserstein distance
$$\mathbb{W}_k(\mu, \nu) := \inf_{\pi \in \mathcal{C}(\mu,\nu)} \left(\int_{\mathbb{R}^d \times \mathbb{R}^d} |x-y|^k \pi(\mathrm{d}x, \mathrm{d}y)\right)^{\frac{1}{k \vee 1}}, \quad \mu, \nu \in \mathcal{P}_k,$$
where $\mathcal{C}(\mu, \nu)$ is the set of all couplings for μ and ν.

(2) V-weighted variation space for a measurable function $V \geq 1$:
$$\mathcal{P}_V := \{\mu \in \mathcal{P} : \mu(V) < \infty\},$$
which is a complete (but not separable) metric space under the V-weighted variation distance $\|\mu - \nu\|_V$.

When $V = 1 + |\cdot|^k$ for some $k > 0$, we denote $\|\cdot\|_V$ by $\|\cdot\|_{k,var}$, i.e.
$$\|\mu - \nu\|_{k,var} := \sup_{f \in \mathcal{B}_b(\mathbb{R}^d), |f| \leq 1 + |\cdot|^k} |\mu(f) - \nu(f)| = |\mu - \nu|(1 + |\cdot|^k).$$

Remark 3.2.1. According to Theorem 6.15 in [Villani (2009)], for any $k > 0$ there exists a constant $c > 0$ such that
$$\|\mu - \nu\|_{var} + \mathbb{W}_k(\mu, \nu)^{1 \vee k} \leq c\|\mu - \nu\|_{k,var}, \quad \mu, \nu \in \mathcal{P}_k. \tag{3.2.1}$$
However, when $k > 1$, for any constant $c > 0$, $\mathbb{W}_k(\mu, \nu) \leq c\|\mu - \nu\|_{k,var}$ does not hold. Indeed, by taking
$$\mu = \delta_0, \quad \nu = (1 - n^{-1-k})\delta_0 + n^{-1-k}\delta_{ne}, \quad n \geq 1, e \in \mathbb{R}^d \text{ with } |e| = 1,$$
we have $\mathbb{W}_k(\mu, \nu) = n^{-\frac{1}{k}}$, while
$$\|\mu - \nu\|_{k,var} = n^{-1-k}\|\delta_0 - \delta_{ne}\|_{k,var}$$
$$\leq n^{-1-k}\{\delta_0(1 + |\cdot|^k) + \delta_{ne}(1 + |\cdot|^k)\} \leq \frac{3}{n}, \quad n \geq 1,$$
so that $\lim_{n \to \infty} \frac{\mathbb{W}_k(\mu,\nu)}{\|\mu-\nu\|_{k,var}} = \infty$ for $k > 1$.

For any $\gamma \in \hat{\mathcal{P}}$, consider the path space over $\hat{\mathcal{P}}$

$$\mathcal{C}^\gamma(\hat{\mathcal{P}}) := \{\mu_. \in C_b^w([0,T]; \hat{\mathcal{P}}) : \mu_0 = \gamma\}.$$

The following is an easy observation reducing the well-posedness of DDSDEs to that of classical SDEs.

Theorem 3.2.1. *Assume that for any $\gamma \in \hat{\mathcal{P}}$ and $\mu \in \mathcal{C}^\gamma(\hat{\mathcal{P}})$, the SDE*

$$\mathrm{d}X_t^\mu = b_t(X_t^\mu, \mu_t)\mathrm{d}t + \sigma_t(X_t^\mu, \mu_t)\mathrm{d}W_t, \quad t \in [0,T] \tag{3.2.2}$$

with $L_{X_0^\mu} = \gamma$ has a unique weak solution, such that the map

$$\mathcal{C}^\gamma(\hat{\mathcal{P}}) \ni \mu \mapsto \Phi^\gamma \mu = (\Phi_t^\gamma \mu)_{t \in [0,T]} := (\mathcal{L}_{X_t^\mu})_{t \in [0,T]} \in \mathcal{C}^\gamma(\hat{\mathcal{P}})$$

has a unique fixed point. Then the DDSDE (3.1.1) is weakly well-posed for distributions in $\hat{\mathcal{P}}$. If moreover (3.2.2) is strongly well-posed, then so is (3.1.1) for distributions in $\hat{\mathcal{P}}$.

The following is a simple consequence of Theorem 3.2.1, where for invertible $\sigma_t \sigma_t^*$, (3.2.4) holds for

$$\Gamma_t(x, \mu, \nu) := \{\sigma_t^*(\sigma_t \sigma_t^*)^{-1}[b_t(\cdot, \nu) - b_t(\cdot, \mu)]\}(x).$$

To this end, we recall the Pinsker's inequality: for any measurable space $(E, \mathcal{B}(E))$,

$$\|\mu - \nu\|_{var}^2 := \sup_{A \in \mathcal{B}(E)} |\mu(A) - \nu(B)|^2 \leq 2\mathrm{Ent}(\mu|\nu), \quad \mu, \nu \in \mathcal{P}(E), \tag{3.2.3}$$

where $\mathcal{P}(E)$ is the set of all probability measures on W.

Corollary 3.2.2. *Let $\sigma_t(x, \mu) = \sigma_t(x)$ for $t \in [0,T]$. Assume that (3.2.2) is weak (respectively strong) well-posed for any $\mu \in C^w([0,T]; \hat{\mathcal{P}})$. If there exists a measurable map*

$$\Gamma : [0,T] \times \mathbb{R}^d \times \mathcal{P} \times \mathcal{P} \to \mathbb{R}^m$$

such that

$$b_t(x, \nu) - b_t(x, \mu) = \sigma_t(x)\Gamma_t(x, \mu, \nu),$$
$$(t, x, \mu, \nu) \in [0,T] \times \mathbb{R}^d \times \mathcal{P} \times \mathcal{P}, \tag{3.2.4}$$

and there exists a constant $K > 0$ such that

$$|\Gamma_t(x, \mu, \nu)| \leq K\|\mu - \nu\|_{var}, \quad (t, x, \mu, \nu) \in [0,T] \times \mathbb{R}^d \times \mathcal{P} \times \mathcal{P}. \tag{3.2.5}$$

Then (3.1.1) is weakly (respectively strongly) well-posed.

Proof. Let $\gamma \in \mathcal{P}$ and let Φ^γ be defined in Theorem 3.2.1. Then it suffices to prove that Φ^γ has a unique fixed point in $\mathcal{C}^\gamma(\mathcal{P})$. Let $\mu, \nu \in \mathcal{C}^\gamma(\mathcal{P})$, and let X_t^μ solve (3.2.2). By (3.2.5),
$$R_t := e^{\int_0^t \langle \Gamma_t(X_r^\mu, \mu_r, \nu_r), dW_r\rangle - \frac{1}{2}\int_0^t |\Gamma_r(X_r^\mu, \mu_r, \nu_r)|^2 dr}, \quad t \in [0, T]$$
is a martingale, such that by Girsanov's theorem,
$$\tilde{W}_t := W_t - \int_0^t \Gamma_s(X_s^\mu, \mu_s, \nu_s) ds, \quad t \in [0, T]$$
is a Brownian motion under the probability $\mathbb{Q}_T := R_T \mathbb{P}$. By (3.2.4), we may reformulate (3.2.2) as
$$dX_t^\mu = b_t(X_t^\mu, \nu_t)dt + \sigma_t(X_t^\mu)d\tilde{W}_t, \quad t \in [0, T], \mathcal{L}_{X_0^\mu} = \gamma.$$
By the weak uniqueness of (3.2.2) we obtain $\Phi_t^\gamma \nu = \mathcal{L}_{X_t^\mu | \mathbb{Q}_T}, t \in [0, T]$. So, (3.2.3) and (3.2.5) yield
$$\|\Phi_t^\gamma \mu - \Phi_t^\gamma \nu\|_{var}^2 = \sup_{|f|\leq 1}\left|\mathbb{E}[(R_t - 1)f(X_t^\mu)]\right|^2 = \left(\mathbb{E}[|R_t - 1|]\right)^2$$
$$\leq 2\mathbb{E}[R_t \log R_t] = 2\mathbb{E}_{\mathbb{Q}_T}[\log R_t] = \mathbb{E}_{\mathbb{Q}_T}\int_0^t |\Gamma_r(X_r^\mu, \mu_r, \nu_r)|^2 dr$$
$$\leq K^2 \int_0^t \|\mu_r - \nu_r\|_{var}^2 dr, \quad t \in [0, T].$$
Considering the complete metrics
$$\rho_\lambda(\mu, \nu) := \sup_{t \in [0,T]} e^{-\lambda t}\|\mu_t - \nu_t\|_{var}, \quad \lambda > 0$$
on $\mathcal{C}^\gamma(\mathcal{P})$, we derive
$$\rho_\lambda(\Phi^\gamma \mu, \Phi^\gamma \nu)^2 = \sup_{t \in [0,T]} e^{-2\lambda t}\|\Phi_t^\gamma \mu - \Phi_t^\gamma \nu\|_{var}^2$$
$$\leq K^2 \rho_\lambda(\mu, \nu)^2 \sup_{t \in [0,T]} \int_0^t e^{-2\lambda(t-r)} dr \leq \frac{K^2}{2\lambda}\rho_\lambda(\mu, \nu)^2.$$
Thus, for large enough $\lambda > 0$ the map Φ^γ is contractive in ρ_λ, so that Φ^γ has a unique fixed point on $\mathcal{C}^\gamma(\mathcal{P})$. $\square$

As shown in Chapters 1 and 2, the Yamada-Watanabe principle is a fundamental tool in the study of well-posedness for SDEs. The following is a modified version for DDSDEs, see Lemma 3.4 in [Huang and Wang (2019)].

Theorem 3.2.3. *Assume that for any $\mu \in C_b^w([0, T]; \hat{\mathcal{P}})$, the classical SDE (3.2.2) has pathwise uniqueness. If (3.1.1) has weak existence and strong uniqueness for distributions in $\hat{\mathcal{P}}$, then it is well-posed for distributions in $\hat{\mathcal{P}}$.*

Proof. Let $\mu. = \mathcal{L}_{\bar{X}.|\bar{\mathbb{P}}} \in C_b^w([0,T]; \hat{\mathcal{P}})$ for a weak solution $(\bar{X}_t, \bar{W}_t)$ of (3.1.1) with distribution $\mu_0 \in \hat{\mathcal{P}}$. Then $(\bar{X}_t, \bar{W}_t)$ is a weak solution to (3.2.2). By Yamada-Watanabe principle in Lemma 1.3.2, the strong uniqueness of (3.2.2) implies the well-posedness. So, given initial value X_0 with distribution μ_0, by the weak uniqueness, the strong solution of (3.2.2) satisfies $\mathcal{L}_{X_t} = \mathcal{L}_{\bar{X}_t|\bar{\mathbb{P}}} = \mu_t$, and hence, X_t is a strong solution to (3.1.1). Combining this with the strong uniqueness as assumed, we conclude that (3.1.1) is strongly well-posed for distributions in $\hat{\mathcal{P}}$.

By the same reason, if (3.1.1) has two weak solutions $(\bar{X}_t^i, \bar{W}_t^i)$ under probabilities $\bar{\mathbb{P}}^i$ ($i = 1, 2$) with common initial distribution $\mu_0 \in \hat{\mathcal{P}}$, then the well-posedness of (3.2.2) for $\mu_t^i := \mathcal{L}_{\bar{X}_t^i|\bar{\mathbb{P}}^i}$ replacing μ_t gives two strong solutions X_t^i of (3.1.1) with the same initial value and with $\mathcal{L}_{X_t^i} = \mu_t^i$ for $t \in [0, T]$, so that the pathwise uniqueness of (3.1.1) implies $\mu_t^1 = \mu_t^2$ for $t \in [0, T]$, and the well-posedness of (3.2.2) yields $\mathcal{L}_{\bar{X}_{[0,T]}^1|\bar{\mathbb{P}}^1} = \mathcal{L}_{\bar{X}_{[0,T]}^2|\bar{\mathbb{P}}^2}$. Hence, (3.1.1) also has weak uniqueness, so that the weak well-posedness holds for distributions in $\hat{\mathcal{P}}$. □

3.3 The monotone case

$(A^{3.1})$ Let $k \in [1, \infty)$.
(1) For any $\mu \in C_b^w([0,T]; \mathcal{P}_k)$, the SDE (3.2.2) is well-posed.
(2) There exists $K \in L^1([0,T]; (0, \infty))$ such that for any $t \in [0,T], x, y \in \mathbb{R}^d$ and $\mu, \nu \in \mathcal{P}_k$,

$$\|\sigma_t(x,\mu) - \sigma_t(y,\nu)\|^2 + \langle b_t(x,\mu) - b_t(y,\nu), x - y \rangle^+$$
$$\leq K(t)\{|x-y|^2 + \mathbb{W}_k(\mu,\nu)^2\}.$$

Theorem 3.3.1. *Assume* $(A^{3.1})$ *for some* $k \in [1, \infty)$.

(1) *The DDSDE (3.1.1) is well-posed for distributions in $\mathcal{P}_k$. Moreover, for any $p \geq k$, there exists a constant $c > 0$ such that for any solution X_t of (3.1.1) with $\mathcal{L}_{X_0} \in \mathcal{P}_k$,*

$$\mathbb{E}\Big[\sup_{t \in [0,T]} |X_t|^p \Big| \mathcal{F}_0\Big] \leq c\big(1 + |X_0|^p + \{\mathbb{E}[|X_0|^k]\}^{\frac{p}{k}}\big). \qquad (3.3.1)$$

(2) *For any $p \geq k$, there exists a constant $c > 0$ such that for any two solutions X_t and Y_t of (3.1.1) with $\mathcal{L}_{X_0}, \mathcal{L}_{Y_0} \in \mathcal{P}_k$,*

$$\mathbb{E}\Big[\sup_{t \in [0,T]} |X_t - Y_t|^p \Big| \mathcal{F}_0\Big] \leq c\big(\mathbb{W}_k(\mathcal{L}_{X_0}, \mathcal{L}_{Y_0}) + |X_0 - Y_0|\big)^p. \qquad (3.3.2)$$

110   *Distribution Dependent Stochastic Differential Equations*

Consequently, there exists a constant $c > 0$ such that

$$\mathbb{W}_k(P_t^*\mu, P_t^*\nu) \leq c\mathbb{W}_k(\mu, \nu), \quad t \in [0, T], \mu, \nu \in \mathcal{P}_k. \tag{3.3.3}$$

Proof. (a) Let X_0 be $\mathcal{F}_0$-measurable with $\gamma := \mathcal{L}_{X_0} \in \mathcal{P}_k$. Then

$$\mathcal{C}_k^\gamma := \{\mu \in C_b^w([0, T]; \mathcal{P}_k) : \mu_0 = \gamma\}$$

is a complete space under the following metric for any $\lambda > 0$:

$$\mathbb{W}_{k,\lambda}(\mu, \nu) := \sup_{t \in [0,T]} e^{-\lambda t} \mathbb{W}_k(\mu_t, \nu_t), \quad \mu, \nu \in \mathcal{C}_k^\gamma. \tag{3.3.4}$$

Let $(X_t^\mu)_{t \in [0,T]}$ be the unique solution of (3.2.2) with $X_0^\mu = X_0$. By Theorem 3.2.1, for the well-posedness of (3.1.1), it suffices to prove the contraction of the map

$$\mathcal{C}_k^\gamma \ni \mu \mapsto \Phi^\gamma_{\cdot}\mu := \mathcal{L}_{X^\mu_\cdot} \in \mathcal{C}_k^\gamma$$

under the metric $\mathbb{W}_{k,\lambda}$ for large enough $\lambda > 0$, where the continuity of $\Phi^\gamma_t\mu$ in t follows from (1.3.1).

By $(A^{3.1})$ and Itô's formula, for any $p \geq k \vee 2 := \max\{k, 2\}$, we find a constant $c_1 > 0$ such that

$$d|X_t^\mu - X_t^\nu|^p \leq c_1 K(t)\{|X_t^\mu - X_t^\nu|^p + \mathbb{W}_k(\mu_t, \nu_t)^p\}dt + dM_t$$

holds for some martingale M_t with

$$d\langle M \rangle_t \leq c_1 K(t)\{|X_t^\mu - X_t^\nu|^{2p} + \mathbb{W}_k(\mu_t, \nu_t)^{2p}\}dt.$$

Let

$$\zeta_t := \sup_{s \in [0,t]} |X_s^\mu - X_s^\nu|^p, \quad t \in [0, T].$$

By BDG inequality in Lemma 1.3.5, we find constants $c_2, c_3 > 0$ such that

$$\mathbb{E}[\zeta_t | \mathcal{F}_0] \leq c_2 \int_0^t K(s)\{\mathbb{E}[\zeta_s | \mathcal{F}_0] + \mathbb{W}_k(\mu_s, \nu_s)^p\}ds$$

$$+ c_2 \mathbb{E}\left[\left(\int_0^t K(s)\{\zeta_s^2 + \mathbb{W}_k(\mu_s, \nu_s)^{2p}\}ds\right)^{\frac{1}{2}} \Big| \mathcal{F}_0\right]$$

$$\leq \frac{1}{2}\mathbb{E}[\zeta_t | \mathcal{F}_0] + c_3 \int_0^t K(s)\mathbb{E}[\zeta_s | \mathcal{F}_0]ds + c_3\left(\int_0^t K(s)\mathbb{W}_k(\mu_s, \nu_s)^{2p}ds\right)^{\frac{1}{2}}$$

for $t \in [0, T]$. Thus,

$$\mathbb{E}[\zeta_t | \mathcal{F}_0] \leq 2c_3 \int_0^t K(s)\mathbb{E}[\zeta_s | \mathcal{F}_0]ds$$

$$+ 2c_3\left(\int_0^t K(s)\mathbb{W}_k(\mu_s, \nu_s)^{2p}ds\right)^{\frac{1}{2}}, \quad t \in [0, T]. \tag{3.3.5}$$

By using $\zeta_{t\wedge\tau_k}$ for $\tau_k := \inf\{t \geq 0 : |X_t^\mu - X_t^\nu| \geq k\}$ replacing ζ_t and letting $k \to \infty$, we may and do assume that $\mathbb{E}[\zeta_t|\mathcal{F}_0] < \infty$, so that by Gronwall's inequality,

$$\mathbb{E}[\zeta_t|\mathcal{F}_0] \leq 2c_3 \mathrm{e}^{2c_3 \int_0^t K(s)\mathrm{d}s} \left(\int_0^t K(s)\mathbb{W}_k(\mu_s, \nu_s)^{2p}\mathrm{d}s\right)^{\frac{1}{2}}$$

$$\leq c_4 \mathrm{e}^{\lambda p t}\mathbb{W}_{k,\lambda}(\mu,\nu)^p \left(\int_0^t K(s)\mathrm{e}^{-2p\lambda(t-s)}\mathrm{d}s\right)^{\frac{1}{2}}, \quad t \in [0,T] \quad (3.3.6)$$

holds for some constant $c_4 > 0$. Since $p \geq k$, by Jensen's inequality, this implies

$$\mathbb{W}_{k,\lambda}(\Phi^\gamma \mu, \Phi^\gamma \nu)^k \leq \sup_{t\in[0,T]} \mathrm{e}^{-k\lambda t}\mathbb{E}\left[\left(\mathbb{E}[\zeta_t|\mathcal{F}_0]\right)^{\frac{k}{p}}\right]$$

$$\leq c_4 \mathbb{W}_{k,\lambda}(\mu,\nu)^k \left(\sup_{t\in[0,T]} \int_0^t K(s)\mathrm{e}^{-2p\lambda(t-s)}\mathrm{d}s\right)^{\frac{k}{2p}}.$$

Noting that

$$\lim_{\lambda\to\infty} \sup_{t\in[0,T]} \int_0^t K(s)\mathrm{e}^{-2p\lambda(t-s)}\mathrm{d}s = 0,$$

we see that Φ^γ is contractive in $\mathbb{W}_{k,\lambda}$ for large $\lambda > 0$.

(b) Next, for any $p \geq k \vee 2$, let X_t solve (3.1.1) with $\mathbb{E}[|X_0|^p] < \infty$. By $(A^{3.1})$ and Itô's formula, there exists a constant $c(p) > 0$ such that

$$\mathrm{d}|X_t|^p \leq c(p)K(t)\big\{1 + |X_t|^p + (\mathbb{E}[|X_t|^k])^{\frac{p}{k}}\big\}\mathrm{d}t + \mathrm{d}M_t$$

for some martingale M_t with

$$\mathrm{d}\langle M\rangle_t \leq c(p)K(t)\big\{1 + |X_t|^{2p} + (\mathbb{E}[|X_t|^k])^{2p/k}\big\}\mathrm{d}t.$$

Then (3.3.1) follows from BDG inequality.

(c) By Jensen's inequality, it suffices to prove (3.3.2) and (3.3.3) for $p \geq k \vee 2$. Let X_t and Y_t be two solutions of (3.1.1), and denote $\mu_t = \mathcal{L}_{X_t}, \nu_t = \mathcal{L}_{Y_t}$. By $(A^{3.1})$ and Itô's formula, we find a constant $c_0 > 0$ such that

$$\mathrm{d}|X_t - Y_t|^p \leq c_0\big\{|X_t - Y_t|^p + \mathbb{W}_k(\mu_t,\nu_t)^p\big\}\mathrm{d}t + \mathrm{d}M_t$$

for some martingale M_t satisfying

$$\mathrm{d}\langle M\rangle_t \leq c_0\big\{|X_t - Y_t|^{2p} + \mathbb{W}_k(\mu_t,\nu_t)^{2p}\big\}\mathrm{d}t.$$

By BDG inequality as in above, we find a constant $c_1 > 0$ such that

$$\mathbb{E}\left[\sup_{t\in[0,s]}|X_t - Y_t|^p \Big| \mathcal{F}_0\right] \leq c_1|X_0 - Y_0|^p$$

$$+ c_1 \int_0^s \big\{\mathbb{E}[|X_t - Y_t|^p|\mathcal{F}_0] + \mathbb{W}_k(\mu_t,\nu_t)^p\big\}\mathrm{d}t, \quad t \in [0,T].$$

By Gronwall's inequality, for some constant $c_2 > 0$,

$$\mathbb{E}\left[\sup_{t\in[0,s]} |X_t - Y_t|^p \Big| \mathcal{F}_0\right] \\ \leq c_2|X_0 - Y_0|^p + c_2 \int_0^t \mathbb{W}_k(\mu_s, \nu_s)^p \mathrm{d}s, \quad t \in [0, T]. \tag{3.3.7}$$

Thus, (3.3.2) follows from (3.3.3).

To prove (3.3.3), we take X_0 and Y_0 such that

$$\mathbb{W}_k(\mu_0, \nu_0) = \left(\mathbb{E}[|X_0 - Y_0|^k]\right)^{\frac{1}{k}}, \quad \mu_0 := \mathcal{L}_{X_0}, \nu_0 := \mathcal{L}_{Y_0}.$$

By (3.7.20), we find some constant $c_3 > 0$ such that

$$\sup_{s\in[0,t]} \mathbb{W}_k(\mu_s, \nu_s)^k \leq \mathbb{E}\left(\mathbb{E}\left[\sup_{s\in[0,t]} |X_s - Y_s|^k \Big| \mathcal{F}_0\right]\right)$$

$$\leq \mathbb{E}\left[\left(\mathbb{E}\left[\sup_{s\in[0,t]} |X_s - Y_s|^p \Big| \mathcal{F}_0\right]\right)^{\frac{k}{p}}\right] \text{(Jensen's inequality)}$$

$$\leq \mathbb{E}\left\{c_2|X_0 - Y_0|^p + c_2 \int_0^t \mathbb{W}_k(\mu_s, \nu_s)^p \mathrm{d}s\right\}^{\frac{k}{p}}$$

$$\leq c_2^{\frac{k}{p}} \mathbb{E}[|X_0 - Y_0|^k]$$

$$+ c_2^{\frac{k}{p}} \left(\sup_{s\in[0,t]} \mathbb{W}_k(\mu_s, \nu_s)^k\right)^{\frac{p-k}{p}} \left(\int_0^t \mathbb{W}_k(\mu_s, \nu_s)^k \mathrm{d}s\right)^{\frac{k}{p}}$$

$$\leq c_3 \mathbb{W}_k(\mu_0, \nu_0)^k + \frac{1}{2}\sup_{s\in[0,t]} \mathbb{W}_k(\mu_s, \nu_s)^k + c_3 \int_0^t \mathbb{W}_k(\mu_s, \nu_s)^k \mathrm{d}s.$$

Then

$$\sup_{s\in[0,t]} \mathbb{W}_k(\mu_s, \nu_s)^k \leq 2c_3 \mathbb{W}_k(\mu_0, \nu_0)^k + 2c_3 \int_0^t \mathbb{W}_k(\mu_s, \nu_s)^k \mathrm{d}s, \quad t \in [0, T].$$

By Gronwall's lemma we derive (3.3.3). $\square$

3.4 Singular case: $\|\cdot\|_V$-Lipschitz in distribution

We consider the case that b is singular and $\sigma_t(x, \mu) = \sigma_t(x)$ does not depend on μ, so that (3.1.1) reduces to

$$\mathrm{d}X_t = b_t(X_t, \mathcal{L}_{X_t})\mathrm{d}t + \sigma_t(X_t)\mathrm{d}W_t, \quad t \in [0, T]. \tag{3.4.1}$$

$(A^{3.2})$ Let $1 \leq V \in C^2(\mathbb{R}^d; [1,\infty))$ be a compact function, and $b_t(x,\mu) = b_t^{(0)}(x) + b_t^{(1)}(x,\mu)$.

(1) $(A^{1.2})(1)$ holds for σ and $b^{(0)}$.

(2) For any $\mu \in C_b^w([0,T]; \mathcal{P}_V)$, $b_t^{(1)}(x,\mu_t)$ is locally bounded in $(t,x) \in [0,T] \times \mathbb{R}^d$. Moreover, there exist a constant $\varepsilon \in (0,1)$ and a function $0 \leq K \in L^1([0,T])$ such that

$$\sup_{B(x,\varepsilon)} \{|\nabla V| + \|\nabla^2 V\|\} \leq \varepsilon^{-1} V(x),$$

$$\langle b_t^{(1)}(x,\mu), \nabla V(x) \rangle + \varepsilon |b_t^{(1)}(x,\mu)| \sup_{B(x,\varepsilon)} \{|\nabla V| + \|\nabla^2 V\|\}$$

$$\leq K_t \{V(x) + \mu(V)\}, \quad x \in \mathbb{R}^d, \mu \in \mathcal{P}_V.$$

The following result is due to [Ren (2023)].

Theorem 3.4.1. *Assume* $(A^{3.2})$.

(1) *If for any* $t \in [0,T]$,

$$|b_t^{(1)}(x,\mu) - b_t^{(1)}(x,\nu)|^2 \leq K_t \|\mu - \nu\|_V^2, \quad x \in \mathbb{R}^d, \mu, \nu \in \mathcal{P}_V, \quad (3.4.2)$$

then $(3.4.1)$ *is well-posed for distributions in* $\mathcal{P}_V$, *and for any* $k \geq 1$ *there exists a constant* $c > 0$ *such that for any solution* X_t,

$$\mathbb{E}\left[\sup_{t \in [0,T]} V(X_t)^k \Big| X_0\right] \leq c\{V(X_0)^k + (\mathbb{E}[V(X_0)])^k\}. \quad (3.4.3)$$

Moreover, for any sequence $\{\gamma_n\}_{n \geq 1} \subset \mathcal{P}_V$ *with bounded* $\gamma_n(V^p)$ *for some* $p > 1$ *such that* $\gamma_n \to \gamma$ *weakly,*

$$\lim_{n \to \infty} \|P_t^* \gamma_n - P_t^* \gamma\|_V = 0, \quad t \in [0,T]. \quad (3.4.4)$$

(2) *If for any* $t \in [0,T]$,

$$|b_t^{(1)}(x,\mu) - b_t^{(1)}(x,\nu)|^2 \leq K_t \|\mu - \nu\|_{var}^2, \quad x \in \mathbb{R}^d, \mu, \nu \in \mathcal{P}_V, \quad (3.4.5)$$

then for any $\{\gamma, \gamma_n\}_{n \geq 1} \subset \mathcal{P}_V$ *with* $\gamma_n \to \gamma$ *weakly,*

$$\lim_{n \to \infty} \|P_t^* \gamma_n - P_t^* \gamma\|_{var} = 0, \quad t \in [0,T]. \quad (3.4.6)$$

Proof. (1) Let X_0 be $\mathcal{F}_0$-measurable with $\gamma := \mathcal{L}_{X_0} \in \mathcal{P}_V$. Simply denote

$$\mathcal{C}_V^\gamma := \{\mu \in C_b^w([0,T]; \mathcal{P}_V) : \mu_0 = \gamma\}.$$

For any $\mu \in \mathcal{C}_V^\gamma$, by Theorem 1.3.1, $(A^{3.2})$ implies that the following SDE is well-posed,

$$dX_t^\mu = b_t(X_t^\mu, \mu_t) dt + \sigma_t(X_t^\mu) dW_t, \quad X_0^\mu = X_0. \quad (3.4.7)$$

Recall that $\Phi_t^\gamma \mu := \mathcal{L}_{X_t^\mu}$. By Theorem 3.2.1, for the well-posedness of (3.4.1), it suffices to prove that Φ^γ has a unique fixed point in $\mathcal{C}_V^\gamma$. To this end, we approximate $\mathcal{C}_V^\gamma$ by bounded subsets

$$\mathcal{C}_V^{\gamma,N} := \left\{\mu \in \mathcal{C}^\gamma : \sup_{t \in [0,T]} \mu_t(V) e^{-Nt} \leq N(1 + \gamma(V))\right\}, \quad N \geq 1.$$

(1a) We aim to find a constant $N_0 \geq 1$, such that $\Phi^\gamma \mathcal{C}_V^{\gamma,N} \subset \mathcal{C}_V^{\gamma,N}$ holds for $N \geq N_0$. By Lemma 1.2.2, we consider Zvonkin's transform of X_t^μ:

$$\begin{aligned} Y_t^\mu &= X_t^\mu + u_t^\lambda(X_t^\mu), \quad u_t^\lambda \in \tilde{H}_{q_0}^{p_0}(T), \\ (\partial_t + L_t^0) u_t^\lambda &= \lambda u_t^\lambda - b_t^{(0)}, \quad t \in [0,T], \quad u_T^\lambda = 0, \end{aligned} \quad (3.4.8)$$

for large $\lambda > 0$ such that $\|u_t^\lambda\|_\infty + \|\nabla u_t^\lambda\|_\infty \leq \frac{1}{2}$, where

$$L_t^0 := \nabla_{b_t^{(0)}} + \frac{1}{2}\mathrm{tr}\{\sigma_t \sigma_t^* \nabla^2\}.$$

By $(A^{3.2})$ and Itô's formula, for any $k \geq 1$, we find a constant $c_1 > 0$ and a martingale M_t such that

$$\begin{aligned} \mathrm{d}\{V(Y_t^\mu)\}^k &\leq c_1 K_t \{V(Y_t^\mu)^k + \mu_t(V)^k\}\mathrm{d}t + \mathrm{d}M_t, \\ \mathrm{d}\langle M\rangle_t &\leq k_1 V(Y_t^\mu)^{2(k-1)}\mathrm{d}t, \quad t \in [0,T]. \end{aligned} \quad (3.4.9)$$

By the condition on V and $|X_t^\mu - Y_t^\mu| \leq \frac{1}{2}$, we find a constant $C > 1$ such that

$$C^{-1}V(X_t^\mu) \leq V(Y_t^\mu) \leq CV(X_t^\mu), \quad (3.4.10)$$

so that (3.4.9) implies that for some constant $c_2 > 0$,

$$\begin{aligned} \mathbb{E}&\big(V(X_t^\mu)^2 \big| X_0^\mu\big) \\ &\leq C^2 e^{c_1 \int_0^t K_s \mathrm{d}s} V(X_0^\mu)^2 + C^2 c_1 \int_0^t K_s e^{c_1 \int_s^t K_r \mathrm{d}r} \mu_s(V)^2 \mathrm{d}s \\ &\leq c_2 V(X_0^\mu)^2 + c_2\{N(1+\gamma(V))\}^2 \int_0^t K_s e^{2Ns} \mathrm{d}s \\ &\leq c_2 V(X_0^\mu)^2 + c_2\{N(1+\gamma(V))\}^2 e^{2Nt} \int_0^t K_s e^{-2N(t-s)} \mathrm{d}s, \end{aligned} \quad (3.4.11)$$

$t \in [0,T]$, $\mu \in \mathcal{C}_V^{\gamma,N}$.

Since $0 \leq K \in L^1([0,T])$ implies

$$\lim_{N \to \infty} \sup_{t \in [0,T]} \int_0^t K_s e^{-2N(t-s)} \mathrm{d}s = 0,$$

we find a constant $N_0 \geq 1$ such that
$$\sup_{t\in[0,T]} \{\Phi_t^\gamma \mu\}(V) e^{-Nt} = \sup_{t\in[0,T]} e^{-Nt} \mathbb{E}[V(X_t^\mu)]$$
$$\leq \sqrt{c_2}\gamma(V) + \frac{1}{2}N(1+\gamma(V)) \leq N(1+\gamma(V)), \quad N \geq N_0.$$
Thus, $\Phi^\gamma \mathcal{C}_V^{\gamma,N} \subset \mathcal{C}_V^{\gamma,N}$ for $N \geq N_0$, where the continuity of $\Phi_t^\gamma \mu$ in t is implied by Theorem 1.3.1.

(1b) Let $N \geq N_0$. We prove that Φ^γ has a unique fixed point in $\mathcal{C}_V^{\gamma,N}$, and hence it has a unique fixed point in $\mathcal{C}_V^\gamma$ by the arbitrariness of $N \geq N_0$. Consider the following complete metric on $\mathcal{C}_V^{\gamma,N}$:
$$\rho_\lambda(\mu,\nu) := \sup_{t\in[0,T]} e^{-\lambda t} \|\mu_t - \nu_t\|_V.$$
Let
$$\xi_s := \{\sigma_s^*(\sigma_s\sigma_s^*)^{-1}[b_s(X_s^\mu, \nu_s) - b_s(X_s^\mu, \mu_s)]\}(X_s^\mu), \quad s \in [0,T].$$
By (3.4.2),
$$R_t := e^{\int_0^t \langle \xi_s, dW_s \rangle - \frac{1}{2}\int_0^t |\xi_s|^2 ds} \tag{3.4.12}$$
is a martingale, such that
$$\tilde{W}_r := W_r - \int_0^r \xi_s ds, \quad r \in [0,t]$$
is a Brownian motion under the probability $\mathbb{Q}_t := R_t \mathbb{P}$. Reformulate (3.4.7) as
$$dX_r^\mu = b_r(X_r^\mu, \nu_r)dr + \sigma_r(X_r^\mu)d\tilde{W}_r, \quad X_0^\mu = X_0, \quad r \in [0,t].$$
By the uniqueness we obtain
$$\Phi_t^\gamma \nu = \mathcal{L}_{X_t^\nu} = \mathcal{L}_{X_t^\mu | \mathbb{Q}_t},$$
where $\mathcal{L}_{X_t^\mu | \mathbb{Q}_t}$ stands for the distribution of X_t^μ under $\mathbb{Q}_t$. Then by (3.4.11), we find a constant $c_1(N) > 0$ such that
$$\|\Phi_t^\gamma \mu - \Phi_t^\gamma \nu\|_V = \sup_{|f|\leq V} |\mathbb{E}[f(X_t^\mu)(1-R_t)]|$$
$$\leq \mathbb{E}\Big[\{\mathbb{E}(V(X_t^\mu)^2 | X_0^\mu)\}^{\frac{1}{2}} \{\mathbb{E}[|R_t - 1|^2 | X_0^\mu]\}^{\frac{1}{2}}\Big] \tag{3.4.13}$$
$$\leq c_1(N)\mathbb{E}\Big[V(X_0)\{\mathbb{E}[R_t^2 - 1 | X_0]\}^{\frac{1}{2}}\Big].$$
Since $\mu,\nu \in \mathcal{C}_V^{\gamma,N}$, by $\|\sigma^*(\sigma\sigma^*)^{-1}\|_\infty < \infty$ and (3.4.2), we find a constant $c_2(N) > 0$ such that
$$|\xi_s|^2 \leq c_2(N) K_s (1 \wedge \|\mu_s - \nu_s\|_V^2), \quad s \in [0,T],$$

so that for some constant $c_3(N) > 0$,

$$\mathbb{E}[R_t^2 - 1|X_0] \leq \mathbb{E}\left[e^{2\int_0^t \langle \xi_s, dW_s \rangle - \int_0^t |\xi_s|^2 ds} - 1\Big|X_0\right]$$

$$\leq \mathbb{E}\left[e^{2\int_0^t \langle \xi_s, dW_s \rangle - 2\int_0^t |\xi_s|^2 ds + c_2(N)\int_0^t K_s(1\wedge\|\mu_s-\nu_s\|_V^2)ds} - 1\Big|X_0\right]$$

$$= e^{c_2(N)\int_0^t K_s(1\wedge\|\mu_s-\nu_s\|_V^2)ds} - 1 \leq c_3(N)\int_0^t K_s\|\mu_s - \nu_s\|_V^2 ds,$$

where the last step follows from the fact that $e^r - 1 \leq re^r$ for $r \geq 0$. Combining this with (3.4.13), we find a constant $c_4(N) > 0$ such that

$$\|\Phi_t^\gamma \mu - \Phi_t^\gamma \nu\|_V \leq c_4(N)\int_0^t K_s\|\mu_s - \nu_s\|_V^2 ds, \quad t \in [0, T]. \quad (3.4.14)$$

So,

$$\rho_\lambda(\Phi^\gamma \mu, \Phi^\gamma \nu) = \sup_{t \in [0,T]} e^{-\lambda t}\|\Phi_t^\gamma \mu - \Phi_t^\gamma \nu\|_V$$

$$\leq c_4(N)\rho_\lambda(\mu, \nu)\sup_{t \in [0,T]}\left(\int_0^t K_s e^{-2\lambda(t-s)}ds\right)^{\frac{1}{2}}, \quad \mu, \nu \in \mathcal{C}_V^{\gamma, N}.$$

Therefore, when $\lambda > 0$ is large enough, Φ^γ is contractive in $\mathcal{C}_V^{\gamma,N}$ under ρ_λ, so that it has a unique fixed point in $\mathcal{C}_V^{\gamma,N}$.

(1c) By (3.4.9) for $k = 1$, (3.4.10), $(A^{3.2})$ and $\mathcal{L}_{X_0^\mu} = \mathcal{L}_{X_0} = \gamma$, we find a constant $k_1 > 0$ such that

$$\mathbb{E}[V(Y_t^\mu)] \leq k_1\gamma(V) + k_1\int_0^t \mathbb{E}[V(Y_s^\mu)]ds,$$

so that Gronwall's inequality yields

$$\mathbb{E}[V(Y_t^\mu)] \leq k_1 e^{k_1 t}\gamma(V).$$

Combining this with (3.4.9), (3.4.10), BDG inequality and Hölder's inequality for $\mathbb{E}[\cdot|X_0]$, we find a constant $k_2 > 0$ such that

$$\eta_t := \mathbb{E}\left[\sup_{s \in [0,t]} V(Y_s^\mu)^k \Big|X_0\right], \quad t \in [0, T]$$

satisfies

$$\eta_t \leq k_2 V(X_0)^k + \int_0^t \eta_s ds + k_2\mathbb{E}\left[\left(\int_0^t V(Y_s^\mu)^{2(k-1)}ds\right)^{\frac{1}{2}}\Big|X_0\right]$$

$$\leq k_2 V(X_0)^k + k_3\int_0^t \eta_s ds + \frac{1}{2}\xi_t, \quad t \in [0, T].$$

By Gronwall's inequality and (3.4.10), this implies (3.4.3).

DDSDEs: Well-posedness 117

(1d) It remains to prove (3.4.4). Let $\hat{P}_t$ be the Markov semigroup of X_t^μ for $\mu_t := P_t^*\gamma$. Then
$$\hat{P}_t^*\gamma = P_t^*\gamma, \quad t \in [0,T]. \tag{3.4.15}$$
By Theorem 1.3.1, we have
$$\lim_{y \to x} \|\hat{P}_t^*\delta_x - \hat{P}_t^*\delta_y\|_{var} = 0, \quad x \in \mathbb{R}^d.$$
Since $\gamma_n \to \gamma$ weakly, we may construct random variables $\{\xi_n\}$ and ξ such that $\mathcal{L}_{\xi_n} = \gamma_n, \mathcal{L}_\xi = \gamma$ and $\xi_n \to \xi$ a.s.. Thus, by the dominated convergence theorem we obtain
$$\lim_{n \to \infty} \|\hat{P}_t^*\gamma_n - \hat{P}_t^*\gamma\|_{var} = \lim_{n \to \infty} \|\mathbb{E}[\hat{P}_t^*\delta_{\xi_n} - \hat{P}_t^*\delta_\xi]\|_{var}$$
$$\leq \lim_{n \to \infty} \mathbb{E}[\|\hat{P}_t^*\delta_{\xi_n} - \hat{P}_t^*\delta_\xi\|_{var}] = 0. \tag{3.4.16}$$
Hence,
$$\limsup_{n \to \infty} \|\hat{P}_t^*\gamma_n - \hat{P}_t^*\gamma\|_V$$
$$\leq k \limsup_{n \to \infty} \|\hat{P}_t^*\gamma_n - \hat{P}_t^*\gamma\|_{var} + \sup_{n \geq 1} \int_{\mathbb{R}^d} \hat{P}_t(V-k)^+ \mathrm{d}(\gamma_n + \gamma) \tag{3.4.17}$$
$$= \sup_{n \geq 1} \{\hat{P}_t^*(\gamma_n + \gamma)\}((V-k)^+), \quad k \geq 1.$$
Since $\gamma_n(V^p)$ is bounded for some $p \in (1,2]$, (3.4.11) implies that
$$\sup_{n \geq 1, t \subset [0,T]} (\hat{P}_t^*\gamma_n)(V^p) < \infty, \tag{3.4.18}$$
so that letting $k \to \infty$ in (3.4.17), we derive
$$\limsup_{n \to \infty} \|\hat{P}_t^*\gamma_n - \hat{P}_t^*\gamma\|_V = 0. \tag{3.4.19}$$
On the other hand, by (3.4.15) and the Girsanov transform in step (1b) for γ_n replacing ν, the argument leading to (3.4.14) implies
$$\|P_t^*\gamma_n - \hat{P}_t^*\gamma_n\|_V^2 \leq c \int_0^t K_s \|P_s^*\gamma_n - P_s^*\gamma\|_V^2 \mathrm{d}s, \quad t \in [0,T] \tag{3.4.20}$$
for some constant $c > 0$. Combining this with (3.4.15), (3.4.19) and Fatou's lemma due to (3.4.18), we derive
$$\limsup_{n \to \infty} \|P_t^*\gamma_n - P_t^*\gamma\|_V^2 \leq 2\limsup_{n \to \infty} \{\|\hat{P}_t^*\gamma_n - \hat{P}_t^*\gamma\|_V^2 + \|P_t^*\gamma_n - \hat{P}_t^*\gamma_n\|_V^2\}$$
$$\leq 2 \int_0^t K_s \limsup_{n \to \infty} \|P_s^*\gamma_n - \hat{P}_s^*\gamma\|_V^2 \mathrm{d}s < \infty, \quad t \in [0,T].$$
By Gronwall's inequality, this implies (3.4.4).

(2) Similarly to eqref*B, by (3.4.5), Girsanov's theorem and Pinsker's inequality (3.2.3), we find a constant $c > 0$ such that
$$\|P_t^*\gamma_n - \hat{P}_t^*\gamma_n\|_{var}^2 \leq c \int_0^t K_s \|P_s^*\gamma_n - P_s^*\gamma\|_{var}^2 \mathrm{d}s, \quad t \in [0,T].$$
Combining this with (3.4.15) and (3.4.16), we derive (3.4.6). □

3.5 Singular case: $(\|\cdot\|_{k,var} + \mathbb{W}_k)$-Lipschitz in distribution

Comparing with $(A^{3.2})$, the following assumption allows $b_t(x,\cdot)$ to be Lipschitz in $\|\cdot\|_{k,var} + \mathbb{W}_k$ with Lipschitz constant singular in (t,x). By Remark 3.2.1, when $k > 1$ this norm is essentially larger than $\|\cdot\|_V$ for $V := 1 + |\cdot|^k$. Let

$$\|\mu\|_k := 1_{\{k>0\}} \mu(|\cdot|^k)^{\frac{1}{k}} + 1_{\{k=0\}}. \tag{3.5.1}$$

$(A^{3.3})$ Let $k \geq 0$ and $b_t^\mu := b_t(\cdot, \mu_t)$ for $\mu \in C_b^w([0,T]; \mathcal{P})$.
(1) There exists $\hat{\mu} \in \mathcal{P}_k$ such that $(A^{1.1})$ holds for $\hat{b} := b(\cdot, \hat{\mu})$ replacing b.
(2) There exist a constant $\alpha \geq 0$ and $1 \leq f \in \tilde{L}_{q_0}^{p_0}(T)$ such that for any $t \in [0,T]$, $x \in \mathbb{R}^d$, and $\mu, \nu \in \mathcal{P}_k$,

$$|b_t^\mu(x) - \hat{b}_t^{(1)}(x)| \leq f_t(x) + \alpha \|\mu\|_k, \tag{3.5.2}$$

$$|b_t^\mu(x) - b_t^\nu(x)| \leq f_t(x)\{\|\mu - \nu\|_{k,var} + \mathbb{W}_k(\mu,\nu)\}. \tag{3.5.3}$$

The following result is due to [Wang (2023b)] for $D = \mathbb{R}^d$.

Theorem 3.5.1. *Assume* $(A^{3.3})$.

(1) (3.4.1) *is weak well-posed for distributions in* $\mathcal{P}_k$. *Moreover, for any* $\gamma \in \mathcal{P}_k$, *and any* $n > 0$, *there exists a constant* $c > 0$, *such that*

$$\mathbb{E}\left[\sup_{t \in [0,T]} |X_t|^n \Big| X_0\right] \leq c(1 + |X_0|^n) \tag{3.5.4}$$

holds for the solution with $\mathcal{L}_{X_0} = \gamma$.
(2) *If moreover* σ *satisfies* $(A^{1.2})(1)$, *then* (3.4.1) *is well-posed for distributions in* $\mathcal{P}_k$.

To prove Theorem 3.5.1, we first present a more general result extending Corollary 3.2.2 for $k = 0$ and $p = q = \infty$, which may also apply to the degenerate situation.

For any $k \geq 0, \gamma \in \mathcal{P}_k, N \geq 2$, let

$$\mathcal{C}_k^{\gamma,N} := \Big\{\mu \in C_b^w([0,T]; \mathcal{P}_k) : \mu_0 = \gamma, \sup_{t \in [0,T]} e^{-Nt}(1 + \mu_t(|\cdot|^k)) \leq N\Big\}. \tag{3.5.5}$$

Then as $N \uparrow \infty$,

$$\mathcal{C}_k^{\gamma,N} \uparrow \mathcal{C}_k^\gamma := \big\{\mu \in C_b^w([0,T]; \mathcal{P}_k) : \mu_0 = \gamma\big\}. \tag{3.5.6}$$

For any $\mu \in \mathcal{C}_k^\gamma$, we will assume that the SDE
$$dX_t^\mu = b_t(X_t^\mu, \mu_t)dt + \sigma_t(X_t^\mu)dW_t, \quad t \in [0,T], \mathcal{L}_{X_0^\mu} = \gamma \quad (3.5.7)$$
has a unique weak solution with
$$\Phi_t^\gamma \mu := \mathcal{L}_{X_t^\mu} \in \mathcal{P}_k, \quad t \in [0,T].$$

$(A^{3.4})$ Let $k \geq 0, T > 0$. For any $\gamma \in \mathcal{P}_k$ and $\mu \in \mathcal{C}_k^\gamma$, (3.5.7) has a unique weak solution, and there exist constants $p, q > 1, N_0 \geq 2$ and increasing maps $C : [N_0, \infty) \to (0, \infty)$ and $F : [N_0, \infty) \times [0, \infty) \to (0, \infty)$ such that for any $N \geq N_0$ and $\mu \in \mathcal{C}_k^{\gamma, N}$, the (weak) solution satisfies
$$\Phi^\gamma \mu := \mathcal{L}_{(X_t^\mu)_{t \in [0,T]}} \in \mathcal{C}_k^{\gamma, N}, \quad (3.5.8)$$
$$\left(\mathbb{E}\left[(1+|X_t^\mu|^k)^2 | X_0^\mu\right]\right)^{\frac{1}{2}} \leq C(N)(1+|X_0^\mu|^k), \quad t \in [0,T], \quad (3.5.9)$$
$$\mathbb{E}\left(\int_0^t g_s(X_s^\mu)ds\right)^2 \leq C(N)\|g\|_{\tilde{L}_q^p(t)}^2, \quad (3.5.10)$$
$$\mathbb{E}\left[e^{\int_0^t g_s(X_s^\mu)ds}\right] \leq F(N, \|g\|_{\tilde{L}_q^p(t)}), \quad t \in [0,T], y \in \tilde{L}_q^p(t).$$

Obviously, when $k = 0$, conditions (3.5.8) and (3.5.9) hold for $N_0 = 2$. So, Corollary 3.2.2 is a special situation of the following result with $k = 0$ and $p = q = \infty$.

Theorem 3.5.2. *Assume* $(A^{3.4})$ *and*
$$b_t(x, \nu) - b_t(x, \mu) = \sigma_t(x)\Gamma_t(x, \nu, \mu), \quad x \in \mathbb{R}^d, t \in [0,T], \nu, \mu \subset \mathcal{P}_k \quad (3.5.11)$$
for some measurable map $\Gamma : [0,T] \times \mathbb{R}^d \times \mathcal{P}(\mathbb{R}^d) \to \mathbb{R}^m$.

(1) *If there exists* $f \geq 1$ *with* $|f|^2 \in \tilde{L}_q^p(T)$ *such that*
$$|\Gamma_t(x, \nu, \mu)| \leq f_t(x)\|\nu - \mu\|_{k, var}, \quad x \in \mathbb{R}^d, t \in [0,T], \nu, \mu \in \mathcal{P}_k, \quad (3.5.12)$$
then (3.4.1) *is weakly well-posed for distributions in* $\mathcal{P}_k$. *If, furthermore, in* $(A^{3.4})$ *the SDE* (3.5.7) *is strongly well-posed for any* $\gamma \in \mathcal{P}_k$ *and* $\mu \in \mathcal{C}_k^\gamma$, *so is* (3.4.1) *for distributions in* $\mathcal{P}_k$.

(2) *Let* $k > 1$ *and assume that*
$$|\Gamma_t(x, \nu, \mu)| \leq f_t(x)\{\|\nu - \mu\|_{k, var} + \mathbb{W}_k(\mu, \nu)\},$$
$$(t, x) \in [0,T] \times \mathbb{R}^d, \mu, \nu \in \mathcal{P}_k \quad (3.5.13)$$
holds for some $f \geq 1$ *with* $|f|^2 \in \tilde{L}_q^p(T)$. *If for any* $\gamma \in \mathcal{P}_k$ *and* $N \geq N_0$, *there exists a constant* $K(N) > 0$ *such that for any* $\mu, \nu \in \mathcal{C}_k^{\gamma, N}$,
$$\mathbb{W}_k(\Phi_t^\gamma \mu, \Phi_t^\gamma \nu)^{2k}$$
$$\leq K(N) \int_0^t \{\|\mu_s - \nu_s\|_{k, var}^{2k} + \mathbb{W}_k(\mu_s, \nu_s)^{2k}\}ds, \quad t \in [0,T], \quad (3.5.14)$$
then assertions in (1) *holds*.

Proof. Let $\gamma \in \mathcal{P}_k$. Then the weak solution to (3.5.7) is a weak solution to (3.4.1) if and only if μ is a fixed point of the map Φ^γ in $\mathcal{C}_k^\gamma$. So, if Φ^γ has a unique fixed point in $\mathcal{C}_k^\gamma$, then the (weak) well-posedness of (3.5.7) implies that of (3.4.1). Thus, by (3.5.6), it suffices to show that for any $N \geq N_0$, Φ^γ has a unique fixed point in $\mathcal{C}_k^{\gamma,N}$. By (3.5.8) and the fixed point theorem, we only need to prove that for any $N \geq N_0$, Φ^γ is contractive with respect to a complete metric on $\mathcal{C}_k^{\gamma,N}$.

(1) For any $\lambda > 0$, consider the metric
$$\mathbb{W}_{k,\lambda,var}(\mu,\nu) := \sup_{t\in[0,T]} e^{-\lambda t} \|\mu_t - \nu_t\|_{k,var}, \quad \mu,\nu \in \mathcal{C}_k^{\gamma,N}.$$

Let X_t^μ solve (3.5.7) for some Brownian motion W_t on a complete probability filtration space $(\Omega, \{\mathcal{F}_t\}, \mathbb{P})$. By (3.5.10), (3.5.12) or (3.5.13) with $|f|^2 \in \tilde{L}_q^p(T)$, we find a constant $c_1 > 0$ depending on N such that

$$\sup_{\mu,\nu\in\mathcal{C}_k^{\gamma,N}} \mathbb{E}\big(e^{12\int_0^T |\Gamma_s(X_s^\mu,\nu_s,\mu_s)|^2 ds} \big| \mathcal{F}_0 \big) \leq c_1^2,$$

$$\sup_{\mu\in\mathcal{C}_k^{\gamma,N}} \mathbb{E}\bigg(\bigg(\int_0^T g_s(X_s^\mu) ds\bigg)^2 \bigg| \mathcal{F}_0\bigg) \quad (3.5.15)$$

$$\leq c_1^2 \|g\|_{\tilde{L}_q^p(T)}^2, \quad g \in \tilde{L}_q^p(T), \mu,\nu \in \mathcal{C}_k^{\gamma,N}.$$

Then by Girsanov's theorem,
$$\tilde{W}_t := W_t - \int_0^t \Gamma_s(X_s^\mu, \nu_s, \mu_s) ds, \quad t \in [0,T]$$

is a Brownian motion under the probability $\mathbb{Q} := R_T \mathbb{P}$, where
$$R_t := e^{\int_0^t \langle \Gamma_s(X_s^\mu,\nu_s,\mu_s), dW_s\rangle - \frac{1}{2}\int_0^t |\Gamma_s(X_s^\mu,\nu_s,\mu_s)|^2 ds}, \quad t \in [0,T]$$

is a $\mathbb{P}$-martingale. By (3.5.11), we may formulate (3.5.7) as
$$dX_t^\mu = b_t(X_t^\mu, \nu_t) dt + \sigma_t(X_t^\mu) d\tilde{W}_t, \quad t \in [0,T], \mathcal{L}_{X_0^\mu} = \gamma.$$

By the weak uniqueness due to $(A^{3.4})$, the definition of $\|\cdot\|_{k,var}$, (3.5.9) and (3.5.11), we obtain

$$\|\Phi_t^\gamma \mu - \Phi_t^\gamma \nu\|_{k,var} = \sup_{|\tilde{f}|\leq 1+|\cdot|^k} \big|\mathbb{E}\big[(R_t - 1)\tilde{f}(X_t^\mu)\big]\big|$$
$$\leq \mathbb{E}\big[(1 + |X_t^\mu|^k)|R_t - 1|\big]$$
$$\leq \mathbb{E}\bigg[\big\{\mathbb{E}((1+|X_t^\mu|^k)^2|\mathcal{F}_0)\big\}^{\frac{1}{2}} \big\{\mathbb{E}(|R_t-1|^2|\mathcal{F}_0)\big\}^{\frac{1}{2}}\bigg] \quad (3.5.16)$$
$$\leq C(N)\mathbb{E}\bigg[(1+|X_0^\mu|^k)\big\{\mathbb{E}(e^{6\int_0^t |\Gamma_s(X_s^\mu,\nu_s,\mu_s)|^2 ds} - 1|\mathcal{F}_0)\big\}^{\frac{1}{2}}\bigg],$$

where we have used the facts that $\mathbb{E}(|R_t - 1|^2|\mathcal{F}_0) = \mathbb{E}(R_t^2|\mathcal{F}_0) - 1$ and for $\xi_s := \Gamma_s(X_s^\mu, \nu_s, \mu_s)$,

$$\mathbb{E}(R_t^2|\mathcal{F}_0) \leq \left(\mathbb{E}\left[e^{4\int_0^t \langle \xi_s, dW_s\rangle - 8\int_0^t |\xi_s|^2 ds}\big|\mathcal{F}_0\right]\right)^{\frac{1}{2}} \left(\mathbb{E}\left[e^{6\int_0^t |\xi_s|^2 ds}\big|\mathcal{F}_0\right]\right)^{\frac{1}{2}}$$
$$\leq \mathbb{E}\left[e^{6\int_0^t |\xi_s|^2 ds}\big|\mathcal{F}_0\right].$$

Moreover, by (3.5.15) we find a constant $c_2 > 0$ such that

$$\mathbb{E}(e^{6\int_0^t |\Gamma_s(X_s^\mu,\nu_s,\mu_s)|^2 ds} - 1|\mathcal{F}_0)$$
$$\leq 6\mathbb{E}\left(e^{6\int_0^t |\Gamma_s(X_s^\mu,\nu_s,\mu_s)|^2 ds} \int_0^t |\Gamma_s(X_s^\mu,\nu_s,\mu_s)|^2 ds\bigg|\mathcal{F}_0\right)$$
$$\leq c_2 \left\{\mathbb{E}\left(\left(\int_0^t |f_s(X_s^\mu)|^2 \|\mu_s - \nu_s\|_{k,var}^2 ds\right)^2 \bigg|\mathcal{F}_0\right)\right\}^{\frac{1}{2}}$$
$$\leq c_2 e^{2\lambda t} \mathbb{W}_{k,\lambda,var}(\mu,\nu)^2 \left\{\mathbb{E}\left(\left(\int_0^t |f_s(X_s^\mu)|^2 e^{-2\lambda(t-s)} ds\right)^2 \bigg|\mathcal{F}_0\right)\right\}^{\frac{1}{2}}$$
$$\leq c_2^2 e^{2\lambda t} \|f^2 e^{-2\lambda(t-\cdot)}\|_{\tilde{L}_q^p(t)} \mathbb{W}_{k,\lambda,var}(\mu,\nu)^2, \quad t \in [0,T].$$

Combining this with (3.5.16) and the definition of $\mathbb{W}_{k,\lambda,var}$, we obtain

$$\mathbb{W}_{k,\lambda,var}(\Phi^\gamma \mu, \Phi^\gamma \nu) \leq C(N)(1 + \gamma(|\cdot|^k))c_2\sqrt{\varepsilon(\lambda)}\mathbb{W}_{k,\lambda,var}(\mu,\nu), \quad \lambda > 0, \tag{3.5.17}$$

where

$$\varepsilon(\lambda) := \sup_{t\in[0,T]} \|f^2 e^{-2\lambda(t-\cdot)}\|_{\tilde{L}_q^p(t)} \downarrow 0 \text{ as } \lambda \uparrow \infty.$$

So, Φ^γ is contractive on $(\mathcal{C}_k^{\gamma,N}, \mathbb{W}_{k,\lambda,var})$ for large enough $\lambda > 0$.

(2) Let $k > 1$. We consider the metric $\tilde{\mathbb{W}}_{k,\lambda,var} := \mathbb{W}_{k,\lambda,var} + \mathbb{W}_{k,\lambda}$, where

$$\mathbb{W}_{k,\lambda}(\mu,\nu) := \sup_{t\in[0,T]} e^{-\lambda t}\mathbb{W}_k(\mu_t,\nu_t), \quad \mu,\nu \in \mathcal{C}_k^{\gamma,N}.$$

By using (3.5.13) replacing (3.5.12), instead of (3.5.17) we find constants $\{C(N,\lambda) > 0\}_{\lambda>0}$ with $C(N,\lambda) \to 0$ as $\lambda \to \infty$ such that

$$\mathbb{W}_{k,\lambda,var}(\Phi^\gamma \mu, \Phi^\gamma \nu) \leq C(N,\lambda)\tilde{\mathbb{W}}_{k,\lambda,var}(\mu,\nu), \quad \lambda > 0, \mu,\nu \in \mathcal{C}_k^{\gamma,N}. \tag{3.5.18}$$

On the other hand, (3.5.14) yields

$$\mathbb{W}_{k,\lambda}(\Phi^\gamma \mu, \Phi^\gamma \nu)$$
$$\leq \sup_{t\in[0,T]} \left(C(N) \mathrm{e}^{-2\lambda kt} \int_0^t \{\|\mu_s - \nu_s\|_{k,var}^{2k} + \mathbb{W}_k(\mu_s, \nu_s)^{2k}\} \mathrm{d}s \right)^{\frac{1}{2k}}$$
$$\leq \tilde{\mathbb{W}}_{k,\lambda,var}(\mu,\nu) \sup_{t\in[0,T]} \left(K(N) \int_0^t \mathrm{e}^{-2\lambda k(t-s)} \mathrm{d}s \right)^{\frac{1}{2k}}$$
$$\leq \frac{K(N)^{\frac{1}{2k}}}{(2\lambda k)^{\frac{1}{2k}}} \tilde{\mathbb{W}}_{k,\lambda,var}(\mu,\nu), \quad \lambda > 0.$$

Combining this with (3.5.18), we conclude that Φ^γ is contractive in $\mathcal{C}_k^{\gamma,N}$ under the metric $\tilde{\mathbb{W}}_{k,\lambda,var}$ when λ is large enough, and hence finish the proof. $\square$

Proof of Theorem 3.5.1. Let $\gamma \in \mathcal{P}_k$ be fixed. By Theorem 1.3.1, $(A^{3.3})$ implies the weak well-posedness of (3.5.7) for distributions in $\mathcal{P}_k$ with

$$\Phi^\gamma \mu := \mathcal{L}_{X^\mu} \in \mathcal{C}_k^\gamma, \quad \mu \in \mathcal{C}_k^\gamma,$$

and also implies the strong well-posedness of (3.5.7) in the situation of Theorem 3.5.1(2). Moreover, by Theorem 1.2.3 and (1.2.17), $(A^{3.3})$ implies that (3.5.10) holds for $(p,q) = (p_0/2, q_0/2)$, (3.5.11) with (3.5.12) holds for $k \leq 1$ due to (3.2.1), and (3.5.11) with (3.5.13) holds for $k > 1$. Therefore, by Theorem 3.5.2, it remains to verify (3.5.4), (3.5.8), (3.5.9), and (3.5.14) for $k > 1$. Since (3.5.9) and (3.5.8) are trivial for $k = 0$, we only need to prove:

- (3.5.4);
- (3.5.9) and (3.5.8) for $k > 0$;
- (3.5.14) for $k > 1$.

(a) We first prove that for some constant $c_0 > 0$ and increasing function $c : [1, \infty) \to (0, \infty)$ such that for any $n \geq 1$ and $\mu \in \mathcal{C}_k^\gamma$,

$$\mathbb{E}\left(\int_0^t |f_s(X_s^\mu)|^2 \mathrm{d}s \right)^n \leq c(n) + c(n) \left(\int_0^t \|\mu_s\|_k^2 \mathrm{d}s \right)^n,$$
$$\mathbb{E}\exp\left[n \int_0^t |f_s(X_s^\mu)|^2 \mathrm{d}s \right] \leq c(n) \exp\left[c_0 \int_0^t \|\mu_s\|_k^2 \mathrm{d}s \right]$$

(3.5.19)

for $t \in [0,T]$, where X_t^μ solves (3.5.7). Consider the SDE

$$\mathrm{d}\hat{X}_s = \hat{b}_s(\hat{X}_s)\mathrm{d}s + \sigma_s(\hat{X}_s)\mathrm{d}W_s, \quad \hat{X}_0 = X_0^\mu, s \in [0,t]. \tag{3.5.20}$$

By ($A^{3.4}$), (1.2.17) applies to this SDE, so that for any $n \geq 1$ we find a constant $c_1(n) > 0$ such that
$$\mathbb{E}\big[e^{n\int_0^t (|\hat{b}_s^{(0)}|^2 + |f_s|^2)(\hat{X}_s)\mathrm{d}s}\big] \leq c_1(n), \quad t \in [0, T]. \tag{3.5.21}$$
Let $\xi_s = \big\{[\sigma_s^*(\sigma_s\sigma_s^*)^{-1}](b_s^\mu - \hat{b}_s)\big\}(\hat{X}_s)$, and
$$R_t := e^{\int_0^t \langle \xi_s, \mathrm{d}W_s\rangle - \frac{1}{2}\int_0^t |\xi_s|^2 \mathrm{d}s}, \quad \tilde{W}_s := W_s - \int_0^s \gamma_r \mathrm{d}r, \quad s \in [0, t].$$
By Girsanov's theorem, $(\tilde{W}_s)_{s \in [0,t]}$ is a Brownian motion under $R_t \mathbb{P}$, and (3.5.20) becomes
$$\mathrm{d}\hat{X}_s = b_s(\hat{X}_s, \mu_s)\mathrm{d}s + \sigma_s(\hat{X}_s)\mathrm{d}\tilde{W}_s, \quad \hat{X}_0 = X_0^\mu, s \in [0, t].$$
So, by (3.5.2), (3.5.21) and Hölder's inequality, we find constants c_0, c_1, $c_2(n) > 0$ such that
$$\mathbb{E}\big[e^{n\int_0^t |f_s(X_s^\mu)|^2 \mathrm{d}s}\big] = \mathbb{E}\big[R_t e^{n\int_0^t |f_s(\hat{X}_s)|^2 \mathrm{d}s}\big]$$
$$\leq \big(\mathbb{E}e^{2n\int_0^t |f_s(\hat{X}_s)|^2 \mathrm{d}s}\big)^{\frac{1}{2}} \big(\mathbb{E}[R_t^2]\big)^{\frac{1}{2}}$$
$$\leq \sqrt{c_1(2n)}\big(\mathbb{E}e^{c_1\int_0^t \{|\hat{b}_s^{(0)}|^2 + (f_s + \alpha\|\mu_s\|_k)^2\}(\hat{X}_s)\mathrm{d}s}\big)^{\frac{1}{2}} \leq c_2(n)e^{c_0\int_0^t \|\mu_s\|_k^2 \mathrm{d}s}.$$
Next, taking $c_3(n) > 0$ large enough such that the function
$$r \mapsto [\log(r + c_3(n))]^n$$
is concave for $r \geq 0$, so that this and Jensen's inequality imply
$$\mathbb{E}\bigg(\int_0^t |f_s(X_s^\mu)|^2 \mathrm{d}s\bigg)^n \leq \mathbb{E}\big(\big[\log(c_3(n) + e^{\int_0^t |f_s(X_s^\mu)|^2 \mathrm{d}s})\big]^n\big)$$
$$\leq \big[\log(c_3(n) + \mathbb{E}e^{\int_0^t |f_s(X_s^\mu)|^2 \mathrm{d}s})\big]^n \leq c(n) + c(n)\bigg(\int_0^t \|\mu_s\|_k^2 \mathrm{d}s\bigg)^n$$
for some constant $c(n) > 0$. Therefore, (3.5.19) holds.

(b) Proof of (3.5.8). Simply denote $X_t = X_t^\mu$. By ($A^{3.4}$) and Itô's formula, we find constants $c_1, c_2 > 0$ such that
$$\mathbb{E}(1 + |X_t|^k) \leq c_1(1 + \|\gamma\|_k^k)$$
$$+ c_1 \mathbb{E}\bigg(\int_0^t \{|X_s| + |f_s(X_s)| + \|\mu_s\|_k\}\mathrm{d}s\bigg)^k \tag{3.5.22}$$
$$\leq c_2 + c_2 \mathbb{E}\bigg(\int_0^t \{|X_s|^2 + \|\mu_s\|_k^2\}\mathrm{d}s\bigg)^{\frac{k}{2}}, \quad t \in [0, T].$$

(b1) When $k \geq 2$, by (3.5.22) we find a constant $k_3 > 0$ such that
$$\mathbb{E}(1 + |X_t|^k) \leq k_2 + k_3 \int_0^t \{\mathbb{E}|X_s|^k + \|\mu_s\|_k^k\}\mathrm{d}s, \quad t \in [0, T].$$

By Gronwall's lemma, and noting that $\mu \in \mathcal{C}_k^{\gamma,N}$, we find a constant $k_4 > 0$ such that

$$\mathbb{E}(1+|X_t|^k) \leq k_4 + k_4 \int_0^t (1+\|\mu_s\|_k^k)\mathrm{d}s$$

$$\leq k_4 + k_4 N \mathrm{e}^{Nt} \int_0^t \mathrm{e}^{-N(t-s)}\mathrm{d}s \leq 2k_4 \mathrm{e}^{Nt}, \ t \in [0,T].$$

Taking $N_0 = 2k_4$ we derive

$$\sup_{t\in[0,T]} \mathrm{e}^{-Nt}(1+\|\Phi_t\mu\|_k^k) = \sup_{t\in[0,T]} \mathrm{e}^{-Nt}\mathbb{E}(1+|X_t|^k)$$

$$\leq N_0 \leq N, \quad N \geq N_0, \mu \in \mathcal{C}_k^{\gamma,N},$$

so that (3.5.8) holds.

(b2) When $k \in (0,2)$, by BDG inequality, and by the same reason leading to (3.5.22), we find constants $k_5, k_6, k_7 > 0$ such that

$$U_t := \mathbb{E}\Big[\sup_{s\in[0,t]}(1+|X_s|^k)\Big] \leq k_5 + k_5\mathbb{E}\bigg(\int_0^t \{|X_s|^2 + \|\mu_s\|_k^2\}\mathrm{d}s\bigg)^{\frac{k}{2}}$$

$$\leq k_6 + k_6\mathbb{E}\bigg\{\Big[\sup_{s\in[0,t]}|X_s|^k\Big]^{1-\frac{k}{2}}\bigg(\int_0^t |X_s|^k \mathrm{d}s\bigg)^{\frac{k}{2}}\bigg\} + k_6\bigg(\int_0^t \|\mu_s\|_k^2 \mathrm{d}s\bigg)^{\frac{k}{2}}$$

$$\leq k_6 + \frac{1}{2}U_t + k_7\int_0^t U_s \mathrm{d}s + k_6\bigg(\int_0^t \|\mu_s\|_k^2 \mathrm{d}s\bigg)^{\frac{k}{2}}, \ t \in [0,T].$$

By Gronwall's lemma, we find constants $k_8, k_9 > 0$ such that for any $\mu \in \mathcal{C}_k^{\gamma,N}$,

$$\mathbb{E}(1+|X_t|^k) \leq U_t \leq k_8 + k_8\bigg(\int_0^t \|\mu_s\|_k^2 \mathrm{d}s\bigg)^{\frac{k}{2}}$$

$$\leq k_8 + k_8 N \mathrm{e}^{Nt}\bigg(\int_0^t \mathrm{e}^{-2N(t-s)/k}\mathrm{d}s\bigg)^{\frac{k}{2}} \leq k_8 + k_9 N^{1-\frac{k}{2}}\mathrm{e}^{Nt}, \ t \in [0,T].$$

Thus, there exists $N_0 > 0$ such that for any $N \geq N_0$,

$$\sup_{t\in[0,T]} \mathrm{e}^{-Nt}(1+\|\Phi_t\mu\|_k^k) = \sup_{t\in[0,T]} \mathrm{e}^{-Nt}\mathbb{E}(1+|X_t|^k)$$

$$\leq k_8 + k_9 N^{1-\frac{k}{2}} \leq N, \quad \mu \in \mathcal{C}_k^{\gamma,N},$$

which implies (3.5.8).

(c) Proofs of (3.5.9) and (3.5.4). By Theorem 1.3.1, $(A^{3.4})$ implies that for any $n \geq 1$ there exists a constant $c > 0$ such that

$$\mathbb{E}\Big[\sup_{t\in[0,T]}|\hat{X}_t|^n \Big| \hat{X}_0\Big] \leq c(1+|\hat{X}_0|^n). \qquad (3.5.23)$$

So, by (1.2.17) and Girsanov's theorem,

$$\tilde{W}_t := W_t - \int_0^t \{\sigma_s^*(\sigma_s\sigma_s^*)^{-1}\}(\hat{X}_s)\{b_s(\hat{X}_s,\mu_s) - \hat{b}_s(\hat{X}_s)\}\mathrm{d}s, \quad t \in [0,T]$$

is a $\mathbb{Q}$-Brownian motion for $\mathbb{Q} := R_T\mathbb{P}$, where

$$\eta_t := \{\sigma_t^*(\sigma_t\sigma_t^*)^{-1}\}(\hat{X}_t)\{b_t(\hat{X}_t,\mu_t) - \hat{b}_t(\hat{X}_t)\},$$
$$R_T := e^{\int_0^T \langle \eta_t, \mathrm{d}W_t\rangle - \frac{1}{2}\int_0^T |\eta_t|^2 \mathrm{d}t}.$$

By $(A^{3.4})$ and (1.2.17), we find an increasing function F such that

$$\mathbb{E}(|R_T|^2|\mathcal{F}_0) \leq \mathbb{E}(e^{\int_0^T |f_s(\hat{X}_s)|^2\{\|\mu_s-\hat{\mu}\|_{k,var}+\mathbb{W}_k(\mu_s,\hat{\mu})\}^2\mathrm{d}s}|\mathcal{F}_0) \leq F(\|\mu\|_{k,T}),$$

where $\|\mu\|_{k,T} := \sup_{t\in[0,T]} \mu_t(|\cdot|^k)$. Reformulating (3.5.20) as

$$\mathrm{d}\hat{X}_t = b_t^\mu(\hat{X}_t)\mathrm{d}t + \sigma_t(\hat{X}_t)\mathrm{d}\tilde{W}_t, \quad \mathcal{L}_{\hat{X}_0} = \gamma,$$

by the weak uniqueness we have $\mathcal{L}_{\hat{X}|\mathbb{Q}} = \mathcal{L}_{X^\mu}$, so that (3.5.23) with $2n$ replacing n implies

$$\mathbb{E}\Big[\sup_{t\in[0,T]} |X_t^\mu|^n \Big|\mathcal{F}_0\Big] = \mathbb{E}_\mathbb{Q}\Big[\sup_{t\in[0,T]} |\hat{X}_t|^n \Big|\mathcal{F}_0\Big]$$
$$\leq \Big(\mathbb{E}\Big[\sup_{t\in[0,T]} |\hat{X}_t|^{2n} \Big|\mathcal{F}_0\Big]\Big)^{\frac{1}{2}} (\mathbb{E}R_T^2|\mathcal{F}_0)^{\frac{1}{2}} \leq c\sqrt{(1+|\hat{X}_0|^n)F(\|\mu\|_{k,T})}.$$

Since $\sup_{\mu\in\mathcal{C}_k^\gamma,N} \|\mu\|_{k,T}$ is a finite increasing function of N, this implies (3.5.9).

Finally, since $X_t := X_t^\mu$ solves (3.4.1) with initial distribution γ and $\mu_t = \mathcal{L}_{X_t}$ (i.e. μ is the fixed point of Φ^γ), and since Φ^γ has a unique fixed point in $\mathcal{C}_k^{\gamma,N}$ for some $N > 0$ depending on γ as shown in the proof of Theorem 3.5.2 using (3.5.10) and (3.5.8), we have $\mathcal{L}_X \subset \mathcal{C}_k^{\gamma,N}$, and hence (3.5.4) follows from (3.5.9).

(d) Proof of (3.5.14) for $k > 1$. Let u_t^λ solve (1.2.3) for $L_t = L_{t,\nu}$ with $b^{(0)} = b^\nu - \hat{b}^{(1)}$ under $(A^{1,1})$, such that

$$\|\nabla u^\lambda\|_\infty \leq \frac{1}{2}.$$

Let $\Theta_t = u_t^\lambda + id$ and

$$\xi_t := \Theta_t(X_t^\mu) - \Theta_t(X_t^\nu) = X_t^\mu + u_t^\lambda(X_t^\mu) - X_t^\nu - u_t^\lambda(X_t^\nu).$$

We have

$$|X_t^\mu - X_t^\nu| \leq 2|\xi_t|. \tag{3.5.24}$$

By Itô's formula we obtain

$$d\xi_t = \left\{(\lambda u_t^\lambda + \hat{b}^{(1)})(X_t^\mu) - (\lambda u_t^\lambda + \hat{b}^{(1)})(X_t^\nu)\right\}dt$$
$$+ \left\{[(\nabla\Theta_t^\lambda)\sigma_t](X_t^\mu) - [(\nabla\Theta_t^\lambda)\sigma_t](X_t^\nu)\right\}dW_t.$$

By $(A^{3.4})$ and Lemma 1.3.4, there exists a constant $c_1 > 0$ such that

$$|X_t^\mu - X_t^\nu|^{2k} \le c_1 \int_0^t \left(\|\mu_s - \nu_s\|_{k,var} + \mathbb{W}_k(\mu_s, \nu_s)\right)^{2k} ds$$
$$+ c_1 \int_0^{t\wedge\tau_m} |X_{s\wedge\tau_m}^\mu - X_{s\wedge\tau_m}^\nu|^{2k} d\tilde{\mathcal{L}}_s + \tilde{M}_t$$

holds for some local martingale $\tilde{M}_t$ and

$$\tilde{\mathcal{L}}_t := A_t + \int_0^t |f_s(X_s^\nu)|^2 ds, \quad t \in [0,T],$$

where A_t is in (1.3.25) for (X^μ, X^ν) replacing (X^1, X^2). By (3.5.24), the stochastic Gronwall inequality in Lemma 1.3.3 and Kasminskii's estimate (1.2.17), we find a constant $c > 0$ such that

$$\mathbb{W}_k(\Phi_t^\gamma \mu, \Phi_t^\gamma \nu)^{2k} \le (\mathbb{E}|X_t^\mu - X_t^\nu|^k)^2$$
$$\le c \int_0^t \left\{\|\mu_s - \nu_s\|_{k,var}^{2k} + \mathbb{W}_k(\mu_s, \nu_s)^{2k}\right\} ds. \qquad (3.5.25)$$

Thus, (3.5.14) holds. □

3.6 Singular case: With distribution dependent noise

In this part, we investigate the well-posedness of (3.1.1) where the noise is distribution dependent, and the drift is singular in the spatial variable and $\mathbb{W}_k$-Lipschitzian in the distribution variable.

For any $\mu \in C_b^w([0,T]; \mathcal{P}_k)$, $x \in \mathbb{R}^d$ and $t \in [0,T]$, denote

$$\sigma_t^\mu(x) := \sigma_t(x, \mu_t), \quad b_t^\mu(x) := b_t(x, \mu_t).$$

Recall that by (3.5.1), $\|\mu\|_k := \mu(|\cdot|^k)^{\frac{1}{k}}$ for $k > 0$.

$(A^{3.5})$ Let $k \ge 1$. There exist constants $K > K_0 \ge 0$, $l \in \mathbb{N}$, $\{(p_i, q_i) : 0 \le i \le l\} \subset \mathcal{K}$ with $p_i > 2$, and $1 \le f_i \in \tilde{L}_{q_i}^{p_i}(T)$ for $0 \le i \le l$, such that $\sigma_t^\mu(x)$ and $b_t^\mu(x) := b_t^{\mu,1}(x) + b_t^{\mu,0}(x)$ satisfy the following conditions for all $\mu \in C_b^w([0,T]; \mathcal{P}_k)$.

(1) $a^\mu := \sigma^\mu(\sigma^\mu)^*$ is invertible with $\|a^\mu\|_\infty + \|(a^\mu)^{-1}\|_\infty \leq K$ and
$$\lim_{\varepsilon \downarrow 0} \sup_{\mu \in C_b^w([0,T];\mathcal{P}_k)} \sup_{t \in [0,T], |x-y| \leq \varepsilon} \|a_t^\mu(x) - a_t^\mu(y)\| = 0.$$

(2) $b^{\mu,1}$ is locally bounded on $[0,T] \times \mathbb{R}^d$, σ_t^μ is weakly differentiable such that
$$|b_t^{\mu,0}(x)| \leq f_0(t,x) + K_0 \|\mu_t\|_k, \quad \|\nabla \sigma_t^\mu(x)\| \leq \sum_{i=1}^l f_i(t,x),$$
$$|b_t^{\mu,1}(x) - b_t^{\mu,1}(y)| \leq K|x-y|, \quad t \in [0,T], x, y \in \mathbb{R}^d.$$

(3) For any $t \in [0,T], x \in \mathbb{R}^d$ and $\mu, \nu \in \mathcal{P}_k$,
$$\|\sigma_t(x,\mu) - \sigma_t(x,\nu)\| + |b_t(x,\mu) - b_t(x,\nu)| \leq \mathbb{W}_k(\mu,\nu) \sum_{i=0}^l f_i(t,x).$$

Theorem 3.6.1. *Assume* $(A^{3.5})$. *Then the following assertions hold.*

(1) (3.1.1) *is well-posed for distributions in* $\mathcal{P}_k$. *Moreover, for any* $j \geq k$ *there exists a constant* $c(j) > 0$ *such that any solution* X_t *of* (3.1.1) *satisfies*
$$\mathbb{E}\Big[\sup_{t \in [0,T]} |X_t|^j \Big| \mathcal{F}_0\Big] \leq c(j)\big\{1 + |X_0|^j + (\mathbb{E}[|X_0|^k])^{\frac{j}{k}}\big\}. \qquad (3.6.1)$$

(2) *For any* $N > 0$ *and* $j \geq k$, *there exists a constant* $C_{j,N} > 0$ *such that for any two solutions* X_t^i *of* (3.1.1) *with* $\mathbb{E}[|X_0^i|^k] \leq N, i = 1,2$,
$$\mathbb{E}\Big(\sup_{t \in [0,T]} |X_t^1 - X_t^2|^j \Big| \mathcal{F}_0\Big)$$
$$\leq C_{j,N}\big\{|X_0^1 - X_0^2|^j + (\mathbb{E}[|X_0^1 - X_0^2|^k])^{\frac{j}{k}}\big\}. \qquad (3.6.2)$$

Consequently,
$$\sup_{t \in [0,T]} \mathbb{W}_k(P_t^* \mu^1, P_t^* \mu^2) \leq 2 C_{k,N} \mathbb{W}_k(\mu^1, \mu^2),$$
$$\mu^1, \mu^2 \in \mathcal{P}_k, \ \mu^1(|\cdot|^k), \mu^2(|\cdot|^k) \leq N. \qquad (3.6.3)$$

When $K_0 = 0$, *this estimate holds for some constant* $C_j > 0$ *replacing* $C_{j,N}$ *for any two solutions with distributions in* $\mathcal{P}_k$.

128   Distribution Dependent Stochastic Differential Equations

3.6.1 Some lemmas

We first explain the main idea of the two-step fixed point argument.

Let X_0 be $\mathcal{F}_0$-measurable with $\gamma := \mathcal{L}_{X_0} \in \mathcal{P}_k$. Let
$$\mathcal{C}_k^\gamma := \{\mu \in C_b^w([0,T]; \mathcal{P}_k) : \mu_0 = \gamma\}.$$
We solve (3.1.1) with a fixed distribution parameter $\mu \in \mathcal{C}_k^\gamma$ in the drift:
$$dX_t^\mu = b_t(X_t^\mu, \mu_t)dt + \sigma_t(X_t^\mu, \mathcal{L}_{X_t^\mu})dW_t, \quad t \in [0,T], X_0^\mu = X_0, \quad (3.6.4)$$
such that the well-posedness of this SDE for distributions in $\mathcal{P}_k$ provides a map
$$\mathcal{C}_k^\gamma \ni \mu \mapsto \Phi^\gamma \mu := \mathcal{L}_{X^\mu} \in \mathcal{C}_k^\gamma. \quad (3.6.5)$$
Then the well-posedness of (3.1.1) follows if the map Φ^γ has a unique fixed point in $\mathcal{C}_k^\gamma$.

To solve (3.6.4), we further fix the distribution parameter $\nu \in \mathcal{C}_k^\gamma$ in σ such that the SDE becomes
$$dX_t^{\mu,\nu} = b_t(X_t^{\mu,\nu}, \mu_t)dt + \sigma_t(X_t^{\mu,\nu}, \nu_t)dW_t, \quad t \in [0,T], X_0^{\mu,\nu} = X_0,$$
which is well-posed under $(A^{3.5})$ according to Theorem 1.3.1(3). This gives a map
$$\mathcal{C}_k^\gamma \ni \nu \mapsto \Phi^{\gamma,\mu}\nu := \mathcal{L}_{X^{\mu,\nu}} \in \mathcal{C}_k^\gamma. \quad (3.6.6)$$
So, we first prove that this map has a unique fixed point such that (3.6.4) is well-posed, then apply the fixed point theorem to Φ^γ to derive the well-posedness of the original SDE (3.1.1).

To apply the fixed point theorem, we will use the following complete metric on $\mathcal{C}_k^\gamma$ for $\theta > 0$:
$$\mathbb{W}_{k,\theta}(\mu,\nu) = \sup_{t \in [0,T]} e^{-\theta t} \mathbb{W}_k(\mu_t, \nu_t), \quad \mu, \nu \in \mathcal{C}_k^\gamma. \quad (3.6.7)$$
To prove that Φ^γ has a unique fixed point in $\mathcal{C}_k^\gamma$, we need to restrict the map to the following bounded subspaces of $\mathcal{C}_k^\gamma$:
$$\mathcal{C}_k^{\gamma,N} := \left\{\mu \in \mathcal{C}_k^\gamma : \sup_{t \in [0,T]} e^{-Nt}(1 + \mu_t(|\cdot|^k)) \leq N\right\}, \quad N > 0, \quad (3.6.8)$$
and to prove that these spaces are Φ^γ-invariant for large N. This enables us to verify the contraction of Φ^γ in $\mathcal{C}_k^{\gamma,N}$ under a suitable complete metric.

For this purpose, we present the following lemmas. The first one ensures the well-posedness of (3.6.4).

Lemma 3.6.2. *Assume* $(A^{3.5})$ *and let* $\mu \in \mathcal{C}_k^\gamma$. *Then* (3.6.4) *is well-posed for distributions in* $\mathcal{P}_k$. *Moreover, there exist* $\theta_0 > 0$ *and decreasing function* $\beta : [\theta_0, \infty) \to (0, \infty)$ *with* $\beta(\theta) \downarrow 0$ *as* $\theta \uparrow \infty$ *such that* Φ^γ *defined in* (3.6.5) *satisfies*
$$\mathbb{W}_{k,\theta}(\Phi^\gamma \mu, \Phi^\gamma \nu) \leq \beta(\theta) \mathbb{W}_{k,\theta}(\mu, \nu), \quad \mu, \nu \in \mathcal{C}_k^{\gamma,N}. \quad (3.6.9)$$

Proof. (a) For the well-posedness, it suffices to prove that $\Phi^{\gamma,\mu}$ defined in (3.6.6) has a unique fixed point in $\mathcal{C}_k^\gamma$.

In general, let $\mu^i \in \mathcal{C}_k^{\gamma^i,N}$ for some $N > 0, \gamma^i \in \mathcal{P}^k, i = 1, 2$. For $\nu^i \in \mathcal{C}_k^{\gamma^i,N}$ and initial value X_0^i with $\mathcal{L}_{X_0^i} = \gamma^i, i = 1, 2$, consider the SDEs

$$dX_t^i = b_t^{\mu^i}(X_t^i)dt + \sigma_t^{\nu^i}(X_t^i)dW_t, \quad t \in [0,T], i = 1, 2. \tag{3.6.10}$$

According to Theorem 1.3.1(3), under $(A^{3.5})$ these SDEs are well-posed, and by Lemma 1.2.2, there exist constants $c_0, \lambda_0 \geq 0$ depending on N via $\mu^1 \in \mathcal{C}_k^{\gamma,N}$ due to

$$|b_t^{\mu^1,0}(x)| \leq f_0(t,x) + K_0\|\mu_t^1\|_k,$$

such that for any $\lambda \geq \lambda_0$, the PDE for

$$\left(\partial_t + \frac{1}{2}\mathrm{tr}\{a_t^{\nu^1}\nabla^2\} + \nabla_{b_t^{\mu^1}}\right)u_t = \lambda u_t - b_t^{\mu^1,0}, \quad t \in [0,T], u_T = 0 \tag{3.6.11}$$

has a unique solution such that

$$\|\nabla^2 u\|_{\tilde{L}_{q_0}^{p_0}(T)} \leq c_0, \quad \|u\|_\infty + \|\nabla u\|_\infty \leq \frac{1}{2}. \tag{3.6.12}$$

Let $Y_t^i := \Theta_t(X_t^i), i = 1, 2, \Theta_t := id + u_t$. By Itô's formula we obtain

$$dY_t^1 = \{b_t^{\mu^1,1} + \lambda u_t\}(X_t^1)dt + (\{\nabla\Theta_t\}\sigma_t^{\nu^1})(X_t^1)dW_t,$$

$$dY_t^2 = \left(\{b_t^{\mu^1,1} + \lambda u_t + (\nabla\Theta_t)(b_t^{\mu^2} - b_t^{\mu^1})\}(X_t^2)\right.$$

$$+ \frac{1}{2}\mathrm{tr}\big[\{(a_t^{\nu^2} - a_t^{\nu^1})\nabla^2 u_t\}\big](X_t^2)\bigg)dt + (\{\nabla\Theta_t\}\sigma_t^{\nu^2})(X_t^2)dW_t.$$

Let $\eta_t := |X_t^1 - X_t^2|$ and

$$g_r := \sum_{i=0}^l f_i(r, X_r^2), \quad \tilde{g}_r := g_r\|\nabla^2 u_r(X_r^2)\|,$$

$$\bar{g}_r := \sum_{i=1}^2 \|\nabla^2 u_r\|(X_r^i) + \sum_{j=1}^2\sum_{i=0}^l f_i(r, X_r^j), \quad r \in [0,T].$$

Since $b_t^{(1)} + \lambda u_t$ is Lipschitz continuous uniformly in $t \in [0,T]$, by $(A^{3.5})$ and the maximal functional inequality in Lemma 1.3.4, there exists a constant $c_1 > 0$ depending on N such that

$$\left|\{b_r^{\mu^1,1} + \lambda u_r\}(X_r^1) - \{b_r^{\mu^1,1} + \lambda u_r\}(X_r^2)\right| \leq c_1\eta_r,$$

$$\left|\{(\nabla\Theta_r)(b_r^{\mu^2} - b_r^{\mu^1})\}(X_r^2)\right| \leq c_1 g_r \mathbb{W}_k(\mu_r^1, \mu_r^2),$$

$$\left|[\mathrm{tr}\{(a_r^{\nu^2} - a_r^{\nu^1})\nabla^2 u_r\}](X_r^2)\right| \leq c_1 \tilde{g}_r \mathbb{W}_k(\nu_r^1, \nu_r^2),$$

$$\left\|\{(\nabla\Theta_r)\sigma_r^{\nu^1}\}(X_r^1) - \{(\nabla\Theta_r)\sigma_r^{\nu^2}\}(X_r^2)\right\|$$

$$\leq c_1\bar{g}_r\eta_r + c_1 g_r \mathbb{W}_k(\nu_r^1, \nu_r^2), \quad r \in [0,T].$$

So, by Itô's formula, for any $j \geq k$ we find a constant $c_2 > 1$ depending on N such that
$$d|Y_t^1 - Y_t^2|^{2j} \leq c_2(g_t^2 + \tilde{g}_t)\{\mathbb{W}_k(\mu_t^1, \mu_t^2)^{2j} + \mathbb{W}_k(\nu_t^1, \nu_t^2)^{2j}\}dt \\ + c_2\eta_t^{2j}dA_t + dM_t \quad (3.6.13)$$
holds for some martingale M_t with $M_0 = 0$ and
$$A_t := \int_0^t \{1 + g_s^2 + \tilde{g}_s + \bar{g}_s^2\}ds.$$
Since $\|\nabla u\|_\infty \leq \frac{1}{2}$ implies $|Y_t^1 - Y_t^2| \geq \frac{1}{2}\eta_t$, this implies
$$\eta_t^{2j} \leq 2^{2j}M_t + 2^{2j}\eta_0^{2j} + 2^{2j}c_2\int_0^t \eta_r^{2j}dA_r \\ + 2^{2j}c_2\int_0^t (g_s^2 + \tilde{g}_s)\{\mathbb{W}_k(\mu_s^1, \mu_s^2)^{2j} + \mathbb{W}_k(\nu_s^1, \nu_s^2)^{2j}\}ds \quad (3.6.14)$$
for some constant $c_2 > 0$ and all $t \in [0, T]$. By (3.6.12), $f_i \in \tilde{L}_{q_i}^{p_i}(T)$ for $(p_i, q_i) \in \mathcal{K}$, Krylov's estimate (1.2.7) and Khasminskii's estimate (1.2.17) for $(p, q) = (p_i/2, q_i/2)$, we find an increasing function $\alpha : (0, \infty) \to (0, \infty)$ and a decreasing function $\varepsilon : (0, \infty) \to (0, \infty)$ with $\varepsilon_\theta \to 0$ as $\theta \to \infty$, such that
$$\mathbb{E}[e^{rA_T}|\mathcal{F}_0] \leq \alpha(r), \quad r > 0,$$
$$\sup_{t \in [0,T]} \mathbb{E}\left(\int_0^t e^{-2k\theta(t-r)}(g_r^2 + \tilde{g}_r)dr \Big| \mathcal{F}_0\right) \leq \varepsilon_\theta, \quad \theta > 0.$$
By the stochastic Gronwall inequality in Lemma 1.3.3 and the maximal inequality in Lemma 1.3.4, we find a constant $c_3 > 0$ depending on N such that (3.6.7) and (3.6.14) yield
$$\left\{\mathbb{E}\left(\sup_{s \in [0,t]} \eta_s^j \Big| \mathcal{F}_0\right)\right\}^2 - c_3\eta_0^{2j} \\ \leq c_3\mathbb{E}\left(\int_0^t (g_s^2 + \tilde{g}_s)\{\mathbb{W}_k(\mu_s^1, \mu_s^2)^{2j} + \mathbb{W}_k(\nu_s^1, \nu_s^2)^{2j}\}ds \Big| \mathcal{F}_0\right) \quad (3.6.15) \\ \leq c_3 e^{2k\theta t}\varepsilon_\theta\{\mathbb{W}_{k,\theta}(\mu^1, \mu^2)^{2j} + \mathbb{W}_{k,\theta}(\nu^1, \nu^2)^{2j}\}.$$
Noting that
$$\mathbb{W}_k(\mathcal{L}_{X_t^1}, \mathcal{L}_{X_t^2})^k \leq \mathbb{E}[|X_t^1 - X_t^2|^k] = \mathbb{E}[\eta_t^k],$$
by taking $j = k$ we obtain
$$\mathbb{W}_{k,\theta}(\mathcal{L}_{X^1}, \mathcal{L}_{X^2})^k \\ \leq \sqrt{c_3}\mathbb{E}[\eta_0^k] + \sqrt{c_3\varepsilon_\theta}\{\mathbb{W}_{k,\theta}(\mu^1, \mu^2)^k + \mathbb{W}_{k,\theta}(\nu^1, \nu^2)^k\}. \quad (3.6.16)$$

By taking $X_0^1 = X_0^2 = X_0$ and $\mu^1 = \mu^2 = \mu \in \mathcal{C}_k^{\gamma,N}$, when $\theta > 0$ is large enough such that $\sqrt{c_3 \varepsilon_\theta} \leq \frac{1}{2}$, $\Phi^{\gamma,\mu} \nu^i = \mathcal{L}_{X^i}$ satisfies

$$\mathbb{W}_{k,\theta}(\Phi^{\gamma,\mu}\nu^1, \Phi^{\gamma,\mu}\nu^2) \leq \frac{1}{2} \mathbb{W}_{k,\theta}(\nu^1, \nu^2), \quad \nu_1, \nu_2 \in \mathcal{C}_k^\gamma.$$

Thus, $\Phi^{\gamma,\mu}$ has a unique fixed point in $\mathcal{C}_k^\gamma$, so that (3.6.4) is well-posed for distributions in $\mathcal{P}_k$.

(b) Taking $\nu^i = \Phi^\gamma \mu^i$, we have $\mathcal{L}_{X^i} = \Phi^\gamma \mu^i$, so that (3.6.16) becomes

$$\mathbb{W}_{k,\theta}(\Phi^\gamma \mu^1, \Phi^\gamma \mu^2) \leq (c_3 \varepsilon_\theta)^{\frac{1}{2k}} \left\{ \mathbb{W}_{k,\theta}(\mu^1, \mu^2) + \mathbb{W}_{k,\theta}(\Phi^\gamma \mu^1, \Phi^\gamma \mu^2) \right\}.$$

Choosing $\theta_0 > 0$ large enough such that $c_3 \varepsilon_{\theta_0} < 1$, we derive (3.6.9) for

$$\beta(\theta) := \frac{(c_3 \varepsilon_\theta)^{\frac{1}{2k}}}{1 - (c_3 \varepsilon_\theta)^{\frac{1}{2k}}}, \quad \theta \geq \theta_0.$$

□

Lemma 3.6.3. *Assume* $(A^{3.5})$. *For any* $(p,q) \in \mathcal{K}$, *there exist a constant* $c_0 \geq 1$ *and a function* $c : [1, \infty) \to (0, \infty)$ *such that for any* $j \geq 1$ *and* $\mu \in \mathcal{C}_k^\gamma$, *any solution* X_t *to* (3.6.4) *satisfies*

$$\mathbb{E}\left[e^{\int_0^t |f_s(X_s^\mu)|^2 ds} \Big| \mathcal{F}_0 \right] \leq e^{c_0 + c_0 \int_0^t \|\mu_s\|_k^2 ds + c_0 \|f\|_{\tilde{L}_q^p(t)}^{c_0}}, \quad (3.6.17)$$

$$\mathbb{E}\left[\left(\int_0^t |f_s(X_s^\mu)|^2 ds \right)^j \Big| \mathcal{F}_0 \right] \leq c(j) \left(1 + \int_0^t \|\mu_s\|_k^2 ds \right)^j \|f\|_{\tilde{L}_q^p(t)}^{2j} \quad (3.6.18)$$

for any $t \in [0, T]$ *and* $f \in \tilde{L}_q^p(t), t \in [0, T]$.

Proof. Let Φ^γ be defined in (3.6.5). Consider the SDE

$$d\bar{X}_t = b_t^{(1)}(\bar{X}_t) dt + \sigma_t(\bar{X}_t, \Phi_t^\gamma \mu) dW_t, \quad \bar{X}_0 = X_0, t \in [0, T].$$

By Krylov's estimate (1.2.7) which implies Khasminskii's estimate in Theorem 1.2.4, there exists a constant $c_1 \geq 1$ depending only on K, T, d, p, q and the continuity modulus of a^μ which is uniform in μ, such that

$$\mathbb{E}\left[e^{\int_0^t |f_s(\bar{X}_s^\mu)|^2 ds} \Big| \mathcal{F}_0 \right] \leq e^{c_1 + c_1 \|f\|_{\tilde{L}_q^p(t)}^{c_1}}, \quad f \in \tilde{L}_q^p(t), t \in [0, T]. \quad (3.6.19)$$

By $(A^{3.5})$,

$$\xi_t := \sigma_t(\bar{X}_t, \Phi_t^\gamma \mu)^* \{\sigma_t(\bar{X}_t, \Phi_t^\gamma \mu) \sigma_t(\bar{X}_t, \Phi_t^\gamma \mu)^*\}^{-1} b_t^{\mu,0}(\bar{X}_t)$$

satisfies

$$|\xi_t| \leq c_2 f_0(t, \bar{X}_t) + c_2 \|\mu_t\|_k, \quad t \in [0, T]$$

for some constant $c_2 > 0$. Combining this with (3.6.19), we conclude that

$$R_t := e^{\int_0^t \langle \xi_s, dW_s \rangle - \frac{1}{2} \int_0^t |\xi_s|^2 ds}, \quad t \in [0, T]$$

is a martingale satisfying

$$\mathbb{E}[R_t^2|\mathcal{F}_0] \leq e^{c_3+c_3\int_0^t \|\mu_s\|_k^2 ds}, \quad t \in [0,T] \tag{3.6.20}$$

for some constant $c_3 > 0$. By Girsanov's theorem

$$\tilde{W}_t := W_t - \int_0^t \xi_s ds, \quad t \in [0,T]$$

is m-dimensional Brownian motion under the probability $\mathbb{Q}_T := R_T \mathbb{P}$. Since $b^\mu = b^{(1)} + b^{\mu,0}$, we may reformulate the SDE for $\bar{X}_t$ as

$$d\bar{X}_t = b_t^\mu(\bar{X}_t)dt + \sigma_t(\bar{X}_t, \Phi_t^\gamma \mu)d\tilde{W}_t, \quad \bar{X}_0 = X_0, t \in [0,T],$$

so that the weak uniqueness of (3.6.4) yields $\mathcal{L}_{\bar{X}|\mathbb{Q}_T} = \mathcal{L}_{X^\mu}$. Combining this with (3.6.19) and (3.6.20), we obtain

$$\mathbb{E}\big[e^{\int_0^t f(s,X_s^\mu)^2 ds}\big|\mathcal{F}_0\big] = \mathbb{E}\big[R_t e^{\int_0^t f(s,\bar{X}_s)^2 ds}\big|\mathcal{F}_0\big]$$
$$\leq \big(\mathbb{E}[|R_t|^2|\mathcal{F}_0]\big)^{\frac{1}{2}} \big(\mathbb{E}[e^{\int_0^t f(s,\bar{X}_s)^2 ds}|\mathcal{F}_0]\big)^{\frac{1}{2}} \leq e^{c_4+c_4\int_0^t \|\mu_s\|_k^2 ds + c_4\|f\|_{\tilde{L}_q^p(t)}^{c_1}}$$

for some constant $c_4 > 0$. This implies (3.6.17) for some constant $c_0 > 1$.

By choosing large enough constant $C_j > 0$ such that $h(r) := \{\log(C_j + r)\}^j$ is concave for $r \geq 0$, using Jensen's inequality and (3.6.17), we find a constant $\tilde{C}_j > 1$ increasing in $j \geq 1$ such that

$$\mathbb{E}\bigg[\bigg(\int_0^t |f_s(X_s^\mu)|^2 ds\bigg)^j\bigg|\mathcal{F}_0\bigg] \leq \mathbb{E}\bigg(\bigg[\log\big(C_j + e^{\int_0^t f_s(X_s^\mu)^2 ds}\big)\bigg]^j\bigg|\mathcal{F}_0\bigg)$$
$$\leq \bigg[\log\big(C_j + \mathbb{E}[e^{\int_0^t f_s(X_s^\mu)^2 ds}]|\mathcal{F}_0\big)\bigg]^j \leq \tilde{C}_j \bigg(1 + \int_0^t \|\mu_s\|_k^2 ds + \|f\|_{\tilde{L}_q^p(t)}^{c_1}\bigg)^j.$$

Using $\frac{f}{\|f\|_{\tilde{L}_q^p(t)}}$ replacing f, we derive

$$\mathbb{E}\bigg[\bigg(\int_0^t |f_s(X_s^\mu)|^2 ds\bigg)^j\bigg|\mathcal{F}_0\bigg] \leq \|f\|_{\tilde{L}_q^p(t)}^{2j} \tilde{C}_j \bigg(1 + \int_0^t \|\mu_s\|_k^2 ds + 1\bigg)^j$$

which implies (3.6.18). □

Lemma 3.6.4. *Assume* $(A^{3.5})$.

(1) *There exists a constant $N_0 > 0$ such that $\Phi^\gamma \mathcal{C}_k^{\gamma,N} \subset \mathcal{C}_k^{\gamma,N}$ holds for $N \geq N_0$.*
(2) *There exists $c : [k,\infty) \to (0,\infty)$ such that (3.6.1) holds for any solution X_t to (3.1.1).*

Proof. (1) Simply denote $M_t = \int_0^t \sigma_s(X_s^\mu, \mathcal{L}_{X_s^\mu}) \mathrm{d}W_s$. Since $\|\sigma\|_\infty < \infty$ due to $(A^{3.5})$, we have
$$\sup_{t\in[0,T]} \mathbb{E}[|M_t|^k] < \infty.$$
Combining this with Lemma 3.6.3 below, we find some constants $c_0, c_1 > 0$ such that
$$\mathbb{E}(1 + |X_t^\mu|^k)$$
$$\leq \mathbb{E}(1 + |X_0|^k) + c_0 \mathbb{E}\left|\int_0^t (K_0\|\mu_s\|_k + f_0(s, X_s^\mu) + |X_s^\mu| + 1)\mathrm{d}s\right|^k + \mathbb{E}|M_t|^k$$
$$\leq c_1 + c_1 \left|\int_0^t \|\mu_s\|_k^2 \mathrm{d}s\right|^{k/2} + c_1 \int_0^t \mathbb{E}(1 + |X_s^\mu|^k)\mathrm{d}s, \quad t \in [0, T].$$
By Gronwall's inequality and (3.6.8), we find constants $c_2, c_3 > 0$ such that
$$\mathbb{E}(1 + |X_t^\mu|^k) \leq c_2 + c_2 \left|\int_0^t \mathrm{e}^{-\frac{2N}{k}s} \|\mu_s\|_k^2 \mathrm{e}^{\frac{2N}{k}s} \mathrm{d}s\right|^{k/2}$$
$$\leq c_3 + c_3 N^{1-k/2} \mathrm{e}^{Nt}, \quad \mu \in \mathcal{C}_k^{\gamma, N}, t \in [0, T].$$
Therefore, we find a constant $N_0 > 0$ such that
$$\sup_{t\in[0,T]} (1 + \|\Phi_t^\gamma \mu\|_k^k)\mathrm{e}^{-Nt} \leq c_3 + c_3 N^{1-k/2} \leq N, \quad N \geq N_0, \mu \in \mathcal{C}_k^{\gamma, N}.$$
That is, $\Phi^\gamma \mathcal{C}_k^{\gamma, N} \subset \mathcal{C}_k^{\gamma, N}$ for $N \geq N_0$.

(2) Let X_t solve (3.1.1) with $\gamma := \mathcal{L}_{X_0} \in \mathcal{P}_k$, and denote $\mu_t := \mathcal{L}_{X_t}$. Then $X_t = X_t^\mu$. By $(A^{3.5})$ and Itô's formula, for any $j \geq 1$ we find a constant $c_1 > 0$ such that
$$|X_t|^{2j} - |X_0|^{2j}$$
$$\leq c_1 \int_0^t \{1 + |X_s|^{2j} + |X_s|^{2j-1} f_0(s, X_s) + \|\mu_s\|_k^{2j}\} \mathrm{d}s + M_t \tag{3.6.21}$$
holds for some martingale M_t with $\mathrm{d}\langle M \rangle_t \leq c_1^2 |X_t|^{2(2j-1)} \mathrm{d}t$. Noting that
$$c_1 \int_0^t |X_s|^{2j-1} f_0(s, X_s) \mathrm{d}s \leq c_1 \left(\sup_{s\in[0,t]} |X_s|^{2j-1}\right) \int_0^t f_0(s, X_s) \mathrm{d}s$$
$$\leq \frac{1}{2} \sup_{s\in[0,t]} |X_s|^{2j} + c_2 \left(\int_0^t f_0(s, X_s) \mathrm{d}s\right)^{2j}$$
holds for some constant $c_2 > 0$, we see that $\eta_t := \sup_{s\in[0,t]} |X_s|^{2j}$ satisfies
$$\eta_t \leq 2|X_0|^{2j} + 2c_1 \int_0^t \{1 + \eta_s + \|\mu_s\|_k^{2j}\} \mathrm{d}s$$
$$+ 2c_2 \left(\int_0^t f_0(s, X_s) \mathrm{d}s\right)^{2j} + 2 \sup_{s\in[0,t]} M_s. \tag{3.6.22}$$

By $\mathrm{d}\langle M\rangle_t \leq c_1^2 |X_t|^{2(2j-1)} \mathrm{d}t$ and BDG inequality in Lemma 1.3.5, we find constants $c_3, c_4 > 0$ such that

$$\mathbb{E}\Big(\sup_{s\in[0,t]} M_s \Big| \mathcal{F}_0\Big) \leq c_3 \mathbb{E}\bigg[\bigg(\int_0^t |X_s|^{2(2j-1)} \mathrm{d}s\bigg)^{\frac{1}{2}} \Big| \mathcal{F}_0\bigg]$$

$$\leq \frac{1}{4}\mathbb{E}\big(\eta_t\big|\mathcal{F}_0\big) + c_4 \int_0^t \big\{1 + \mathbb{E}(\eta_s|\mathcal{F}_0)\big\} \mathrm{d}s.$$

Combining this with (3.6.22) and (3.6.18), we find a constant $c_5 > 0$ such that

$$\mathbb{E}\big(\eta_t\big|\mathcal{F}_0\big) \leq c_5 + c_5 |X_0|^{2j} + c_5 \int_0^t \big\{\mathbb{E}(\eta_s|\mathcal{F}_0) + \|\mu_s\|_k^{2j}\big\} \mathrm{d}s \qquad (3.6.23)$$

holds for $t \in [0,T]$. By Gronwall's inequality, there exists a constant $c_6 > 0$ such that

$$\mathbb{E}\big(\eta_t\big|\mathcal{F}_0\big) \leq c_6 + c_6 |X_0|^{2j} + c_6 \int_0^t \|\mu_s\|_k^{2j} \mathrm{d}s, \quad t \in [0,T]. \qquad (3.6.24)$$

In particular, choosing $j = k$ and applying Jensen's inequality, we derive

$$\mathbb{E}\bigg[\sup_{s\in[0,t]} |X_s|^k \Big| \mathcal{F}_0\bigg] \leq \big\{\mathbb{E}(\eta_t|\mathcal{F}_0)\big\}^{\frac{1}{2}}$$

$$\leq \sqrt{c_6}\big(1 + |X_0|^k\big) + \frac{c_6}{2} \int_0^t \|\mu_s\|_k^k \mathrm{d}s + \frac{1}{2} \sup_{s\in[0,t]} \|\mu_s\|_k^k.$$

Noting that $\|\mu_s\|_k^k = \mathbb{E}[|X_s|^k]$, by taking expectation we obtain

$$\|\mu_t\|_k^k \leq \mathbb{E}\bigg[\sup_{s\in[0,t]} |X_s|^k\bigg] \leq 2\sqrt{c_6}\big(1 + \mathbb{E}[|X_0|^k]\big) + c_6 \int_0^t \|\mu_s\|_k^k, \quad t \in [0,T].$$

By Gronwall's inequality, we find a constant $c > 0$ such that

$$\|\mu_t\|_k^k \leq c(1 + \mathbb{E}[|X_0|^k]), \quad t \in [0,T].$$

Substituting into (3.6.24) we derive (3.6.1). $\square$

3.6.2 Proof of Theorem 3.6.1

(1) Since (3.6.1) is included in Lemma 3.6.4, it remains to prove that Φ^γ has a unique fixed point in $\mathcal{C}_k^{\gamma,N}$ for $N > N_0$.

Taking large enough θ such that $\beta(\theta) < 1$, by (3.6.9) we prove the contraction of Φ^γ on the complete metric space $(\mathcal{C}_k^{\gamma,N}, \mathbb{W}_{k,\theta})$, so that Φ^γ has a unique fixed point in $\mathcal{C}_k^{\gamma,N}$.

(2) Let $N > 0$. For any two solutions X_t^i of (3.1.1) with $\mathbb{E}[|X_0^i|^k] \leq N$, they solve (3.6.4) for $\mu_t^i = \nu_t^i = \mathcal{L}_{X_t^i}, i = 1, 2$. By (3.6.1), there exists a constant $K_N > 0$ depending on N such that $\mu, \nu \in \mathcal{C}_k^{\gamma, K_N}$. By (3.6.16) for large θ such that $\sqrt{c_3 \varepsilon_\theta} \leq \frac{1}{4}$, where θ and c_3 depend on N, we obtain

$$\mathbb{W}_{k,\theta}(\mu_t^1, \mu_t^2)^k \leq 2\sqrt{c_3}\mathbb{E}[|X_0^1 - X_0^2|^k].$$

Substituting into (3.6.15) yields the estimate (3.6.2) for some constant $C_{j,N} > 0$. When $K_0 = 0$ we have $|b^{\mu,0}| \leq f_0$ for any $\mu \in C_b^w([0,T]; \mathcal{P}_k)$, so that all the above constants are uniformly bounded in N, hence (3.6.2) holds for some constant $C_{j,N} = C_j$ independent of N.

Finally, by taking $j = k$ and X_0^1, X_0^2 such that

$$\mathcal{L}_{X_0^1} = \mu^1, \quad \mathcal{L}_{X_0^2} = \mu^2, \quad \mathbb{E}[|X_0^1 - X_0^2|^k] = \mathbb{W}_k(\mu^1, \mu^2)^k,$$

we deduce (3.6.3) from (3.6.2).

3.7 Singular density dependent SDEs

In this section, we study the following density dependent SDE on $\mathbb{R}^d$:

$$dX_t = b_t(X_t, \ell_{X_t}(X_t), \ell_{X_t})dt + \sigma_t(X_t, \ell_{X_t})dW_t, \quad t \in [0,T], \quad (3.7.1)$$

where ℓ_ξ is the distribution density function of an absolutely continuous random variable ξ on $\mathbb{R}^d$,

$$b : [0,T] \times \mathbb{R}^d \times [0,\infty) \times \mathcal{D}_+^1 \to \mathbb{R}^d, \quad \sigma : [0,T] \times \mathbb{R}^d \times \tilde{L}^1 \to \mathbb{R}^d \otimes \mathbb{R}^m$$

are measurable, and

$$\mathcal{D}_+^1 := \left\{ f \in L^1(\mathbb{R}^d) : f \geq 0, \int_{\mathbb{R}^d} f(x)dx \leq 1 \right\}$$

is a closed subspace of $L^1(\mathbb{R}^d)$. We take $\mathcal{D}_+^1$ as the set of sub-probability densities rather than probability densities, to ensure the completeness of the set in L^k and $\tilde{L}^k$ for $k > 1$, which is crucial in the proof of well-posedness.

A continuous adapted process $(X_t)_{t \in [0,T]}$ on $\mathbb{R}^d$ is called a (strong) solution of (3.7.1), if

$$\int_0^T \mathbb{E}\big[|b_s(X_s, \ell_{X_s}(X_s), \ell_{X_t})| + \|\sigma_s(X_s, \ell_{X_t})\|^2\big]ds < \infty$$

and $\mathbb{P}$-a.s.

$$X_t = X_0 + \int_0^t b_s(X_s, \ell_{X_s}(X_s), \ell_{X_t})ds + \int_0^t \sigma_s(X_s, \ell_{X_t})dW_s, \quad t \in [0,T].$$

A pair $(X_t, W_t)_{t\in[0,T]}$ is called a weak solution of (3.7.1), if $(W_t)_{t\in[0,T]}$ is an m-dimensional Brownian under a complete filtration probability space $(\Omega, \{\mathcal{F}_t\}_{t\in[0,T]}, \mathbb{P})$ such that $(X_t)_{t\in[0,T]}$ solves (3.7.1). We identify any two weak solutions $(X_t, W_t)_{t\in[0,T]}$ and $(\bar{X}_t, \bar{W}_t)_{t\in[0,T]}$ if $(X_t, l_t)_{t\in[0,T]}$ and $(\bar{X}_t, \bar{l}_t)_{t\in[0,T]}$ have the same distribution under the corresponding probability spaces.

When $m = d$, $\sigma = I_d$ (the $d \times d$ identity matrix), and $b_t(x, r, \rho) = b_t(x, r)$ does not depend on ρ, the weak solutions are studied in [Hao et al. (2021a); Issoglio and Russo (2023)]. In [Hao et al. (2021a)], the initial distribution is not necessarily absolutely continuous, where the weak existence is proved for $b_t(x, r)$ bounded and continuous in (t, r) locally uniformly in x, and the weak and strong uniqueness holds when $b_t(x, r)$ is furthermore Lipschitz continuous in r uniformly in (t, x). In [Issoglio and Russo (2023)], the initial density is in $C^{\beta+} := \cup_{p>\beta} C^p$ for some $\beta \in (0, \frac{1}{2})$, the weak well-posedness is proved for $b_t(x, r) := F(r)\tilde{b}_t(x)$, where $\tilde{b} \in C_b^w([0,T]; C^{-\beta})$ and F is bounded and Lipschitz continuous such that $rF(r)$ is Lipschitz continuous in $r \geq 0$. See [Izydorczyk et al. (2019)] and references within for the case with better drift.

In (3.7.1) the noise does not point-wisely depend on the density. It seems that to solve SDEs with point-wisely density dependent noise, one needs stronger regularity for the initial density and the coefficients. For instance, [Jourdain and Méléard (1998)] proved the well-posedness and studied the propagation of chaos for the following SDE with point-wisely density dependent noise:

$$dX_t = b(\ell_{X_t}(X_t))dt + \sigma(\ell_{X_t}(X_t))dW_t,$$

where the initial distribution density is C^{2+}-smooth, b is C^2-smooth, and σ is uniformly elliptic and C^3-smooth.

For $k > 1$ and a signed measure μ with density function $\ell_\mu(x) := \frac{\mu(dx)}{dx}$, let

$$\|\mu\|_{L^k} := \|\ell_\mu\|_{L^k}, \quad \|\mu\|_{\tilde{L}^k} := \|\ell_\mu\|_{\tilde{L}^k}.$$

When $k = 1$, we define

$$\|\mu\|_{L^1} := \sup_{\|f\|_\infty \leq 1} |\mu(f)|, \quad \|\mu\|_{\tilde{L}^1} := \sup_{z \in \mathbb{R}^d} \sup_{\|f\|_\infty \leq 1} |\mu(1_{B(z,1)}f)|,$$

where $\mu(f) := \int_{\mathbb{R}^d} f d\mu$. Note that $\|\cdot\|_{L^1}$ is the total variation norm.

We will solve (3.7.1) with initial distributions in the classes

$$\mathcal{P}^k := \left\{\nu \in \mathcal{P} : \|\nu\|_{L^k} < \infty\right\}, \quad \tilde{\mathcal{P}}^k := \left\{\nu \in \mathcal{P} : \|\nu\|_{\tilde{L}^k} < \infty\right\}, \quad k \in [1, \infty],$$

which are complete metric spaces under distances $\|\nu_1 - \nu_2\|_{L^k}$ and $\|\nu_1 - \nu_2\|_{\tilde{L}^k}$ respectively. The main results in this part come from [Wang (2023e)].

3.7.1 Density free noise

$(A^{3.6})$ $a_t(x) := (\sigma_t \sigma_t^*)(x)$ and $b_t(x, r, \rho) = b_t^{(1)}(x) + b_t^{(0)}(x, r, \rho)$ satisfy the following conditions for some $k \in [1, \infty]$.

(1) $a_t(x)$ is invertible with $\|a\|_\infty + \|a^{-1}\|_\infty < \infty$, and there exist constants $\alpha \in (0, 1)$ and $C > 0$ such that

$$\sup_{t \in [0,T]} \|a_t(x) - a_t(y)\| \leq C|x - y|^\alpha, \quad x, y \in \mathbb{R}^d.$$

(2) There exist $(p_0, q_0) \in \mathcal{K}$, $\theta > \frac{2}{q_0} + \frac{d}{p_0} - 1$, and $1 \leq f_0 \in \tilde{L}_{q_0}^{p_0}(T)$ such that

$$|b_t^{(0)}(x, r, \rho) - b_t^{(0)}(x, \tilde{r}, \tilde{\rho})| \leq f_0(t, x) t^\theta \big(|r - \tilde{r}| + \|\rho - \tilde{\rho}\|_{\tilde{L}^k}\big),$$

$|b_t^{(0)}(x, r, \rho)| \leq f_0(t, x), (t, x) \in (0, T] \times \mathbb{R}^d, r, \tilde{r} \in [0, \infty), \rho, \tilde{\rho} \in \tilde{L}^k \cap \mathcal{D}_+^1$.

(3) $b_t^{(1)}(0)$ is bounded in $t \in [0, T]$ and

$$\|\nabla b^{(1)}\|_\infty := \sup_{t \in [0,T]} \sup_{x \neq y} \frac{|b_t^{(1)}(x) - b_t^{(1)}(y)|}{|x - y|} < \infty.$$

To ensure $\ell_{X_t} \in L^k$ for $\ell_{X_0} \in L^k$, we replace $(A^{3.6})(2)$ by the following condition.

($2'$) There exist a constant $C > 0, (p_0, q_0) \in \mathcal{K}$, $\theta > \frac{2}{q_0} + \frac{d}{p_0} - 1$, and $0 \leq f_0 \in L_{q_0}^{p_0}(T)$ such that

$$|b_t^{(0)}(x, r, \rho) - b_t^{(0)}(x, s, \tilde{\rho})| \leq \big(C + f_0(t, x)\big)\big(|r - s| + \|\rho - \tilde{\rho}\|_{L^k}\big),$$

$|b_t^{(0)}(x, r, \rho)| \leq f_0(t, x), (t, x) \in (0, T] \times \mathbb{R}^d, r, s \in [0, \infty), \rho, \tilde{\rho} \in L^k \cap \mathcal{D}_+^1$.

Under the above assumptions, the following result ensures the well-posedness of (3.7.1) for initial distributions in $\tilde{\mathcal{P}}^k$ or $\mathcal{P}^k$ for

$$k \in \Big[\frac{p_0}{p_0 - 1}, \infty\Big] \cap (k_0, \infty], \quad k_0 := \frac{d}{2\theta + 1 - 2q_0^{-1} - dp_0^{-1}}.$$

This explains the role played by the quantity θ in $(A^{3.6})(2)(2')$: for bigger θ, (2) and ($2'$) provide stronger conditions on $|b_t^{(0)}(x, r, \rho) - b_t^{(0)}(x, \tilde{r}, \tilde{\rho})|$ for small t, so that the SDE is solvable for initial distributions in larger classes $\mathcal{P}^k$ and $\tilde{\mathcal{P}}^k$. In particular, when $p_0 = \infty$ and θ is large enough such that $k_0 < 1$, we may take $k = 1$ so that the SDE is well-posed for any initial distribution $\nu \in \mathcal{P}$.

Theorem 3.7.1. Let $k \in [\frac{p_0}{p_0 - 1}, \infty]$ with $k > k_0 := \frac{d}{2\theta + 1 - 2q_0^{-1} - dp_0^{-1}}$.

(1) Under $(A^{3.6})$, for any $\gamma \in \tilde{\mathcal{P}}^k$, (3.7.1) has a unique weak solution with $\mathcal{L}_{X_0} = \gamma$ satisfying $\ell_{X.} \in \tilde{L}_\infty^k(T)$, and there exists an increasing function $\Lambda : [0, \infty) \to (0, \infty)$ such that for any two weak solutions $\{X_t^i\}_{i=1,2}$ of (3.7.1) with $\ell_{X^i} \in \tilde{L}_\infty^k$,
$$\sup_{t \in [0,T]} \|\ell_{X_t^1} - \ell_{X_t^2}\|_{\tilde{L}^k} \le \Lambda\big(\|\mathcal{L}_{X_0^1}\|_{\tilde{L}^k} \wedge \|\mathcal{L}_{X_0^1}\|_{\tilde{L}^k}\big)\|\mathcal{L}_{X_0^1} - \mathcal{L}_{X_0^2}\|_{\tilde{L}^k}. \quad (3.7.2)$$
If moreover σ_t is weakly differentiable with
$$\|\nabla \sigma\| \le \sum_{i=1}^l f_i \text{ for some } l \in \mathbb{N},$$
$$0 \le f_i \in \tilde{L}_{q_i}^{p_i}(T), (p_i, q_i) \in \mathcal{K}, 1 \le i \le l, \quad (3.7.3)$$
then for any X_0 with $\mathcal{L}_{X_0} \in \tilde{L}^k$, (3.7.1) has a unique strong solution with $\ell_{X.} \in \tilde{L}_\infty^k(T)$.

(2) Under $(A^{3.6})$ with $(2')$ replacing $(A^{3.6})(2)$, assertions in (1) hold for $(\mathcal{P}^k, L_\infty^k, L^k)$ replacing $(\tilde{\mathcal{P}}^k, \tilde{L}_\infty^k(T), \tilde{L}^k)$.

For fixed $k \ge 1$ and $\gamma \in \tilde{\mathcal{P}}^k$, let $\tilde{\mathcal{P}}_{\gamma,T}^k$ be the set of all bounded measurable maps
$$\rho : (0, T] \to \tilde{L}^k \cap \mathcal{D}_+^1, \quad \rho_0 = \gamma.$$
When $k = 1$, the initial value γ may be singular, and if it is absolutely continuous we regard it as its density function.

Then $\tilde{\mathcal{P}}_{\gamma,T}^k$ is complete under the metric
$$\tilde{d}_{k,\lambda}(\gamma^1, \gamma^2) := \sup_{t \in [0,T]} e^{-\lambda t} \|\rho_t^1 - \rho_t^2\|_{\tilde{L}^k}, \quad \rho^1, \rho^2 \in \tilde{\mathcal{P}}_{\gamma,T}^k$$
for $\lambda > 0$. We define $(\mathcal{P}_{\gamma,T}^k, d_{k,\lambda})$ in the same way with $(L^k, \mathcal{P}^k)$ replacing $(\tilde{L}^k, \tilde{\mathcal{P}}^k)$.

To prove the well-posedness of (3.7.1), we will use the fixed point theorem for the map induced by the SDE for $\rho \in \tilde{\mathcal{P}}_{\gamma,T}^k$ replacing $\ell_{X.}$. For any $\rho \in \tilde{\mathcal{P}}_{\gamma,T}^k$, let
$$b_t^\rho(x) := b_t(x, \rho_t(x), \rho_t), \quad \sigma_t^\rho(x) := \sigma_t(x, \rho_t), \quad t \in [0, T], x \in \mathbb{R}^d.$$
Then for $\gamma := \mathcal{L}_{X_0} \in \tilde{L}^k$, (3.7.1) has a unique (weak or strong) solution with $\ell_{X.} \in \tilde{L}_\infty^k$ if we could verify the following two things:

(1) For any $\rho \in \tilde{\mathcal{P}}_{\gamma,T}^k$, the SDE
$$dX_t^\rho = b_t^\rho(X_t^\rho)dt + \sigma_t^\rho(X_t^\rho)dW_t, \quad t \in [0,T], \ X_0^\rho = X_0 \quad (3.7.4)$$
is (weakly or strongly) well-posed, and
$$\rho \mapsto \Phi_t^\gamma \rho := \mathcal{L}_{X_t^\rho}, \quad t \in (0, T]$$
provides a map $\Phi^\gamma : \tilde{\mathcal{P}}_{\gamma,T}^k \to \tilde{\mathcal{P}}_{\gamma,T}^k$.

(2) Φ^γ has a unique fixed point $\bar\rho$ in $\tilde{\mathcal{P}}^k_{\gamma,T}$. Indeed, from these we see that $X_t := X_t^{\bar\rho}$ is the unique (weak or strong) solution of (3.7.1) with $\mathcal{L}_{X.} \in \tilde{L}^k_\infty(T)$.

To verify (1) and (2), we recall some heat kernel upper bounds of [Menozzi et al. (2021)], and estimate the $\tilde{L}^p_q$-$\tilde{L}^{p'}_q$ norm for time inhomogeneous semigroups.

We consider heat kernel estimates for the time dependent second order differential operator

$$L^{a,b}_t := \frac{1}{2}\mathrm{tr}\{a_t \nabla^2\} + \nabla_{b_t},$$

where

$$a: [0,T] \times \mathbb{R}^d \to \mathbb{R}^d \otimes \mathbb{R}^d, \quad b: [0,T] \times \mathbb{R}^d \to \mathbb{R}^d$$

satisfy the following conditions.

($H^{a,b}$) $a_t(x)$ is invertible and there exist constants $C > 0$ and $\alpha \in (0,1)$ such that

$$\|b.(0)\|_\infty + \|a\|_\infty + \|a^{-1}\|_\infty \leq C,$$
$$\sup_{t \in [0,T]} \|a_t(x) - a_t(y)\| \leq C|x-y|^\alpha,$$
$$\sup_{t \in [0,T]} |b_t(x) - b_t(y)| \leq C(|x-y| + |x-y|^\alpha), \quad x,y \in \mathbb{R}^d.$$

(H^a) $a_t(x)$ is differentiable in x, and there exist constants $C \in (0,\infty)$ and $\alpha \in (0,1)$ such that

$$\|\nabla a\|_\infty \leq C, \quad \sup_{t \in [0,T]} \|\nabla a_t(x) - \nabla a_t(y)\| \leq C|x-y|^\alpha, \quad x,y \in \mathbb{R}^d.$$

Under ($H^{a,b}$), for any $s \in [0,T)$, the SDE

$$\mathrm{d}X^x_{s,t} = b_s(X^x_{s,t})\mathrm{d}s + \sqrt{a_s}(X^x_{s,t})\mathrm{d}W_s, \quad t \in [s,T], X^x_{s,s} = x \in \mathbb{R}^d$$

is weakly well-posed with semigroup $\{P^{a,b}_{s,t}\}_{0 \leq s < t \leq T}$ and transition density $\{p^{a,b}_{s,t}\}_{0 \leq s < t \leq T}$ given by

$$P^{a,b}_{s,t}f(x) = \int_{\mathbb{R}^d} p^{a,b}_{s,t}(x,y)f(y)\mathrm{d}y = \mathbb{E}[f(X^x_{s,t})], \quad f \in \mathcal{B}_b(\mathbb{R}^d),$$

and we have the following Kolmogorov backward equation (see Remark 2.2 in [Menozzi et al. (2021)])

$$\partial_s P^{a,b}_{s,t}f = -L_s P^{a,b}_{s,t}f, \quad f \in C^\infty_b(\mathbb{R}^d), s \in [0,t], t \in (0,T]. \tag{3.7.5}$$

Next, we denote $\psi_{s,t} = \theta^{(1)}_{t,s}$ presented in [Menozzi et al. (2021)]. Then $(\psi_{s,t})_{0 \le s \le t \le T}$ is a family of diffeomorphisms on $\mathbb{R}^d$ satisfying

$$\sup_{0 \le s \le t \le T} \left\{ \|\nabla \psi_{s,t}\|_\infty + \|\nabla \psi_{s,t}^{-1}\|_\infty \right\} \le \delta \tag{3.7.6}$$

for some constant $\delta > 0$ depending on α, C. For any $\kappa > 0$, consider the Gaussian heat kernel

$$p_t^\kappa(x) := (\kappa \pi t)^{-\frac{d}{2}} e^{-\frac{|x|^2}{\kappa t}}, \quad t > 0, \ x \in \mathbb{R}^d.$$

The following result is taken from [Menozzi et al. (2021)].

Theorem 3.7.2. *Assume* $(H^{a,b})$. *Then there exist constants* $c, \kappa > 0$ *depending on* C, α *such that*

$$\begin{aligned}|\nabla^i p_{s,t}^{a,b}(\cdot, y)(x)| &\le c(t-s)^{-\frac{i}{2}} p_{t-s}^\kappa(\psi_{s,t}(x) - y), \\ i &= 0,1,2, \ 0 \le s < t \le T, \ x, y \in \mathbb{R}^d.\end{aligned} \tag{3.7.7}$$

If moreover (H^a) *holds, then*

$$\begin{aligned}|\nabla p_{s,t}^{a,b}(x, \cdot)(y)| &\le c(t-s)^{-\frac{1}{2}} p_{t-s}^\kappa(\psi_{s,t}(x) - y), \\ 0 &\le s < t \le T, x, y \in \mathbb{R}^d,\end{aligned} \tag{3.7.8}$$

and for any $\beta \in (0,1)$ *there exists a constant* $c' > 0$ *depending on* C, α, β *such that*

$$\begin{aligned}&|\nabla p_{s,t}^{a,b}(\cdot, y)(x) - \nabla p_{s,t}^{a,b}(\cdot, y')(x)| + |\nabla p_{s,t}^{a,b}(x, \cdot)(y) - \nabla p_{s,t}^{a,b}(x, \cdot)(y')| \\ &\le c'|y - y'|^\beta (t-s)^{-\frac{1+\beta}{2}} \left\{ p_{t-s}^\kappa(\psi_{s,t}(x) - y) + p_{t-s}^\kappa(\psi_{s,t}(x) - y') \right\}, \\ &0 \le s < t \le T, \ x, x', y \in \mathbb{R}^d.\end{aligned} \tag{3.7.9}$$

For any $f \in \mathcal{B}_b(\mathbb{R}^d)$, $0 \le s < t \le T$ and $x \in \mathbb{R}^d$, let

$$\begin{aligned}P_t^\kappa f(x) &:= \int_{\mathbb{R}^d} p_t^\kappa(x-y) f(y) \mathrm{d}y, \\ \hat{P}_{s,t}^\kappa f(x) &:= \int_{\mathbb{R}^d} p_{t-s}^\kappa(\psi_{s,t}(x) - y) f(y) \mathrm{d}y, \\ \tilde{P}_{s,t}^\kappa f(x) &:= \int_{\mathbb{R}^d} p_{t-s}^\kappa(\psi_{s,t}(y) - x) f(y) \mathrm{d}y.\end{aligned} \tag{3.7.10}$$

It is well known that for some constant $c > 0$,

$$\begin{aligned}\|P_t^\kappa\|_{L^p \to L^{p'}} &:= \sup_{\|f\|_p \le 1} \|P_t^\kappa f\|_{L^{p'}} \le c t^{-\frac{d(p'-p)}{2pp'}}, \\ t &> 0, 1 \le p \le p' \le \infty.\end{aligned} \tag{3.7.11}$$

Combining this with (3.7.6) we obtain
$$\|\hat{P}^\kappa_{s,t}\|_{L^p \to L^{p'}} + \|\tilde{P}^\kappa_{s,t}\|_{L^p \to L^{p'}} \le c(t-s)^{-\frac{d(p'-p)}{2pp'}}, \quad (3.7.12)$$
$$0 \le s < t \le T, 1 \le p \le p' \le \infty$$
for some different constant $c > 0$. Below we extend this estimate to the $\tilde{L}^p_q \tilde{L}^{p'}_q$ norm.

Lemma 3.7.3. *There exists a constant $c > 0$ such that for any $0 \le s < t \le T$, $1 \le p \le p' \le \infty$ and $q \in [1, \infty]$,*
$$\|\hat{P}^\kappa_{\cdot,t}f\|_{\tilde{L}^{p'}_q(t)} + \|\tilde{P}^\kappa_{\cdot,t}f\|_{\tilde{L}^{p'}_q(t)} \le c\|(t-\cdot)^{-\frac{d(p'-p)}{2pp'}}f\|_{\tilde{L}^p_q(t)}, \quad f \in \mathcal{B}_b([0,T] \times \mathbb{R}^d), \quad (3.7.13)$$
where and in the sequel, $(t-\cdot)(s) := t-s$ is a function on $[0,t]$, and
$$\sup_{z \in \mathbb{R}^d} \|g\hat{P}^\kappa_{s,t}(1_{B(z,1)}f)\|_{L^1}$$
$$\le c(t-s)^{-\frac{d(p'-p)}{2pp'}}\|g\|_{L^{\frac{p'}{p'-1}}}\|f\|_{\tilde{L}^p}, \quad f,g \in \mathcal{B}_b(\mathbb{R}^d). \quad (3.7.14)$$

Proof. Let $\mathcal{B}_n := \{v \subset \mathbb{Z}^d : |v|_1 := \sum_{i=1}^d |v_i| = n\}, n \ge 0$. By (3.7.6), we find a constant $c_1 > 1$ such that $|\psi_{s,t}(x)-y|^2 \ge (c_1^{-1}n^2-c_1)^+, x \in B(\psi_{s,t}^{-1}(z),1), y \in \cup_{v \in \mathcal{B}_n} B(z+v,d), z \in \mathbb{R}^d$. Combining this with (3.7.11), we find constants $c_2, c_3, c_4 > 0$ such that for any $z \in \mathbb{R}^d$, $0 \le s < t \le T$, and $f,g \in \mathcal{B}_b^+(\mathbb{R}^d)$,

$$\|1_{B(\psi_{s,t}^{-1}(z),1)}g\hat{P}^\kappa_{s,t}f\|_{L^1} \le \sum_{n=0}^\infty \sum_{v \in \mathbb{Z}^d: |v|_1 = n} \|1_{B(\psi_{s,t}^{-1}(z),1)}g\hat{P}^\kappa_{s,t}(1_{B(z+v,d)}f)\|_{L^1}$$

$$\le \sum_{n=0}^\infty \sum_{v \in \mathcal{B}_n} \int_{\mathbb{R}^d \times \mathbb{R}^d} |1_{B(\psi_{s,t}^{-1}(z),1)}g|(x)p^\kappa_{t-s}(\psi_{s,t}(x)-y)|1_{B(z+v,d)}f|(y) \mathrm{d}x\mathrm{d}y$$

$$\le c_2 \sum_{n=0}^\infty \sum_{v \in \mathcal{B}_n} e^{-\frac{n^2}{c_3(t-s)}} \int_{\mathbb{R}^d \times \mathbb{R}^d} |1_{B(\psi_{s,t}^{-1}(z),1)}g|(x)p^\kappa_{2(t-s)}(\psi_{s,t}(x)-y)$$
$$|1_{B(z+v,d)}f|(y)\mathrm{d}x\mathrm{d}y$$

$$\le c_3 \sum_{n=0}^\infty \sum_{v \in \mathcal{B}_n} e^{-\frac{n^2}{c_3(t-s)}} \|\{P^\kappa_{2(t-s)}(1_{B(\psi_{s,t}^{-1}(z),1)}g)\}1_{B(z+v,d)}f\|_{L^1}$$

$$\le c_3 \sum_{n=0}^\infty \sum_{v \in \mathcal{B}_n} e^{-\frac{n^2}{c_3(t-s)}} \|P^\kappa_{2(t-s)}(1_{B(\psi_{s,t}^{-1}(z),1)}g)\|_{L^{\frac{p}{p-1}}} \|1_{B(z+v,d)}f\|_{L^p}$$

$$\le c_4(t-s)^{-\frac{d(p'-p)}{2pp'}} \|g\|_{L^{\frac{p'}{p'-1}}} \sum_{n=0}^\infty \sum_{v \in \mathcal{B}_n} e^{-\frac{n^2}{c_3(t-s)}} \|1_{B(z+v,d)}f\|_{L^p}.$$

Since
$$\sup_{z\in\mathbb{R}^d}\left(\int_0^t \|1_{B(z,d)}f\|_{L^p}^q \mathrm{d}s\right)^{\frac{1}{q}} \le c_5 \|f\|_{\tilde{L}_q^p(t)} \tag{3.7.15}$$
holds for some constant $c_5 > 0$, we find a constant $c_6 > 0$ such that this and Hölder's inequality imply

$$\sup_{z\in\mathbb{R}^d}\left(\int_0^t \|1_{B(z,1)}\hat{P}_{s,t}^\kappa f_s\|_{L^{p'}}^q \mathrm{d}s\right)^{\frac{1}{q}} = \sup_{z\in\mathbb{R}^d}\left(\int_0^t \|1_{B(\psi_{s,t}^{-1}(z),1)}\hat{P}_{s,t}^\kappa f_s\|_{L^{p'}}^q \mathrm{d}s\right)^{\frac{1}{q}}$$

$$\le \sup_{z\in\mathbb{R}^d}\left(\int_0^t \left\{c_4\|1_{B(z,d)}f_s\|_{L^p}(t-s)^{-\frac{d(p'-p)}{2pp'}}\right\}^q \mathrm{d}s\right)^{\frac{1}{q}} \sup_{r\in(0,T]}\sum_{n=0}^\infty\sum_{v\in\mathcal{B}_n} e^{-\frac{n^2}{c_3 r}}$$

$$\le c_6 \|(t-\cdot)^{-\frac{d(p'-p)}{2pp'}}f\|_{\tilde{L}_q^p(t)}\sum_{n=0}^\infty\sum_{v\in\mathcal{B}_n} e^{-\frac{n^2}{c_3 T}}.$$

This implies the upper bound for $\hat{P}^\kappa$ in (3.7.13), by noting that for some constant $K > 0$,
$$\sum_{n=0}^\infty\sum_{v\in\mathcal{B}_n} e^{-\frac{n^2}{c_3 T}} \le \sum_{n=0}^\infty K(1+n^{d-1})e^{-\frac{n^2}{c_3 T}} < \infty. \tag{3.7.16}$$

By (3.7.6) and integral transforms, the estimate on $\tilde{P}_{s,t}$ follows from that of $\hat{P}_{s,t}^\kappa$.

Similarly, we find a constant $K > 1$ such that

$$\|g\hat{P}_{s,t}^\kappa(1_{B(\psi_{s,t}(z),1)}f)\|_{L^1} \le \sum_{n=0}^\infty\sum_{v\in\mathbb{Z}^d:|v|_1=n} \|1_{B(z+v,d)}g\hat{P}_{s,t}^\kappa(1_{B(\psi_{s,t}(z),1)}f)\|_{L^1}$$

$$\le \sum_{n=0}^\infty\sum_{v\in\mathcal{B}_n}\int_{\mathbb{R}^d\times\mathbb{R}^d}|1_{B(z+v,d)}g|(x)p_{t-s}^\kappa(\psi_{s,t}(x)-y)|1_{B(\psi_{s,t}(z),1)}f|(y)\mathrm{d}x\mathrm{d}y$$

$$\le K(t-s)^{-\frac{d(p'-p)}{2pp'}}\sum_{n=0}^\infty\sum_{v\in\mathcal{B}_n} e^{-\frac{n^2}{K(t-s)}}\|g\|_{\tilde{L}^{\frac{p'}{p'-1}}}\|1_{B(z,1)}f\|_{L^p}.$$

This together with (3.7.15) and (3.7.16) implies (3.7.14) for some $c > 0$. $\square$

For $\rho \in \tilde{\mathcal{P}}_{\gamma,T}^k$, we denote
$$\sigma_t^\rho(x) := \sigma_t(x,\rho_t), \quad b_t^\rho(x) := b_t(x,\rho_t(x),\rho_t) = b_t^{(1)}(x) + b_t^{\rho,0}(x),$$
$$b_t^{\rho,0}(x) := b_t^{(0)}(x,\rho_t(x),\rho_t), \quad t\in[0,T], x\in\mathbb{R}^d.$$

Lemma 3.7.4. *Assume* $(A^{3.6})$ *with* $(A^{3.6})(1)$ *holding for* σ^ρ *replacing* σ *uniformly in* $\rho \in \tilde{L}^k \cap \mathcal{D}_+^1$, *where* $k \in [\frac{p_0}{p_0-1},\infty]$. *Then* (3.7.4) *is weakly* well-posed for any $\gamma \in \tilde{\mathcal{P}}^k$ and $\rho \in \tilde{\mathcal{P}}^k_{\gamma,T}$, and for any $\beta \in (0,1)$ there exists a constant $c > 1$ independent of γ and ρ such that $\Phi^\gamma_t \rho := \ell_{X^\rho_t}$ satisfies

$$\|\Phi^\gamma \rho\|_{\tilde{L}^k_\infty} \le c\|\gamma\|_{\tilde{L}^k}. \tag{3.7.17}$$

Moreover, under $(2')$ replacing $(A^{3.6})(2)$, this estimate holds for (L^k, L^k_∞) in place of $(\tilde{L}^k, \tilde{L}^k_\infty(T))$.

Proof. (a) By $(A^{3.6})(2)$,

$$\|b^{\rho,0}\|_{\tilde{L}^{p_0}_{q_0}(T)} \le \|f_0\|_{\tilde{L}^{p_0}_{q_0}(T)} < \infty, \quad \rho \in \tilde{\mathcal{P}}^k_{\gamma,T}. \tag{3.7.18}$$

According to Theorem 1.3.1, this together with $(A^{3.6})$ imply the well-posedness of (3.7.4). Moreover, by Theorem 6.2.7(ii)–(iii) in [Bogachev et al. (2015)], the distribution density function $\ell_{X^\rho_t}$ exists.

(b) To estimate $\Phi^\gamma_t \rho$ for $\rho \in \tilde{L}^k_\infty \cap \mathcal{D}^1_+$, consider the SDE

$$\begin{aligned}\mathrm{d}\bar{X}^\rho_s &= b^{(1)}_s(\bar{X}^\rho_s)\mathrm{d}s + \sigma^\rho_s(\bar{X}^\rho_s)\mathrm{d}W_s,\\ s &\in [0,t],\ \bar{X}^\rho_0 = X^\rho_0 = X_0 \text{ with } \ell_{X_0} = \gamma.\end{aligned} \tag{3.7.19}$$

Let $a^\rho := \sigma^\rho(\sigma^\rho)^*$. Then

$$\mathbb{E}[f(\bar{X}^\rho_t)] = \mathbb{E}[(P^{a^\rho,b^{(1)}}_{0,t} f)(X_0)] = \int_{\mathbb{R}^d \times \mathbb{R}^d} p^{a^\rho,b^{(1)}}_{0,t}(x,y) f(y)\gamma(\mathrm{d}x)\mathrm{d}y,\ f \in \mathcal{B}^+(\mathbb{R}^d),$$

and (3.7.7) holds for $p^{a^\rho,b^{(1)}}_{s,t}$ with constants $c, \kappa > 0$ uniformly in ρ. So, we find a constant $c_1 > 0$ such that

$$\mathbb{E}[f(\bar{X}^\rho_t)] \le c_1 \int_{\mathbb{R}^d} (\hat{P}^\kappa_{0,t} f)(x)\gamma(\mathrm{d}x) = c_1(\hat{P}^{\kappa*}_{0,t}\gamma)(f),\ f \in \mathcal{B}^+(\mathbb{R}^d), \tag{3.7.20}$$

where

$$(\hat{P}^{\kappa*}_{0,t}\gamma)(\mathrm{d}y) := \left(\int_{\mathbb{R}^d} \hat{p}^\kappa_{0,t}(x,y)\gamma(\mathrm{d}x)\right)\mathrm{d}y,\ t \in (0,T], \gamma \in \mathcal{P}. \tag{3.7.21}$$

On the other hand, let

$$R_t := e^{\int_0^t \langle \xi_s, \mathrm{d}W_s\rangle - \frac{1}{2}\int_0^t |\xi_s|^2 \mathrm{d}s},\ \xi_s := \{\sigma^\rho_s(\sigma_s(\sigma^\rho_s)^*)^{-1}b^{\rho,0}_s\}(\bar{X}_s).$$

By (3.7.18), the uniform boundedness of $\|\sigma^\rho(\sigma^\rho(\sigma^\rho)^*)^{-1}\|_\infty$, and Khasminskii's estimate in Theorem 1.2.4 and the Krylov's estimate (1.2.7), we find a map $K_\rho : [1,\infty) \to (0,\infty)$ such that

$$K_\rho(p) := (\mathbb{E}[R^p_t])^{\frac{1}{p}} < \infty,\ p \ge 1. \tag{3.7.22}$$

By Girsanov's theorem,

$$\tilde{W}_s := W_s - \int_0^s \xi_r \mathrm{d}r,\ s \in [0,t]$$

is an m-dimensional Brownian motion under the probability measure $\mathbb{Q}_t := R_t \mathbb{P}$, with which the SDE (3.7.19) reduces to
$$d\bar{X}_s = b_s^\rho(\bar{X}_s)ds + \sigma_s^\rho(\bar{X}_s)d\tilde{W}_s, \quad s \in [0,t], \bar{X}_0 = X_0^\rho.$$
By the weak uniqueness, the law of X_t^ρ under $\mathbb{P}$ coincides with that of $\bar{X}_t$ under $\mathbb{Q}_t$. Combining this with (3.7.20), (3.7.22) and (3.7.14), for any $p > 1$ and $k' \geq k$ we find constants $c_1(p), c_2(p) > 0$ such that

$$\int_{\mathbb{R}^d} \{(\Phi_t^\gamma \rho) 1_{B(z,1)} f\}(y) dy = \mathbb{E}\big[(1_{B(z,1)} f)(X_t^\rho)\big] = \mathbb{E}\big[R_t (1_{B(z,1)} f)(\bar{X}_t^\rho)\big]$$
$$\leq \big(\mathbb{E}[R_t^{\frac{p}{p-1}}]\big)^{\frac{p-1}{p}} \big(\mathbb{E}[(1_{B(z,1)} f^p)(\bar{X}_t^\rho)]\big)^{\frac{1}{p}}$$
$$\leq c_1(p) \left(\int_{\mathbb{R}^d} \{(\hat{P}_{0,t}^\kappa (1_{B(z,1)} f^p)\}(x) \gamma(dx) \right)^{\frac{1}{p}}$$
$$\leq c_2(p) \|\gamma\|_{\tilde{L}^k}^{\frac{1}{p}} t^{-\frac{d(k'-k)}{2kk'p}} \|f\|_{\tilde{L}^{\frac{pk'}{k'-1}}}, \quad t \in (0,T], f \in \mathcal{B}^+(\mathbb{R}^d).$$

Therefore, for any probability density $\gamma \in \tilde{L}^k$,

$$\|\Phi_t^\gamma \rho\|_{\tilde{L}^{\frac{pk'}{pk'-k'+1}}} \leq c_2(p) \|\gamma\|_{\tilde{L}^k}^{\frac{1}{p}} t^{-\frac{d(k'-k)}{2kk'p}},$$
$$p > 1, k' \geq k, \rho \in \tilde{\mathcal{P}}_{\gamma,T}^k, t \in (0,T],$$
(3.7.23)

where for $k' = k = \infty$ we set $\frac{pk'}{pk'-k'+1} := \frac{p}{p-1}$, $\frac{d(k'-k)}{2kk'p} := 0$. Using (3.7.12) replacing the estimate in Lemma 3.7.3, we find a map $c: (1,\infty) \to (0,\infty)$ such that

$$\|\Phi_t^\gamma \rho\|_{L^{\frac{pk'}{pk'-k'+1}}} \leq c(p) \|\gamma\|_{\tilde{L}^k}^{\frac{1}{p}} t^{-\frac{d(k'-k)}{2kk'p}},$$
$$p > 1, k' \geq k, \rho \in \tilde{\mathcal{P}}_{\gamma,T}^k, t \in (0,T].$$
(3.7.24)

(c) By the backward Kolmogorov equation (3.7.5) and Itô's formula, for any $f \in C_0^\infty(\mathbb{R}^d)$ we have

$$d\{(P_{s,t}^{a^\rho,b^{(1)}} f)(X_s^\rho)\} = \{(\partial_s + L_s^{a^\rho,b^{(1)}} + \nabla_{b_s^\rho,0}) P_{s,t}^{a^\rho,b^{(1)}} f\}(X_s^\rho) ds + dM_s$$
$$= \{\nabla_{b_s^\rho,0} P_{s,t} f\}(X_s^\rho) ds + dM_s, \quad s \in [0,t]$$

for some martingale M_s. Then

$$\mathbb{E}[f(X_t^\rho)] = \mathbb{E}[P_{t,t}^{a^\rho,b^{(1)}} f(X_t^\rho)]$$
$$= \mathbb{E}[P_{0,t}^{a^\rho,b^{(1)}} f(X_0)] + \int_0^t \mathbb{E}[(\nabla_{b_s^\rho,0} P_{s,t}^{a^\rho,b^{(1)}} f)(X_s^\rho)] ds, \quad s \in [0,t].$$
(3.7.25)

We explain that the last term in (3.7.25) exists. Indeed, by (1.4.1), there exists a constant $c_2 > 0$ such that

$$\|\nabla P_{s,t}^{a^\rho,b^{(1)}} f\|_\infty \leq c_2 \|\nabla f\|_\infty, \quad 0 \leq s \leq t, f \in C_b^1(\mathbb{R}^d),$$

so that (3.7.18) and Krylov's estimate (1.2.7) yield
$$\mathbb{E}\int_0^t |(\nabla_{b_s^{\rho,0}} P_{s,t}^{a^\rho,b^{(1)}} f)(X_s^\rho)|\mathrm{d}s \leq \mathbb{E}\int_0^t c_2\|\nabla f\|_\infty |b_s^{\rho,0}|(X_s^\rho)\mathrm{d}s < \infty.$$
Noting that $\Phi_s^\gamma \rho := \ell_{X_s^\rho}$ and
$$P_{s,t}^{a^\rho,b^{(1)}} f(x) = \int_{\mathbb{R}^d} p_{s,t}^{a^\rho,b^{(1)}}(x,y)f(y)\mathrm{d}y,$$
(3.7.25) is equivalent to
$$\int_{\mathbb{R}^d} \{\Phi_s^\gamma f\}(y)\mathrm{d}y = \int_{\mathbb{R}^d \times \mathbb{R}^d} \gamma(x) p_{0,t}^{a^\rho,b^{(1)}}(x,y) f(y)\mathrm{d}x\mathrm{d}y$$
$$+ \int_0^t \mathrm{d}s \int_{\mathbb{R}^d \times \mathbb{R}^d} (\Phi_s^\gamma \rho)(x) \{\nabla_{b_s^{\rho,0}} p_{s,t}^{a^\rho,b^{(1)}}(\cdot,y)(x)\} f(y)\mathrm{d}y,$$
$$f \in C_0^\infty(\mathbb{R}^d), s \in [0,t].$$
Thus,
$$(\Phi_t^\gamma \rho)(y) = \int_{\mathbb{R}^d} p_{0,t}^{a^\rho,b^{(1)}}(x,y)\gamma(\mathrm{d}x)$$
$$+ \int_0^t \mathrm{d}s \int_{\mathbb{R}^d} (\Phi_s^\gamma \rho)(x) \{\nabla_{b_s^{\rho,0}} p_{s,t}^{a^\rho,b^{(1)}}(\cdot,y)(x)\}\mathrm{d}x, \quad t \in [0,T]. \quad (3.7.26)$$
By (3.7.12) for $p = p'$, $\|\hat{P}_t^{\kappa*}\gamma\|_{\tilde{L}^l} \leq K\|\gamma\|_{\tilde{L}^l}$ holds for some constant $K > 0$. Combining this with (3.7.7), (3.7.18) and (3.7.26), we find a constant $c_3 > 0$ such that for any $l \in [1,\infty]$,
$$\|\Phi_t^\gamma \rho\|_{\tilde{L}^l} \leq c_3 \|\gamma\|_{\tilde{L}^l}$$
$$+ c_3 \sup_{z \in \mathbb{R}^d} \int_0^t (t-s)^{-\frac{1}{2}} \|1_{B(z,1)} \hat{P}_{t-s}^\kappa \{(\Phi_s^\gamma \rho) f_0(s,\cdot)\}\|_{L^l} \mathrm{d}s. \quad (3.7.27)$$
By $k > k_0$ and $k \geq \frac{p_0}{p_0-1}$, for any $l \in (k_0, k] \cap [\frac{p_0}{p_0-1}, k]$ we have
$$q_l := \frac{p_0 l}{p_0 + l} \in (1, l], \quad \frac{1}{q_l} = \frac{1}{p_0} + \frac{1}{l}, \quad (3.7.28)$$
and $(p_0, q_0) \in \mathcal{K}$ implies
$$\frac{1}{2} + \frac{d(l-q_l)}{2lq_l} = \frac{1}{2} + \frac{d}{2p_0} =: \delta' < \frac{q_0 - 1}{q_0}. \quad (3.7.29)$$
Combining these with (3.7.13) for $(p',p) = (l, q_l)$ and applying Hölder's inequality, we find a constant $c_4 > 0$ such that
$$\int_0^t (t-s)^{-\frac{1}{2}} \|\hat{P}_{t-s}^\kappa \{(\Phi_s^\gamma \rho) f_0(s,\cdot)\}\|_{\tilde{L}^l} \mathrm{d}s \leq c_4 \|(t-\cdot)^{-\delta'} f_0 \Phi_\cdot^\gamma \rho\|_{\tilde{L}_1^{q_l}(t)}$$
$$\leq c_4 \|f_0\|_{\tilde{L}_{q_0}^{p_0}(t)} \|(t-\cdot)^{-\delta'} \Phi_\cdot^\gamma \rho\|_{\tilde{L}_{\frac{q_0}{q_0-1}}^l(t)}, \quad l \in (k_0, k] \cap \left[\frac{p_0}{p_0-1}, k\right],$$

where $\{(t-\cdot)^{-\delta'}\Phi_\cdot^\gamma\}(s,x) := (t-s)^{-\delta'}\Phi_s^\gamma(x)$. This together with (3.7.27) implies that for some constant $c_5 > 0$,

$$\|\Phi_t^\gamma \rho\|_{\tilde{L}^l} \leq c_5 \|\gamma\|_{\tilde{L}^l}$$
$$+ c_5 \|f_0\|_{\tilde{L}_{q_0}^{p_0}(t)} \left(\int_0^t \left\{ (t-s)^{-\delta'} \|\Phi_s^\gamma \rho\|_{\tilde{L}^l} \right\}^{\frac{q_0}{q_0-1}} \mathrm{d}s \right)^{\frac{q_0-1}{q_0}}, \quad (3.7.30)$$
$$t \in [0,T], l \in (k_0, k] \cap \left[\frac{p_0}{p_0-1}, k\right].$$

Similarly, using (3.7.12) replacing Lemma 3.7.3, we derive

$$\|\Phi_t^\gamma \rho\|_{L^l} \leq c_5 \|\gamma\|_{L^l}$$
$$+ c_5 \|f_0\|_{L_{q_0}^{p_0}(t)} \left(\int_0^t \left\{ (t-s)^{-\delta'} \|\Phi_s^\gamma \rho\|_{\tilde{L}^l} \right\}^{\frac{q_0}{q_0-1}} \mathrm{d}s \right)^{\frac{q_0-1}{q_0}}, \quad (3.7.31)$$
$$t \in [0,T], l \in (k_0, k] \cap \left[\frac{p_0}{p_0-1}, k\right].$$

Below we prove (3.7.17) by considering two different situations.

(c_1) $k < \infty$. For any $k' \in (k, \infty)$ we have

$$p_{k,k'} := \frac{k(k'-1)}{k'(k-1)} > 1, \quad \frac{p_{k,k'}k'}{p_{k,k'}k' - k' + 1} = k.$$

Noting that

$$\lim_{k' \downarrow k} \frac{d(k'-k)}{2kk' p_{k,k'}} = 0,$$

by (3.7.29) we find $k' > k$ such that

$$\varepsilon_{k,k'} := \frac{d(k'-k)}{2kk' p_{k,k'}} \subset \left(0, 1 - \frac{\delta' q_0}{q_0 - 1}\right).$$

Combining this with (3.7.23) and (3.7.30) for $l = k$, we find a constant $K > 0$ such that

$$\sup_{t \in [0,T]} \|\Phi_t^\gamma \rho\|_{\tilde{L}^k} \leq K \|\gamma\|_{\tilde{L}^k}$$
$$+ K \sup_{t \in [0,T]} \left(\int_0^t (t-s)^{-\frac{q_0 \delta'}{q_0-1}} s^{-\varepsilon_{k,k'}} \|\gamma\|_{\tilde{L}^k}^{\frac{q_0}{p_{k,k'}(q_0-1)}} \mathrm{d}s \right)^{\frac{q_0-1}{q_0}} < \infty.$$

Therefore, by the generalized Gronwall inequality [Ye et al. (2007)], (3.7.29) and (3.7.30) implies (3.7.17).

When $f_0 \in L_{q_0}^{p_0}$, by using (3.7.24) and (3.7.31) replacing (3.7.23) and (3.7.30), we obtain this estimate for L replacing $\tilde{L}$.

(c_2) $k = \infty$. We take $k' = k = \infty$, so that by (3.7.23), for any $p > 1$ we find a constant $c(p) > 0$ such that

$$\|\Phi_t^\gamma \rho\|_{\tilde{L}^{\frac{p}{p-1}}} \le c(p)\|\gamma\|_{\tilde{L}^k}^{\frac{1}{p}}.$$

Combining this with (3.7.30) for $l \in (\frac{p_0}{p_0-1} \vee k_0, \infty)$ and $p := \frac{l}{l-1} > 1$, we obtain

$$\sup_{t \in [0,T]} \|\Phi_t^\gamma \rho\|_{\tilde{L}^l} < \infty,$$

so that by the generalized Gronwall inequality [Ye et al. (2007)], (3.7.30) implies (3.7.17) for $l \in (\frac{p_0}{p_0-1} \vee k_0, \infty)$ replacing $k = \infty$ with a uniform constant $c > 0$. By letting $l \uparrow k = \infty$, we derive (3.7.17).

Noting that a probability density function $\rho \in L^\infty$ implies $\rho \in L^l$ for any $l \geq 1$, when $f_0 \in L^{p_0}_{q_0}$ we derive (3.7.17) for L replacing $\tilde{L}$ by using (3.7.24) and (3.7.31) replacing (3.7.23) and (3.7.30) respectively. □

Proof of Theorem 3.7.1(1). By Lemma 3.7.3, (3.7.4) is weakly well-posed. By Theorem 1.3.1, it is also strongly well-posed provided (3.7.3) holds. Thus, for the weak or strong well-posedness of (3.7.1), it suffices to prove that Φ^γ has a unique fixed point in $\tilde{\mathcal{P}}^k_{\gamma,T}$. In general, for probability density functions $\gamma^1, \gamma^2 \in \tilde{L}^k$ and $\rho^1, \rho^2 \in \tilde{\mathcal{P}}^k_{\gamma,T}$, we estimate $\tilde{d}_{k,\lambda}(\Phi^{\gamma^1}\rho^1, \Phi^{\gamma^2}\rho^2)$ for $\lambda > 0$.

By (3.7.26) for a^ρ independent of ρ, $(A^{3.6})(2)$ and (3.7.7), we find a constant $c_1 > 0$ such that

$$\|\Phi_t^{\gamma_1}\rho^1 - \Phi_t^{\gamma_2}\rho^2\|_{\tilde{L}^l} - c_1\|\gamma_1 - \gamma_0\|_{\tilde{L}^l}$$
$$\le c_1 \int_0^t (t-s)^{-\frac{1}{2}} \Big\|\hat{P}^\kappa_{s,t}\Big\{f_0(s,\cdot)\big[|\Phi_s^{\gamma_1}\rho^1 - \Phi_s^{\gamma_2}\rho^2| \qquad (3.7.32)$$
$$+ s^\theta(\Phi_s^{\gamma_1}\rho^1)\big(|\rho_s^1 - \rho_s^2| + \|\rho_s^1 - \rho_s^2\|_{\tilde{L}^k}\big)\big]\Big\}\Big\|_{\tilde{L}^k} \mathrm{d}s.$$

Letting

$$F_l(s,x) := (t-s)^{-\frac{d(k-l)-kl}{2kl}} s^\theta \big[(\Phi_s^{\gamma_1}\rho^1)(|\rho_s^1 - \rho_s^2| + \|\rho_s^1 - \rho_s^2\|_{\tilde{L}^k})\big](x)$$
$$+ (t-s)^{-\frac{d(k-l)-kl}{2kl}} |\Phi_s^{\gamma_1}\rho^1 - \Phi_s^{\gamma_2}\rho^2|(x), \quad l \in \Big[1, \frac{kp_0}{k+p_0}\Big],$$

by (3.7.13) for $q = 1$ and $(p',p) = (k,l)$, and applying Hölder's inequality, we find a constant $c_2 > 0$ such that

$$\int_0^t (t-s)^{-\frac{1}{2}} \Big\| \hat{P}_{s,t}^\kappa \Big\{ f_0(s,\cdot) \big[|\Phi_s^{\gamma_1}\rho^1 - \Phi_s^{\gamma_2}\rho^2| $$
$$+ s^\theta (\Phi_s^{\gamma_1}\rho^1)(|\rho_s^1 - \rho_s^2| + \|\rho_s^1 - \rho_s^2\|_{\tilde{L}^k}) \big] \Big\} \Big\|_{\tilde{L}^k} ds$$
$$\leq c_2 \|f_0 F_l\|_{\tilde{L}_1^l(t)} \leq c \|f_0\|_{\tilde{L}_{q_0}^{p_0}(t)} \|F_l\|_{\tilde{L}^{\frac{p_0 l}{p_0 - l}}_{\frac{q_0}{q_0 - 1}}(t)}$$
$$\leq c_2 \|f_0\|_{\tilde{L}_{q_0}^{p_0}(t)} \left(\int_0^t \Big\{ (t-\cdot)^{-\frac{d(k-l)}{2kl} - \frac{1}{2}} \Big[\|\Phi^{\gamma_1}\rho^1 - \Phi^{\gamma_2}\rho^2\|_{\tilde{L}^{\frac{p_0 l}{p_0 - l}}} \right.$$
$$\left. + s^\theta \|(\Phi_s^{\gamma_1}\rho^1)(|\rho_s^1 - \rho_s^2| + \|\rho_s^1 - \rho_s^2\|_{\tilde{L}^k})\|_{\tilde{L}^{\frac{p_0 k}{p_0 - k}}} \Big] \Big\}^{\frac{q_0}{q_0 - 1}} ds \right)^{\frac{q_0 - 1}{q_0}}.$$

Since $l \in [1, \frac{kp_0}{k+p_0}]$ implies $\frac{p_0 l}{p_0 - l} \leq k$, combining this with (3.7.32) and applying Hölder's inequality, we find a constant $c_3 > 0$ such that

$$\|\Phi_t^{\gamma_1}\rho^1 - \Phi_t^{\gamma_2}\rho^2\|_{\tilde{L}^k} - c_1\|\gamma_1 - \gamma_2\|_{\tilde{L}^k}$$
$$\leq c_3 \left(\int_0^t \Big\{ (t-s)^{-\frac{d(k-l)}{2kl} - \frac{1}{2}} \Big[\|\Phi_s^{\gamma_1}\rho^1 - \Phi_s^{\gamma_2}\rho^2\|_{\tilde{L}^k} \right. \quad (3.7.33)$$
$$\left. + s^\theta \|\Phi_s^{\gamma_1}\rho^1\|_{\tilde{L}^{\frac{kp_0 l}{k(p_0-l)-p_0 l}}} \|\rho_s^1 - \rho_s^2\|_{\tilde{L}^k} \Big] \Big\}^{\frac{q_0}{q_0 - 1}} ds \right)^{\frac{q_0 - 1}{q_0}}$$

holds for $l \in [1, \frac{kp_0}{k+p_0}]$. Letting

$$\alpha_l := \frac{q_0}{q_0 - 1}\Big(\frac{d(k-l)}{2kl} + \frac{1}{2}\Big), \quad \beta_l := \frac{kp_0 l}{k(p_0 - l) - p_0 l}, \quad (3.7.34)$$

by the definition of $\tilde{d}_{k,\lambda}$, this implies that for any $\lambda > 0$ and $l \in [1, \frac{kp_0}{k+p_0}]$,

$$\tilde{d}_{k,\lambda}(\Phi^{\gamma^1}\rho^1, \Phi^{\gamma^2}\rho^2) \leq c_1 \|\gamma_1 - \gamma_2\|_{\tilde{L}^k}$$
$$+ c_3 \big\{ \tilde{d}_{k,\lambda}(\Phi^{\gamma^1}\rho^1, \Phi^{\gamma^2}\rho^2) + \tilde{d}_{k,\lambda}(\rho^1, \rho^2) \big\}$$
$$\times \sup_{t \in (0,T]} \left\{ \left(\int_0^t (t-s)^{-\alpha_l} e^{-\frac{\lambda q_0}{q_0 - 1}(t-s)} ds \right)^{\frac{q_0 - 1}{q_0}} \right. \quad (3.7.35)$$
$$\left. + \left(\int_0^t (t-s)^{-\alpha_l} \big(s^\theta e^{-\lambda(t-s)} \|\Phi_s^{\gamma_1}\rho^1\|_{\tilde{L}^{\beta_l}} \big) \right)^{\frac{q_0}{q_0 - 1}} ds \right)^{\frac{q_0 - 1}{q_0}} \right\}.$$

Below we complete the proof by considering two different situations respectively.

(a) Let $k < \infty$. By $(p_0, q_0) \in \mathcal{K}$ and $k > k_0 := \frac{d}{2\theta+1-dp_0^{-1}-2q_0^{-1}}$, α_l in (3.7.34) satisfies
$$\lim_{l \uparrow \frac{kp_0}{k+p_0}} \alpha_l + \frac{q_0}{q_0-1}\left(\frac{d}{2k} - \theta\right)^+ = \frac{q_0}{q_0-1}\left\{\frac{d}{2p_0} + \frac{1}{2} + \left(\frac{d}{2k} - \theta\right)^+\right\} < 1.$$
So, we may take $l \in (1, \frac{kp_0}{k+p_0})$ such that
$$\alpha_l + \frac{q_0}{q_0-1}\left(\frac{d}{2k} - \theta\right)^+ < 1, \quad \beta_l \in (1, \infty). \tag{3.7.36}$$
By (3.7.23) for $k' = \infty$ and $p = \frac{\beta_l}{\beta_l-1}$, there exists a constant $c_4 > 0$ such that
$$\|\Phi_s^{\gamma_1} \rho^1\|_{L^{\beta_l}} \leq c_4 \|\gamma^1\|_{\tilde{L}^k} s^{-\frac{d}{2k}}.$$
Combining this with (3.7.35) and (3.7.36), for large enough $\lambda > 0$ which is increasing in $\|\gamma^1\|_{\tilde{L}^k} (\leq \|\gamma^2\|_{\tilde{L}^k})$, we have
$$\tilde{d}_{k,\lambda}(\Phi^{\gamma^1}\rho^1, \Phi^{\gamma^2}\rho^2) \leq c_1 \|\gamma_1 - \gamma_2\|_{\tilde{L}^k} + \frac{1}{4}\tilde{d}_{k,\lambda}(\Phi^{\gamma^1}\rho^1, \Phi^{\gamma^2}\rho^2) + \frac{1}{4}\tilde{d}_{k,\lambda}(\rho^1, \rho^2).$$
Taking $\gamma^1 = \gamma^2 = \gamma$ we derive the contraction of Φ^γ on the complete metric space $(\mathcal{P}_{\gamma,T}^k, d_{k,\lambda})$, and hence Φ^γ has a unique fixed point. This implies the weak (also strong under (3.7.3)) well-posedness of (3.7.1). Moreover, for two solutions $(X^i)_{i=1,2}$ of this SDE with initial distribution densities $(\gamma^i)_{i=1,2}$, by taking $\rho^i = \mathcal{L}_{X^i}$ so that $\rho^i = \Phi^{\gamma^i}\rho^i$, we deduce (3.7.2) for some increasing function Λ.

(b) Let $k = \infty$. By taking $l = p_0$, we have $\beta_l = \infty$ and $\theta > \frac{2}{q_0} + \frac{d}{p_0} - 1$ in $(A^{3.6})(2)$ implies
$$\alpha_l + \frac{q_0}{q_0-1}\left(\frac{d}{2k} - \theta\right)^+ = \frac{q_0}{q_0-1}\left\{\frac{d}{2p_0} + \frac{1}{2} + \theta^-\right\} < 1.$$
Combining (3.7.35) with (3.7.17) for $k = \infty$, we derive that for a large enough $\lambda > 0$ increasing in $\|\gamma^1\|_{\tilde{L}^\infty} (\leq \|\gamma^2\|_{\tilde{L}^\infty})$,
$$\tilde{d}_{k,\lambda}(\Phi^{\gamma^1}\rho^1, \Phi^{\gamma^2}\rho^2) \leq c_1\|\gamma_1 - \gamma_2\|_{\tilde{L}^\infty}$$
$$+ c_3 \tilde{d}_{k,\lambda}(\Phi^{\gamma^1}\rho^1, \Phi^{\gamma^2}\rho^2) \sup_{t \in (0,T]} \left(\int_0^t (t-s)^{-\alpha_l} e^{-\frac{\lambda q_0}{q_0-1}(t-s)} ds\right)^{\frac{q_0-1}{q_0}}$$
$$+ c_3 \tilde{d}_{k,\lambda}(\rho^1, \rho^2) \sup_{t \in (0,T]} \left(\int_0^t (t-s)^{-\alpha_l} e^{-\frac{\lambda q_0}{q_0-1}(t-s)} (s^\theta \|\gamma^1\|_{\tilde{L}^\infty})^{\frac{q_0}{q_0-1}} ds\right)^{\frac{q_0-1}{q_0}}$$
$$\leq c_1 \|\gamma_1 - \gamma_2\|_{\tilde{L}^\infty} + \frac{1}{4}\tilde{d}_{k,\lambda}(\Phi^{\gamma^1}\rho^1, \Phi^{\gamma^2}\rho^2) + \frac{1}{4}\tilde{d}_{k,\lambda}(\rho^1, \rho^2).$$
Then we finish the proof as in step (a). □

Proof of Theorem 3.7.1(2). Let $(A^{3.6})$ hold with condition $(2')$ replacing $(A^{3.6})(2)$. By (3.7.12) and Hölder's inequality, we find constants $c_1, c_2 > 0$ such that for any $0 \le s < t \le T$ and $l \in [1, \frac{kp_0}{k+p_0}]$,

$$\left\| \hat{P}_{s,t}^{\kappa} \left\{ (C + f_0(s,\cdot)) \left(|\Phi_s^{\gamma_1} \rho^1 - \Phi_s^{\gamma_2} \rho^2| + s^\theta (\Phi_s^{\gamma_1} \rho^1) |\rho_s^1 - \rho_s^2| \right) \right\} \right\|_{L^k}$$

$$\le c_1 (t-s)^{-\frac{d(k-l)}{2kl}} \left\{ \|\Phi_s^{\gamma_1} \rho^1 - \Phi_s^{\gamma_2} \rho^2\|_{L^l} + s^\theta \|(\Phi_s^{\gamma_1} \rho^1)|\rho_s^1 - \rho_s^2|\|_{L^l} \right.$$

$$\left. + \|f_0(s,\cdot)(\Phi_s^{\gamma_1} \rho^1 - \Phi_s^{\gamma_2} \rho^2)\|_{L^l} + s^\theta \|f_0(s,\cdot)(\Phi_s^{\gamma_1} \rho^1)|\rho_s^1 - \rho_s^2|\|_{L^l} \right\}$$

$$\le c_1 (t-s)^{-\frac{d(k-l)}{2kl}} \left\{ \|\Phi_s^{\gamma_1} \rho^1 - \Phi_s^{\gamma_2} \rho^2\|_{L^l} + s^\theta \|\Phi_s^{\gamma_1} \rho^1\|_{L^{\frac{kl}{k-l}}} \||\rho_s^1 - \rho_s^2|\|_{L^k} \right.$$

$$+ \|f_0(s,\cdot)\|_{L^{p_0}} \|\Phi_s^{\gamma_1} \rho^1 - \Phi_s^{\gamma_2} \rho^2\|_{L^{\frac{p_0 l}{p_0 - l}}}$$

$$\left. + s^\theta \|f_0(s,\cdot)\|_{L^{p_0}} \|\Phi_s^{\gamma_1} \rho^1\|_{L^{\frac{p_0 kl}{p_0 k - kl - p_0 l}}} \||\rho_s^1 - \rho_s^2|\|_{L^k} \right\}.$$

Noting that $l \in [1, \frac{kp_0}{k+p_0}]$ implies $l \vee \frac{p_0 l}{p_0 - l} \le k$ and $\frac{kl}{k-l} \le \frac{p_0 kl}{p_0 k - kl - p_0 l}$, by combining this with $(2')$, (3.7.7), (3.7.26) and Hölder's inequality, we find constants $c_3, c_4 > 0$ such that

$$\|\Phi_t^{\gamma_1} \rho^1 - \Phi_t^{\gamma_2} \rho^2\|_{L^k} - c_1 \|\gamma_1 - \gamma_2\|_{L^k}$$

$$\le c_3 \int_0^t (t-s)^{-\frac{1}{2}} \left\| \hat{P}_{s,t}^{\kappa} \left\{ (C + f_0(s,\cdot)) \right. \right.$$

$$\left. \left. \times \left(|\Phi_s^{\gamma_1} \rho^1 - \Phi_s^{\gamma_2} \rho^2| + s^\theta (\Phi_s^{\gamma_1} \rho^1)|\rho_s^1 - \rho_s^2| \right) \right\} \right\|_{L^k} ds$$

$$\le c_4 \left(1 + \|f_0\|_{\tilde{L}_{q_0}^{p_0}}(t) \right) \left(\int_0^t \left\{ (t-s)^{-\frac{d(k-l)}{2kl} - \frac{1}{2}} \left[\|\Phi_s^{\gamma_1} \rho^1 - \Phi_s^{\gamma_2} \rho^2\|_{L^k} \right. \right. \right.$$

$$\left. \left. \left. + s^\theta \|\Phi_s^{\gamma_1} \rho^1\|_{L^{\frac{kp_0 l}{k(p_0 - l) - p_0 l}}} \|\rho_s^1 - \rho_s^2\|_{L^k} \right] \right\}^{\frac{q_0}{q_0 - 1}} ds \right)^{\frac{q_0 - 1}{q_0}}, \quad l \in \left[1, \frac{kp_0}{k+p_0}\right].$$

Then the remainder of the proof is similar to that of Theorem 3.7.1(1) from (3.7.33) with L replacing $\tilde{L}$. □

3.7.2 Density dependent noise

In this part we allow σ to be density dependent but make stronger assumptions for the coefficients in the spatial variable.

$(A^{3.7})$ There exist $1 \le f_0 \in \tilde{L}_{q_0}^{p_0}(T)$, $C \in (0, \infty)$ and $\alpha \in (0, 1)$, such that the following conditions hold for all $t \in (0, T]$, $x, y \in \mathbb{R}^d$, $r, \tilde{r} \in [0, \infty)$ and $\rho, \tilde{\rho} \in L^\infty$:

$$|b_t(x, r, \rho)| \le f_0(t, x),$$

$$|b_t(x,r,\rho) - b_t(x,\tilde{r},\tilde{\rho})| \leq C(|r-\tilde{r}| + \|\rho-\tilde{\rho}\|_\infty),$$

$$\|\sigma\|_\infty + \|\nabla\sigma\|_\infty + \|(\sigma\sigma^*)^{-1}\|_\infty \leq C,$$

$$\|\nabla\sigma_t(\cdot,\rho)(x) - \nabla\sigma_t(\cdot,\rho)(y)\| \leq C|x-y|^\alpha,$$

$$\|\sigma_t(\cdot,\rho) - \sigma_t(\cdot,\tilde{\rho})\|_{C_b^\alpha} \leq C\|\rho-\tilde{\rho}\|_\infty.$$

Theorem 3.7.5. *Assume* $(A^{3.7})$ *and let* $\beta \in (0, 1 - \frac{d}{p_0} - \frac{2}{q_0})$. *For any initial value (initial density) with* $\ell_{X_0} \in C_b^\beta(\mathbb{R}^d)$, (3.7.1) *has a unique strong (weak) solution satisfying* $\ell_{X_\cdot} \in L_\infty^\infty$, *and there exists a constant* $c > 0$ *such that*

$$\sup_{t\in[0,T]} \|\ell_{X_t}\|_{C_b^\beta} \leq c\|\ell_{X_0}\|_{C_b^\beta}. \qquad (3.7.37)$$

Moreover, there exists an increasing function $\Lambda : (0,\infty) \to (0,\infty)$ *such that for any two solutions* $\{X_t^i\}_{i=1,2}$ *with* $\ell_{X_0^i} \in C_b^\beta(\mathbb{R}^d)$ *and* $\ell_{X_\cdot^i} \in L_\infty^\infty$,

$$\sup_{t\in[0,T]} \|\ell_{X_t^1} - \ell_{X_t^2}\|_\infty \leq \Lambda\big(\|\ell_{X_0^1}\|_{C_b^\beta} \wedge \|\ell_{X_0^1}\|_{C_b^\beta}\big)\|\ell_{X_0^1} - \ell_{X_0^2}\|_\infty. \qquad (3.7.38)$$

Let $\gamma \in \mathcal{P}^\infty$ with $\ell_\gamma \in C_b^\beta$. By Theorem 3.7.1 and $(A^{3.7})$, for any $\rho \in \mathcal{P}_{\gamma,T}^\infty$, the following density dependent SDE has a unique (weak and strong) solution with $\ell_{X^{\rho,\gamma}} \in L_\infty^\infty$:

$$\begin{aligned}\mathrm{d}X_t^{\rho,\gamma} &= b_t(X_t^{\rho,\gamma}, \ell_{X_t^{\rho,\gamma}}(X_t^{\rho,\gamma}), \ell_{X_t^{\rho,\gamma}})\mathrm{d}t + \sigma_t^\rho(X_t^{\rho,\gamma}),\\ \mathcal{L}_{X_0^{\rho,\gamma}} &= \gamma, t \in [0,T],\end{aligned} \qquad (3.7.39)$$

and there exists a constant $c > 0$ depending on C, α such that

$$\|\ell_{X_t^{\rho,\gamma}}\|_\infty \leq c\|\ell_\gamma\|_\infty, \quad \rho \in \mathcal{P}_{\gamma,T}^\infty. \qquad (3.7.40)$$

We aim to show that the map

$$\rho \mapsto \ell_{X_\cdot^{\rho,\gamma}}$$

has a unique fixed point in $\mathcal{P}_{\gamma,T}^\infty$, such that the (weak and strong) well-posedness of (3.7.39) implies that of (3.7.1). As shown in the proof of Theorem 3.7.1, we will need heat kernel estimates presented in Section 2 for the operator $L_t^{a^\rho, b^{\rho,\gamma}}$, where

$$a_t^\rho := \frac{1}{2}\sigma_t^\rho(\sigma_t^\rho)^*, \quad b_t^{\rho,\gamma} := b_t\big(\cdot, \ell_{X_t^{\rho,\gamma}}(\cdot), \ell_{X_t^{\rho,\gamma}}\big), \quad t \in [0,T].$$

To this end, we first prove the Hölder continuity of $b_t^{\rho,\gamma}$. By $(A^{3.7})$, this follows from the Hölder continuity of $\ell_{X_t^{\rho,\gamma}}$.

Lemma 3.7.6. *Assume* $(A^{3.7})$ *and let* $\beta \in (0, 1 - \frac{d}{p_0} - \frac{2}{q_0})$. *Then there exists a constant* $c > 0$ *such that for any* $\rho \in \mathcal{P}_{\gamma,T}^\infty$ *and* $\gamma \in \mathcal{P}^\infty$ *with* $\ell_\gamma \in C_b^\beta$,

$$\|\ell_{X_t^{\rho,\gamma}}\|_{C_b^\beta} \le c\|\ell_\gamma\|_{C_b^\beta}, \quad t \in (0,T]. \tag{3.7.41}$$

Proof. Simply denote $\ell_t = \ell_{X_t^{\rho,\gamma}}$. Let $p_{s,t}^\rho$ be the heat kernel for the operator

$$L_t^\rho := \frac{1}{2}\mathrm{div}\{a_t^\rho \nabla\} = L_t^{a^\rho, \bar{b}^\rho},$$

where

$$a_t^\rho := \frac{1}{2}\sigma_t^\rho (\sigma_t^\rho)^*, \quad (\bar{b}_t^\rho)_i := \sum_{j=1}^d \partial_j (a_t^\rho)_{ij}.$$

Then $p_{s,t}^\rho(x,y) = p_{s,t}^\rho(y,x)$, and by Theorem 3.7.2, there exist constants $c, \kappa > 0$ depending on C, α, β such that for some diffeomorphisms $\psi_{s,t}$ satisfying (3.7.6),

$$\begin{aligned}&|\nabla^i p_{s,t}^\rho(\cdot,y)(x)| \le c_1 (t-s)^{-\frac{i}{2}} p_{t-s}^\kappa(\psi_{s,t}(x) - y), \quad i = 0, 1, 2,\\ &|\nabla p_{s,t}^\rho(x,\cdot)(y)| \le c_1(t-s)^{-\frac{1}{2}} p_{t-s}^\kappa(\psi_{s,t}(x) - y),\\ &|\nabla p_{s,t}^\rho(\cdot,y)(x) - \nabla p_{s,t}^\rho(\cdot,y')(x)|\\ &\le c_1 |y-y'|^\beta (t-s)^{-\frac{1+\beta}{2}} p_{t-s}^\kappa(\psi_{s,t}(x) - y)\end{aligned} \tag{3.7.42}$$

hold for all $0 \le s < t \le T$, $x, y \in \mathbb{R}^d$. By the argument leading to (3.7.26), we obtain

$$\begin{aligned}\ell_t(y) = &\int_{\mathbb{R}^d} p_{0,t}^\rho(x,y)\ell_\gamma(x)\mathrm{d}x\\ &+ \int_0^t \mathrm{d}s \int_{\mathbb{R}^d} \ell_s(x)\{\nabla_{b_s(x,\ell_s(x),\ell_s) - \bar{b}_s^\rho(x)} p_{s,t}^\rho(\cdot,y)\}(x)\mathrm{d}x.\end{aligned} \tag{3.7.43}$$

By the symmetry of $p_{0,t}^\rho(x,y)$ we have

$$\int_{\mathbb{R}^d} p_{0,t}^\rho(x,y)\ell_\gamma(x)\mathrm{d}x = \int_{\mathbb{R}^d} p_{0,t}^\rho(y,x)\ell_\gamma(x)\mathrm{d}x =: (P_{0,t}^\rho \ell_\gamma)(y). \tag{3.7.44}$$

Let X_t^x solve the SDE

$$\mathrm{d}X_t^x = \bar{b}_t^\rho(X_t^x)\mathrm{d}t + \sigma_t^\rho(X_t)\mathrm{d}W_t, \quad t \in [0,T], X_0 = x.$$

By Theorem 1.3.1(3), we find a constant $c_1 > 0$ depending on C, α in $(A^{3.7})$ such that
$$\mathbb{E}\Big[\sup_{t\in[0,T]} |X_t^x - X_t^y|\Big] \leq c_1 |x-y|, \quad x, y \in \mathbb{R}^d.$$
Then (3.7.44) implies
$$|(P_{0,t}^\rho \ell_\gamma)(y) - (P_{0,t}^\rho \ell_\gamma)(y')| = |\mathbb{E}[\ell_\gamma(X_t^y) - \ell_\gamma(X_t^{y'})]|$$
$$\leq \|\ell_\gamma\|_{C_b^\beta} \mathbb{E}[|X_t^y - X_t^{y'}|^\beta] \leq \|\ell_\gamma\|_{C_b^\beta} (c_1|y-y'|)^\beta.$$
Since $(A^{3.7})$ implies $|b| + |\bar{b}^\rho| \leq c f_0$ for some constant $c > 0$, by combining this with (3.7.40), the last inequality in (3.7.42), and (3.7.43), we find a constant $c_2 > 0$ independent of ρ, γ such that
$$|\ell_t(y) - \ell_t(y')| - c_2|y - y'|^\beta$$
$$\leq c_2 \|\ell_\gamma\|_\infty |y-y'|^\beta \int_0^t (t-s)^{-\frac{1+\beta}{2}} \{\tilde{P}_{s,t}^\kappa f_0(s,\cdot)(y) + \tilde{P}_{s,t}^\kappa f_0(s,\cdot)(y')\} \mathrm{d}s,$$
where $\tilde{P}_{s,t}$ is in (3.7.10). By (3.7.13) for $(p,q) = (p_0, q_0)$ and $p' = \infty$, we find a constant $c_3 > 0$ such that this implies
$$\frac{|\ell_t(y) - \ell_t(y')| - c_2|y-y'|^\beta}{c_2\|\ell_\gamma\|_\infty |y-y'|^\beta}$$
$$\leq \int_0^t (t-s)^{-(\frac{1+\beta}{2} + \frac{d}{2p_0})} \Big(\tilde{P}_{s,t}^\kappa \{(t-s)^{\frac{d}{2p_0}} f_0(s,\cdot)\}(y)$$
$$+ \tilde{P}_{s,t}^\kappa \{(t-s)^{\frac{d}{2p_0}} f_0(s,\cdot)\}(y') \Big) \mathrm{d}s$$
$$\leq 2c_2 \Big(\int_0^t (t-s)^{-(\frac{1+\beta}{2} + \frac{d}{2p_0})\frac{q_0}{q_0-1}} \mathrm{d}s \Big)^{\frac{q_0-1}{q_0}} \|\tilde{P}_{\cdot,t}^\kappa \{(t-\cdot)^{\frac{d}{2p_0}} f_0\}\|_{L_{q_0}^\infty(t)}$$
$$\leq c_3 \|f\|_{\tilde{L}_{q_0}^{r_0}(t)}, \quad y \neq y', t \in (0, T],$$
where we have used the fact that $\|\cdot\|_{\tilde{L}_{q_0}^\infty(t)} = \|\cdot\|_{L_{q_0}^\infty(t)}$ and $(\frac{1+\beta}{2} + \frac{d}{2p_0})\frac{q_0}{q_0-1} < 1$ due to $\beta \in (0, 1 - \frac{2}{q_0} - \frac{d}{p_0})$. Combining this with (3.7.40), we finish the proof. $\square$

The next lemma contains two classical estimates on the operator $1 - \Delta$ and the heat semigroup $P_t = \mathrm{e}^{t\Delta}$.

Lemma 3.7.7. *Let $P_t = \mathrm{e}^{t\Delta}$.*

(1) *For any $\beta > 0$, there exists a constant $c > 0$ such that*
$$\|(1-\Delta)^{\frac{\beta}{2}} f\|_\infty \leq c\|f\|_{C_b^\beta}.$$

(2) For any $\alpha, \beta, k \geq 0$, there exists a constant $c > 0$ such that
$$\|(1-\Delta)^{-k}P_t f\|_{C_b^{\alpha+\beta}} \leq ct^{-(\frac{\alpha}{2}-k)^+}\|f\|_{C_b^\beta}, \quad t > 0.$$

Proof of Theorem 3.7.5. For $\mathscr{L}_{X_0^i} = \gamma^i$ with $\ell_{\gamma^i} \in C_b^\beta(\mathbb{R}^d)$ and $\rho^i \in \mathcal{P}_{\gamma,T}^\infty$, simply denote
$$\ell_t^i = \ell_{X_t^{\rho^i,\gamma^i}}, \quad b_t^{\ell^i} := b_t(\cdot, \ell_t^i(\cdot), \ell_t^i), \quad t \in [0,T], \, i = 1,2.$$
Without loss of generality, let $\|\ell_{\gamma^2}\|_{C_b^\beta} \leq \|\ell_{\gamma^1}\|_{C_b^\beta}$.
By (3.7.43) with $(\gamma,\rho) = (\gamma^1, \rho^1)$, we obtain
$$\ell_t^1(y) = P_{0,t}^{\rho^1}\ell_{\gamma^1}(y) + \int_0^t ds \int_{\mathbb{R}^d} \ell_s^1(x)\{\nabla_{b_s^{\ell^1}(x) - \bar{b}_s^{\rho^1}(x)} p_{s,t}^{\rho^1}(\cdot, y)\}(x) dx.$$
By the argument leading to (3.7.26) for $(p_{s,t}^{\rho^1}, X_s^{\rho^2,\gamma^2})$ replacing $(p_{s,t}^{a^\rho, b^{(1)}}, X_s^\rho)$, we derive
$$\ell_t^2(y) = P_{0,t}^{\rho^1}\ell_{\gamma^2}(y) + \int_0^t ds \int_{\mathbb{R}^d} \ell_s^2(x)\{\nabla_{b_s^{\ell^2}(x) - \bar{b}_s^{\rho^1}(x)} p_{s,t}^{\rho^1}(\cdot, y)\}(x) dx$$
$$+ \frac{1}{2}\sum_{i,j=1}^d \int_0^t ds \int_{\mathbb{R}^d} \{\ell_s^2(a_s^{\rho^2} - a_s^{\rho^1})_{ij}\partial_i\partial_j p_{s,t}^{\rho^1}(\cdot, y)\}(x) dx.$$
Thus,
$$\|\ell_t^1 - \ell_t^2\|_\infty \leq I_1 + I_2 + \sum_{i,j=1}^d I_{ij}, \qquad (3.7.45)$$
where
$$I_1 := \|P_{0,t}^{\rho^1}\ell_{\gamma^1} - P_{0,t}^{\rho^1}\ell_{\gamma^2}\|_\infty \leq \|\ell_{\gamma^1} - \ell_{\gamma^2}\|_\infty, \qquad (3.7.46)$$
and
$$I_2 := \int_0^t ds \int_{\mathbb{R}^d} \left|\left\{[\ell_s^2(b_s^{\ell^2} - \bar{b}_s^{\rho^1}) - \ell_s^1(b_s^{\ell^1} - \bar{b}_s^{\rho^1})]\nabla p_{s,t}^{\rho^1}(\cdot, y)\right\}(x)\right| dx,$$
$$I_{ij} := \frac{1}{2}\sup_{y\in\mathbb{R}^d}\left|\int_0^t ds \int_{\mathbb{R}^d} \{\ell_s^2(a_s^{\rho^2} - a_s^{\rho^1})_{ij}\partial_i\partial_j p_{s,t}^{\rho^1}(\cdot, y)\}(x) dx\right|.$$
Below we estimate I_2 and I_{ij} respectively.
Firstly, by $(A^{3.7})$ and (3.7.40), we find a constant $c_1 > 0$ such that
$$|\ell_s^2\{b_s^{\ell^2}(x) - \bar{b}_s^{\rho^1}(x)\} - \ell_s^1(x)\{b_s^{\ell^1}(x) - \bar{b}_s^{\rho^1}(x)\}|$$
$$\leq \|\ell_s^1 - \ell_s^2\|_\infty |b_s^{\ell^2}(x) - \bar{b}_s^{\rho^1}(x)| + \|\ell_s^2\|_\infty |b_s^{\ell^2}(x) - b_s^{\ell^1}(x)|$$
$$\leq c_1 \|\ell_{\gamma^2}\|_\infty \|\ell_s^1 - \ell_s^2\|_\infty f_0(s,x), \quad s \in [0,T], x \in \mathbb{R}^d.$$

Combining this with (3.7.42) for $i = 1$, (3.7.13) for $(p, q) = (p_0, q_0)$ and $p' = \infty$, and applying Hölder's inequality, we find constant $c_2, c_3 > 0$ such that for any $t \in [0, T]$,

$$I_2 \leq c_2 \|\ell_{\gamma^2}\|_\infty \int_0^t (t-s)^{-\frac{1}{2}} \|\ell_s^1 - \ell_s^2\|_\infty \tilde{P}_{s,t}^\kappa f_0(s, \cdot)(y) ds$$

$$\leq c_2 \|\ell_{\gamma^2}\|_\infty \|(t - \cdot)^{\frac{d}{2p_0}} f_0\|_{\tilde{L}_{q_0}^\infty(t)}$$

$$\times \left(\int_0^t \{(t-s)^{-(\frac{1}{2} + \frac{d}{2p_0})} \|\ell_s^1 - \ell_s^2\|_\infty\}^{\frac{q_0}{q_0-1}} ds \right)^{\frac{q_0-1}{q_0}} \quad (3.7.47)$$

$$\leq c_3 \|\ell_{\gamma^2}\|_\infty \|f_0\|_{\tilde{L}_{q_0}^{p_0}(t)} \left(\int_0^t (t-s)^{-\frac{q_0(p_0+d)}{2p_0(q_0-1)}} \|\ell_s^1 - \ell_s^2\|_\infty^{\frac{q_0}{q_0-1}} ds \right)^{\frac{q_0-1}{q_0}}.$$

Next, by integration by parts formula, $(A^{3.7})$, (3.7.41), (3.7.42) for $i = 1$ and Lemma 3.7.7, for any $\delta := \alpha \wedge \beta$, we find constants $c_4, c_5 > 0$ such that

$$\left| \int_{\mathbb{R}^d} \{\ell_s^2 (a_s^{\rho^2} - a_s^{\rho^1})_{ij} \partial_i \partial_j p_{s,t}^{\rho^1}(\cdot, y)\}(x) dx \right|$$

$$= \left| \int_{\mathbb{R}^d} \left[(1-\Delta)^{\frac{\delta}{2}} \{\ell_s^2 (a_s^{\rho^2} - a_s^{\rho^1})_{ij}\}(x)\right] \cdot \left[\partial_i \partial_j (1-\Lambda)^{-\frac{\delta}{2}} p_{s,t}^{\rho^1}(\cdot, y)(x)\right] dx \right|$$

$$\leq \left\| (1-\Delta)^{\frac{\delta}{2}} \{\ell_s^2 (a_s^{\rho^2} - a_s^{\rho^1})_{ij}\} \right\|_\infty \int_{\mathbb{R}^d} |\partial_i \partial_j (1-\Delta)^{-\frac{\delta}{2}} p_{s,t}^{\rho^1}(\cdot, y)(x)| dx$$

$$\leq c_4 \|\ell_s^2 (a_s^{\rho^2} - a_s^{\rho^1})_{ij}\|_{C_b^{\beta \wedge \alpha}} (t-s)^{\frac{\delta}{2}-1}$$

$$\leq c_5 \|\ell_{\gamma^2}\|_{C_b^\beta} (t-s)^{\frac{\delta}{2}-1} \|\rho_s^1 - \rho_s^2\|_\infty.$$

By combining this with (3.7.45), (3.7.46) and (3.7.47), we arrive at

$$\|\ell_t^1 - \ell_t^2\|_\infty \leq \|\ell_{\gamma^1} - \ell_{\gamma^2}\|_\infty$$

$$+ c_3 \|\ell_{\gamma^2}\|_\infty \left(\int_0^t (t-s)^{-\frac{q_0(p_0+d)}{2p_0(q_0-1)}} \|\ell_s^1 - \ell_s^2\|_\infty^{\frac{q_0}{q_0-1}} ds \right)^{\frac{q_0-1}{q_0}}$$

$$+ \frac{d^2 c_5}{2} \|\ell_{\gamma^2}\|_{C_b^\beta} \int_0^t (t-s)^{\frac{\delta}{2}-1} \|\rho_s^1 - \rho_s^2\|_\infty ds, \quad t \in [0, T].$$

Consequently, for any $\lambda > 0$,

$$d_{\infty,\lambda}(\ell_{X^{\rho^1},\gamma^1}, \ell_{X^{\rho^2},\gamma^2}) := \sup_{t \in [0,T]} e^{-\lambda t} \|\ell_t^1 - \ell_t^2\|_\infty$$

$$\leq \|\ell_{\gamma^1} - \ell_{\gamma^2}\|_\infty + \varepsilon(\lambda)\{d_{\infty,\lambda}(\ell_{X^{\rho^1},\gamma^1}, \ell_{X^{\rho^2},\gamma^2}) + d_{\infty,\lambda}(\rho^1, \rho^2)\}$$

holds for

$$\varepsilon(\lambda) := \sup_{t \in [0,T]} \left\{ c_3 \|\ell_{\gamma^2}\|_\infty \left(\int_0^t (t-s)^{-\frac{q_0(p_0+d)}{2p_0(q_0-1)}} e^{-\frac{q_0 \lambda (t-s)}{q_0-1}} ds \right)^{\frac{q_0-1}{q_0}} \right.$$

$$\left. + \frac{d^2 c_5}{2} \|\ell_{\gamma^2}\|_{C_b^\beta} \int_0^t (t-s)^{\frac{\delta}{2}-1} e^{-\lambda(t-s)} ds \right\}.$$

Since $(p_0, q_0) \in \mathcal{K}$ implies $\frac{q_0(p_0+d)}{2p_0(q_0-1)} < 1$, and since $1 - \frac{\delta}{2} < 1$, by taking large enough $\lambda > 0$ increasing in $\|\ell_{\gamma^2}\|_{C_b^\beta}$, we obtain

$$\begin{aligned} d_{\infty,\lambda}(\ell_{X^{\rho^1},\gamma^1}, \ell_{X^{\rho^2},\gamma^2}) &\leq \|\ell_{\gamma^1} - \ell_{\gamma^2}\|_\infty \\ &+ \frac{1}{4}\{d_{\infty,\lambda}(\ell_{X^{\rho^1},\gamma^1}, \ell_{X^{\rho^2},\gamma^2}) + d_{\infty,\lambda}(\rho^1, \rho^2)\}. \end{aligned} \quad (3.7.48)$$

Taking $\gamma^1 = \gamma^2 = \gamma$, we see that the map $\rho \mapsto \ell_{X^{\rho,\gamma}}$ is contractive on the complete metric space $(\mathcal{P}_{\gamma,T}^\infty, d_{\infty,\lambda})$, so that it has a unique fixed point. Therefore, (3.7.1) is well-posed. Estimate (3.7.37) follows from Lemma 3.7.6 for $\rho_t = \ell_{X_t}$ for the solution to (3.7.1), while (3.7.38) follows from (3.7.48) for $\rho_t^i := \ell_{X_t^i}, \gamma^i = \mathcal{L}_{X_0^i}, i = 1, 2$.

3.8 Notes and further results

There exist many other results on DDSDEs and related topics, see for instance [Wang (2018)], [Hammersley et al. (2021)], [Huang and Wang (2019)], [Huang and Wang (2021a)], [Huang and Wang (2022)], [Röckner and Zhang (2021)], [Chaudru de Raynal (2017)], [Chaudru de Raynal (2019)], [Zhao (2020)] and references within. In the following, we introduce some further results on singular DDSDEs and path-distribution dependent models. See [Hong et al. (2024)] for distribution dependent SDEs/SPDEs under local monotone conditions.

3.8.1 DDSDEs with linear functional derivative of noise coefficients

Let f be a function on $\mathcal{P}$, $\partial_\mu f(\mu) \in \mathcal{B}_b(\mathbb{R}^d)$ is called the linear functional derivative of f at $\mu \in \mathcal{P}$, if for any $\nu \in \mathcal{P}$ we have

$$\lim_{\varepsilon \downarrow 0} \frac{f((1-\varepsilon)\mu + \varepsilon\nu) - f(\mu)}{\varepsilon} = \int_{\mathbb{R}^d} \partial_\mu f(\mu)(y)(\nu - \mu)(dy). \quad (3.8.1)$$

If $\partial_\mu f(\mu)(y)$ has linear functional derivative in μ, we say that f has second order linear functional derivative, and denote

$$\partial_\mu^2 f(\mu)(y, z) = \partial_\mu\{\partial_\mu f(\mu)(y)\}(z). \quad (3.8.2)$$

($A^{3.8}$) $\|b\|_\infty + \|\sigma\|_\infty + \|(\sigma\sigma^*)^{-1}\|_\infty + \|\nabla\sigma\|_\infty < \infty$, $b_t(x,\cdot)$ and $a_t(x,\cdot) := (\sigma_t \sigma_t^*)(x,\cdot)$ have linear functional derivatives, and there exist constants $K > 0$ and $\alpha \in (0,1]$ such that for any $t \in [0,T], \nu, \mu \in \mathcal{P}$ and $x, x', y, y', z, z' \in \mathbb{R}^d$,

$$|b_t(x,\mu) - b_t(x,\nu)| \leq K\|\mu - \nu\|_{var},$$
$$\|a_t(x,\mu) - a_t(x',\mu)\| \leq K|x - x'|^\alpha,$$
$$\|\partial_\mu a_t(x,\mu)(y) - \partial_\mu a_t(x',\mu)(y')\| \leq K\bigl(|x-x'|^\alpha + |y-y'|^\alpha\bigr).$$

The following result is due to Theorem 3.4 and Corollary 3.5 in [Chaudru de Raynal and Frikha (2022)]. See also [Zhao (2020)] for further result where $|b| \in L_q^p$ for some $(p,q) \in \mathcal{K}$ and $b_t(x,\cdot)$ is Lipschitz continuous in a weighted variation distance.

Theorem 3.8.1. *Assume* ($A^{3.8}$). *Then* (3.1.1) *is weak well-posed, and it is well-posed if in addition that* $\sigma_t(\cdot,\mu)$ *is Lispchitz continuous uniformly in* $(t,\mu) \in [0,T] \times \mathcal{P}$.

3.8.2 Singular DDSDEs with integral type, Kato class and critical drifts

Consider
$$dX_t = \int_{\mathbb{R}^d} \tilde{b}_t(X_t, y)\mathcal{L}_{X_t}(dy) + dW_t, \quad t \in [0,T], \tag{3.8.3}$$
where $\tilde{b}$ satisfies $|\tilde{b}_t(x,y)| \leq f_t(x-y)$ for some $f \in \tilde{L}_q^p$ with $(p,q) \in \mathcal{K}, p > 2$. The well-posedness of (3.8.3) is proved in [Röckner and Zhang (2021)] for initial values with $\mathbb{E}[|X_0|^k] < \infty$ for some $k > 2$.

Next, consider
$$dX_t = b_t(X_t, \mathcal{L}_{X_t})dt + dW_t, \quad t \in [0,T]. \tag{3.8.4}$$
Combining Theorem 1.7.4 with Theorem 3.5.2 for $k = 0$ and f being a constant, we have the following result.

Theorem 3.8.2. *Assume that there exists a constant* $K > 0$ *such that*
$$|b_t(x,\mu) - b_t(x,\nu)| \leq K\|\mu - \nu\|_{var}, \quad t \in [0,T], x \in \mathbb{R}^d, \mu, \nu \in \mathcal{P}. \tag{3.8.5}$$
If for any $\mu \in C^w([0,T]; \mathcal{P})$, $|b^\mu|^2 \in \mathbf{K}_{d,\alpha}$ *holds for some* $\alpha > 0$, *where* $b_t^\mu(x) := b_t(x, \mu_t)$, *then* (3.8.4) *is well-posed. If* $|b^\mu| \in \mathbf{K}_{d,1}$ *for any* $\mu \in C^w([0,T]; \mathcal{P})$, *then* (3.8.4) *is weakly well-posed, and for some constants* $c_1, c_2 > 0$:
$$\frac{(P_{s,t}^*\nu)(dx)}{dx} \leq \frac{c_1}{(t-s)^{\frac{d}{2}}} \int_{\mathbb{R}^d} e^{-\frac{c_2|x-y|^2}{t-s}} \nu(dy), \quad 0 \leq s < t \leq T, x \in \mathbb{R}^d, \nu \in \mathcal{P}.$$

Moreover, Theorem 1.7.5 and Theorem 3.5.2 imply the following result.

Theorem 3.8.3. *Assume that* (3.8.5) *holds for some constant* $K > 0$. *Let* $b_t^\mu(x) := b_t(x, \mu_t)$ *for* $\mu \in C^w([0, T]; \mathcal{P})$. *If for any* $\mu \in C^w([0, T]; \mathcal{P})$ *we have either* $b^\mu \in C_b^w([0, T]; L^d(\mathbb{R}^d))$ *or* $|b| \in L_{q_0}^{p_0}(T)$ *for some* $(p_0, q_0) \in (2, \infty)$ *with* $\frac{d}{p_0} + \frac{2}{q_0} = 1$, *then* (3.8.4) *is well-posed. If* $|b^\mu| \in L_\infty^d(T)$ *holds for any* $\mu \in C^w([0, T]; \mathcal{P})$, *then* (3.8.4) *is weakly well-posed.*

3.8.3 Singular distribution dependent semilinear SPDEs

Let $\mathbb{H}, \mathbb{U}$ be two separable Hilbert spaces, and let $\mathcal{L}(\mathbb{U}; \mathbb{H})$ be the space of bounded linear operators from $\mathbb{U}$ to $\mathbb{H}$. Consider the following distribution dependent semilinear SPDE on $\mathbb{H}$:

$$dX_t = \{b_t(X_t, \mathcal{L}_{X_t}) + AX_t\}dt + \sigma_t(X_t, \mathcal{L}_{X_t})dW_t, \quad t \in [0, T], \quad (3.8.6)$$

where W_t is the cylindrical Brownian motion on $\mathbb{U}$, $(A, \mathcal{D}(A))$ is a negative definite self-adjoint operator on $\mathbb{H}$, and

$$b : [0, T] \times \mathbb{H} \times \mathcal{P}_2(\mathbb{H}) \to \mathbb{H}, \quad \sigma : [0, T] \times \mathbb{H} \times \mathcal{P}_2(\mathbb{H}) \to \mathcal{L}(\mathbb{U}; \mathbb{H})$$

are measurable, for $\mathcal{P}_2(\mathbb{H})$ being the class of probability measures on $\mathbb{H}$ having finite second moment.

The following result is due to [Huang and Song (2021)] extending the corresponding result of [Wang (2016)] for singular semilinear SPDEs, see also [Criens (2023)] for the weak existence under a growth condition and further study on propagation of chaos.

Theorem 3.8.4. (3.8.6) *is well-posed for distributions in* $\mathcal{P}_2(\mathbb{H})$ *provided the following conditions hold.*

(1) *A has discrete spectrum with eigenvalues $\{-\lambda_n\}_{n \geq 0}$ satisfying*

$$\sum_{n=1}^\infty \lambda_n^{-\varepsilon} < \infty$$

for some $\varepsilon \in (0, 1)$.

(2) *There exists a constant* $K > 0$ *and an increasing* $\phi : [0, \infty) \to [0, \infty)$ *with* ϕ^2 *concave and* $\int_0^1 \frac{\phi(s)}{s} ds < \infty$ *such that*

$\|(\sigma\sigma^*)^{-1}\|_\infty + \|\sigma\|_\infty + \|\nabla\sigma\|_\infty + \|\nabla^2\sigma\|_\infty < \infty$,

$\lim_{n \to \infty} \|\sigma_t(x, \mu) - \sigma_t(\pi_n x, \mu)\|_{HS} = 0, \quad t \in [0, T], x \in \mathbb{H}, \mu \in \mathcal{P}_2(\mathbb{H})$,

$\|\sigma_t(x, \mu) - \sigma_t(x, \nu)\|_{HS} \leq K\mathbb{W}_2(\mu, \nu), \quad t \in [0, T], x \in \mathbb{H}, \mu, \nu \in \mathcal{P}_2(\mathbb{H})$,

$|b_t(x, \mu) - b_t(y, \nu)| \leq \phi(|x - y|) + K\mathbb{W}_2(\mu, \nu)$,

$t \in [0, T], x, y \in \mathbb{H}, \mu, \nu \in \mathcal{P}_2(\mathbb{H})$, where π_n is the projection onto the eigenspace corresponding to the first n eigenvalues of $-A$.

See also [Hong and Liu (2021)] for distribution dependent quasi-linear SPDEs.

3.8.4 Path-distribution dependent SDEs

The well-posedness has been studied in [Huang *et al.* (2019)] for the following path-dependent DDSDEs on $\mathbb{R}^d$:

$$dX_t = b_t(X_{r_0,t}, \mathcal{L}_{X_{r_0,t}})dt + \sigma_t(X_{r_0,t}, \mathcal{L}_{X_{r_0,t}})dW_t, \quad X_{r_0,0} \in \mathcal{C}, t \in [0,T], \tag{3.8.7}$$

where $r_0 > 0$ is a fixed constant, $\mathcal{C} := C([-r_0,0];\mathbb{R}^d)$, $X_{r_0,t} \in \mathcal{C}$ with $X_{r_0,t}(\theta) := X_{t-\theta}$ for $\theta \in [-r_0,0]$, and

$$b : [0,T] \times \mathcal{C} \times \mathcal{P}(\mathcal{C}) \to \mathbb{R}^d, \quad \sigma : [0,T] \times \mathcal{C} \times \mathcal{P}(\mathcal{C}) \to \mathbb{R}^d \otimes \mathbb{R}^m$$

satisfy some monotone conditions.

3.8.5 Singular path-distribution dependent nonlinear SPDEs

Let $\mathbb{H}, \mathbb{U}$ be two separable Hilbert spaces, and let $\mathcal{L}_2(\mathbb{U};\mathbb{H})$ be the space of Hilbert-Schmidt operators from $\mathbb{U}$ to $\mathbb{H}$ with Hilbert-Schmidt norm $\|\cdot\|_{\mathcal{L}_2(\mathbb{U};\mathbb{H})}$. For a Banach space $\mathbb{M}$, let $\mathcal{P}_{T,\mathbb{M}}$ be the set of probability measures on the path space $\mathcal{C}_{T,\mathbb{M}} := C([0,T];\mathbb{M})$. We also consider the weakly continuous path space

$$\mathcal{C}^w_{T,\mathbb{M}} := \{\xi : [0,T] \to \mathbb{M} \text{ is weak continuous}\}.$$

Both $\mathcal{C}_{T,\mathbb{M}}$ and $\mathcal{C}^w_{T,\mathbb{M}}$ are Banach spaces under the uniform norm

$$\|\xi\|_{T,\mathbb{M}} := \sup_{t \in [0,T]} \|\xi(t)\|_{\mathbb{M}}.$$

Let $\mathcal{P}^w_{T,\mathbb{M}}$ be the space of all probability measures on $\mathcal{C}^w_{T,\mathbb{M}}$ equipped with the weak topology. Denote $\mathcal{P}_{T,\mathbb{M}} = \{\mu \subset \mathcal{P}^w_{T,\mathbb{M}} : \mu(\mathcal{C}_{T,\mathbb{M}}) = 1\}$.

For any map $\xi : [0,T] \to \mathbb{M}$ and $t \in [0,T]$, the path $\pi_t(\xi)$ of ξ before time t is given by

$$\pi_t(\xi) := \xi_t : [0,T] \to \mathbb{M}, \quad \xi_t(s) := \xi(t \wedge s), \quad s \in [0,T].$$

Then the marginal distribution before time t of a probability measure $\mu \in \mathcal{P}^w_{T,\mathbb{M}}$ reads

$$\mu_t := \mu \circ \pi_t^{-1}.$$

The well-posedness is studied in [Ren et al. (2020)] for the following distribution-path dependent nonlinear SPDE on $\mathbb{H}$:

$$dX_t = \{B(t, X_t) + b_t(X_{\cdot \wedge t}, \mathcal{L}_{X_{\cdot \wedge t}})\}dt + \sigma_t(X_{\cdot \wedge t}, \mathcal{L}_{X_{\cdot \wedge t}})dW(t), \quad t \in [0, T],$$

where $X_{\cdot \wedge t}$ is a random variable on $\mathcal{C}^w_{T, \mathbb{H}}$ with $X_{\cdot \wedge t}(s) := X_{s \wedge t}$ for $s \in [0, T]$, and for some separable Hilbert space $\mathbb{B}$ with $\mathbb{H} \hookrightarrow\hookrightarrow \mathbb{B}$ (" $\hookrightarrow\hookrightarrow$ " means the embedding is compact),

$$B : [0, T] \times \mathbb{H} \times \Omega \to \mathbb{B},$$
$$b : [0, T] \times \mathcal{C}^w_{T, \mathbb{H}} \times \mathcal{P}^w_{T, \mathbb{H}} \times \Omega \to \mathbb{H}, \quad (3.8.8)$$
$$\sigma : [0, T] \times \mathcal{C}^w_{T, \mathbb{H}} \times \mathcal{P}^w_{T, \mathbb{H}} \times \Omega \to \mathcal{L}_2(\mathbb{U}; \mathbb{H})$$

are progressively measurable maps.

In applications, $B(t, \cdot)$ is a singular nonlinear term which may not take values in the state space $\mathbb{H}$. For instance, for the stochastic transport SPDE, we take $B(t, X) = -(X \cdot \nabla)X$ for X in a functional space over a Riemannian manifold, while b and σ are regular terms which are locally Lipschitz continuous in the variables (ξ, μ).

3.8.6 Singular degenerate DDSDEs

As extensions to (1.7.5) and (1.7.6), consider the following distribution dependent degenerate SDE for $(X_t, Y_t) \in \mathbb{R}^{2d}$:

$$\begin{cases} dX_t = Z_t(X_t, Y_t)dt, \\ dY_t = b_t(X_t, Y_t, \mathcal{L}_{(X_t, Y_t)})dt + \sigma_t(X_t, Y_t)dW_t, \quad t \in [0, T]. \end{cases} \quad (3.8.9)$$

Combining Theorems 1.7.2 and 1.7.3 with Theorem 3.5.2 for $k = 0$ and f being a constant, we have the following results.

Theorem 3.8.5. *The SDE* (3.8.9) *is well-posed, if for any* $\mu \in C^w_b([0, T]; \mathcal{P}(\mathbb{R}^{2d}))$, *the conditions in Theorem 1.7.2 holds for* $b_t(x, y, \mu_t)$ *replacing* $b_t(x, y)$, *and there exists a constant* $K > 0$ *such that*

$$|b_t(z, \mu) - b_t(z, \nu)| \leq K\|\mu - \nu\|_{var},$$
$$t \in [0, T], z \in \mathbb{R}^{2d}, \mu, \nu \in \mathcal{P}(\mathbb{R}^{2d}). \quad (3.8.10)$$

Theorem 3.8.6. *The SDE* (3.8.9) *with* $Z_t(x, y) = y$ *is well-posed if the conditions in Theorem 1.7.3 holds for* $b_t(x, y, \mu_t)$ *replacing* $b_t(x, y)$, *and there exists a constant* $K > 0$ *such that* (3.8.10) *holds.*

Next, we consider the weak existence for the following degenerate SDE with distribution dependent noise:

$$\begin{cases} dX_t = Y_t dt, \\ dY_t = b_t(X_t, Y_t, \mathcal{L}_{(X_t,Y_t)})dt + \sigma_t(X_t, \mathcal{L}_{(X_t,Y_t)})dW_t \end{cases} \quad (3.8.11)$$

for $t \in [0, T]$, where for $\rho_\mu(z) := \frac{\mu(dz)}{dz}$ and $\rho_{\mu_1}(x) := \int_{\mathbb{R}^d} \rho_\mu(x, y) dy$,

$$b_t(z, \mu) := \int_{\mathbb{R}^{2d}} \tilde{b}_t(z, \rho_\mu(z), z') \mu(dz'),$$

$$\sigma_t(x, \mu) := \left(2 \int_{\mathbb{R}^{2d}} a_t(x, \rho_{\mu_1}(x), z') \mu(dz') \right)^{\frac{1}{2}}.$$

For any $q > 1$ and $\mathbf{p} \in (1, \infty)^{2d}$, we write $f \in \tilde{L}_q^{\mathbf{p}}(T)$ if f is a measurable function on $[0, T] \times \mathbb{R}^{2d}$ such that

$$\int_0^T \| \cdots \| \| f_t \|_{\tilde{L}^{p_{2d}}(dx_{2d})} \|_{\tilde{L}^{p_{2d-1}}(dx_{2d-1})} \cdots \|_{L^{p_1}(dx_1)} dt < \infty.$$

The following result is taken from Theorem 1.3 in [Zhang (2021)].

Theorem 3.8.7. *For any initial distribution,* (3.8.11) *has at least one weak solution if* $a_t(x, r, z)$ *is continuous in* r, $\|a\|_\infty + \|a^{-1}\|_\infty < \infty$,

$$\lim_{\varepsilon \downarrow 0} \sup_{r,r' \in [0,n], |r-r'| \leq \varepsilon} \|\tilde{b}.(\cdot, r, \cdot) - \tilde{b}.(\cdot, r', \cdot)\|_{L^1(K)} = 0$$

for any compact set $K \subset [0, T] \times \mathbb{R}^{2d} \times \mathbb{R}^{2d}$, *and there exists* $f \in \tilde{L}_q^{\mathbf{p}}(T)$ *for some* $q \in (2, 4)$ *and* $\mathbf{p} \in (1, \infty)^{2d}$ *with*

$$\sum_{i=1}^{d} 3 p_i^{-1} + \sum_{i=d+1}^{2d} p_i^{-1} + \frac{2}{q} < \infty,$$

$$|\tilde{b}_t(z, r, z')| \leq f_t(z - z'), \quad t \in [0, T], r \geq 0, z, z' \in \mathbb{R}^{2d}.$$

Theorem 1.5 in [Zhang (2021)] also presents the existence and uniqueness of the "generalized martingale solution" of (3.8.11) when the initial distribution density is in C_b^1, $a_t(x, r, z) = a_t(x)$ and $\tilde{b}_t(z, \cdot, z')$ is Lipschitz continuous uniformly in (t, z, z'). See also [Hao et al. (2021b)] for the study of martingale solutions to the SDE

$$\begin{cases} dX_t = Y_t dt, \\ dY_t = \{b_t(X_t, Y_t) + (K * \mathcal{L}_{X_t})(X_t)\} dt + dW_t, \quad t \in [0, T], \end{cases}$$

where b and K are singular functions and $(K * \mu)(x) := \int_{\mathbb{R}^d} K(x - y) \mu(dy)$.

Chapter 4

DDSDEs: Harnack Inequality and Derivative Estimates

In this chapter we study the regularity of the maps

$$\mu \mapsto P_t^*\mu, \quad t \in (0,T],$$

where $P_t^*\mu := \mathcal{L}_{X_t}$ for X_t solving (3.1.1) with $\mathcal{L}_{X_0} = \mu$. Since a probability measure is determined by integrals of $f \in \mathcal{B}_b(\mathbb{R}^d)$, it suffices to study the regularity of the functionals

$$\mu \mapsto P_t f(\mu) := \int_{\mathbb{R}^d} f \mathrm{d}(P_t^*\mu), \quad f \in \mathcal{B}_b(\mathbb{R}^d), t \in (0,T]. \tag{4.0.1}$$

We will establish dimension-free Harnack inequalities and Bismut formulas for $P_t f$ when the noise is distribution free. For distribution dependent noise, these inequalities and formulas are still open except for a very special situation considered in [Bai and Huang (2023)] and [Huang and Wang (2022b)], where the noise only depends on the time and distribution variables. Derivative estimates are derived in Subsections 4.5.2 and 4.5.3 for the case with distribution dependent noise.

4.1 Log-Harnack inequality

In this part, we study the following type of log-Harnack inequality for P_t defined in (4.0.1):

$$P_t \log f(\nu) \leq \log P_t f(\mu) + c(t)\mathbb{W}_2(\mu,\nu)^2, \quad f \in \mathcal{B}_b^+(\mathbb{R}^d), \mu, \nu \in \mathcal{P}_2, t \in (0,T]$$

for some function $c : (0,T] \to (0,\infty)$. This is equivalent to the entropy-cost inequality

$$\mathrm{Ent}(P_t^*\mu|P_t^*\nu) \leq c(t)\mathbb{W}_2(\mu,\nu)^2, \quad \mu,\nu \subset \mathcal{P}_2, \ t \in (0,T].$$

4.1.1 Monotone and non-degenerate case

$(A^{4.1})$ σ, b are bounded on bounded subsets of $[0,T] \times \mathbb{R}^d \times \mathcal{P}_2$, $\sigma\sigma^*$ is invertible, and there exists a constant $L > 0$ such that $\|\sigma^*(\sigma\sigma^*)^{-1}\|_\infty \leq L$ and

$$\|\sigma_t(x,\mu) - \sigma_t(y,\nu)\|^2 + \langle b_t(x,\mu) - b_t(y,\nu), x - y \rangle^+$$
$$\leq L\{|x-y|^2 + \mathbb{W}_2(\mu,\nu)^2\}, \quad t \in [0,T], \ x,y \in \mathbb{R}^d, \ \mu,\nu \in \mathcal{P}_2.$$

By Theorem 3.3.1, $(A^{4.1})$ implies that (3.1.1) is well-posed for distributions in $\mathcal{P}_2$, and there exists a constant $c > 0$ such that

$$\mathbb{W}_2(P_t^*\mu, P_t^*\nu) \leq c\mathbb{W}_2(\mu,\nu), \quad \mu \in \mathcal{P}_2. \tag{4.1.1}$$

The following result is due to [Wang (2018)].

Theorem 4.1.1. *Assume* $(A^{4.1})$. *Then there exists a constant* $C > 0$ *such that the following inequalities hold for all* $t \in (0,T]$ *and* $\mu, \nu \in \mathcal{P}_2$:

$$P_t \log f(\nu) \leq \log P_t f(\mu) + \frac{C}{t} \mathbb{W}_2(\mu,\nu)^2, \quad f \in \mathcal{B}_b^+(\mathbb{R}^d), \tag{4.1.2}$$

$$\frac{1}{2}\|P_t^*\mu - P_t^*\nu\|_{var}^2 \leq \mathrm{Ent}(P_t^*\nu|P_t^*\mu) \leq \frac{C}{t}\mathbb{W}_2(\mu,\nu)^2, \tag{4.1.3}$$

$$|P_t f(\mu) - P_t f(\nu)| \leq \frac{\sqrt{2C}\|f\|_\infty}{\sqrt{t}} \mathbb{W}_2(\mu,\nu), \quad f \in \mathcal{B}_b(\mathbb{R}^d). \tag{4.1.4}$$

Proof. Noting that (4.1.3) and (4.1.4) are simple consequences of (4.1.2) and Pinsker's inequality (3.2.3), we only prove (4.1.2).

(a) For $\mu_0, \nu_0 \in \mathcal{P}_2$, let (X_0, Y_0) be $\mathcal{F}_0$-measurable such that

$$\mathcal{L}_{X_0} = \mu_0, \quad \mathcal{L}_{Y_0} = \nu_0, \quad \mathbb{E}|X_0 - Y_0|^2 = \mathbb{W}_2(\mu_0, \nu_0)^2. \tag{4.1.5}$$

Denote

$$\mu_t := P_t^*\mu_0, \quad \nu_t := P_t^*\nu_0, \quad t \geq 0.$$

Let X_t solve (3.4.1) with initial value X_0. We have

$$dX_t = b_t(X_t, \mu_t)dt + \sigma_t(X_t)dW_t, \quad t \in [0,T]. \tag{4.1.6}$$

Next, we use the coupling as in the proof of Theorem 1.5.2. For any $t_0 \in (0,T]$ consider the SDE

$$dY_t = \Big\{b_t(Y_t, \nu_t) + \frac{\sigma_t(Y_t)\{\sigma_t^*(\sigma_t\sigma_t^*)^{-1}\}(X_t)(X_t - Y_t)}{\xi_t}\Big\}dt$$
$$+ \sigma_t(Y_t)dW_t, \quad t \in [0,t_0). \tag{4.1.7}$$

For the constant $L > 0$ in $(A^{4.1})$, let
$$\xi_t := \frac{1}{L}\left(1 - e^{L(t-t_0)}\right), \quad t \in [0, t_0). \tag{4.1.8}$$
By $(A^{4.1})$, (4.1.7) has a unique solution up to times
$$\tau_{n,k} := \frac{t_0 n}{n+1} \wedge \inf\{t \in [0, t_0) : |Y_t| \geq k\}, \quad n, k \geq 1.$$
By Itô's formula and $(A^{4.1})$, for any $n \geq 1$ we find a constant $c(n) > 0$ such that
$$d|Y_t|^2 \leq c(n)(1 + |Y_t|^2)dt + dM_t, \quad t \in [0, \tau_{n,k}], \; n, k \geq 1$$
holds for some martingale M_t. This implies
$$\lim_{n \to \infty} \lim_{k \to \infty} \tau_{n,k} = \lim_{n \to \infty} \frac{t_0 n}{n+1} = t_0,$$
and hence (4.1.7) has a unique solution up to time t_0.

(b) For any $n \geq 1$, let
$$\tau_n := \frac{t_0 n}{n+1} \wedge \inf\{t \in [0, t_0) : |X_t - Y_t| \geq n\}. \tag{4.1.9}$$
By $(A^{4.1})$,
$$\eta_s := \{\sigma_s^*(\sigma_s \sigma_s^*)^{-1}\}(X_s)(X_s - Y_s)$$
satisfies $|\eta_s| \leq L|X_s - Y_s|$. By Girsanov's theorem,
$$\tilde{W}_t := W_t + \int_0^t \frac{\eta_s}{\xi_s} ds, \quad t \in [0, \tau_n]$$
is an m-dimensional Brownian motion under the probability $\mathbb{Q}_n := R_n \mathbb{P}$, where
$$R_n := e^{-\int_0^{\tau_n} \frac{1}{\xi_s} \langle \eta_s, dW_s \rangle - \frac{1}{2} \int_0^{\tau_n} \frac{|\eta_s|^2}{|\xi_s|^2} ds}. \tag{4.1.10}$$
Then (4.1.6) and (4.1.7) imply
$$dX_t = \left\{b_t(X_t, \mu_t) - \frac{X_t - Y_t}{\xi_t}\right\}dt + \sigma_t(X_t)d\tilde{W}_t,$$
$$dY_t = b_t(Y_t, \nu_t)dt + \sigma_t(Y_t)d\tilde{W}_t, \quad t \in [0, \tau_n], n \geq 1. \tag{4.1.11}$$
Combining this with $(A^{4.1})$, (4.1.1), (4.1.8) and Itô's formula, we obtain
$$d\frac{|X_t - Y_t|^2}{\xi_t} - dM_t$$
$$\leq \left\{\frac{L|X_t - Y_t|^2 + L|X_t - Y_t|\mathbb{W}_2(\mu_t, \nu_t)}{\xi_t} - \frac{|X_t - Y_t|^2(2 + \xi_t')}{\xi_t^2}\right\}dt$$
$$\leq \left\{\frac{L^2 \mathbb{W}_2(\mu_t, \nu_t)^2}{2} - \frac{|X_t - Y_t|^2(2 + \xi_t' - L\xi_t - \frac{1}{2})}{\xi_t^2}\right\}dt \tag{4.1.12}$$
$$\leq \left\{\frac{L^2 e^{2Lt} \mathbb{W}_2(\mu_0, \nu_0)^2}{2} - \frac{|X_t - Y_t|^2}{2\xi_t^2}\right\}dt, \quad t \in [0, \tau_n],$$

where $\mathrm{d}M_t := \frac{2}{\xi_t}\langle X_t - Y_t, \{\sigma_t(X_t) - \sigma_t(Y_t)\}\mathrm{d}\tilde{W}_t\rangle$ is a $\mathbb{Q}_n$-martingale. Combining this with (4.1.1) and ($A^{4.1}$), we derive

$$\mathbb{E}[R_n \log R_n] = \mathbb{E}_{\mathbb{Q}_n}[\log R_n] = \frac{1}{2}\mathbb{E}_{\mathbb{Q}_n}\int_0^{\tau_n}\frac{|\eta_s|^2}{\xi_s^2}\mathrm{d}s$$
$$\leq \frac{L^2}{2}\mathbb{E}_{\mathbb{Q}_n}\int_0^{\tau_n}\frac{|X_s - Y_s|^2}{\xi_s^2}\mathrm{d}s \leq \frac{c}{t_0}\mathbb{W}_2(\mu_0,\nu_0)^2, \quad n \geq 1 \qquad (4.1.13)$$

for some constant $c > 0$ uniformly in $t_0 \in (0, T]$. Therefore, by the martingale convergence theorem, $R_\infty := \lim_{n\to\infty} R_n$ exists, and

$$N_t := e^{-\int_0^t \frac{1}{\xi_s}\langle \eta_s, \mathrm{d}W_s\rangle - \frac{1}{2}\int_0^t \frac{|\eta_s|^2}{|\xi_s|^2}\mathrm{d}s}, \quad t \in [0, t_0]$$

is a $\mathbb{P}$-martingale.

(c) Finally, let $\mathbb{Q} := N_{t_0}\mathbb{P}$. By Girsanov's theorem, $(\tilde{W}_t)_{t\in[0,t_0]}$ is an m-dimensional Brownian motion under the probability $\mathbb{Q}$, and $(X_t)_{t\in[0,t_0]}$ solves the SDE

$$\mathrm{d}X_t = \Big\{b_t(X_t, \mu_t) - \frac{X_t - Y_t}{\xi_t}\Big\}\mathrm{d}t + \sigma_t(X_t)\mathrm{d}\tilde{W}_t, \quad t \in [0, t_0]. \qquad (4.1.14)$$

Let $(Y_t)_{t\in[0,t_0]}$ solve

$$\mathrm{d}Y_t = b_t(Y_t, \nu_t)\mathrm{d}t + \sigma_t(Y_t)\mathrm{d}\tilde{W}_t, \quad t \in [0, t_0]. \qquad (4.1.15)$$

By the well-posedness of (3.4.1), this extends the second equation in (4.1.11) with $\mathcal{L}_{Y_{t_0}|\mathbb{Q}} = \nu_{t_0}$. Moreover, (4.1.13) and Fatou's lemma imply

$$\frac{1}{2}\mathbb{E}_\mathbb{Q}\int_0^{t_0}\frac{|\{\sigma_s^*(\sigma_s\sigma_s^*)^{-1}\}(X_s)(X_s - Y_s)|^2}{|\xi_s|^2}\mathrm{d}s$$
$$= \mathbb{E}[N_{t_0}\log N_{t_0}] \leq \liminf_{n\to\infty}\mathbb{E}[R_n \log R_n] \leq \frac{c}{t_0}\mathbb{W}_2(\mu_0,\nu_0)^2, \qquad (4.1.16)$$

which in particular implies $\mathbb{Q}(X_{t_0} = Y_{t_0}) = 1$ as explained in the proof of Theorem 1.5.2. Combining this with the Young inequality (see Lemma 2.4 in [Arnaudon et al. (2009)])

$$\mu(fg) \leq \mu(f \log f) + \log \mu(e^g), \quad f, g \geq 0, \mu(f) = 1, \mu \in \mathcal{P}, \qquad (4.1.17)$$

we arrive at

$$P_{t_0}\log f(\nu_0) = \mathbb{E}[N_{t_0}\log f(Y_{t_0})] = \mathbb{E}[N_{t_0}\log f(X_{t_0})]$$
$$\leq \mathbb{E}[N_{t_0}\log N_{t_0}] + \log \mathbb{E}[f(X_{t_0})]$$
$$\leq \log P_{t_0}f(\mu_0) + \frac{c}{t_0}\mathbb{W}_2(\mu_0,\nu_0)^2, \quad t_0 \in (0, T].$$

Hence, (4.1.2) holds. □

4.1.2 Degenerate case

Consider the following distribution dependent stochastic Hamiltonian system for $(X_t, Y_t) \in \mathbb{R}^d := \mathbb{R}^{d_1} \times \mathbb{R}^{d_2}$:

$$\begin{cases} \mathrm{d}X_t = (AX_t + BY_t)\mathrm{d}t, \\ \mathrm{d}Y_t = Z(t,(X_t,Y_t),\mathcal{L}_{(X_t,Y_t)})\mathrm{d}t + \sigma_t \mathrm{d}W_t, \quad t \in [0,T], \end{cases} \quad (4.1.18)$$

where A is a $d_1 \times d_1$-matrix, B is a $d_1 \times d_2$-matrix, σ is a $d_2 \times d_2$-matrix, W_t is the d_2-dimensional Brownian motion on a complete filtration probability space $(\Omega, \{\mathcal{F}_t\}_{t \geq 0}, \mathbb{P})$, and

$$Z : [0,\infty) \times \mathbb{R}^d \times \mathcal{P}_2 \to \mathbb{R}^{d_2}, \quad \sigma : [0,\infty) \to \mathbb{R}^{d_2} \otimes \mathbb{R}^{d_2}$$

are measurable. We assume

$(A^{4.2})$ σ_t is invertible, there exists a constant $K > 0$ such that

$$\|\sigma(t)^{-1}\| \leq K, \quad |Z(t,x,\mu) - Z(t,y,\nu)| \leq K\{|x-y| + \mathbb{W}_2(\mu,\nu)\}$$

holds for all $t \geq 0, \mu, \nu \in \mathcal{P}_2$ and $x, y \in \mathbb{R}^d$, and A, B satisfy the following Kalman's rank condition for some $k \geq 1$:

$$\operatorname{Rank}[A^0 B, \ldots, A^{k-1}B] = d_1, \quad A^0 := I_{d_1 \times d_1}.$$

This assumption implies $(A^{3.1})$ for $k = 2$. By Theorem 3.3.1, (4.1.18) is well-posed for distributions in $\mathcal{P}_2$ and

$$\mathbb{W}_2(P_t^* \mu, P_t^* \nu) \leq \mathrm{e}^{K't}\mathbb{W}_2(\mu,\nu), \quad t \geq 0, \mu, \nu \in \mathcal{P}_2 \quad (4.1.19)$$

holds for some constant $K' > 0$. The following result is due to [Ren and Wang (2021b)].

Theorem 4.1.2. *Assume* $(A^{4.2})$ *and let* P_t *be associated with* (4.1.18). *Then there exists a constant* $c > 0$ *such that for any* $t \in (0,T]$,

$$P_t \log f(\nu) \leq P_t \log f(\mu) + \frac{c}{t^{4k-1}} \mathbb{W}_2(\mu,\nu)^2, \quad (4.1.20)$$
$$t \in (0,T], \mu, \nu \in \mathcal{P}_2, f \in \mathcal{B}_b^+(\mathbb{R}^d).$$

Proof. By the Kalman rank condition in $(A^{4.2})$, for any $t_0 \in (0,T]$,

$$Q_{t_0} := \int_0^{t_0} t(t_0 - t)\mathrm{e}^{(t_0-t)A} BB^* \mathrm{e}^{(t_0-t)A^*} \mathrm{d}t$$

is invertible and there exists a constant $c_1 > 0$ such that

$$\|Q_{t_0}^{-1}\| \leq \frac{c_1 \mathrm{e}^{c_1 t_0}}{t_0^{2k+1}}, \quad t_0 \in (0,T], \quad (4.1.21)$$

see for instance Theorem 4.2(1) in [Wang and Zhang (2013)].

168   *Distribution Dependent Stochastic Differential Equations*

Let $(X_0, Y_0), (\bar{X}_0, \bar{Y}_0) \in L^2(\Omega \to \mathbb{R}^d, \mathcal{F}_0, \mathbb{P})$ such that $\mathcal{L}_{(X_0,Y_0)} = \mu$, $\mathcal{L}_{(\bar{X}_0,\bar{Y}_0)} = \nu$ and

$$\mathbb{E}\big(|X_0 - \bar{X}_0|^2 + |Y_0 - \bar{Y}_0|^2\big) = \mathbb{W}_2(\mu,\nu)^2. \qquad (4.1.22)$$

Next, let (X_t, Y_t) solve (4.1.18). Then $\mathcal{L}_{(X_t,Y_t)} = P_t^*\mu$. Consider the modified equation with initial value $(\bar{X}_0, \bar{Y}_0)$:

$$\begin{cases} \mathrm{d}\bar{X}_t = \big(A\bar{X}_t + B\bar{Y}_t\big)\mathrm{d}t, \\ \mathrm{d}\bar{Y}_t = \Big\{ Z(t,(X_t,Y_t),P_t^*\mu) + \frac{Y_0 - \bar{Y}_0}{t_0} \\ \qquad\qquad + \frac{\mathrm{d}}{\mathrm{d}t}\big[t(t_0-t)B^*\mathrm{e}^{(t_0-t)A^*}v\big]\Big\}\mathrm{d}t + \sigma_t \mathrm{d}W_t, \end{cases} \qquad (4.1.23)$$

where

$$v := Q_{t_0}^{-1}\Big\{ \mathrm{e}^{t_0 A}(X_0 - \bar{X}_0) + \int_0^{t_0} \frac{t - t_0}{t_0} \mathrm{e}^{(t_0-t)A} B(\bar{Y}_0 - Y_0)\mathrm{d}t \Big\}. \qquad (4.1.24)$$

Then

$$\begin{aligned} \bar{Y}_t - Y_t &= \bar{Y}_0 - Y_0 + \int_0^t \Big\{ \frac{Y_0 - \bar{Y}_0}{t_0} + \frac{\mathrm{d}}{\mathrm{d}r}\big[r(t_0-r)B^*\mathrm{e}^{(t_0-r)A^*}v\big]\Big\}\mathrm{d}r \\ &= \frac{t_0 - t}{t_0}(\bar{Y}_0 - Y_0) + t(t_0 - t)B^*\mathrm{e}^{(t_0-t)A^*}v, \quad t \in [0,t_0]. \end{aligned} \qquad (4.1.25)$$

Consequently, $Y_{t_0} = \bar{Y}_{t_0}$, and combining with Duhamel's formula, we obtain

$$\begin{aligned} \bar{X}_t - X_t &= \mathrm{e}^{tA}(\bar{X}_0 - X_0) \\ &\quad + \int_0^t \mathrm{e}^{(t-r)A} B\Big\{ \frac{t_0 - r}{t_0}(\bar{Y}_0 - Y_0) + r(t_0 - r)B^*\mathrm{e}^{(t_0-r)A^*}v \Big\}\mathrm{d}r \end{aligned} \qquad (4.1.26)$$

for $t \in [0,t_0]$. This and (4.1.24) imply

$$\bar{X}_{t_0} - X_{t_0} = \mathrm{e}^{t_0 A}(\bar{X}_0 - X_0) + \int_0^{t_0} \frac{t_0 - r}{t_0}\mathrm{e}^{(t_0-r)A}B(\bar{Y}_0 - Y_0)\mathrm{d}r + Q_{t_0}v = 0,$$

which together with $Y_{t_0} = \bar{Y}_{t_0}$ observed above yields

$$(X_{t_0}, Y_{t_0}) = (\bar{X}_{t_0}, \bar{Y}_{t_0}). \qquad (4.1.27)$$

On the other hand, let

$$\begin{aligned} \xi_t &= \sigma^{-1}\Big\{ \frac{1}{t_0}(Y_0 - \bar{Y}_0) + \frac{\mathrm{d}}{\mathrm{d}t}\big[t(t_0-t)B^*\mathrm{e}^{(t_0-t)A^*}v\big] \\ &\quad + Z\big(t,(X_t,Y_t),P_t^*\mu\big) - Z\big(t,(\bar{X}_t,\bar{Y}_t),P_t^*\nu\big)\Big\}, \quad t \in [0,t_0]. \end{aligned}$$

By ($A^{4.2}$), (4.1.19), (4.1.21), (4.1.24), (4.1.25), and (4.1.26), we find a constant $c_2 > 0$ such that
$$|\xi_t|^2 \leq \frac{c_2}{t_0^{4k}} e^{c_2 t_0}\big\{|X_0 - \bar{X}_0|^2 + |Y_0 - \bar{Y}_0|^2 + \mathbb{W}_2(\mu,\nu)^2\big\}, \quad t \in [0, t_0]. \quad (4.1.28)$$
So, the Girsanov theorem implies that
$$\tilde{W}_t := W_t + \int_0^t \xi_s ds, \quad t \in [0, t_0]$$
is a d_2-dimensional Brownian motion under the probability measure $\mathbb{Q} := R\mathbb{P}$, where
$$R := e^{-\int_0^{t_0} \langle \xi_t, dW_t \rangle - \frac{1}{2}\int_0^{t_0} |\xi_t|^2 dt}. \quad (4.1.29)$$
Reformulating (4.1.23) as
$$\begin{cases} d\bar{X}_t = (A\bar{X}_t + B\bar{Y}_t)dt, \\ d\bar{Y}_t = Z(t, (\bar{X}_t, \bar{Y}_t), P_t^*\nu)dt + \sigma_t d\tilde{W}_t, \quad t \in [0, t_0], \end{cases}$$
by the weak uniqueness of (4.1.18) and that the distribution of $(\bar{X}_0, \bar{Y}_0)$ under $\mathbb{Q}$ coincides with $\mathcal{L}_{(\bar{X}_0, \bar{Y}_0)} = \nu$, we obtain $\mathcal{L}_{(X_t, \bar{Y}_t)|\mathbb{Q}} = P_t^*\nu$ for $t \in [0, T]$. Combining this with (4.1.27) and using the Young inequality (4.1.17), for any $f \in \mathcal{B}_b^+(\mathbb{R}^d)$ we have
$$\begin{aligned}(P_{t_0} \log f)(\nu) &= \mathbb{E}[R \log f(\bar{X}_{t_0}, \bar{Y}_{t_0})] = \mathbb{E}[R \log f(X_{t_0}, Y_{t_0})] \\ &\leq \log \mathbb{E}[f(X_{t_0}, Y_{t_0})] + \mathbb{E}[R \log R] = \log(P_{t_0} f)(\mu) + \mathbb{E}_{\mathbb{Q}}[\log R].\end{aligned} \quad (4.1.30)$$
By (4.1.28), and (4.1.29), $\tilde{W}_t$ is a Brownian motion under $\mathbb{Q}$, and noting that $\mathbb{Q}|_{\mathcal{F}_0} = \mathbb{P}|_{\mathcal{F}_0}$ and (4.1.22) imply
$$\mathbb{E}_{\mathbb{Q}}\big(|X_0 - \bar{X}_0|^2 + |Y_0 - \bar{Y}_0|^2\big) = \mathbb{W}_2(\mu,\nu)^2,$$
we find a constant $c > 0$ such that
$$\mathbb{E}_{\mathbb{Q}}[\log R] = \frac{1}{2}\mathbb{E}_{\mathbb{Q}}\int_0^{t_0} |\xi_t|^2 dt \leq \frac{c e^{ct_0}}{t_0^{4k-1}} \mathbb{W}_2(\mu,\nu)^2.$$
Therefore, (4.1.20) follows from (4.1.30). □

4.1.3 Singular case

Let $\|\mu\|_2 := \sqrt{\mu(|\cdot|^2)}$ for $\mu \in \mathcal{P}_2$. We make the following assumption.

($A^{4.3}$) $(\sigma_t(x), b_t(x, \delta_0))$ satisfies ($A^{1.3}$), and there exist a constant $\alpha \geq 0$ and a function $1 \leq f \in \tilde{L}_{q_0}^{p_0}(T)$ such that
$$|b_t(x,\mu) - b_t(x,\nu)| \leq \min\big\{f_t(x)\mathbb{W}_2(\mu,\nu),\ f_t(x) + \alpha\|\mu\|_2 + \alpha\|\nu\|_2\big\},$$
$$\mu,\nu \in \mathcal{P}_2, (t,x) \in [0,T] \times \mathbb{R}^d.$$

According to Theorem 3.5.1, $(A^{4.3})$ implies the well-posedness of (3.4.1) for distributions in $\mathcal{P}_2$. Let $P_t^*\mu = \mathcal{L}_{X_t}$ for the solution to (3.4.1) with $\mathcal{L}_{X_0} = \mu \in \mathcal{P}_2$ and let $P_t f(\mu)$ be in (4.0.1). The following result is due to [Wang (2023b)].

Theorem 4.1.3. *Assume* $(A^{4.3})$. *For any* $N > 0$, *let* $\mathcal{P}_{2,N}(\mathbb{R}^d) := \{\mu \in \mathcal{P}_2 : \|\mu\|_2 \leq N\}$.

(1) *For any* $N > 0$, *there exists a constant* $C(N) > 0$ *such that for any* $\mu, \nu \in \mathcal{P}_{2,N}(\mathbb{R}^d)$ *and any* $t \in [0,T]$, *the following inequalities hold:*

$$\mathbb{W}_2(P_t^*\mu, P_t^*\nu)^2 \leq C(N)\mathbb{W}_2(\mu,\nu)^2, \tag{4.1.31}$$

$$\mathrm{Ent}(P_t^*\nu | P_t^*\mu) \leq \frac{C(N)}{t}\mathbb{W}_2(\mu,\nu)^2, \tag{4.1.32}$$

$$|P_t g(\nu) - P_t g(\mu)| \leq \frac{\sqrt{2C(N)}}{\sqrt{t}}\|g\|_\infty \mathbb{W}_2(\mu,\nu), \quad g \in \mathcal{B}_b(\mathbb{R}^d). \tag{4.1.33}$$

(2) *If* $(A^{4.3})$ *holds for* $\alpha = 0$, *then there exists a constant* $C > 0$ *such that*

$$\mathbb{W}_2(P_t^*\mu, P_t^*\nu)^2 \leq C\mathbb{W}_2(\mu,\nu)^2, \quad \mu,\nu \in \mathcal{P}_2. \tag{4.1.34}$$

Moreover, if $\|f\|_\infty < \infty$, *then* (4.1.32)-(4.1.33) *hold for some constant* C *replacing* $C(N)$ *and all* $\mu,\nu \in \mathcal{P}_2$.

Proof. (1) By Pinsker's inequality (3.2.3), we only need to prove (4.1.31) and (4.1.32). For any $\mu, \nu \in \mathcal{P}_2$, let X_t solve (3.4.1) for $\mathcal{L}_{X_0} = \mu$, and denote

$$\mu_t := P_t^*\mu = \mathcal{L}_{X_t}, \quad \nu_t := P_t^*\nu, \quad \bar{\mu}_t := \mathcal{L}_{\bar{X}_t}, \quad t \in [0,T],$$

where $\bar{X}_t$ solves

$$d\bar{X}_t = b_t(\bar{X}_t, \nu_t)dt + \sigma_t(\bar{X}_t)dW_t, \quad t \in [0,T], \bar{X}_0 = X_0.$$

Let σ and $\hat{b} := b(\cdot, \delta_0) = \hat{b}^{(1)} + \hat{b}^{(0)}$ satisfy $(A^{1.3})$. Consider the decomposition

$$b_t^\nu := b_t(\cdot, \nu_t) = \hat{b}_t^{(1)} + b_t^{\nu,0}.$$

By $(A^{4.3})$, there exists a constant $K(N) > 0$ such that

$$|b_t^{\nu,0}| \leq |\hat{b}_t^{(0)}| + K(N)f_t, \quad \|\nu\|_2 \leq N, \quad t \in [0,T]. \tag{4.1.35}$$

So, by Theorem 1.3.1(3) and Theorem 1.5.1, there exists a constant $c_1(N) > 0$ such that

$$\mathbb{W}_2(\bar{\mu}_t, \nu_t)^2 \leq c_1(N)\mathbb{W}_2(\mu,\nu)^2, \quad t \in [0,T], \mu \in \mathcal{P}_2, \tag{4.1.36}$$

$$\mathrm{Ent}(\nu_t|\bar{\mu}_t) \le \frac{c_1(N)}{t} \mathbb{W}_2(\mu,\nu)^2, \quad t \in (0,T], \ \mu \in \mathcal{P}_2. \tag{4.1.37}$$

Moreover, repeating step (d) in the proof of Theorem 3.5.1 for $k = 2$ and $(X_t, \bar{X}_t)$ replacing $(X_t^{\mu,\gamma}, X_t^{\nu,\gamma})$, and using $(A^{4.3})$, instead of (3.5.25) where $\|\mu_s - \nu_s\|_{k,var}^2$ disappears in the present case, we derive

$$\mathbb{W}_2(\mu_t, \bar{\mu}_t)^4 \le (\mathbb{E}|X_t - \bar{X}_t|^2)^2 \le c_2(N) \int_0^t \mathbb{W}_2(\mu_s, \nu_s)^4 \mathrm{d}s, \quad t \in [0,T]$$

for some constant $c_2(N) > 0$. This together with (4.1.36) yields

$$\mathbb{W}_2(\mu_t, \nu_t)^4 \le 8\mathbb{W}_2(\mu_t, \bar{\mu}_t)^4 + 8\mathbb{W}_2(\bar{\mu}_t, \nu_t)^2$$
$$\le 8c_1(N)^2 \mathbb{W}_2(\mu,\nu)^4 + 8c_2(N) \int_0^t \mathbb{W}_2(\mu_s, \mu_s)^4 \mathrm{d}s, \quad t \in [0,T].$$

By Gronwall's inequality, (4.1.31) holds for some constant $C(N) > 0$. On the other hand, let $\|\mu\|_2 \le N$ and define

$$R_t := \exp\left[-\int_0^t \langle \gamma_s, \mathrm{d}W_s \rangle - \frac{1}{2}\int_0^t |\gamma_s|^2 \mathrm{d}s\right],$$
$$\gamma_s := \{\sigma_s^*(\sigma_s \sigma_s^*)^{-1}\}(X_s)[b_s(X_s, \mu_s) - b_s(X_s, \nu_s)].$$

By Girsanov's theorem, we obtain

$$\mathbb{E}\left[\left(f \frac{\mathrm{d}\bar{\mu}_t}{\mathrm{d}\mu_t}\right)(X_t)\right] = \int_{\mathbb{R}^d} f \frac{\mathrm{d}\bar{\mu}_t}{\mathrm{d}\mu_t} \mathrm{d}\mu_t$$
$$= \int_{\mathbb{R}^d} f \mathrm{d}\bar{\mu}_t = \mathbb{E}[f(\bar{X}_t)] = \mathbb{E}[R_t f(X_t)], \quad f \in \mathcal{B}_b(\mathbb{R}^d).$$

This implies $\frac{\mathrm{d}\bar{\mu}_t}{\mathrm{d}\mu_t}(X_t) = \mathbb{E}[R_t|X_t]$, so that by Jensen's inequality,

$$\int_{\mathbb{R}^d} \left(\frac{\mathrm{d}\bar{\mu}_t}{\mathrm{d}\mu_t}\right)^2 \mathrm{d}\mu_t = \mathbb{E}\left\{\left(\frac{\mathrm{d}\bar{\mu}_t}{\mathrm{d}\mu_t}(X_t)\right)^2\right\} = \mathbb{E}\left\{\left(\mathbb{E}[R_t|X_t]\right)^2\right\} \le \mathbb{E}[R_t^2]$$

By combining this with the Young inequality (4.1.17), we derive

$$\mathrm{Ent}(\nu_t|\mu_t) = \int_{\mathbb{R}^d} \log\left(\frac{\mathrm{d}\nu_t}{\mathrm{d}\mu_t}\right) \mathrm{d}\nu_t = \int_{\mathbb{R}^d} \left\{\log \frac{\mathrm{d}\nu_t}{\mathrm{d}\bar{\mu}_t} + \log \frac{\mathrm{d}\bar{\mu}_t}{\mathrm{d}\mu_t}\right\} \mathrm{d}\nu_t$$
$$= \mathrm{Ent}(\nu_t|\bar{\mu}_t) + \int_{\mathbb{R}^d} \left(\frac{\mathrm{d}\nu_t}{\mathrm{d}\bar{\mu}_t}\right) \log \frac{\mathrm{d}\bar{\mu}_t}{\mathrm{d}\mu_t} \mathrm{d}\bar{\mu}_t$$
$$\le 2\mathrm{Ent}(\nu_t|\bar{\mu}_t) + \log \int_{\mathbb{R}^d} \frac{\mathrm{d}\bar{\mu}_t}{\mathrm{d}\mu_t} \mathrm{d}\bar{\mu}_t$$
$$= 2\mathrm{Ent}(\nu_t|\bar{\mu}_t) + \log \int_{\mathbb{R}^d} \left(\frac{\mathrm{d}\bar{\mu}_t}{\mathrm{d}\mu_t}\right)^2 \mathrm{d}\mu_t \le 2\mathrm{Ent}(\nu_t|\bar{\mu}_t) + \log \mathbb{E}[R_t^2]. \tag{4.1.38}$$

By ($A^{4.3}$), (4.1.31), $\|\sigma^*(\sigma\sigma^*)^{-1}\|_\infty < \infty$ and (1.2.17), we find constants $c_3(N), c_4(N) > 0$ such that

$$\mathbb{E}[R_t^2] - 1 \leq \left(\mathbb{E}[R_t^2]\right)^2 \leq \mathbb{E}\mathrm{e}^{c_3(N)\mathbb{W}_2(\mu,\nu)^2 \int_0^t f_s(X_s)^2 \mathrm{d}s} - 1$$
$$\leq \mathbb{E}\left[c_3(N)\mathbb{W}_2(\mu,\nu)^2 \mathrm{e}^{c_3(N)\mathbb{W}_2(\mu,\nu)^2 \int_0^t f_s(X_s)^2 \mathrm{d}s} \int_0^t f_s(X_s)^2 \mathrm{d}s\right]$$
$$\leq c_3(N)\mathbb{W}_2(\mu,\nu)^2 \left[\mathbb{E}\left(\int_0^t f_s(X_s)^2 \mathrm{d}s\right)^2\right]^{\frac{1}{2}} \qquad (4.1.39)$$
$$\times \left[\mathbb{E}\mathrm{e}^{2c_3(N)\mathbb{W}_2(\mu,\nu)^2 \int_0^t f_s(X_s)^2 \mathrm{d}s}\right]^{\frac{1}{2}}$$
$$\leq c_4(N)\mathbb{W}_2(\mu,\nu)^2.$$

Combining this with (4.1.37) and (4.1.38), we derive (4.1.32) for some constant $C(N) > 0$.

(2) When $\alpha = 0$, (4.1.35) holds for $K(N) = K$ independent of N, so that (4.1.36) and (4.1.37) hold for some constant $C_1(N) = C_1 > 0$ independent of N and all $\mu, \nu \in \mathcal{P}_2$, and in (4.1.39) the constant $C_3(N) = C_3$ is independent of N as well. Consequently, when $\|f\|_\infty < \infty$ we find a constant $C' > 0$ such that

$$\mathbb{E}[R_t^2] \leq \mathbb{E}\mathrm{e}^{C_3\mathbb{W}_2(\mu,\nu)^2 \int_0^t f_s(X_s)^2 \mathrm{d}s} \leq \mathrm{e}^{C'\mathbb{W}_2(\mu,\nu)^2}.$$

Combining this with (4.1.37) and (4.1.38) we derive (4.1.32) for some constant $C(N) = C$ independent of N. □

4.2 Power Harnack inequality

The power Harnack inequality was established in [Wang (2018)] for the monotone case by using the coupling constructed in the proof of Theorem 4.1.1. In this section we only consider the singular case.

Let $k \geq 0$. For any $\mu \in C_b^w([0,T]; \mathcal{P}_k)$ let $b_t^\mu(x) := b_t(x, \mu_t)$, $(t,x) \in [0,T] \times \mathbb{R}^d$.

($A^{4.4}$) Let $k \geq 0$. Assumption ($A^{1.3}$) holds for (σ, b^μ) uniformly in $\mu \in C_b^w([0,T]; \mathcal{P}_k)$. Moreover, there exist constants $K > 0$ and $\kappa \geq 0$ such that

$$|b_t(x,\mu) - b_t(x,\nu)| \leq K\mathbb{W}_k(\mu,\nu),$$
$$(t,x,\mu,\nu) \in [0,T] \times \mathbb{R}^d \times \mathcal{P}_k \times \mathcal{P}_k.$$

By Theorem 3.5.1, $(A^{4.4})$ implies the well-posedness of (3.4.1) for distributions in $\mathcal{P}_k$, and

Theorem 4.2.1. *Assume* $(A^{4.4})$. *Let* κ_0, κ_1 *and* p^* *be in Theorem 1.5.2. Then for any* $p > p^*$ *there exists a constant* $c > 0$ *such that*

$$|P_t f|^p(\mu) \leq \{P_t |f|^p(\nu)\} e^{c \mathbb{W}_k(\mu,\nu)^2}$$
$$\times \inf_{\pi \in \mathcal{C}(\mu,\nu)} \int_{\mathbb{R}^d \times \mathbb{R}^d} e^{\frac{c}{t}|x-y|^2} \pi(\mathrm{d}x, \mathrm{d}y), \qquad (4.2.1)$$
$$t \in (0,T], \ \mu, \nu \in \mathcal{P}_k.$$

When $k \in [1,2]$, *the term* $e^{c \mathbb{W}_k(\mu,\nu)^2}$ *can be dropped.*

Proof. By Theorem 3.6.1, $(A^{4.4})$ implies (3.6.3) for $C_{k,N} = C_k$ independent of N, i.e.

$$\mathbb{W}_k(P_t^* \mu, P_t^* \nu) \leq C_k \mathbb{W}_k(\mu,\nu), \quad t \in [0,T], \mu,\nu \in \mathcal{P}_k. \qquad (4.2.2)$$

Next, let P_t^μ be the Markov semigroup associated to the SDE

$$\mathrm{d}X_t^{\mu,x} = b_t(X_t^{\mu,x}, P_t^* \mu)\mathrm{d}t + \sigma_t(X_t^{\mu,x})\mathrm{d}W_t, \quad X_0^{\mu,x} = x,$$

i.e. $P_t^\mu f(x) := \mathbb{E}[f(X_t^{\mu,x})]$. We have

$$P_t f(\mu) = \int_{\mathbb{R}^d} P_t^\mu f(x) \mu(\mathrm{d}x), \quad t \in [0,T], f \in \mathcal{B}_b(\mathbb{R}^d), \mu \in \mathcal{P}_k. \qquad (4.2.3)$$

By Theorem 1.5.2, for any $p > p^*$, there exists a constant $c_1 = c_1(p) > 0$, which is independent of μ since $(A^{1.3})$ holds for (σ, b^μ) uniformly in μ, such that for $p' := \frac{p+p^*}{2} > p^*$,

$$|P_t^\mu f(x)|^{p'} \leq (P_t^\mu |f|^{p'}(y)) e^{c_1 t^{-1} |x-y|^2}, \quad t \in (0,T], x, y \in \mathbb{R}^d, f \in \mathcal{B}_b(\mathbb{R}^d). \qquad (4.2.4)$$

On the other hand, for fixed $t \in (0,T]$, let

$$\xi_s := \{\sigma_s^*(\sigma_s \sigma_s^*)^{-1} [b_s(\cdot, P_s^* \mu) - b_s(\cdot, P_s^* \nu)]\}(X_s^{\nu,y}),$$
$$R := e^{\int_0^t \langle \xi_s, \mathrm{d}W_s \rangle - \frac{1}{2} \int_0^t |\xi_s|^2 \mathrm{d}s}.$$

By $(A^{4.4})$ and (4.2.2), we find a constant $c_0 > 0$ such that

$$|\xi_s| \leq c_0 \mathbb{W}_k(\mu,\nu), \quad s \in [0,t]. \qquad (4.2.5)$$

By Girsanov's theorem,

$$\tilde{W}_s := W_s - \int_0^s \xi_r \mathrm{d}r, \quad s \in [0,t]$$

is an m-dimensional Brownian motion under $\mathbb{Q} := R\mathbb{P}$. Moreover, the SDE for $X_s^{\nu,y}$ can be formulated as

$$\mathrm{d}X_s^{\nu,y} = b_s(X_s^{\nu,y}, P_s^*\nu)\mathrm{d}t + \sigma_s(X_s^{\nu,y})\mathrm{d}W_s$$
$$= b_s(X_s^{\nu,y}, P_s^*\mu)\mathrm{d}t + \sigma_s(X_s^{\nu,y})\mathrm{d}\tilde{W}_s, \quad X_0^{\nu,y} = y, \quad s \in [0,t],$$

so that by the uniqueness, $\mathcal{L}_{X_t^{\nu,y}|\mathbb{Q}} = \mathcal{L}_{X_t^{\mu,y}|\mathbb{P}}$. Consequently, by (4.2.5), we find a constant $c_2 > 0$ such that

$$P_t^\mu |f|^{p'}(y) = \mathbb{E}[R|f|^{p'}(X_t^{\nu,y})] \leq \left(\mathbb{E}[|f|^p(X_t^{\nu,y})]\right)^{\frac{p'}{p}} \left(\mathbb{E}[R^{\frac{p}{p-p'}}]\right)^{\frac{p-p'}{p}}$$
$$\leq \mathrm{e}^{c_2 \mathbb{W}_k(\mu,\nu)^2} (P_t^\nu |f|^p(y))^{\frac{p'}{p}}.$$

Combining this with (4.2.4), we obtain

$$|P_t^\mu f(x)|^p \leq (P_t^\nu |f|^p(y))\mathrm{e}^{c_1 p(p't)^{-1}|x-y|^2 + c_2 p(p')^{-1}\mathbb{W}_k(\mu,\nu)^2}, t \in (0,T], x, y \in \mathbb{R}^d.$$

By (4.2.3), integrating both sides with respect to $\pi \in \mathcal{C}(\mu,\nu)$ and applying Jensen's inequality, we derive

$$|P_t f(\mu)|^p \leq (P_t|f|^p(\nu))\mathrm{e}^{c_2 p(p')^{-1}\mathbb{W}_k(\mu,\nu)^2}$$
$$\times \inf_{\pi \in \mathcal{C}(\mu,\nu)} \int_{\mathbb{R}^d \times \mathbb{R}^d} \mathrm{e}^{c_1 p(p't)^{-1}|x-y|^2} \pi(\mathrm{d}x, \mathrm{d}y).$$

Thus, (4.2.1) holds. When $k \in [1,2]$, we have $\mathbb{W}_k \leq \mathbb{W}_2$, so that for any $\pi \in \mathcal{C}(\mu,\nu)$, by the definition of $\mathbb{W}_k$ and Jensen's inequality, we obtain

$$\mathrm{e}^{c\mathbb{W}_k^2(\mu,\nu)} \leq \mathrm{e}^{c\mathbb{W}_2(\mu,\nu)^2} \leq \int_{\mathbb{R}^d \times \mathbb{R}^d} \mathrm{e}^{c|x-y|^2} \pi(\mathrm{d}x, \mathrm{d}y)$$

for any constant $c > 0$, so that the term $\mathrm{e}^{c\mathbb{W}_k^2(\mu,\nu)}$ can be dropped from (4.2.1) by taking a large constant in the other term. □

4.3 Chain rule for intrinsic/L-derivatives

The intrinsic derivative for measures was introduced in [Albeverio *et al.* (1996)] to construct diffusion processes on configuration spaces over a Riemannian manifold, and was used in [Otto (2001)] to study the geometry of dissipative evolution equations, see [Ambrosio *et al.* (2005)] for analysis and geometry on the Wasserstein space over a metric measure space.

In this part, we introduce the intrinsic and L-derivatives for probability measures on a separable Banach space, and establish the chain rule.

Let $(\mathbb{B}, \|\cdot\|_\mathbb{B})$ be a separable Banach space, and let $(\mathbb{B}^*, \|\cdot\|_{\mathbb{B}^*})$ be its dual space. For any $k \in [1,\infty)$, denote $k^* = \frac{k}{k-1}$ when $k > 1$ and $k^* = \infty$

for $k = 1$. Let $\mathcal{P}(\mathbb{B})$ be the class of all probability measures on $\mathbb{B}$ equipped with the weak topology. Then

$$\mathcal{P}_k(\mathbb{B}) := \left\{\mu \in \mathcal{P}(\mathbb{B}) : \|\mu\|_k := \{\mu(\|\cdot\|_{\mathbb{B}}^k)\}^{\frac{1}{k}} < \infty\right\}$$

is a Polish space under the L^k-Wasserstein distance

$$\mathbb{W}_k(\mu_1, \mu_2) := \inf_{\pi \in \mathcal{C}(\mu_1, \mu_2)} \left(\int_{\mathbb{B} \times \mathbb{B}} \|x - y\|_{\mathbb{B}}^k \pi(\mathrm{d}x, \mathrm{d}y)\right)^{\frac{1}{k}},$$

where $\mathcal{C}(\mu_1, \mu_2)$ is the set of all couplings of μ_1 and μ_2.

For any $\mu \in \mathcal{P}_k(\mathbb{B})$, the tangent space at μ is given by

$$T_{\mu,k} = L^k(\mathbb{B} \to \mathbb{B}; \mu) := \left\{\phi : \mathbb{B} \to \mathbb{B} \text{ is measurable with } \mu(\|\phi\|_{\mathbb{B}}^k) < \infty\right\},$$

which is a Banach space under the norm $\|\phi\|_{T_{\mu,k}} := \{\mu(\|\phi\|_{\mathbb{B}}^k)\}^{\frac{1}{k}}$, and its dual space is

$$T_{\mu,k}^* = L^{k^*}(\mathbb{B} \to \mathbb{B}^*; \mu)$$
$$:= \left\{\psi : \mathbb{B} \to \mathbb{B}^* \text{ is measurable with } \|\psi\|_{T_{\mu,k}^*} := \left\|\|\psi\|_{\mathbb{B}^*}\right\|_{L^{k^*}(\mu)} < \infty\right\}.$$

The following definitions and chain rule are taken from [Bao et al. (2021)].

Definition 4.3.1. Let $f : \mathcal{P}_k(\mathbb{B}) \to \mathbb{R}$ be a continuous function for some $p \in [1, \infty)$, and let id be the identity map on $\mathbb{B}$.

(1) f is called intrinsically differentiable at $\mu \in \mathcal{P}_k(\mathbb{B})$, if

$$T_{\mu,k} \ni \phi \mapsto D_\phi^I f(\mu) := \lim_{\varepsilon \downarrow 0} \frac{f(\mu \circ (id + \varepsilon\phi)^{-1}) - f(\mu)}{\varepsilon} \in \mathbb{R}$$

is a well-defined bounded linear functional. In this case, the unique element $D^I f(\mu) \in T_{\mu,k}^*$ such that

$$_{T_{\mu,k}^*}\langle D^L f(\mu), \phi\rangle_{T_{\mu,k}} := \int_{\mathbb{B}} {}_{\mathbb{B}^*}\langle D^I f(\mu)(x), \phi(x)\rangle_{\mathbb{B}}\mu(\mathrm{d}x)$$
$$= D_\phi^I f(\mu), \quad \phi \in T_{\mu,k}$$

is called the intrinsic derivative of f at μ.
If moreover

$$\lim_{\|\phi\|_{T_{\mu,k}} \downarrow 0} \frac{|f(\mu \circ (id + \phi)^{-1}) - f(\mu) - D_\phi^I f(\mu)|}{\|\phi\|_{T_{\mu,k}}} = 0,$$

f is called L-differentiable at μ with the L-derivative $D^L f(\mu) := D^I f(\mu)$.

(2) We write $f \in C^1(\mathcal{P}_k(\mathbb{B}))$ if f is L-differentiable at any $\mu \in \mathcal{P}_k(\mathbb{B})$, and the L-derivative has a version $D^L f(\mu)(x)$ jointly continuous in $(x, \mu) \in \mathbb{B} \times \mathcal{P}_k(\mathbb{B})$. If moreover $D^L f(\mu)(x)$ is bounded, we denote $f \in C_b^1(\mathcal{P}_k(\mathbb{B}))$.

Definition 4.3.2. A probability space $(\Omega, \mathcal{F}, \mathbb{P})$ is called Polish, if $\mathcal{F}$ is the $\mathbb{P}$-completeness of the Borel σ-field induced by a Polish metric on Ω. $\mathbb{P}$ is called atomless if $\mathbb{P}(A) = 0$ holds for any atom $A \in \mathcal{F}$.

Noting that when $\mathbb{B} = \mathbb{R}^d$ and $k = 2$, the L-derivative $D^L f(\mu)$ named after Lions is defined as the unique element in $T_{\mu,2}$ such that for any atomless probability space $(\Omega, \mathcal{F}, \mathbb{P})$ and any random variables X, Y with $\mathcal{L}_X = \mu$,

$$\lim_{\|Y-X\|_{L^2(\mathbb{P})} \downarrow 0} \frac{|f(\mathcal{L}_Y) - f(\mathcal{L}_X) - \mathbb{E}[\langle D^L f(\mu)(X), Y - X \rangle]|}{\|Y - X\|_{L^2(\mathbb{P})}} = 0.$$

Since $D^L f(\mu)$ does not depend on the choice of probability space, when μ is atomless we may choose $(\Omega, \mathcal{F}, \mathbb{P}) = (\mathbb{R}^d, \mathcal{B}^d, \mu)$ such that $D^L f(\mu) = D^I f(\mu)$, see for instance Chapter 5 in [Carmona and Delarue (2019)]. Since by approximations one may drop the atomless condition, the above notion of L-derivative coincides with the Lions' derivative.

Example 4.3.1. Let $\mathbb{B} = \mathbb{R}^d$. We denote $f \in \mathcal{F}C_b^1(\mathcal{P}_k)$ if

$$f(\mu) = g(\mu(h_1), \ldots, \mu(h_n))$$

for some $n \geq 1, g \in C^2(\mathbb{R}^n)$ and $h_i \in C_b^1(\mathbb{R}^d), 1 \leq i \leq n$. Then it is easy to see that $f \in C_b^1(\mathcal{P}_k)$ with

$$D^L f(\mu) = \sum_{i=1}^d (\partial_i g)(\mu(h_1), \ldots, \mu(h_n)) \nabla h_i.$$

We call $\mathcal{F}C_b^1(\mathcal{P}_k)$ the class of C_b^1-cylindrical functions on $\mathcal{P}_k$.

To establish the chain rule for functions of distributions of random variables, we need the following lemma, which extends Lemma A.2 in [Hammersley et al. (2021)] for $\mathbb{B} = \mathbb{R}^d$.

Lemma 4.3.1. *Let $\{(\Omega_i, \mathcal{F}_i, \mathbb{P}_i)\}_{i=1,2}$ be two atomless, Polish probability spaces, and let $X_i, i = 1, 2$, be $\mathbb{B}$-valued random variables on these two probability spaces respectively such that $\mathcal{L}_{X_1|\mathbb{P}_1} = \mathcal{L}_{X_2|\mathbb{P}_2}$. Then for any $\varepsilon > 0$, there exist measurable maps*

$$\tau : \Omega_1 \to \Omega_2, \quad \tau^{-1} : \Omega_2 \to \Omega_1$$

such that
$$\mathbb{P}_1(\tau^{-1}\circ\tau = id_{\Omega_1}) = \mathbb{P}_2(\tau\circ\tau^{-1} = id_{\Omega_2}) = 1,$$
$$\mathbb{P}_1 = \mathbb{P}_2\circ\tau,\quad \mathbb{P}_2 = \mathbb{P}_1\circ\tau^{-1},$$
$$\|X_1 - X_2\circ\tau\|_{L^\infty(\mathbb{P}_1)} + \|X_2 - X_1\circ\tau^{-1}\|_{L^\infty(\mathbb{P}_2)} \leq \varepsilon,$$
where id_{Ω_i} stands for the identity map on $\Omega_i, i = 1, 2$.

Proof. Since $\mathbb{B}$ is separable, there is a measurable partition $(A_n)_{n\geq 1}$ of $\mathbb{B}$ such that $\mathrm{diam}(A_n) < \varepsilon$, $n \geq 1$. Let $A_n^i = \{X_i \in A_n\}, n \geq 1, i = 1, 2$. Then $(A_n^i)_{n\geq 1}$ forms a measurable partition of Ω_i so that $\sum_{n\geq 1} A_n^i = \Omega_i, i = 1, 2$, and, due to $\mathcal{L}_{X_1}|\mathbb{P}_1 = \mathcal{L}_{X_2}|\mathbb{P}_2$,
$$\mathbb{P}_1(A_n^1) = \mathbb{P}_2(A_n^2),\quad n \geq 1.$$
Since the probabilities $(\mathbb{P}_i)_{i=1,2}$ are atomless, according to Theorem C in Section 41 of [Halmos (1950)], for any $n \geq 1$ there exist measurable sets $\tilde{A}_n^i \subset A_n^i$ with $\mathbb{P}_i(A_n^i \setminus \tilde{A}_n^i) = 0, i = 1, 2$, and a measurable bijective map
$$\tau_n : \tilde{A}_n^1 \to \tilde{A}_n^2$$
such that
$$\mathbb{P}_1|_{\tilde{A}_n^1} = \mathbb{P}_2\circ\tau_n|_{\tilde{A}_n^1},\quad \mathbb{P}_2|_{\tilde{A}_n^2} = \mathbb{P}_1\circ\tau_n^{-1}|_{\tilde{A}_n^2}.$$
By $\mathrm{diam}(A_n) < \varepsilon$ and $\mathbb{P}_i(A_n^i \setminus \tilde{A}_n^i) = 0$, we have
$$\|(X_1 - X_2\circ\tau_n)1_{\tilde{A}_n^1}\|_{L^\infty(\mathbb{P}_1)} \vee \|(X_2 - X_1\circ\tau_n^{-1})1_{\tilde{A}_n^2}\|_{L^\infty(\mathbb{P}_2)} \leq \varepsilon.$$
Then the proof is finished by taking, for fixed points $\hat{\omega}_i \in \Omega_i, i = 1, 2$,
$$\tau(\omega_1) := \begin{cases} \tau_n(\omega_1), & \text{if } \omega_1 \in \tilde{A}_n^1 \text{ for some } n \geq 1, \\ \hat{\omega}_2, & \text{otherwise,} \end{cases}$$
$$\tau^{-1}(\omega_2) := \begin{cases} \tau_n^{-1}(\omega_2), & \text{if } \omega_2 \in \tilde{A}_n^2 \text{ for some } n \geq 1, \\ \hat{\omega}_1, & \text{otherwise.} \end{cases} \qquad \square$$

The following chain rule is taken from Theorem 2.1 in [Bao et al. (2021)], which extends the corresponding formulas for functions on $\mathcal{P}_2$ presented in [Carmona and Delarue (2019); Hammersley et al. (2021)] and references therein.

Theorem 4.3.2. *Let $f : \mathcal{P}_k(\mathbb{B}) \to \mathbb{R}$ be continuous for some $k \in [1,\infty)$, and let $(\xi_\varepsilon)_{\varepsilon\in[0,1]}$ be a family of $\mathbb{B}$-valued random variables on a complete probability space $(\Omega, \mathcal{F}, \mathbb{P})$ such that $\dot{\xi}_0 := \lim_{\varepsilon\downarrow 0}\frac{\xi_\varepsilon - \xi_0}{\varepsilon}$ exists in $L^k(\Omega \to \mathbb{B}, \mathbb{P})$. We assume that either ξ_ε is continuous in $\varepsilon \in [0,1]$, or the probability space is Polish.*

(1) Let $\mu_0 = \mathcal{L}_{\xi_0}$ be atomless. If f is L-differentiable such that $D^L f(\mu_0)$ has a continuous version satisfying

$$\|D^L f(\mu_0)(x)\|_{\mathbb{B}^*} \leq C(1 + \|x\|_{\mathbb{B}}^{k-1}), \quad x \in \mathbb{B} \qquad (4.3.1)$$

for some constant $C > 0$, then

$$\lim_{\varepsilon \downarrow 0} \frac{f(\mathcal{L}_{\xi_\varepsilon}) - f(\mathcal{L}_{\xi_0})}{\varepsilon} = \mathbb{E}[_{\mathbb{B}^*}\langle D^L f(\mu_0)(\xi_0), \dot{\xi}_0 \rangle_{\mathbb{B}}]. \qquad (4.3.2)$$

(2) If f is L-differentiable in a neighborhood O of μ_0 such that $D^L f$ has a version jointly continuous in $(x, \mu) \in \mathbb{B} \times O$ satisfying

$$\|D^L f(\mu)(x)\|_{\mathbb{B}^*} \leq C(1 + \|x\|_{\mathbb{B}}^{k-1}), \quad (x, \mu) \in \mathbb{B} \times O \qquad (4.3.3)$$

for some constant $C > 0$, then (4.3.2) holds.

Proof. Without loss of generality, we may and do assume that $\mathbb{P}$ is atomless. Otherwise, by taking

$$(\tilde{\Omega}, \tilde{\mathcal{F}}, \tilde{\mathbb{P}}) := (\Omega \times [0,1], \mathcal{F} \times \mathcal{B}([0,1]), \mathbb{P} \times ds), (\tilde{\xi}_\varepsilon)(\omega, s) := \xi_\varepsilon(\omega) \text{ for } (\omega, s) \in \tilde{\Omega},$$

where $\mathcal{B}([0,1])$ is the completion of the Borel σ-algebra on $[0,1]$ w.r.t. the Lebesgue measure ds, we have

$$\mathcal{L}_{\tilde{\xi}_\varepsilon | \tilde{\mathbb{P}}} = \mathcal{L}_{\xi_\varepsilon | \mathbb{P}}, \quad \mathbb{E}[_{\mathbb{B}^*}\langle D^L f(\mu_0)(\xi_0), \dot{\xi}_0 \rangle_{\mathbb{B}}] = \tilde{\mathbb{E}}[_{\mathbb{B}^*}\langle D^L f(\mu_0)(\tilde{\xi}_0), \dot{\tilde{\xi}}_0 \rangle_{\mathbb{B}}].$$

In this way, we go back to the atomless situation. Moreover, it suffices to prove for the Polish probability space case. Indeed, when ξ_ε is continuous in ε, we may take $\bar{\Omega} = C([0,1]; \mathbb{R}^d)$, let $\bar{\mathbb{P}}$ be the distribution of $\xi_\cdot$, let $\bar{\mathcal{F}}$ be the $\bar{\mathbb{P}}$-complete Borel σ-field on $\bar{\Omega}$ induced by the uniform norm, and consider the coordinate random variable $\bar{\xi}_\cdot(\omega) := \omega, \omega \in \bar{\Omega}$. Then $\mathcal{L}_{\bar{\xi}_\cdot | \bar{\mathbb{P}}} = \mathcal{L}_{\xi_\cdot | \mathbb{P}}$, so that $\mathcal{L}_{\bar{\xi}_\varepsilon | \bar{\mathbb{P}}} = \mathcal{L}_{\xi_\varepsilon | \mathbb{P}}$ for any $\varepsilon \in [0,1]$ and $\mathcal{L}_{\bar{\xi}'_0 | \bar{\mathbb{P}}} = \mathcal{L}_{\xi'_0 | \mathbb{P}}$. Hence we have reduced the situation to the Polish setting.

(1) Let $\mathcal{L}_{\xi_0} = \mu_0 \in \mathcal{P}_k(\mathbb{B})$ be atomless. In this case, $(\mathbb{B}, \mathcal{B}(\mathbb{B}), \mu_0)$ is an atomless Polish complete probability space, where $\mathcal{B}(\mathbb{B})$ is the μ_0-complete Borel σ-algebra of $\mathbb{B}$. By Lemma 3.6.2, for any $n \geq 1$ we find measurable maps

$$\tau_n : \Omega \to \mathbb{B}, \quad \tau_n^{-1} : \mathbb{B} \to \Omega$$

such that

$$\mathbb{P}(\tau_n^{-1} \circ \tau_n = id_\Omega) = \mu_0(\tau_n \circ \tau_n^{-1} = id) = 1,$$
$$\mathbb{P} = \mu_0 \circ \tau_n, \quad \mu_0 = \mathbb{P} \circ \tau_n^{-1}, \qquad (4.3.4)$$
$$\|\xi_0 - \tau_n\|_{L^\infty(\mathbb{P})} + \|id - \xi_0 \circ \tau_n^{-1}\|_{L^\infty(\mu_0)} \leq \frac{1}{n},$$

where id_Ω is the identity map on Ω.

Since f is L-differentiable at μ_0, there exists a decreasing function $h: [0,1] \to [0,\infty)$ with $h(r) \downarrow 0$ as $r \downarrow 0$ such that

$$\sup_{\|\phi\|_{L^k(\mu_0)} \leq r} \left| f(\mu_0 \circ (id + \phi)^{-1}) - f(\mu_0) - D^L_\phi f(\mu_0) \right| \leq rh(r), \quad r \in [0,1] \tag{4.3.5}$$

By $\mathcal{L}_{\xi_\varepsilon - \xi_0} \in \mathcal{P}_k(\mathbb{B})$ and (4.3.4), we have

$$\phi_{n,\varepsilon} := (\xi_\varepsilon - \xi_0) \circ \tau_n^{-1} \in T_{\mu,k}, \quad \|\phi_{n,\varepsilon}\|_{T_{\mu,k}} = \|\xi_\varepsilon - \xi_0\|_{L^k(\mathbb{P})}. \tag{4.3.6}$$

Next, (4.3.4) implies

$$\mathcal{L}_{\tau_n + \xi_\varepsilon - \xi_0} = \mathbb{P} \circ (\tau_n + \xi_\varepsilon - \xi_0)^{-1}$$
$$= (\mu_0 \circ \tau_n) \circ (\tau_n + \xi_\varepsilon - \xi_0)^{-1} = \mu_0 \circ (id + \phi_{n,\varepsilon})^{-1}. \tag{4.3.7}$$

Moreover, by $\frac{\xi_\varepsilon - \xi_0}{\varepsilon} \to \dot{\xi}_0$ in $L^k(\mathbb{P})$ as $\varepsilon \downarrow 0$, we find a constant $c \geq 1$ such that

$$\|\xi_\varepsilon - \xi_0\|_{L^k(\mathbb{P})} \leq c\varepsilon, \quad \varepsilon \in [0,1]. \tag{4.3.8}$$

Combining (4.3.4)–(4.3.8) leads to

$$\left| f(\mathcal{L}_{\tau_n + \xi_\varepsilon - \xi_0}) - f(\mathcal{L}_{\xi_0}) - \mathbb{E}[_{\mathbb{B}^*}\langle (D^L f)(\mu_0)(\tau_n), (\xi_\varepsilon - \xi_0)\rangle_\mathbb{B}] \right|$$
$$= \left| f(\mu_0 \circ (id + \phi_{n,\varepsilon})^{-1}) - f(\mu_0) - D^L_{\phi_{n,\varepsilon}} f(\mu_0) \right|$$
$$\leq \|\phi_{n,\varepsilon}\|_{T_{\mu,k}} h(\|\phi_{n,\varepsilon}\|_{T_{\mu,k}}) \tag{4.3.9}$$
$$= \|\xi_\varepsilon - \xi_0\|_{L^k(\mathbb{P})} h(\|\xi_\varepsilon - \xi_0\|_{L^k(\mathbb{P})}), \quad \varepsilon \in [0, c^{-1}].$$

Since $f(\mu)$ is continuous in μ and $D^L f(\mu_0)(x)$ is continuous in x, by (4.3.1) and (4.3.4), we may apply the dominated convergence theorem to deduce from (4.3.9) with $n \to \infty$ that

$$\left| f(\mathcal{L}_{\xi_\varepsilon}) - f(\mathcal{L}_{\xi_0}) - \mathbb{E}[_{\mathbb{B}^*}\langle (D^L f)(\mu_0)(\xi_0), (\xi_\varepsilon - \xi_0)\rangle_\mathbb{B}] \right|$$
$$\leq \|\xi_\varepsilon - \xi_0\|_{L^k(\mathbb{P})} h(\|\xi_\varepsilon - \xi_0\|_{L^k(\mathbb{P})}), \quad \varepsilon \in [0, c^{-1}].$$

Combining this with (4.3.8) and $h(r) \to 0$ as $r \to 0$, we derive (4.3.2).

(2) When μ_0 has an atom, we take a $\mathbb{B}$-valued bounded random variable X which is independent of $(\xi_\varepsilon)_{\varepsilon \in [0,1]}$ and $\mathcal{L}_X$ does not have an atom. Then $\mathcal{L}_{\xi_0 + sX + r(\xi_\varepsilon - \xi_0)} \in \mathcal{P}_k(\mathbb{B})$ does not have an atom for any $s > 0, \varepsilon \in [0,1]$. By conditions in Theorem 4.3.2(2), there exists a small constant $s_0 \in (0,1)$ such that for any $s, \varepsilon \in (0, s_0]$, we may apply (4.3.2) to the family $\xi_0 + sX + (r+\delta)(\xi_\varepsilon - \xi_0)$ for small $\delta > 0$ to conclude

$$f(\mathcal{L}_{\xi_\varepsilon + sX}) - f(\mathcal{L}_{\xi_0 + sX}) = \int_0^1 \frac{\mathrm{d}}{\mathrm{d}\delta} f(\mathcal{L}_{\xi_0 + sX + (r+\delta)(\xi_\varepsilon - \xi_0)})\big|_{\delta=0}\, \mathrm{d}r$$
$$= \int_0^1 \mathbb{E}[_{\mathbb{B}^*}\langle D^L f(\mathcal{L}_{\xi_0 + sX + r(\xi_\varepsilon - \xi_0)})(\xi_0 + sX + r(\xi_\varepsilon - \xi_0)), \xi_\varepsilon - \xi_0\rangle_\mathbb{B}]\, \mathrm{d}r.$$

By conditions in Theorem 4.3.2(2), we may let $s \downarrow 0$ to derive

$$f(\mathcal{L}_{\xi_\varepsilon}) - f(\mathcal{L}_{\xi_0})$$
$$= \int_0^1 \mathbb{E}[_{\mathbb{B}^*}\langle D^L f(\mathcal{L}_{\xi_0 + r(\xi_\varepsilon - \xi_0)})(\xi_0 + r(\xi_\varepsilon - \xi_0)), \xi_\varepsilon - \xi_0\rangle_\mathbb{B}]\,\mathrm{d}r, \quad \varepsilon \in (0, s_0).$$

Multiplying both sides by ε^{-1} and letting $\varepsilon \downarrow 0$, we finish the proof. □

As a consequence of the chain rule, we have the following Lipschitz estimate for L-differentiable functions on $\mathcal{P}_k(\mathbb{B})$.

Corollary 4.3.3. *Let f be L-differentiable on $\mathcal{P}_k(\mathbb{B})$ such that for any $\mu \in \mathcal{P}_k(\mathbb{B})$, $D^L f(\mu)(\cdot)$ has a continuous version satisfying*

$$|D^L f(\mu)(x)| \leq c(\mu)(1 + |x|^{k-1}), \quad x \in \mathbb{B}, \tag{4.3.10}$$

which holds for some constant $c(\mu) > 0$, and

$$K_0 := \sup_{\mu \in \mathcal{P}_k(\mathbb{B})} \|D^L f(\mu)\|_{L^{k^*}(\mu)} < \infty. \tag{4.3.11}$$

Then

$$|f(\mu_1) - f(\mu_2)| \leq K_0 \mathbb{W}_k(\mu_1, \mu_2), \quad \mu_1, \mu_2 \in \mathcal{P}_k(\mathbb{B}). \tag{4.3.12}$$

Proof. Let ξ_1, ξ_2 be two random variables with

$$\mathcal{L}_{\xi_1} = \mu_1, \quad \mathcal{L}_{\xi_2} = \mu_2, \quad \mathbb{W}_k(\mu_1, \mu_2) = (\mathbb{E}[|\xi_1 - \xi_2|^k])^{\frac{1}{k}}.$$

Let η be a normal random variable independent of (ξ_1, ξ_2) such that $\mathcal{L}_\eta$ is atomless. Then

$$\gamma_\varepsilon(r) := \varepsilon\eta + r\xi_1 + (1-r)\xi_2, \quad r \in [0,1], \varepsilon \in (0,1]$$

are absolutely continuous with respect to the Lebesgue measure and hence atomless. By Theorem 4.3.2, (4.3.10) and the continuity of $D^L f(\mu)(\cdot)$ imply

$$|f(\mathcal{L}_{\gamma_\varepsilon(1)}) - f(\mathcal{L}_{\gamma_\varepsilon(0)})| = \left|\int_0^1 \mathbb{E}\big[\langle D^L f(\mathcal{L}_{\gamma_\varepsilon(r)})(\gamma_\varepsilon(r)), \xi_1 - \xi_2\rangle\big]\mathrm{d}r\right|$$
$$\leq \big(\mathbb{E}[|\xi_1 - \xi_2|^k]\big)^{\frac{1}{k}} \int_0^1 \|D^L f(\mathcal{L}_{\gamma_\varepsilon(r)})\|_{L^{k^*}(\mathcal{L}_{\gamma_\varepsilon(r)})}\mathrm{d}r$$
$$\leq K\mathbb{W}_k(\mu_1, \mu_2), \quad \varepsilon \in (0, 1].$$

Letting $\varepsilon \to 0$ we derive (4.3.12). □

4.4 Bismut formula for singular DDSDEs

Let $k \in [1, \infty)$. We aim to establish Bismut formula for the intrinsic/L derivative of $P_t f(\mu)$ for $\mu \in \mathcal{P}_k$, where P_t is defined in (4.0.1) for the singular DDSDE (3.4.1). To this end, we will assume that $b_t(x, \mu)$ includes a singular term in $\tilde{L}_{q_0}^{p_0}(T)$ and a regular term in the following class $\mathcal{D}_k$, such that the chain rule in Theorem 4.3.2 applies. This part is organized from [Wang (2023d)].

Definition 4.4.1. $\mathcal{D}_k$ is the class of continuous functions g on $\mathbb{R}^d \times \mathcal{P}_k$ such that $g(x, \mu)$ is differentiable in x, L-differentiable in μ, and $D^L g(x, \mu)(y)$ has a version jointly continuous in $(x, y, \mu) \in \mathbb{R}^d \times \mathbb{R}^d \times \mathcal{P}_k$ such that

$$|D^L g(x, \mu)(y)| \leq c(x, \mu)(1 + |y|^{k-1}), \quad x, y \in \mathbb{R}^d, \mu \in \mathcal{P}_k$$

holds for some positive function c on $\mathbb{R}^d \times \mathcal{P}_k$.

According to Example 4.3.1, we have $g \in \mathcal{D}_k$ for

$$g(x, \mu) := F(x, \mu(h_1), \ldots, \mu(h_n)),$$

where $F \in C^1(\mathbb{R}^d \times \mathbb{R}^n)$ and $\{h_i\}_{1 \leq i \leq n} \subset C^1(\mathbb{R}^d)$ such that

$$\sup_{1 \leq i \leq n} |\nabla h_i(y)| \leq c(1 + |y|^{k-1}), \quad y \in \mathbb{R}^d$$

holds for some constant $c > 0$.

$(A^{4.5})$ $b_t(x, \mu) = b_t^{(0)}(x) + b_t^{(1)}(x, \mu)$ such that the following conditions hold.
(1) $(A^{1.2})(1)$ holds; i.e. $a := \sigma \sigma^*$ is invertible with $\|a\|_\infty + \|a^{-1}\|_\infty < \infty$,

$$\lim_{\varepsilon \to 0} \sup_{|x-y| \leq \varepsilon, t \in [0,T]} \|a_t(x) - a_t(y)\| = 0,$$

and there exist $l \in \mathbb{N}$, $\{(p_i, q_i)\}_{0 \leq i \leq l} \subset \mathcal{K}$ with $p_i > 2$, and $0 \leq f_i \in \tilde{L}_{q_i}^{p_i}(T)$ such that

$$|b^{(0)}| \leq f_0, \quad \|\nabla \sigma\| \leq \sum_{i=1}^{l} f_i$$

(2) $b_t^{(1)} \in \mathcal{D}_k$ such that $\sup_{t \in [0,T]} |b_t^{(1)}(0, \delta_0)| < \infty$ and

$$\sup_{(t,x,\mu) \in [0,T] \times \mathbb{R}^d \times \mathcal{P}_k} \left\{ \|\nabla b_t^{(1)}(x, \mu)\| + \|D^L b_t^{(1)}(x, \mu)\|_{L^{k^*}(\mu)} \right\} < \infty,$$

where δ_0 is the Dirac measure at $0 \in \mathbb{R}^d$, ∇ is the gradient in the space variable $x \in \mathbb{R}^d$, and D^L is the L-derivative in the distribution variable $\mu \in \mathcal{P}_k$.

We will show that $(A^{4.5})$ implies the well-posedness of the DDSDE (3.4.1) for distributions in $\mathcal{P}_k$. To calculate the intrinsic derivative $D^I P_t f(\mu)$, for any $\varepsilon \in [0,1]$ and $\phi \in T_{\mu,k}$, we consider the following DDSDE:

$$\begin{aligned} \mathrm{d}X_t^{\mu,\varepsilon\phi} &= b_t(X_t^{\mu,\varepsilon\phi}, \mathcal{L}_{X_t^{\mu,\varepsilon\phi}})\mathrm{d}t + \sigma_t(X_t^{\mu,\varepsilon\phi})\mathrm{d}W_t, \\ t &\in [0,T], \quad X_0^{\mu,\varepsilon\phi} = X_0^\mu + \varepsilon\phi(X_0^\mu). \end{aligned} \quad (4.4.1)$$

We will prove that the derivative process

$$\nabla_\phi X_t^\mu := \lim_{\varepsilon \downarrow 0} \frac{X_t^{\mu,\varepsilon\phi} - X_t^\mu}{\varepsilon}, \quad t \in [0,T] \quad (4.4.2)$$

exists in $L^k(\Omega \to C([0,T]; \mathbb{R}^d), \mathbb{P})$. We also need the derivative of the decoupled SDE

$$\begin{aligned} \mathrm{d}X_t^{\mu,x} &= b_t(X_t^{\mu,x}, P_t^*\mu)\mathrm{d}t + \sigma_t(X_t^{\mu,x})\mathrm{d}W_t, \\ t &\in [0,T], X_0^{\mu,x} = x, x \in \mathbb{R}^d, \mu \in \mathcal{P}_k. \end{aligned} \quad (4.4.3)$$

By Theorem 1.4.2, $(A^{4.5})$ implies the well-posedness of (4.4.3) and that for any $v \in \mathbb{R}^d$,

$$\nabla_v X_t^{\mu,x} := \lim_{\varepsilon \downarrow 0} \frac{X_t^{\mu,x+\varepsilon v(x)} - X_t^{\mu,x}}{\varepsilon}, \quad t \in [0,T] \quad (4.4.4)$$

exists in $L^k(\Omega \to C([0,T]; \mathbb{R}^d), \mathbb{P})$.

4.4.1 Main results

Theorem 4.4.1. *Assume* $(A^{4.5})$. *Then* (3.4.1) *is well-posed for distributions in* $\mathcal{P}_k$ *and the following assertions hold for* P_t *defined in* (4.0.1).

(1) *For any* $\mu \in \mathcal{P}_k$, $\phi \in T_{\mu,k}$ *and* $v, x \in \mathbb{R}^d$, $\nabla_\phi X_t^\mu$ *and* $\nabla_v X_t^{\mu,x}$ *exist in* $L^k(\Omega \to C([0,T]; \mathbb{R}^d), \mathbb{P})$. *Moreover, for any* $j \geq 1$ *there exists a constant* $c > 0$ *such that*

$$\mathbb{E}\Big[\sup_{t\in[0,T]} |\nabla_\phi X_t^\mu|^j \Big| \mathcal{F}_0\Big] \leq c\big\{\|\phi\|_{L^k(\mu)}^j + |\phi(X_0^\mu)|^j\big\}, \quad (4.4.5)$$

$$\mu \in \mathcal{P}_k, \phi \in T_{\mu,k},$$

$$\mathbb{E}\Big[\sup_{t\in[0,T]} |\nabla_v X_t^{\mu,x}|^j\Big] \leq c|v|^j, \quad \mu \in \mathcal{P}_k, x, v \in \mathbb{R}^d. \quad (4.4.6)$$

(2) Denote $\zeta = \sigma(\sigma\sigma^*)^{-1}$. For any $t \in (0,T]$ and $f \in \mathcal{B}_b(\mathbb{R}^d)$, $P_t f$ is intrinsically differentiable on $\mathcal{P}_k$. Moreover, for any $\phi \in T_{\mu,k}$ and $\beta \in C^1([0,t])$ with $\beta_0 = 0$ and $\beta_t = 1$,

$$D^I_\phi P_t f(\mu) = \int_{\mathbb{R}^d} \mathbb{E}\big[f(X^{\mu,x}_t) M^{\mu,x}_{\beta,t}\big] \mu(\mathrm{d}x) + \mathbb{E}\big[f(X^\mu_t) N^\mu_t\big] \quad (4.4.7)$$

holds for

$$M^{\mu,x}_{\beta,t} := \int_0^t \beta'_s \langle \zeta_s(X^{\mu,x}_s) \nabla_{\phi(x)} X^{\mu,x}_s, \mathrm{d}W_s \rangle,$$

$$N^\mu_t := \int_0^t \Big\langle \zeta_s(X^\mu_s) \mathbb{E}\big[\langle D^L b^{(1)}_s(z, P^*_s \mu)(X^\mu_s), \nabla_\phi X^\mu_s \rangle\big]\big|_{z=X^\mu_s}, \mathrm{d}W_s \Big\rangle.$$

The following is a direct consequence of Theorem 4.4.1.

Corollary 4.4.2. *Assume* $(A^{4.5})$. *Then for any* $p > 1$ *there exists a constant* $c > 0$ *such that*

$$\|D^I P_t f(\mu)\|_{L^{k^*}(\mu)} \le \frac{c}{\sqrt{t}} \Big\|\big(\mathbb{E}[|f|^p(X^\mu_t)|\mathcal{F}_0]\big)^{\frac{1}{p}}\Big\|_{L^{k^*}(\mathbb{P})},$$

$$t \in (0,T], f \in \mathcal{B}_b(\mathbb{R}^d), \mu \in \mathcal{P}_k.$$

In particular, there exists a constant $c > 0$ *such that*

$$\|D^I P_t f(\mu)\|_{L^{k^*}(\mu)} \le \frac{c}{\sqrt{t}} \|f(X^\mu_t)\|_{L^{k^*}(\mathbb{P})}, \quad t \in (0,T], f \in \mathcal{B}_b(\mathbb{R}^d), \mu \in \mathcal{P}_k.$$

For the L-differentiability of $P_t f$, we need the uniform continuity of $\sigma_t(x)$, $\nabla b^{(1)}_t(x,\mu)$ and $D^L b_t(x,\mu)(y)$ in (x,y,μ):

$$\limsup_{\varepsilon \downarrow 0} \Big\{\|\sigma_t(x) - \sigma_t(x')\| + \|\nabla b^{(1)}_t(x,\mu) - \nabla b^{(1)}_t(x',\nu)\|$$

$$+ \|D^L b^{(1)}_t(x,\mu)(y) - D^L b^{(1)}_t(x',\nu)(y')\| : t \in [0,T], \quad (4.4.8)$$

$$|x - x'| \vee \mathbb{W}_k(\mu,\nu) \vee |y - y'| \le \varepsilon\Big\} = 0.$$

Under this condition and $(A^{4.5})$, the following result ensures the L-differentiability of $P_t f$ in $\mathcal{P}_k$ for $k > 1$. See also [Huang et al. (2021)] for the case that $k = 2$, σ_t is Lipschitz continuous and $b^{(0)}_t$ is Dini continuous.

Theorem 4.4.3. *Assume* $(A^{4.5})$ *and* $(4.4.8)$ *for* $k \in (1,\infty)$. *Then for any* $t \in (0,T]$ *and* $f \in \mathcal{B}_b(\mathbb{R}^d)$, $P_t f$ *is L-differentiable on* $\mathcal{P}_k$.

4.4.2 Some lemmas

Lemma 4.4.4. *Assume* $(A^{4.5})$. *Then the following assertions hold.*

(1) (3.4.1) *is well-posed for distributions in* $\mathcal{P}_k$, *and for any* $j \geq 1$ *there exists a constant* $c > 0$ *such that any solution* X_t *satisfies*

$$\mathbb{E}\left[\sup_{t\in[0,T]} |X_t|^j \Big| \mathcal{F}_0\right] \leq c\left\{1 + (\mathbb{E}[|X_0|^k])^{\frac{j}{k}} + |X_0|^j\right\}. \quad (4.4.9)$$

In particular, there exists a constant $c > 0$ *such that*

$$\mathbb{E}\left[\sup_{t\in[0,T]} |X_t|^k\right] \leq c\big(1 + \mathbb{E}[|X_0|^k]\big). \quad (4.4.10)$$

(2) *For any* $j \geq 1$ *there exists a constant* $c > 0$ *such that for any two solutions* (X_t^1, X_t^2) *of* (3.4.1) *with initial distributions in* $\mathcal{P}_k$,

$$\mathbb{E}\left[\sup_{t\in[0,T]} |X_t^1 - X_t^2|^j \Big| \mathcal{F}_0\right] \leq c\left\{\big(\mathbb{E}[|X_0^1 - X_0^2|^k]\big)^{\frac{j}{k}} + |X_0^1 - X_0^2|^j\right\}. \quad (4.4.11)$$

In particular, there exists a constant $c > 0$ *such that*

$$\mathbb{E}\left[\sup_{t\in[0,T]} |X_t^1 - X_t^2|^k\right] \leq c\mathbb{E}[|X_0^1 - X_0^2|^k]. \quad (4.4.12)$$

(3) *There exists a constant* $c > 0$ *such that*

$$\|P_t^*\mu - P_t^*\nu\|_{var} \leq \frac{c}{\sqrt{t}} \mathbb{W}_k(\mu, \nu), \quad t \in (0, T], \mu, \nu \in \mathcal{P}_k. \quad (4.4.13)$$

Proof. (1) By $(A^{4.5})$, we have $b_t^{(1)} \in \mathcal{D}_k$ with $\|D^L b_t^{(1)}(x, \mu)\|_{L^{k^*}(\mu)} \leq K$ for some constant $K > 0$. Then Corollary 4.3.3 implies

$$|b_t^{(1)}(x, \mu) - b_t^{(1)}(x, \nu)| \leq K\mathbb{W}_k(\mu, \nu), \quad (4.4.14)$$

so that the well-posedness of (3.4.1) follows from Theorem 3.5.1.

To prove (4.4.9) and (4.4.10), we use Zvonkin's transform. Consider the differential operator

$$L_t^\mu := L_{t,\mu_t} = \frac{1}{2}\mathrm{tr}\{\sigma_t\sigma_t^*\nabla^2\} + \nabla_{b_t(\cdot,\mu_t)}, \quad t \in [0,T]. \quad (4.4.15)$$

By Lemma 1.2.2, $(A^{4.5})$ implies that for some constant $\lambda_0 > 0$ uniformly in μ_0, when $\lambda \geq \lambda_0$, the PDE

$$(\partial_t + L_t^\mu)u_t = \lambda u_t - b_t^{(0)}, \quad t \in [0,T], u_T = 0 \quad (4.4.16)$$

has a unique solution $u \in \tilde{H}_{q_0}^{2,p_0}(T)$ such that

$$f_0 := \|\nabla^2 u\| + |(\partial_t + \nabla_{b^{(1)}})u| \in \tilde{L}_{q_0}^{p_0}(T), \quad \|u\|_\infty + \|\nabla u\|_\infty \leq \frac{1}{2}. \quad (4.4.17)$$

Let $\Theta_t := id + u_t$. By Itô's formula,
$$Y_t := \Theta_t(X_t) = X_t + u_t(X_t)$$
solves the SDE
$$\begin{aligned} dY_t &= \{b_t^{(1)}(X_t, \mu_t) + \lambda u_t(X_t)\}dt + \{(\nabla \Theta_t)\sigma_t\}(X_t)dW_t, \\ Y_0 &= \Theta_0(X_0), \quad t \in [0, T]. \end{aligned} \quad (4.4.18)$$

By (4.4.17), there exists a constant $c_1 > 1$ such that
$$|X_t| \leq c_1(1 + |Y_t|) \leq c_1^2(1 + |X_t|), \quad t \in [0, T]. \quad (4.4.19)$$
For any $n \geq 1$, let
$$\gamma_{t,n} := \sup_{s \in [0, t \wedge \tau_n]} |Y_s|, \quad \tau_n := \inf\{s \geq 0 : |Y_s| \geq n\}, \quad t \in [0, T].$$

By BDG's inequality in Lemma 1.3.5, $(A^{4.5})$ and (4.4.17), for any $j \geq 1$ there exists a constant $c(j) > 0$ such that
$$\mathbb{E}(\gamma_{t,n}^j | \mathcal{F}_0) \leq 2|Y_0|^j + c(j) \int_0^t \{\mathbb{E}(\gamma_{s,n}^j | \mathcal{F}_0) + (\mathbb{E}[|Y_s|^k])^{\frac{j}{k}} + 1\}ds + c(j),$$
$$n \geq 1, t \in [0, T].$$

By Gronwall's inequality, we obtain
$$\mathbb{E}(\gamma_{t,n}^j | \mathcal{F}_0)$$
$$\leq \left(2|Y_0|^j + c(j) \int_0^t \{(\mathbb{E}[|Y_s|^k])^{\frac{j}{k}} + 1\}ds + c(j)\right)e^{c(j)t}, \quad (4.4.20)$$
$$n \geq 1, t \in [0, T].$$

Taking expectations with $j = k$ and letting $n \to \infty$, we find a constant $c_2 > 0$ such that
$$\mathbb{E}[\gamma_t^k] := \mathbb{E}\left[\sup_{s \in [0, t]} |Y_s|^k\right] \leq c_2(1 + \mathbb{E}[|Y_0|^k]) + c_2 \int_0^t \mathbb{E}[|Y_s|^k]ds, \quad t \in [0, T].$$

Noting that $\sup_{t \in [0,T]} \mathbb{E}[|X_t|^k] < \infty$ as X_t is the solution of (3.4.1) for distributions in $\mathcal{P}_k$, by combining this with (4.4.19) and $\mathbb{E}[\gamma_t^k] \geq \mathbb{E}[|Y_s|^k]$ we obtain
$$\mathbb{E}[\gamma_t^k] := \mathbb{E}\left[\sup_{s \in [0, t]} |Y_s|^k\right] \leq c_2 + c_2 \int_0^t \mathbb{E}[\gamma_s^k]ds < \infty, \quad t \in [0, T],$$

so that by Gronwall's inequality and (4.4.19), we derive (4.4.10) for some constant $c > 0$. Substituting this into (4.4.66) and letting $n \to \infty$, we prove (4.4.9).

(2) Denote $\mu_t^i := \mathcal{L}_{X_t^i}, i = 1, 2, t \in [0, T]$. Let u solve (4.4.16) for $L_t^{\mu^1}$ replacing L_t^μ such that (4.4.17) holds. Let $\Theta_t = id + u_t$ and

$$Y_t^i = \Theta_t(X_t^i), \quad t \in [0, T], i = 1, 2.$$

By (4.4.16) and Itô's formula we obtain

$$dY_t^1 = \{b_t^{(1)}(X_t^1, \mu_t^1) + \lambda u_t(X_t^1)\}dt + \{(\nabla\Theta_t)\sigma_t\}(X_t^1)dW_t,$$

$$dY_t^2 = \{b_t^{(1)}(X_t^2, \mu_t^2) + \lambda u_t(X_t^2) + \nabla_{b_t^{(1)}(X_t^2, \mu_t^2) - b_t^{(1)}(X_t^2, \mu_t^1)} u_t(X_t^2)\}dt$$
$$+ \{(\nabla\Theta_t)\sigma_t\}(X_t^2)dW_t, \quad t \in [0, T].$$

So, by Itô's formula, the process

$$v_t := Y_t^2 - Y_t^1, \quad t \in [0, T]$$

satisfies the SDE

$$dv_t = \left\{b_t^{(1)}(X_t^2, \mu_t^2) + \lambda u_t(X_t^2) - b_t^{(1)}(X_t^1, \mu_t^1)\right.$$
$$\left. - \lambda u_t(X_t^1) + \nabla_{b_t^{(1)}(X_t^2, \mu_t^2) - b_t^{(1)}(X_t^2, \mu_t^1)} u_t(X_t^2)\right\}dt$$
$$+ \left\{[(\nabla\Theta_t)\sigma_t](X_t^2) - [(\nabla\Theta_t)\sigma_t](X_t^1)\right\}dW_t,$$
$$v_0 = \Theta_0(X_0^2) - \Theta_0(X_0^1).$$

By (4.4.17) and (4.4.14), we obtain

$$|b_t^{(1)}(x, \mu_t^2) - b_t^{(1)}(x, \mu_t^1)|^k \le K^k \mathbb{E}[|X_t^2 - X_t^1|^k] \le (2K)^k \mathbb{E}[|Y_t^2 - Y_t^1|^k].$$

Combining this with $(A^{4.5})$, (4.4.17), Lemma 1.3.4, and applying Itô's formula, for any $j \ge k$ we find a constant $c_1 > 0$ such that

$$|v_t|^{2j} \le |v_0|^{2j} + c_1 \int_0^t |v_s|^{2j}\Big\{1 + \sum_{i=0}^l \mathcal{M}f_i^2(s, X_s)\Big\}ds$$
$$+ c_1 \int_0^t (\mathbb{E}[|v_s|^k])^{\frac{2j}{k}}ds + M_t, \quad t \in [0, T] \tag{4.4.21}$$

holds for some local martingale M_t with $M_0 = 0$. Since (4.4.17) implies

$$|v_0| \le 2|X_0^1 - X_0^2|,$$

by stochastic Gronwall's inequality in Lemma 1.3.3 and Khasminskii's estimate in Theorem 1.2.4, we find a constant $c_2 > 0$ such that

$$\gamma_t := \sup_{s \in [0, t]} |v_s|, \quad t \in [0, T]$$

satisfies

$$\mathbb{E}\big[|\gamma_t|^j \big| \mathcal{F}_0\big] \leq c_2 \left(|X_0^1 - X_0^2|^{2j} + \int_0^t (\mathbb{E}[|v_s|^k])^{\frac{2j}{k}} ds \right)^{\frac{1}{2}}$$
$$\leq c_2 |X_0^1 - X_0^2|^j + \frac{1}{2} \sup_{s \in [0,t]} \left(\mathbb{E}[|v_s|^k] \right)^{\frac{j}{k}} \quad (4.4.22)$$
$$+ \frac{c_2^2}{2} \int_0^t \left(\mathbb{E}[|v_s|^k] \right)^{\frac{j}{k}} ds < \infty, \quad t \in [0,T].$$

Noting that $\sup_{s \in [0,t]} \mathbb{E}[|v_s|^k] \leq \mathbb{E}[|\gamma_t|^k]$, by taking expectation in (4.4.22) with $j = k$, we derive

$$\mathbb{E}\big[|\gamma_t|^k\big] \leq 2c_2 \mathbb{E}[|X_0^1 - X_0^2|^k] + c_2^2 \int_0^t \mathbb{E}[|\gamma_s|^k] ds, \quad t \in [0,T].$$

Since $\mathbb{E}\big[|\gamma_t|^k\big] < \infty$ due to (4.4.10), by Gronwall's inequality we find a constant $c > 0$ such that

$$\sup_{t \in [0,T]} \mathbb{E}[|v_s|^k] \leq \mathbb{E}[|\gamma_T|^k] \leq c\mathbb{E}[|X_0^1 - X_0^2|^k].$$

Substituting this into (4.4.22) implies (4.4.11).

(3) Let $\nu \in \mathcal{P}_k$ and take $\mathcal{F}_0$-measurable random variables $X_0, \tilde{X}_0$ such that

$$\mathcal{L}_{X_0} = \mu, \quad \mathcal{L}_{\tilde{X}_0} = \nu, \quad \mathbb{E}[|X_0 - \tilde{X}_0|^k] = \mathbb{W}_k(\mu,\nu)^k. \quad (4.4.23)$$

Let X_t and $\tilde{X}_t$ solve (3.4.1) with initial values X_0 and $\tilde{X}_0$ respectively, and denote

$$\mu_t := P_t^* \mu = \mathcal{L}_{X_t}, \quad \nu_t := P_t^* \nu = \mathcal{L}_{\tilde{X}_t}, \quad t \in [0,T].$$

Let P_t^μ be the semigroup associated with $X_t^{\mu,x}$. According to Remark 1.4.1, (1.4.3) holds for P_t^μ replacing P_t and some constant $c > 0$ independent of μ. Then

$$\|P_t^* \mu - (P_t^\mu)^* \nu\|_{var} = \|(P_t^\mu)^* \mu - (P_t^\mu)^* \nu\|_{var} < \frac{c}{\sqrt{t}} \mathbb{W}_1(\mu,\nu). \quad (4.4.24)$$

On the other hand, let

$$R_t := e^{\int_0^t \langle \zeta_s(X_s)\{b_s(\tilde{X}_s,\mu_s) - b_s(\tilde{X}_s,\nu_s)\}, dW_s\rangle - \frac{1}{2}\int_0^s |\zeta_s(X_s)\{b_s(\tilde{X}_s,\nu_s) - b_s(\tilde{X}_s,\mu_s)\}|^2 ds}.$$

By $(A^{4.5})$ and Girsanov's theorem, $\mathbb{Q}_t := R_t \mathbb{P}$ is a probability measure under which

$$\tilde{W}_s := W_s - \int_0^s \zeta_s(X_s)\{b_s(\tilde{X}_s,\mu_s) - b_s(\tilde{X}_s,\nu_s)\} ds, \quad r \in [0,t]$$

is a Brownian motion. Reformulating the SDE for $\tilde{X}_s$ as

$$\mathrm{d}\tilde{X}_s = b_s(\tilde{X}_s, \mu_s)\mathrm{d}s + \sigma_s(\tilde{X}_s)\mathrm{d}\tilde{W}_s, \quad \mathcal{L}_{\tilde{X}_0} = \nu,$$

by the uniqueness we obtain $\mathcal{L}_{\tilde{X}_t|\mathbb{Q}_t} = (P_t^\mu)^*\nu$, so that by Pinsker's inequality (3.2.3) and ($A^{4.5}$), we find a constant $c_1 > 0$ such that

$$\|P_t^*\nu - (P_t^\mu)^*\nu\|_{var}^2 = \sup_{|f|\leq 1} |\mathbb{E}[f(\tilde{X}_t)(R_t - 1)]|^2 \leq 2\mathbb{E}[R_t \log R_t]$$

$$\leq c_1 \mathbb{E}_{\mathbb{Q}_t} \int_0^t \mathbb{W}_k(\mu_s, \nu_s)^2 \mathrm{d}s = c_1 \int_0^t \mathbb{W}_k(\mu_s, \nu_s)^2 \mathrm{d}s.$$

Combining this with (4.4.25) and (4.4.24), we derive (4.4.13) for some constant $c > 0$. □

Remark 4.4.1. By taking X_0^1 and X_0^2 such that

$$\mathcal{L}_{X_0^1} = \mu, \quad \mathcal{L}_{X_0^2} = \nu, \quad \mathbb{E}[|X_0^1 - X_0^2|^k] = \mathbb{W}_k(\mu, \nu)^k,$$

we deduce from (4.4.12) that

$$\mathbb{W}_k(P_t^*\mu, P_t^*\nu) \leq c\mathbb{W}_k(\mu, \nu), \quad t \in [0, T], \mu, \nu \in \mathcal{P}_k \qquad (4.4.25)$$

holds for some constant $c > 0$.

Next, we calculate $\nabla_\phi X_t^\mu$. In general, let X_t^μ solve (3.4.1) for $\mathcal{L}_{X_0^\mu} = \mu \in \mathcal{P}_k$, and for any $\varepsilon \in [0,1]$ and $\mathcal{F}_0$-measurable random variable η with $\mathcal{L}_\eta \in \mathcal{P}_k$, let X_t^ε solve (3.4.1) with $X_0^r = X_0^\mu + \varepsilon\eta$. We intend to calculate

$$\nabla_\eta X_t^\mu := \lim_{\varepsilon \downarrow 0} \frac{X_t^\varepsilon - X_t^\mu}{\varepsilon}, \quad t \in [0, T] \qquad (4.4.26)$$

in $L^k(\Omega \to C([0,T]; \mathbb{R}^d), \mathbb{P})$. In particular, taking $\eta := \phi(X_0^\mu)$ for $\phi \in \mathcal{T}_{\mu,k}$, we have

$$\nabla_\phi X_t^\mu = \nabla_\eta X_t^\mu, \quad t \in [0, T]. \qquad (4.4.27)$$

Choosing general η instead of $\phi(X_0^\mu)$ is useful in the proof of Theorem 4.4.3.

Let u solve (4.4.16) such that (4.4.17) and (1.4.9) hold as explained in the proof of Theorem 1.4.2. Let $\Theta_t = id + u_t$ and

$$Y_t^r := \Theta_t(X_t^r) = X_t^r + u_t(X_t^r), \quad t \in [0,T], \; r \in [0,1]. \qquad (4.4.28)$$

By (4.4.16) and Itô's formula, for any $r \in [0,1]$ we have

$$\mathrm{d}Y_t^r = \Big\{b_t^{(1)}(X_t^r, \mu_t^r) + \lambda u_t(X_t^r)$$
$$+ \nabla_{b_t^{(1)}(X_t^r, \mu_t^r) - b_t^{(1)}(X_t^r, \mu_t)} u_t(X_t^r)\Big\}\mathrm{d}t + \{(\nabla\Theta_t)\sigma_t\}(X_t^r)\mathrm{d}W_t, \qquad (4.4.29)$$
$$Y_0^r = \Theta_0(X_0^r) = X_0^\mu + r\eta + u_0(X_0^\mu + r\eta).$$

For any $t \in [0, T]$ and $v \in L^k(\Omega \to \mathbb{R}^d, \mathbb{P})$, let

$$\psi_t(v) := \mathbb{E}\big[\langle D^L b_t^{(1)}(z, \mu_t)(X_t^\mu), (\nabla \Theta_t(X_t^\mu))^{-1} v \rangle\big]\big|_{z=X_t^\mu}. \qquad (4.4.30)$$

By $(A^{4.5})(2)$, there exists a constant $K > 0$ such that for any $v, \tilde{v} \in L^k(\Omega \to \mathbb{R}^d, \mathbb{P})$,

$$\psi_t(0) = 0, \quad |\psi_t(v) - \psi_t(\tilde{v})| \leq K(\mathbb{E}[|v - \tilde{v}|^k])^{\frac{1}{k}}, \quad t \in [0, T]. \qquad (4.4.31)$$

If

$$v_t^\eta := \nabla_\eta Y_t^0 := \lim_{\varepsilon \downarrow 0} \frac{Y_t^\varepsilon - Y_t^0}{\varepsilon} \qquad (4.4.32)$$

exists in $L^k(\Omega \to C([0,T]; \mathbb{R}^d), \mathbb{P})$, by (4.4.17), (1.4.9) and (4.4.28) we see that $\nabla_\eta X_t^\mu$ exists in the same sense and

$$\nabla_\eta X_t^\mu = (\nabla \Theta_t(X_t^\mu))^{-1} \nabla_\eta Y_t^0 = (\nabla \Theta_t(X_t^\mu))^{-1} v_t^\eta. \qquad (4.4.33)$$

Combining this with $(A^{4.5})$, applying the chain rule Theorem 4.3.2, and noting that μ_t is absolutely continuous due to Theorem 6.3.1 in [Bogachev et al. (2015)], we obtain that for $\zeta_r := r X_t^\varepsilon + (1-r) X_t^\mu$,

$$\lim_{\varepsilon \to 0} \frac{b_t^{(1)}(X_t^\varepsilon, \mu_t^\varepsilon) - b_t^{(1)}(X_t^\varepsilon, \mu_t)}{\varepsilon} = \lim_{\varepsilon \to 0} \int_0^1 \frac{1}{\varepsilon} \frac{d}{dr} b_t^{(1)}(X_t^\varepsilon, \mathcal{L}_{\zeta_r}) dr$$
$$= \lim_{\varepsilon \to 0} \int_0^1 \mathbb{E}\Big[\Big\langle D^L b_t^{(1)}(z, \mathcal{L}_{\zeta_r})(\zeta_r), \frac{X_t^\varepsilon - X_t^\mu}{\varepsilon}\Big\rangle dr\Big]\Big|_{z=X_t^\varepsilon} \qquad (4.4.34)$$
$$= \psi_t(v_t^\eta),$$

which together with (4.4.33) yields

$$\lim_{\varepsilon \to 0} \frac{b_t^{(1)}(X_t^\varepsilon, \mu_t^\varepsilon) - b_t^{(1)}(X_t^\mu, \mu_t)}{\varepsilon} = \psi_t(v_t^\eta) + \nabla_{(\nabla \Theta_t(X_t^\mu))^{-1} v_t^\eta} b_t^{(1)}(X_t^\mu, \mu_t),$$

$$\lim_{\varepsilon \to 0} \frac{\{(\nabla \Theta_t)\sigma_t\}(X_t^\varepsilon) - \{(\nabla \Theta_t)\sigma_t\}(X_t^\mu)}{\varepsilon} = \nabla_{(\nabla \Theta_t(X_t^\mu))^{-1} v_t^\eta} \{(\nabla \Theta_t)\sigma_t\}(X_t^\mu),$$

$$\lim_{\varepsilon \to 0} \frac{u_t(X_t^\varepsilon) - u_t(X_t^\mu)}{\varepsilon} = \nabla_{(\nabla \Theta_t(X_t^\mu))^{-1} v_t^\eta} u_t(X_t^\mu).$$

Thus, if v_t^η in (4.4.32) exists, by (4.4.29) it should solve the SDE

$$\begin{aligned} dv_t^\eta &= \big\{\psi_t(v_t^\eta) + \nabla_{(\nabla \Theta_t(X_t^\mu))^{-1} v_t^\eta} b_t^{(1)}(X_t^\mu, \mu_t) \\ &\quad + \nabla_{\psi_t(v_t^\eta) + \lambda(\nabla \Theta_t(X_t^\mu))^{-1} v_t^\eta} u_t(X_t^\mu)\big\} dt \\ &\quad + \nabla_{(\nabla \Theta_t(X_t^\mu))^{-1} v_t^\eta} \{(\nabla \Theta_t)\sigma_t\}(X_t^\mu) dW_t, \\ v_0^\eta &= \eta + (\nabla_\eta u_0)(X_0). \end{aligned} \qquad (4.4.35)$$

Therefore, in terms of (4.4.33), to study $\nabla_\eta X_t^\mu$ we first consider the SDE (4.4.35).

Lemma 4.4.5. *Assume* $(A^{4.5})$. *For any* $\eta \in L^k(\Omega \to \mathbb{R}^d, \mathcal{F}_0, \mathbb{P})$, *the SDE* (4.4.35) *has a unique solution, and for any* $j \geq 1$ *there exists a constant* $c > 0$ *such that*

$$\mathbb{E}\left[\sup_{t\in[0,T]} |v_t^\eta|^j \Big| \mathcal{F}_0\right] \leq c\big\{(\mathbb{E}[|\eta|^k])^{\frac{j}{k}} + |\eta|^j\big\}, \tag{4.4.36}$$

$$\mu \in \mathcal{P}_k, \eta \in L^k(\Omega \to \mathbb{R}^d, \mathcal{F}_0, \mathbb{P}).$$

Proof. Let $X_t (= X_t^\mu)$ solve (3.4.1) with $\mathcal{L}_{X_0} = \mu$.

(1) Well-posedness of (4.4.35). Consider the space

$$\mathcal{C}_k := \left\{(v_t)_{t\in[0,T]} \text{ is continuous adapted}, v_0 = v_0^\eta, \mathbb{E}\left[\sup_{t\in[0,T]} |v_t|^k\right] < \infty\right\},$$

which is complete under the metric

$$\rho_\lambda(v^1, v^2) := \left(\mathbb{E}\left[\sup_{t\in[0,T]} e^{-\lambda t}|v_t^1 - v_t^2|^k\right]\right)^{\frac{1}{k}}, \quad v^1, v^2 \in \mathcal{C}_k$$

for $\lambda > 0$. By $(A^{4.5})$, (4.4.17) and (4.4.31), there exist a constant $K > 0$ and a function $1 \leq f_0 \in \tilde{L}_{q_0}^{p_0}$ such that for any random variable v,

$$\begin{aligned}&\big|\nabla_{(\nabla\Theta_t(X_t))^{-1}v} b_t^{(1)}(X_t, \mu_t) + \nabla_{\psi_t(v_t) + \lambda(\nabla\Theta_t(X_t))^{-1}v} u_t(X_t)\big| \\ &\leq K|v|,\\ &\big\|\nabla_{(\nabla\Theta_t(X_t))^{-1}v}\{(\nabla\Theta_t)\sigma_t\}(X_t)\big\| \leq K|v|\sum_{i=0}^{l} f_i(t, X_t),\\ &|\psi_t(v)| \leq K\big(\mathbb{E}[|v|^k]\big)^{\frac{1}{k}}, \quad t \in [0,T].\end{aligned} \tag{4.4.37}$$

Let $f = \sum_{i=0}^{l} f_i$. Let $\theta > 1$ such that $(\theta^{-1}p_i, \theta^{-1}q_i) \in \mathcal{K}, 0 \leq i \leq l$. By Krylov's estimate Theorem 1.2.3(2), we find a constant $c > 0$ such that

$$\mathbb{E}\int_0^T f_t(X_t)^{2\theta} dt \leq c\sum_{i=0}^{l}\|f^{2\theta}\|_{\tilde{L}_{q_i/\theta}^{p_i/\theta}} = c\sum_{i=0}^{l}\|f_i\|_{\tilde{L}_{q_i}^{p_i}}^{2\theta} < \infty.$$

So,

$$\tau_n := T \wedge \inf\left\{t \geq 0 : \int_0^t |f_t(X_t)|^{2\theta} ds \geq n\right\} \to T \text{ as } n \to \infty. \tag{4.4.38}$$

Thus,

$$\begin{aligned}H_t(v) := v_0^\eta &+ \int_0^t \big\{\psi_s(v_s) + \nabla_{(\nabla\Theta_s(X_s))^{-1}v_s} b_s^{(1)}(X_s, \mu_s)\\ &+ \nabla_{\psi_s(v_s)+\lambda(\nabla\Theta_s(X_s))^{-1}v_s} u_s(X_s)\big\} ds \\ &+ \int_0^t \nabla_{(\nabla\Theta_s(X_s))^{-1}v_s}\{(\nabla\Theta_s)\sigma_s\}(X_s) dW_s, \quad t \in [0,T]\end{aligned} \tag{4.4.39}$$

is an adapted continuous process on $\mathbb{R}^d$, and for any $n \geq 1$,
$$H_{\cdot \wedge \tau_n} : \mathcal{C}_{k,n} \to \mathcal{C}_{k,n}, \quad \mathcal{C}_{k,n} := \big\{(v_{\cdot \wedge \tau_n}) : v \in \mathcal{C}_k\big\}.$$
So, it remains to prove that H has a unique fixed point $v^\eta \in \mathcal{C}_k$ satisfying (4.4.36), which is then the unique solution of (4.4.35). In the following we explain that it suffices to prove

$$H_{\cdot \wedge \tau_n} \text{ has a unique fixed point in } \mathcal{C}_{k,n}, \quad n \geq 1. \tag{4.4.40}$$

Indeed, if (4.4.40) holds, then the unique fixed point $v^{\eta,n}_{\cdot \wedge \tau_n}$ satisfies
$$v^{\eta,n}_{\cdot \wedge \tau_n} = v^{\eta,n+k}_{\cdot \wedge \tau_n}, \quad n, k \geq 1,$$
so that
$$v^\eta_t := \lim_{n \to \infty} v^{\eta,n}_{t \wedge \tau_n}$$
is a continuous adapted process on $\mathbb{R}^d$, and
$$H_{\cdot \wedge \tau_n}(v^\eta) = v^\eta_{\cdot \wedge \tau_n} \in \mathcal{C}_{k,n}, \quad n \geq 1.$$
By this and (4.4.37), for any $j \geq k$ we find a constant $c > 0$ such that
$$d|v^\eta_t|^{2j} \leq c\big(\{\mathbb{E}[|v^\eta_{t \wedge \tau_n}|^k]\}^{\frac{2j}{k}} + |v^\eta_t|^{2j}\big)(1 + f_t^2(X_t))dt + d\tilde{M}_t, \quad t \in [0, \tau_n],$$
holds for some local martingale $\tilde{M}_t$. By the stochastic Gronwall inequality in Lemma 1.3.3, we find constants $k_1, k_2 > 0$ such that
$$\mathbb{E}\Big[\sup_{t \in [0,T]} |v^\eta_{t \wedge \tau_n}|^j \Big| \mathcal{F}_0\Big]$$
$$\leq k_1 \bigg(\int_0^t \{\mathbb{E}[|v^\eta_{t \wedge \tau_n}|^k]\}^{\frac{2j}{k}} ds + \mathbb{E}[|v^\eta_0|^{2j}|\mathcal{F}_0]\bigg)^{\frac{1}{2}} \tag{4.4.41}$$
$$\leq k_2 |\eta|^j + k_1 \bigg(\int_0^t \{\mathbb{E}[|v^\eta_{t \wedge \tau_n}|^k]\}^{\frac{2j}{k}} ds\bigg)^{\frac{1}{2}}.$$
Choosing $j = k$ we obtain
$$\mathbb{E}\Big[\sup_{t \in [0,T]} |v^\eta_{t \wedge \tau_n}|^k \Big| \mathcal{F}_0\Big]$$
$$\leq k_2 |\eta|^k + \frac{k_1^2}{2} \int_0^t \mathbb{E}[|v^\eta_{t \wedge \tau_n}|^k] ds + \frac{1}{2} \mathbb{E}\Big[\sup_{t \in [0,T]} |v^\eta_{t \wedge \tau_n}|^k\Big], \quad t \in [0,T].$$
Taking expectation and applying Gronwall's inequality, we find a constant $k_3 > 0$ such that
$$\sup_{n \geq 1} \mathbb{E}\Big[\sup_{t \in [0,T]} |v^\eta_{t \wedge \tau_n}|^k\Big] \leq k_3 \mathbb{E}[|\eta|^k],$$

so that (4.4.41) with $n \to \infty$ implies (4.4.36), and v_t^η is the unique solution of (4.4.35) in $\mathcal{C}_k$, since for each $n \geq 1$, $v_{t \wedge \tau_n}^\eta$ is the unique fixed point of $H_{\cdot \wedge \tau_n}$ in $\mathcal{C}_{k,n}$.

(2) We now verify (4.4.40). By (4.4.31), (4.4.37) and (4.4.38), we find constants $c_1, c_2 > 0$ such that

$$\rho_\lambda(H_{\cdot \wedge \tau_n}(v^1), H_{\cdot \wedge \tau_n}(v^2))^k := \mathbb{E}\left[\sup_{t \in [0,\tau_n]} e^{-\lambda t} |H_t(v^1) - H_t(v^2)|^k\right]$$

$$\leq c_1 \mathbb{E}\left[\sup_{t \in [0,\tau_n]} e^{-\lambda t} \left\{\left(\int_0^t \{|v_s^1 - v_s^2| + (\mathbb{E}|v_s^1 - v_s^2|^k)^{\frac{1}{k}}\}ds\right)^k \right.\right.$$

$$\left.\left. + \left(\int_0^t |v_s^1 - v_s^2|^2 f_s(X_s)^2 ds\right)^{\frac{k}{2}}\right\}\right]$$

$$\leq 2c_1 T^{k-1} \rho_{\lambda,n}(v^1, v^2) \sup_{t \in [0,T]} \int_0^t e^{-\lambda(t-s)} ds$$

$$+ c_1 \mathbb{E}\left[\sup_{t \in [0,\tau_n]} \left(e^{-\lambda t}|v_t^1 - v_t^2|^k\right)\left(\int_0^t e^{-\frac{2\lambda(t-s)}{k}} f_s(X_s)^2 ds\right)^{\frac{k}{2}}\right]$$

$$\leq \rho_\lambda(v^1, v^2)\left\{\frac{c_2}{\lambda}\right.$$

$$\left. + c_1 \sup_\Omega \sup_{t \in [0,\tau_n]} \left(\int_0^t f_s(X_s)^{2\theta} ds\right)^{\frac{k}{2\theta}} \left(\int_0^t e^{-\frac{2\theta^* \lambda(t-s)}{k}} ds\right)^{\frac{k}{2\theta^*}}\right\}$$

$$\leq \left\{\frac{c_2}{\lambda} + c_1 n^{\frac{k}{2\theta}} \left(\frac{k}{2\lambda \theta^*}\right)^{\frac{k}{2\theta^*}}\right\} \rho_\lambda(v^1, v^2), \quad v^1, v^2 \in \mathcal{C}_{k,n}, \quad \theta^* := \frac{\theta}{\theta - 1}.$$

Therefore, when $\lambda > 0$ is large enough, $H_{\cdot \wedge \tau_n}$ is contractive in ρ_λ for large $\lambda > 0$, and hence has a unique fixed point on $\mathcal{C}_{k,n}$. □

4.4.3 Proof of Theorem 4.4.1(1)

Theorem 4.4.1(1) is implied by the following result for $\eta = \phi(X_0^\mu)$.

Proposition 4.4.6. *Assume* $(A^{4.5})$. *For any* $v \in \mathbb{R}^d$ *and* $\eta \in L^k(\Omega \to \mathbb{R}^d, \mathcal{F}_0, \mathbb{P})$, $\nabla_\eta X_t^\mu$ *in* (4.4.26) *and* $\nabla_v X_t^{\mu,x}$ *in* (4.4.4) *exist in* $L^k(\Omega \to C([0,T]; \mathbb{R}^d), \mathbb{P})$, *and for any* $j \geq 1$ *there exists a constant* $c > 0$ *such that*

$$\mathbb{E}\left[\sup_{t \in [0,T]} |\nabla_\eta X_t^\mu|^j \Big| \mathcal{F}_0\right] \leq c\{\mathbb{E}[|\eta|^k]\}^{\frac{j}{k}} + c|\eta|^j,$$
$$\mu \in \mathcal{P}_k, \eta \in L^k(\Omega \to \mathbb{R}^d, \mathcal{F}_0, \mathbb{P}),$$
(4.4.42)

$$\mathbb{E}\Big[\sup_{t\in[0,T]}|\nabla_v X_t^{\mu,x}|^j\Big] \leq c|v|^j, \quad x,v \in \mathbb{R}^d, \mu \in \mathcal{P}_k. \tag{4.4.43}$$

Proof. The existence of $\nabla_v X_t^{\mu,x}$ and (4.4.43) follow from Theorem 1.4.2(1) for $b_t(x) := b_t(x,\mu_t)$ where the constant in (1.4.1) is uniformly in μ_t according to Remark 1.4.1. So, it suffices to prove the existence of $\nabla_\eta X_t^\mu$ and to verify (4.4.42). We simply denote
$$X_t = X_t^\mu, \quad v_t = v_t^\eta, \quad t \in [0,T].$$
For any $r \in (0,1]$ let Y_t^r be in (4.4.28). We have $Y_t := Y_t^0 = \Theta_t(X_t)$. Let
$$\tilde{v}_t^\varepsilon := \frac{Y_t^\varepsilon - Y_t}{\varepsilon}, \quad t \in [0,T], \varepsilon \in (0,1). \tag{4.4.44}$$
By Lemma 4.4.4(2) and (4.4.17), for any $j \geq 1$ there exists $c(j) > 0$ such that
$$\mathbb{E}\Big[\sup_{t\in[0,T]}|\tilde{v}_t^\varepsilon|^j \Big| \mathcal{F}_0\Big] \leq c(j)\big(\{\mathbb{E}[|\eta|^k]\}^{\frac{j}{k}} + |\eta|^j\big), \quad \varepsilon \in (0,1). \tag{4.4.45}$$
We claim that it suffices to prove
$$\lim_{\varepsilon\downarrow 0} \mathbb{E}\Big[\sup_{t\in[0,T]}|\tilde{v}_t^\varepsilon - v_t|^k\Big] = 0. \tag{4.4.46}$$
Indeed, this implies that
$$\nabla_\eta Y_t := \lim_{\varepsilon\downarrow 0} \tilde{v}_t^\varepsilon = v_t$$
exists in $L^k(\Omega \to C([0,T];\mathbb{R}^d),\mathbb{P})$, so that (4.4.17), (1.4.9) and $\Theta_t := id + u_t$ yield
$$\nabla_\eta X_t := \lim_{\varepsilon\downarrow 0} \frac{X_t^\varepsilon - X_t}{\varepsilon} = (\nabla\Theta_t(X_t))^{-1} v_t$$
exists in the same space, and (4.4.42) follows from (4.4.36).

Recall that $\mu_s^\varepsilon = \mathcal{L}_{X_s^\varepsilon}, \varepsilon \in [0,1]$. By (4.4.16) and Itô's formula, we obtain
$$d\tilde{v}_t^\varepsilon = \frac{1}{\varepsilon}\Big\{b_t^{(1)}(X_t^\varepsilon,\mu_t^\varepsilon) - b_t^{(1)}(X_t,\mu_t) + \nabla_{b_t^{(1)}(X_t^\varepsilon,\mu_t^\varepsilon) - b_t^{(1)}(X_t^\varepsilon,\mu_t)} u_t(X_t^\varepsilon)\Big\}dt$$
$$+ \frac{1}{\varepsilon}\Big\{[(\nabla\Theta_t)\sigma_t](X_t^\varepsilon) - [(\nabla\Theta_t)\sigma_t](X_t)\Big\}dW_t, \quad \tilde{v}_0^\varepsilon = \frac{\Theta_0(X_0^\varepsilon) - \Theta_0(X_0)}{\varepsilon}.$$
Then
$$\tilde{v}_t^\varepsilon = \tilde{v}_0^\varepsilon + \int_0^t \Big\{\nabla_{(\nabla\Theta_s(X_s))^{-1}\tilde{v}_s^\varepsilon} b_s^{(1)}(X_s,\mu_s)$$
$$+ \psi_s(\tilde{v}_s^\varepsilon) + \nabla_{\psi_s(\tilde{v}_s^\varepsilon)} u_s(X_s)\Big\}ds \tag{4.4.47}$$
$$+ \int_0^t \nabla_{(\nabla\Theta_s(X_s))^{-1}\tilde{v}_s^\varepsilon}\{(\nabla\Theta_s)\sigma_s\}(X_s)dW_s + \alpha_t^\varepsilon, \quad t \in [0,T],$$

where $\psi_t(v)$ is in (4.4.30), and for $t \in [0, T]$,

$$\alpha_t^\varepsilon := \int_0^t \xi_s^\varepsilon \mathrm{d}s + \int_0^t \eta_s^\varepsilon \, \mathrm{d}W_s,$$

$$\xi_s^\varepsilon := \frac{1}{\varepsilon}\Big\{ b_s^{(1)}(X_s^\varepsilon, \mu_s^\varepsilon) - b_s^{(1)}(X_s, \mu_s) + \nabla_{b_s^{(1)}(X_s^\varepsilon, \mu_s^\varepsilon) - b_s^{(1)}(X_s, \mu_s)} u_t(X_s^\varepsilon) \Big\}$$
$$- \Big\{ \nabla_{(\nabla \Theta_s(X_s))^{-1}\tilde{v}_s^\varepsilon} b_s^{(1)}(X_s, \mu_s) + \psi_s(\tilde{v}_s^\varepsilon) + \nabla_{\psi_s(\tilde{v}_s^\varepsilon)} u_s(X_s) \Big\},$$

$$\eta_s^\varepsilon := \frac{\{(\nabla \Theta_s)\sigma_s\}(X_s^\varepsilon) - \{(\nabla \Theta_s)\sigma_s\}(X_s)}{\varepsilon} - \nabla_{(\nabla \Theta_s(X_s))^{-1}\tilde{v}_s^\varepsilon}\{(\nabla \Theta_s)\sigma_s\}(X_s).$$

We claim

$$\lim_{\varepsilon \to 0} \mathbb{E}\Big[\sup_{t\in[0,T]} |\alpha_t^\varepsilon|^n \Big| \mathcal{F}_0 \Big] = 0, \quad n \geq 1. \qquad (4.4.48)$$

This can be proved by the argument leading to (1.4.13), but with the conditional expectation $\mathbb{E}[\cdot|\mathcal{F}_0]$ replacing the expectation.

Firstly, by (4.4.45), $Y_t^\varepsilon = X_t^\varepsilon + u_t(X_t^\varepsilon)$ and (4.4.17), for any $j \geq 1$ there exists $c(j) > 0$ such that

$$\sup_{\varepsilon \in (0,1]} \mathbb{E}\Big[\sup_{t\in[0,T]} \Big|\frac{X_t^\varepsilon - X_t}{\varepsilon}\Big|^j \Big| \mathcal{F}_0 \Big] \leq c(j)\big(\{\mathbb{E}[|\eta|^k]\}^{\frac{j}{k}} + |\eta|^j \big). \qquad (4.4.49)$$

Since $\{(\nabla \Theta_s)\sigma_s\}, b_s^{(1)}(\cdot, \mu_s)$ and ∇u_s are a.e. differentiable, by the same reason leading to (1.4.14), (4.4.49) implies that for any $s \in (0, T]$, $\mathbb{P}$-a.s.

$$\lim_{\varepsilon \to 0} \bigg\{ \bigg| \frac{\{(\nabla \Theta_s)\sigma_s\}(X_s^\varepsilon) - \{(\nabla \Theta_s)\sigma_s\}(X_s)}{\varepsilon}$$
$$- \nabla_{(\nabla \Theta_s(X_s))^{-1}\tilde{v}_s^\varepsilon}\{(\nabla \Theta_s)\sigma_s\}(X_s) \bigg|$$
$$+ \bigg| \frac{b_s^{(1)}(X_s^\varepsilon, \mu_s) - b_s^{(1)}(X_s, \mu_s)}{\varepsilon} \nabla_{(\nabla \Theta_s(X_s))^{-1}\tilde{v}_s^\varepsilon} h^{(1)}(X_s, \mu_s) \bigg| \bigg\} = 0.$$

Next, as in (4.4.34), by the chain rule in Theorem 4.3.2 and $b_t^{(1)} \in \mathcal{D}_k$, we obtain

$$\lim_{\varepsilon \to 0} \bigg| \frac{b_s^{(1)}(X_s^\varepsilon, \mu_s^\varepsilon) - b_s^{(1)}(X_s^\varepsilon, \mu_s)}{\varepsilon} - \psi_s(\tilde{v}_s^\varepsilon) \bigg| = 0, \quad s \in (0, T].$$

Thus, for any $s \in (0, T]$,

$$\lim_{\varepsilon \to 0} \big\{ |\xi_s^\varepsilon| + \|\eta_s^\varepsilon\| \big\} = 0, \quad \mathbb{P}\text{-a.s.}. \qquad (4.4.50)$$

Moreover, by $(A^{4.5})$ and Lemma 1.3.4, we find a constant $c > 0$ such that

$$|\xi_s^\varepsilon| + \|\eta_s^\varepsilon\| \leq c|\tilde{v}_s^\varepsilon|\Big(1 + \sum_{i=0}^{l} \big\{ \mathcal{M}f_i(s,\cdot)(X_s) + \mathcal{M}f_i(s,\cdot)(X_s^\varepsilon) \big\} \Big), \quad s \in [0, T].$$

Finally, let $\theta > 1$ be in the proof of (1.4.13) such that (1.4.15) holds for X_t^ε replacing $X_t^{x+\varepsilon v}$. By (1.4.15) for X_t^ε, (4.4.45), and Lemma 1.3.4, for any $n \geq 1$ there exist constants $c_1(n), c_2(n) > 0$ such that

$$\mathbb{E}\left[\left(\int_0^T \{|\xi_s^\varepsilon|^{2\theta} + \|\eta_s^\varepsilon\|^{2\theta}\}\mathrm{d}s\right)^n \bigg| \mathcal{F}_0\right]$$

$$\leq c_1(n)\mathbb{E}\left[\left(\sup_{s\in[0,T]} |\tilde{v}_s^\varepsilon|^{2\theta n}\right) \right.$$

$$\times \left. \left(\int_0^T \left(1 + \sum_{i=0}^l \{\mathcal{M}f_i^{2\theta}(s, X_s) + \mathcal{M}f_i^{2\theta}(s, X_s^\varepsilon)\}\right)\mathrm{d}s\right)^n \bigg| \mathcal{F}_0\right]$$

$$\leq c_1(n)\left(\mathbb{E}\left[\sup_{s\in[0,T]} |\tilde{v}_s^\varepsilon|^{4\theta n} \bigg| \mathcal{F}_0\right]\right)^{\frac{1}{2}}$$

$$\times \left(\mathbb{E}\left[\left(\int_0^T \left(1 + \sum_{i=0}^l \{\mathcal{M}f_i^{2\theta}(s, X_s) + \mathcal{M}f_i^{2\theta}(s, X_s^\varepsilon)\}\right)\mathrm{d}s\right)^{2n} \bigg| \mathcal{F}_0\right]\right)^{\frac{1}{2}}$$

$$\leq c_2(n)(1 + |\eta|^{2\theta n}) < \infty.$$

By BDG's inequality in Lemma 1.3.5 and the dominated convergence theorem, this and (4.4.50) imply (4.4.48).

Now, by (4.4.35) and (4.4.47), the argument leading to (4.4.21) gives

$$|v_t - \tilde{v}_t^\varepsilon|^{2k} \leq |v_0 - \tilde{v}_0^\varepsilon|^{2k} + \int_0^t \{|v_s - v_s^\varepsilon|^{2k}\gamma_t + (\mathbb{E}|v_s - \tilde{v}_s^\varepsilon|^k)^2\}\mathrm{d}t$$

$$+ K \sup_{r\in[0,t]} |\alpha_r^\varepsilon|^{2k} + M_t, \quad t \in [0,T],$$

where $K > 0$ is a constant and γ_t is a positive process satisfying

$$\mathbb{E}[e^{N\int_0^T \gamma_t \mathrm{d}t}] < \infty, \quad N > 0.$$

Therefore, by the stochastic Gronwall inequality in Lemma 1.3.3, we find a constant $c > 0$ such that for any $t \in [0,T]$,

$$\mathbb{E}\left[\sup_{s\in[0,t]} |\tilde{v}_s^\varepsilon - v_s|^k \bigg| \mathcal{F}_0\right]$$

$$\leq c|v_0 - \tilde{v}_0^\varepsilon|^k + c\left(\mathbb{E}\left[\sup_{s\in[0,t]} |\alpha_s^\varepsilon|^{2k} \bigg| \mathcal{F}_0\right]\right)^{\frac{1}{2}} + c\left(\int_0^t (\mathbb{E}[|\tilde{v}_s^\varepsilon - v_s|^k])^2 \mathrm{d}s\right)^{\frac{1}{2}}.$$

Combining this with (4.4.48) and $\lim_{\varepsilon\to 0} |v_0 - \tilde{v}_0^\varepsilon| = 0$, we obtain

$$\limsup_{\varepsilon\to 0} \mathbb{E}\left[\sup_{s\in[0,t]} |\tilde{v}_s^\varepsilon - v_s|^k \bigg| \mathcal{F}_0\right]$$

$$\leq c\limsup_{\varepsilon\to 0} \left(\int_0^t (\mathbb{E}[|\tilde{v}_s^\varepsilon - v_s|^k])^2 \mathrm{d}s\right)^{\frac{1}{2}}. \quad (4.4.51)$$

Taking $j = k$, by (4.4.36), (4.4.45) and (4.4.49), we conclude that
$$\left\{ \mathbb{E}\left[\sup_{t \in [0,T]} \{|\tilde{v}_t^\varepsilon|^k + |v_t|^k\} \Big| \mathcal{F}_0 \right] : \varepsilon \in (0,1] \right\}$$
is uniformly integrable with respect to $\mathbb{P}$, so that by Fatou's lemma, (4.4.51) implies
$$h_t := \limsup_{\varepsilon \to 0} \mathbb{E}\left[\sup_{s \in [0,t]} |\tilde{v}_s^\varepsilon - v_s|^k \right] = \limsup_{\varepsilon \to 0} \mathbb{E}\left\{ \mathbb{E}\left[\sup_{s \in [0,t]} |\tilde{v}_s^\varepsilon - v_s|^k \Big| \mathcal{F}_0 \right] \right\}$$
$$\leq \mathbb{E}\left\{ \limsup_{\varepsilon \to 0} \mathbb{E}\left[\sup_{s \in [0,t]} |\tilde{v}_s^\varepsilon - v_s|^k \Big| \mathcal{F}_0 \right] \right\} \leq c \left(\int_0^t h_s^2 \mathrm{d}s \right)^{\frac{1}{2}}, \quad t \in [0,T]$$
and $h_t < \infty$, so that $h_t = 0$ for all $t \in [0,T]$. Therefore, (4.4.46) holds and hence the proof is finished. □

4.4.4 Proof of Theorem 4.4.1(2)

For any $\eta \in L^k(\Omega \to \mathbb{R}^d, \mathcal{F}_0, \mathbb{P})$, $\mu \in \mathcal{P}_k$, and $\varepsilon \in [0,1]$, let X_t^ε solve (3.4.1) for $X_0^\varepsilon = X_0^\mu + \varepsilon \eta$, where $\mathcal{L}_{X_0^\mu} = \mu$. Then $X_t^0 = X_t^\mu$. Consider
$$\Gamma_\eta(f(X_t^\mu)) := \lim_{\varepsilon \downarrow 0} \frac{\mathbb{E}[f(X_t^\varepsilon) - f(X_t^\mu)]}{\varepsilon}, \quad t \in (0,T], f \in \mathcal{B}_b(\mathbb{R}^d).$$
Theorem 4.4.1(2) is implied by the following result for $\eta = \phi(X_0)$.

Proposition 4.4.7. *Assume* $(A^{4.5})$. *$D_\eta^I P_t f(\mu)$ exists for any $t \in (0,T]$, $f \in \mathcal{B}_b(\mathbb{R}^d)$, $\eta \in L^k(\Omega \to \mathbb{R}^d, \mathcal{F}_0, \mathbb{P})$ and $\mu \in \mathcal{P}_k$. Moreover, for any $\beta \in C^1([0,t])$ with $\beta_0 = 0$ and $\beta_t = 1$, the formula*
$$\Gamma_\eta(f(X_t^\mu)) = \int_{\mathbb{R}^d \times \mathbb{R}^d} \mathbb{E}\big[f(X_t^{\mu,x}) M_{\beta,t}^{\mu,x}(v)\big] \mathcal{L}_{(X_0^\mu, \eta)}(\mathrm{d}x, \mathrm{d}v) \qquad (4.4.52)$$
$$+ \mathbb{E}\big[f(X_t^\mu) N_t^\mu(\eta)\big],$$
holds for
$$M_{\beta,t}^{\mu,x}(v) := \int_0^t \beta_s' \langle \zeta_s(X_s^{\mu,x}) \nabla_v X_s^{\mu,x}, \mathrm{d}W_s \rangle,$$
$$N_t^\mu(\eta) := \int_0^t \Big\langle \zeta_s(X_s^\mu) \mathbb{E}\big[\langle D^L b_s^{(1)}(z, P_s^*\mu)(X_s^\mu), \nabla_\eta X_s^\mu \rangle\big]\big|_{z=X_s^\mu}, \mathrm{d}W_s \Big\rangle.$$
Consequently, there exists a constant $c > 0$ such that
$$|\Gamma_\eta(f(X_t^\mu))| \leq \frac{c}{\sqrt{t}} \big(P_t|f|^{\frac{k}{k-1}}(\mu)\big)^{\frac{k-1}{k}} (\mathbb{E}[|\eta|^k])^{\frac{1}{k}}, \qquad (4.4.53)$$
$$t \in (0,T], f \in \mathcal{B}_b(\mathbb{R}^d), \mu \in \mathcal{P}_k, \eta \in L^k(\Omega \to \mathbb{R}^d, \mathcal{F}_0, \mathbb{P}).$$

Proof. Let $X_t^{\mu,x}$ solve (4.4.3). Since X_t^μ solves the same SDE with initial value X_0^μ replacing x, the pathwise uniqueness implies

$$X_t^\mu = X_t^{\mu,X_0^\mu}, \quad t \in [0,T]. \tag{4.4.54}$$

Let $(P_{s,t}^\mu)_{0 \le s \le t \le T}$ be the semigroup associated with (4.4.3), i.e. for $(X_{s,t}^{\mu,x})_{t \in [s,T]}$ solving (4.4.3) from time s with $X_{s,s}^{\mu,x} = x$,

$$P_{s,t}^\mu f(x) := \mathbb{E}[f(X_{s,t}^{\mu,x})], \quad t \in [s,T], x \in \mathbb{R}^d. \tag{4.4.55}$$

Simply denote $P_t^\mu = P_{0,t}^\mu$. Then (4.4.54) implies

$$P_t f(\mu) = \mathbb{E}[f(X_t^\mu)] = \int_{\mathbb{R}^d} P_t^\mu f(x) \mu(\mathrm{d}x), \quad t \in [0,T], f \in \mathcal{B}_b(\mathbb{R}^d). \tag{4.4.56}$$

By Theorem 1.4.2, $(A^{4.5})$ implies that for any $t \in (0,T]$ and $\beta \in C^1([0,t])$ with $\beta_0 = 0$ and $\beta_t = 1$,

$$\nabla_v P_t^\mu f(x) = \mathbb{E}\big[f(X_t^{\mu,x}) M_{t,\beta}^{\mu,x}(v)\big], \quad v \in \mathbb{R}^d, f \in \mathcal{B}_b(\mathbb{R}^d). \tag{4.4.57}$$

Next, denote $\mu_t = P_t^* \mu = \mathcal{L}_{X_t^\mu}$ and let $\bar{X}_s^\varepsilon$ solve (1.1.3) for $\bar{X}_0^\varepsilon = X_0^\varepsilon$, i.e.

$$\mathrm{d}\bar{X}_s^\varepsilon = b_s(X_s^\varepsilon, \mu_s)\mathrm{d}s + \sigma_s(\bar{X}_s^\varepsilon)\mathrm{d}W_s, \quad s \in [0,t], \bar{X}_0^\varepsilon = X_0^\varepsilon. \tag{4.4.58}$$

We have

$$\mathbb{E}[f(\bar{X}_t^\varepsilon)] = \int_{\mathbb{R}^d} (P_t^\mu f)(x) \mathcal{L}_{X_0^\mu + \varepsilon \eta}(\mathrm{d}x)$$

$$= \int_{\mathbb{R}^d \times \mathbb{R}^d} P_t^\mu f(x + \varepsilon v) \mathcal{L}_{(X_0^\mu, \eta)}(\mathrm{d}x, \mathrm{d}v), \quad f \in \mathcal{B}_b(\mathbb{R}^d).$$

Combining this with (4.4.56) and (4.4.57), and applying the dominated convergence theorem, we obtain

$$\lim_{\varepsilon \downarrow 0} \frac{\mathbb{E}[f(\bar{X}_t^\varepsilon)] - P_t f(\mu)}{\varepsilon} = \int_{\mathbb{R}^d \times \mathbb{R}^d} \nabla_v P_t^\mu f(x) \mathcal{L}_{(X_0^\mu, \eta)}(\mathrm{d}x, \mathrm{d}v)$$
$$= \int_{\mathbb{R}^d \times \mathbb{R}^d} \mathbb{E}\big[f(X_t^{\mu,x}) M_{\beta,t}^{\mu,x}(v)\big] \mathcal{L}_{(X_0^\mu, \eta)}(\mathrm{d}x, \mathrm{d}v). \tag{4.4.59}$$

On the other hand, denote $\mu_t^\varepsilon = \mathcal{L}_{X_t^\varepsilon}$ and let

$$R_t^\varepsilon := \mathrm{e}^{\int_0^t \langle \xi_s, \mathrm{d}W_s \rangle - \frac{1}{2} \int_0^t |\xi_s|^2 \mathrm{d}s},$$

$$\xi_s := \zeta_s(X_s^\varepsilon)\{b_s^{(1)}(X_s^\varepsilon, \mu_s) - b_s^{(1)}(X_s^\varepsilon, \mu_s^\varepsilon)\}.$$

By $(A^{4.5})$, $\zeta_s = \sigma_s^*(\sigma_s \sigma_s^*)^{-1}$ and Girsanov's theorem, $\mathbb{Q}_t^\varepsilon := R_t^\varepsilon \mathbb{P}$ is a probability measure under which

$$\tilde{W}_r^\varepsilon := W_r - \int_0^r \zeta_s(X_s^\varepsilon)\{b_s^{(1)}(X_s^\varepsilon, \mu_s) - b_s^{(1)}(X_s^\varepsilon, \mu_s^\varepsilon)\}\mathrm{d}s, \quad r \in [0,t]$$

is a Brownian motion, and

$$\sup_{r\in[0,T],\varepsilon\in(0,1]} \mathbb{E}\left[\frac{|R_r^\varepsilon - 1|^j}{\varepsilon^j}\right] < \infty, \quad j \geq 1. \quad (4.4.60)$$

Reformulate the SDE for X_s^ε as

$$dX_s^\varepsilon = b_s(X_s^\varepsilon, \mu_s)+\sigma_s(X_s^\varepsilon)d\tilde{W}_s^\varepsilon, \quad X_0^\varepsilon = \bar{X}_0^\varepsilon.$$

By the well-posedness we obtain $\mathcal{L}_{X_t^\varepsilon|\mathbb{Q}_t^\varepsilon} = \mathcal{L}_{\bar{X}_t^\varepsilon|\mathbb{P}}$, so that

$$\mathbb{E}[f(\bar{X}_t^\varepsilon)] = \mathbb{E}[R_t^\varepsilon f(X_t^\varepsilon)], \quad f \in \mathcal{B}_b(\mathbb{R}^d).$$

Thus,

$$\frac{\mathbb{E}[f(X_t^\varepsilon)] - \mathbb{E}[f(\bar{X}_t^\varepsilon)]}{\varepsilon} = \frac{\mathbb{E}[f(X_t^\varepsilon)(1-R_t^\varepsilon)]}{\varepsilon} = I_1(\varepsilon) + I_2(\varepsilon),$$

$$I_1(\varepsilon) := \mathbb{E}\left[f(X_t^\mu)\frac{1-R_t^\varepsilon}{\varepsilon}\right], \quad I_2(\varepsilon) := \mathbb{E}\left[\{f(X_t^\varepsilon) - f(X_t^\mu)\}\frac{1-R_t^\varepsilon}{\varepsilon}\right].$$

By (4.4.14), (4.4.34) and the dominated convergence theorem, we obtain

$$\lim_{\varepsilon \to 0} I_1(\varepsilon) = \mathbb{E}\left[f(X_t^\mu)\int_0^t \langle \zeta_s(X_s^\mu)\mathbb{E}[\langle D^L b_s^{(1)}(z,\mu_s), \nabla_\eta X_s^\mu\rangle]|_{z=X_s^\mu}, dW_s\rangle\right].$$

So, to prove (4.4.52) it suffices to verify

$$\lim_{\varepsilon \to 0} I_2(\varepsilon) = 0. \quad (4.4.61)$$

By (4.4.14), we obtain

$$\lim_{r\uparrow t} \sup_{\varepsilon\in(0,1]} \mathbb{E}\left[\frac{|R_t^\varepsilon - R_r^\varepsilon|}{\varepsilon}\right] = 0. \quad (4.4.62)$$

Since $(A^{4.5})$ holds for $[r,T]$ replacing $[0,T]$, we have (4.4.13) for (r,T) replacing $(0,T)$. Similarly, (1.4.3) holds for $P_{r,t}^{\mu^\varepsilon}$ and $P_{r,t}^\mu$ replacing P_{t-r}, where $P_{r,t}^{\mu^\varepsilon}$ and $P_{r,t}^\mu$ are defined in (4.4.55). Therefore, by the Markov property,

$$|\mathbb{E}[f(X_t^\varepsilon) - f(X_t^\mu)|\mathcal{F}_r]| = |(P_{r,t}^{\mu^\varepsilon}f)(X_r^\varepsilon) - (P_{r,t}^\mu f)(X_r^\mu)|$$

$$\leq |(P_{r,t}^{\mu^\varepsilon}f)(X_r^\varepsilon) - (P_{r,t}^{\mu^\varepsilon}f)(X_r^\mu)| + |(P_{r,t}^{\mu^\varepsilon}f)(X_r^\mu) - (P_{r,t}^\mu f)(X_r^\mu)| \quad (4.4.63)$$

$$\leq c\|f\|_\infty \left(\frac{|X_r^\varepsilon - X_r^\mu|}{\sqrt{t-s}} \wedge 1\right) + |(P_{r,t}^{\mu^\varepsilon}f)(X_r^\mu) - (P_{r,t}^\mu f)(X_r^\mu)|.$$

On the other hand, let $(\tilde{X}_{r,s}^\varepsilon)_{s\in[r,t]}$ solve the SDE

$$d\tilde{X}_{r,s}^\varepsilon = b_s(\tilde{X}_{r,s}^\varepsilon, \mu_s^\varepsilon)ds + \sigma_s(\tilde{X}_{r,s}^\varepsilon)dW_s, \quad \tilde{X}_{r,r}^\varepsilon = X_r^\mu, s \in [r,t].$$

We have
$$P^{\mu^\varepsilon}_{r,t}f(X^\mu_r) = \mathbb{E}\big[f(\tilde{X}^\varepsilon_{r,t})\big|\mathcal{F}_r\big], \quad P^\mu_{r,t}f(X^\mu_r) = \mathbb{E}\big[f(X^\mu_t)\big|\mathcal{F}_r\big].$$
Noting that (4.4.14) and (4.4.25) imply
$$|b(x,\mu^\varepsilon_t) - h_t(x,\mu_t)| \leq c_1 \mathbb{W}_k(\mu^\varepsilon_0, \mu_0) \leq c_1\varepsilon(\mathbb{E}[|\eta|^k])^{\frac{1}{k}} \tag{4.4.64}$$
for some constant $c_1 > 0$, by Girsanov's theorem, for any $r \in [0,t)$,
$$R^\varepsilon_{r,t} := e^{\int_r^t \langle \xi_s, dW_s\rangle - \frac{1}{2}\int_r^t |\xi_s|^2 ds} = \frac{R^\varepsilon_t}{R^\varepsilon_r}$$
is a probability density such that under $\mathbb{Q}_{r,t} := R^\varepsilon_{r,t}\mathbb{P}$,
$$\tilde{W}_s := W_s - \int_r^s \zeta(X_\theta)\{b_s(X^\mu_\theta, \mu^\varepsilon_\theta) - b_\theta(X^\mu_\theta, \mu_\theta)\}d\theta, \quad s \in [r,t]$$
is a Brownian motion. Reformulating the SDE for $(X^\mu_s)_{s\in[r,t]}$ as
$$dX^\mu_s = b_s(X^\mu_s, \mu^\varepsilon_s)ds + \sigma_s(X^\mu_s)d\tilde{W}_s, \quad X^\mu_r = \tilde{X}^\varepsilon_{r,r}, \quad s \in [r,t],$$
by the weak uniqueness, we obtain
$$P^{\mu^\varepsilon}_{r,t}f(X^\mu_r) = \mathbb{E}\big[R^\varepsilon_{r,t}f(X^\mu_t)\big|\mathcal{F}_r\big],$$
so that by Pinsker's inequality (3.2.3) and (4.4.64), we find a constant $c_2 > 0$ such that
$$\begin{aligned}|(P^{\mu^\varepsilon}_{r,t}f)(X^\mu_r) - (P^\mu_{r,t}f)(X^\mu_r)|^2 &\leq \|f\|^2_\infty \big|\mathbb{E}[|1-R^\varepsilon_{r,t}|\big|\mathcal{F}_r]\big|^2 \\ &\leq 2\|f\|^2_\infty \mathbb{E}_{\mathbb{Q}_{r,t}}\big[\log R^\varepsilon_{r,t}\big|\mathcal{F}_0\big] \\ &= \|f\|^2_\infty \int_r^t \mathbb{E}_{\mathbb{Q}_{r,t}}\big[|\zeta(X^\mu_s)\{b_s(X^\mu_s, \mu^\varepsilon_s) - b_s(X^\mu_s, \mu_s)\}|^2\big|\mathcal{F}_r\big]ds \\ &\leq c_2\|f\|^2_\infty(t-r)\varepsilon^2\|\eta\|^2_{L^k(\mathbb{P})}.\end{aligned} \tag{4.4.65}$$
Combining this with (4.4.12), (4.4.60), (4.4.63) and that $(s\wedge 1)^2 \leq s$ for $s \geq 0$, we find constants $c_3, c_4 > 0$ such that
$$\begin{aligned}\bigg|\mathbb{E}\Big[\{f(X^\varepsilon_t) - f(X_t)\}\frac{1-R^\varepsilon_r}{\varepsilon}\Big]\bigg| &\leq \bigg(\mathbb{E}\big|\mathbb{E}[f(X^\varepsilon_t) - f(X_t)|\mathcal{F}_r]\big|^2\bigg)^{\frac{1}{2}} \bigg(\mathbb{E}\Big[\frac{|1-R^\varepsilon_r|^2}{\varepsilon^2}\Big]\bigg)^{\frac{1}{2}} \\ &\leq c_4\|f\|_\infty \bigg(\frac{\mathbb{E}[|X^\varepsilon_r - X^\mu_r|]}{\sqrt{t-r}}\bigg)^{\frac{1}{2}} + c_4\|f\|_\infty \varepsilon \\ &\leq c_5\sqrt{T}\|f\|_\infty \Big(\frac{\varepsilon}{t-r}\Big)^{\frac{1}{2}}, \quad \varepsilon \in (0,1], t \in [0,T].\end{aligned}$$

Combining this with (4.4.62) we obtain

$$\lim_{\varepsilon \downarrow 0} I_2(\varepsilon)$$
$$\leq \lim_{r \uparrow t} \lim_{\varepsilon \downarrow 0} \left\{ \left| \mathbb{E}\left[\{f(X_t^\varepsilon) - f(X_t)\} \frac{1 - R_r^\varepsilon}{\varepsilon} \right] \right| + 2\|f\|_\infty \mathbb{E}\left[\frac{|R_t^\varepsilon - R_r^\varepsilon|}{\varepsilon} \right] \right\} = 0.$$

Therefore, (4.4.62) holds.

It remains to prove (4.4.53). By Jensen's inequality, we only need to consider $p \in (1,2]$. By (4.4.52), we have

$$|\Gamma_\eta(f(X_t^\mu))| \leq \mathbb{E}(|J_1(X_0^\mu, \eta)|) + |J_2|, \qquad (4.4.66)$$

where

$$J_1(x,v) := \mathbb{E}\left[f(X_t^{\mu,x}) \int_0^t \beta'_s \langle \zeta_s(X_s^{\mu,x}) \nabla_v X_s^{\mu,x}, \mathrm{d}W_s \rangle \right], \quad x, v \in \mathbb{R}^d,$$

$$J_2 := \mathbb{E}\left[f(X_t^\mu) \int_0^t \langle \zeta_s(X_s^\mu) \mathbb{E}[\langle D^L b_s^{(1)}(z, P_s^* \mu)(X_s^\mu), \nabla_\eta X_s^\mu \rangle]|_{z=X_s^\mu}, \mathrm{d}W_s \rangle \right].$$

Taking $\beta_s = \frac{s}{t}$, by $\|\zeta\|_\infty < \infty$, (4.4.6) and Hölder's inequality, we find constants $c_1, c_2 > 0$ such that

$$|J_1(x,v)| \leq \frac{c_1}{t} (P_t^\mu |f|^p(x))^{\frac{1}{p}} \left\{ \mathbb{E}\left[\left(\int_0^t |\nabla_v X_s^{\mu,x}|^2 \mathrm{d}s \right)^{\frac{p^*}{2}} \right] \right\}^{\frac{1}{p^*}}$$
$$\leq \frac{c_2 |v|}{\sqrt{t}} (P_t^\mu |f|^p(x))^{\frac{1}{p}}, \quad t \in (0,T], \ x, v \in \mathbb{R}^d.$$

Combining this with (4.4.56) and $P_t^\mu |f|^p(X_0^\mu) = \mathbb{E}[|f(X_t^\mu)|^p | \mathcal{F}_0]$, we derive

$$\mathbb{E}[|J_1(X_0^\mu, \eta)|] \leq \frac{c_2}{\sqrt{t}} \mathbb{E}\left[|\eta| (P_t^\mu |f|^p (X_0^\mu))^{\frac{1}{p}} \right]$$
$$\leq \frac{c_2 \|\eta\|_{L^k(\mathbb{P})}}{\sqrt{t}} \| (\mathbb{E}[|f(X_t^\mu)|^p | \mathcal{F}_0])^{\frac{1}{p}} \|_{L^{k^*}(\mathbb{P})}, \quad t \in (0,T]. \qquad (4.4.67)$$

On the other hand, by $(A^{4.5})$, Hölder's inequality and (4.4.42) for $j = k$, we find constants $c_3, c_4 > 0$ such that

$$I_s(z) := \left| \zeta_s(X_s^\mu) \mathbb{E}[\langle D^L b_s^{(1)}(z, P_s^* \mu)(X_s^\mu), \nabla_\eta X_s^\mu \rangle] \right|$$
$$\leq c_3 \|\nabla_\eta X_s^\mu\|_{L^k(\mathbb{P})} \leq c_4 \|\eta\|_{L^k(\mathbb{P})},$$

so that

$$|J_2| \leq \mathbb{E}\left[(\mathbb{E}[|f(X_t^\mu)|^p | \mathcal{F}_0])^{\frac{1}{p}} \left(\mathbb{E}\left[\int_0^t I_s(X_s^\mu)^2 \mathrm{d}s \right]^{\frac{p^*}{2}} \right)^{\frac{1}{p^*}} \right]$$
$$\leq c_4 \sqrt{t} \|\eta\|_{L^k(\mathbb{P})} \mathbb{E}\left[(\mathbb{E}[|f(X_t^\mu)|^p | \mathcal{F}_0])^{\frac{1}{p}} \right].$$

This and (4.4.67) imply (4.4.53). □

4.4.5 Proof of Theorem 4.4.3

Let $X_t(=X_t^\mu)$ solve (3.4.1) with $\mathcal{L}_{X_0} = \mu$. For any $\varepsilon \in [0,1]$, let X_t^ε solve (3.4.1) with $X_0^\varepsilon = X_0 + \varepsilon\phi(X_0)$, $\mu^\varepsilon := \mathcal{L}_{X_0+\varepsilon\phi(X_0)}$ and $\mu_t^\varepsilon := P_t^*\mu^\varepsilon = \mathcal{L}_{X_t^\varepsilon}$.
We have
$$P_t f(\mu \circ (id + \varepsilon\phi^{-1})) = \mathbb{E}[f(X_t^\varepsilon)].$$
It suffices to prove
$$\lim_{\varepsilon \downarrow 0} \sup_{\|\phi\|_{L^k(\mu)} \leq 1} \left| \frac{\mathbb{E}[f(X_t^\varepsilon)] - f(X_t)]}{\varepsilon} - D_\phi^I P_t f(\mu) \right| = 0. \tag{4.4.68}$$

By applying (4.4.52) with $\beta_s = \frac{s}{t}$ for $(\mu^r, \phi(X_0))$ replacing (μ, η), we obtain
$$\frac{\mathrm{d}}{\mathrm{d}r}\mathbb{E}[f(X_t^r)] := \lim_{\varepsilon \downarrow 0} \frac{\mathbb{E}[f(X_t^{r+\varepsilon})] - f(X_t^r)]}{\varepsilon} = \Gamma_{\phi(X_0)}(f(X_t^{\mu^r}))$$
$$= \frac{1}{t} \int_{\mathbb{R}^d} \mathbb{E}\left[f(X_t^{\mu^r, id+r\phi}) \int_0^t \langle \zeta_s(X_s^{\mu^r, id+r\phi}) \nabla_\phi X_s^{\mu^r, id+r\phi}, \mathrm{d}W_s \rangle \right] \mathrm{d}\mu$$
$$+ \mathbb{E}\left[f(X_t^r) \int_0^t \langle \zeta_s(X_s^r) \mathbb{E}[\langle D^L b_s^{(1)}(z, \mu_s^r)(X_s^r), \nabla_{\phi(X_0)} X_s^{\mu^r} \rangle]|_{z=X_s^r}, \mathrm{d}W_s \rangle \right].$$

Combining this with (4.4.7) for $\beta_s = \frac{s}{t}$, we derive
$$\sup_{\|\phi\|_{L^k(\mu)} \leq 1} \left| \frac{\mathbb{E}[f(X_t^\varepsilon)] - f(X_t)]}{\varepsilon} - D_\phi^I P_t f(\mu) \right|$$
$$= \sup_{\|\phi\|_{L^k(\mu)} \leq 1} \left| \frac{1}{\varepsilon} \int_0^\varepsilon \left\{ \frac{\mathrm{d}}{\mathrm{d}r} \mathbb{E}[F(X_t^r)] - D_\phi^I P_t f(\mu) \right\} \mathrm{d}r \right| \leq \frac{c}{t\varepsilon} \int_0^\varepsilon \sum_{i=1}^4 \alpha_i(r) \mathrm{d}r$$

for some constant $c > 0$, where letting
$$G_\phi(r) := \int_0^t \langle \zeta_s(X_s^r) \mathbb{E}[\langle D^L b_s^{(1)}(z, \mu_s^r)(X_s^r), \nabla_{\phi(X_0)} X_s^{\mu^r} \rangle]|_{z=X_s^r}, \mathrm{d}W_s \rangle,$$
$$F_\phi(r,x) := \int_0^t \langle \zeta_s(X_s^{\mu^r, x+r\phi(x)}) \nabla_{\phi(x)} X_s^{\mu^r, x+r\phi(x)}, \mathrm{d}W_s \rangle, \quad r \in [0,1], x \in \mathbb{R}^d,$$

we set
$$\alpha_1(r) := \sup_{\|\phi\|_{L^k(\mu)} \leq 1} \left| \int_{\mathbb{R}^d} \mathbb{E}\left[f(X_t^{\mu^r, x+r\phi(x)}) \{F_\phi(r,x) - F_\phi(0,x)\} \right] \mu(\mathrm{d}x) \right|,$$
$$\alpha_2(r) := \sup_{\|\phi\|_{L^k(\mu)} \leq 1} \left| \int_{\mathbb{R}^d} \mathbb{E}\left[\{f(X_t^{\mu^r, x+r\phi(x)}) - f(X_t^{\mu,x})\} F_\phi(0,x) \right] \mu(\mathrm{d}x) \right|,$$
$$\alpha_3(r) := \sup_{\|\phi\|_{L^k(\mu)} \leq 1} \left| \mathbb{E}\left[f(X_t^r)\{G_\phi(r) - G_\phi(0)\} \right] \right|,$$
$$\alpha_4(r) := \sup_{\|\phi\|_{L^k(\mu)} \leq 1} \left| \mathbb{E}\left[\{f(X_t^r) - f(X_t)\} G_\phi(0) \right] \right|.$$

Since $\|f\|_\infty < \infty$, by $(A^{4.5})$, (4.4.5) and (4.4.6), we conclude that $\{\alpha_i\}_{1\leq i\leq 4}$ are bounded on $[0,1]$. So, (4.4.68) follows if
$$\lim_{r\downarrow 0}\alpha_i(r) = 0, \quad 1 \leq i \leq 4.$$
To prove these limits, we need the following two lemmas.

Lemma 4.4.8. *Assume* $(A^{4.5})$. *For any* $j \geq 1$ *there exists a constant* $c > 0$ *such that for any* $\mu \in \mathcal{P}_k$ *and* $\phi \in T_{\mu,k}$ *with* $\|\phi\|_{L^k(\mu)} \leq 1$,
$$\mathbb{E}\left[\sup_{t\in[0,T]}|X_t^{\mu^r, x+r\phi(x)} - X_t^{\mu,x}|^j\right] \leq cr^j(1+|\phi(x)|^j), \quad r \in [0,1].$$

Proof. By (4.4.6), we have
$$\mathbb{E}[|X_t^{\mu^r, x+r\phi(x)} - X_t^{\mu^r, x}|^j] \leq cr^j|\phi(x)|^j, \quad r \in [0,1], x \in \mathbb{R}^d.$$
Combining this with $\mathbb{W}_k(\mu^r, \mu) \leq r\|\phi\|_{L^k(\mu)} \leq r$, we need only to prove
$$\sup_{x\in\mathbb{R}^d}\mathbb{E}[|X_t^{\mu,x} - X_t^{\nu,x}|^j] \leq c\mathbb{W}_k(\mu,\nu)^j, \quad \mu,\nu \in \mathcal{P}_k \quad (4.4.69)$$
for some constant $c > 0$, where $X_t^{\nu,x}$ solves (4.4.3) for $\nu_t := P_t^*\nu$ replacing $\mu_t := P_t^*\mu$. Let u solve (4.4.16) such that (4.4.17) holds. Let $\Theta_t = id + u_t$ and
$$Y_t^{\mu,x} := \Theta_t(X_t^{\mu,x}), \quad Y_t^{\nu,x} := \Theta_t(X_t^{\nu,x}), \quad t \in [0,T].$$
By Itô's formula we obtain
$$d(Y_t^{\mu,x} - Y_t^{\nu,x}) = \left\langle\{(\nabla\Theta_t)\sigma_t\}(X_t^{\mu,x}) - \{(\nabla\Theta_t)\sigma_t\}(X_t^{\nu,x}), dW_t\right\rangle$$
$$+ \left\{b_t^{(1)}(X_t^{\mu,x}, \mu_t) + \lambda u_t(X_t^{\mu,x}) - b_t^{(1)}(X_t^{\nu,x}, \nu_t) - \lambda u_t(X_t^{\nu,x})\right.$$
$$\left. + \nabla_{b_t^{(1)}(X_t^{\mu,x},\mu_t) - b_t^{(1)}(X_t^{\nu,x},\nu_t)} u_t(X_t^{\nu,x})\right\}dt.$$
By $(A^{4.5})$, (4.4.17), Lemma 1.3.4 and Itô's formula, for any $j \geq 1$ we find a constant $c > 0$ such that
$$|Y_t^{\mu,x} - Y_t^{\nu,x}|^{2j}$$
$$\leq c\int_0^t |Y_s^{\mu,x} - Y_s^{\nu,x}|^{2j}\sum_{i=0}^l\{1 + \mathcal{M}f_i^2(x, X_s^{\mu,x}) + \mathcal{M}f_i^2(x, X_s^{\nu,x})\}ds$$
$$+ c\int_0^t \mathbb{W}_k(\mu_s,\nu_s)^{2j}ds + M_t, \quad t \in [0,T]$$
holds for some local martingale M_t with $M_0 = 0$. Since $\mathbb{W}_k(\mu_s,\nu_s) \leq c\mathbb{W}_k(\mu,\nu)$ due to (4.4.25), (4.4.69) follows from the stochastic Gronwall inequality in Lemma 1.3.3, the maximal inequalities in Lemma 1.3.4, and Khasminskii's estimate in Theorem 1.2.4 for $X_s^{\mu,x}$ and $X_s^{\nu,x}$ replacing X_s respectively. □

Lemma 4.4.9. *Assume* $(A^{4.5})$ *and* (4.4.8). *For any* $j \geq 1$ *there exist a constant* $c > 0$ *and a positive function* $\varepsilon(\cdot)$ *on* $[0,1]$ *with* $\varepsilon(r) \downarrow 0$ *as* $r \downarrow 0$, *such that for any* $\phi \in T_{\mu,k}$ *with* $\|\phi\|_{L^k(\mu)} \leq 1$ *and* $r \in [0,1]$,

$$\sup_{|v|\leq 1} \mathbb{E}\left[\sup_{t\subset [0,T]} |\nabla_v X_t^{\mu^r,x+r\phi(x)} - \nabla_v X_t^{\mu,x}|^j\right]$$
$$\leq \min\{c, \varepsilon(r)(1+|\phi(x)|^j)\}, \quad x \in \mathbb{R}^d, \tag{4.4.70}$$

$$\mathbb{E}\left[\sup_{t\in [0,T]} |\nabla_{\phi(X_0)} X_t^{\mu^r} - \nabla_{\phi(X_0)} X_t^{\mu}|^j \Big| \mathcal{F}_0\right]$$
$$\leq |\phi(X_0)|^j \min\{c, \varepsilon(r)(1+|\phi(X_0)|^j)\}. \tag{4.4.71}$$

Proof. We only prove (4.4.70), since (4.4.71) can be proved in the same way by using (4.4.42) and (4.4.11) replacing (4.4.6) and Lemma 4.4.8 respectively. We simply denote

$$X_t^x := X_t^{\mu,x}, \quad X_t^{r,x} := X_t^{\mu^r, x+r\phi(x)},$$
$$\tilde{v}_t := \nabla_v X_t^{\mu,x}, \quad \tilde{v}_t^r := \nabla_v X_t^{\mu^r, x+r\phi(x)}. \tag{4.4.72}$$

Let u solve (4.4.16) such that (4.4.17) holds. We may also assume that u satisfies (1.4.9) as explained before. Let $\Theta_t = id + u_t$ and denote

$$Y_t^x := \Theta_t(X_t^x), \quad Y_t^{r,x} := \Theta_t(X_t^{r,x}),$$
$$v_t := (\nabla \Theta_t(X_t^x))^{-1} \tilde{v}_t, \quad v_t^r := (\nabla \Theta_t(X_t^{r,x}))^{-1} \tilde{v}_t^r. \tag{4.4.73}$$

By (4.4.6) and (4.4.17), to prove (4.4.70) it suffices to find $\varepsilon(r) \downarrow 0$ as $r \downarrow 0$ such that

$$\sup_{|v|\leq 1} \mathbb{E}\left[\sup_{t\in [0,T]} |v_t^r - v_t|^j\right] \leq \varepsilon(r)(1+|\phi(x)|^j), \quad r \in [0,1], x \in \mathbb{R}^d. \tag{4.4.74}$$

By Jensen's inequality, we only need to prove for $j \geq 4$.

To calculate v_t and v_t^r, for any $\varepsilon \in [0,1]$ we let

$$Y_t^{r,x}(\varepsilon) := \Theta_t(X_t^{\mu^r, x+r\phi(x)+\varepsilon v}), \quad Y_t^x(\varepsilon) := \Theta_t(X_t^{\mu, x+\varepsilon v}).$$

Then the argument leading to (1.4.8) implies that

$$v_t = \lim_{\varepsilon \downarrow 0} \frac{Y_t^x(\varepsilon) - Y_t^x}{\varepsilon}, \quad v_t^r = \lim_{\varepsilon \downarrow 0} \frac{Y_t^{r,x}(\varepsilon) - Y_t^{r,x}}{\varepsilon}. \tag{4.4.75}$$

By (4.4.16) and Itô's formula, we obtain

$$dY_t^x(\varepsilon) = \left\{b_t^{(1)}(X_t^{\mu,x+\varepsilon v}, \mu_t) + \lambda u_t(X_t^{\mu,x+\varepsilon v})\right\}dt$$
$$+ \{(\nabla \Theta_t)\sigma_t\}(X_t^{\mu,x+\varepsilon v})dW_t, \quad Y_0^x(\varepsilon) = x + \varepsilon v,$$

$$\mathrm{d}Y_t^{r,x}(\varepsilon) = \left\{b_t^{(1)}(X_t^{\mu^r,x+r\phi(x)+\varepsilon v},\mu_t^r) + \lambda u_t(X_t^{\mu^r,x+r\phi(x)+\varepsilon v})\right.$$
$$\left. + \nabla_{b_t^{(1)}(X_t^{\mu^r,x+r\phi(x)+\varepsilon v},\mu_t^r) - b_t^{(1)}(X_t^{\mu^r,x+r\phi(x)+\varepsilon v},\mu_t)} u_t(X_t^{\mu^r,x+r\phi(x)+\varepsilon v})\right\}\mathrm{d}t$$
$$+ \left\{(\nabla\Theta_t)\sigma_t\right\}(X_t^{\mu^r,x+r\phi(x)+\varepsilon v})\mathrm{d}W_t, \quad Y_0^{r,x}(\varepsilon) = x + r\phi(x) + \varepsilon x.$$

Combining this with (4.4.72) and (4.4.75), we conclude that (v_t, v_t^r) solves the SDEs

$$\mathrm{d}v_t = \left\{\nabla_{\tilde{v}_t} b_t^{(1)}(X_t^x,\mu_t) + \lambda \nabla_{\tilde{v}_t} u_t(X_t^x)\right\}\mathrm{d}t + \nabla_{\tilde{v}_t}\left\{(\nabla\Theta_t)\sigma_t\right\}(X_t^x)\mathrm{d}W_t,$$
$$v_0 = (\nabla\Theta_0(x))^{-1}v,$$

$$\mathrm{d}v_t^r = \left\{\nabla_{\tilde{v}_t^r} b_t^{(1)}(X_t^{r,x},\mu_t^r) + \lambda \nabla_{\tilde{v}_t^r} u_t(X_t^{r,x}) + \nabla_{\tilde{v}_t^r - \tilde{v}_t} u_t(X_t^{r,x})\right\}\mathrm{d}t$$
$$+ \nabla_{\tilde{v}_t^r}\left\{(\nabla\Theta_t)\sigma_t\right\}(X_t^{r,x})\mathrm{d}W_t, \quad v_0^r = \left\{\nabla\Theta_0(x+r\phi(x))\right\}^{-1}v.$$

Therefore, by (4.4.73),
$$z_t^r := v_t^r - v_t, \quad t \in [0,T]$$
solves the SDE

$$\mathrm{d}z_t^r = \left\{\nabla_{(\nabla\Theta_t(X_t^x))^{-1}z_t^r}\left[b_t^{(1)}(\cdot,\mu_t) + \lambda u_t\right](X_t^x)\right.$$
$$\left. + \nabla_{(\nabla\Theta_t(X_t^x))^{-1}z_t^r} u_t(X_t^{r,x})\right\}\mathrm{d}t \qquad (4.4.76)$$
$$+ \nabla_{(\nabla\Theta_t(X_t^x))^{-1}z_t^r}\left\{(\nabla\Theta_t)\sigma_t\right\}(X_t^x)\mathrm{d}W_t - \eta_t^r\mathrm{d}t - \xi_t^r\mathrm{d}W_t,$$
$$z_0^r = \left\{(\nabla\Theta_0(x+r\phi(x))^{-1} - (\nabla\Theta_0)(x))^{-1}\right\}v,$$

where for any $t \in [0,T]$,
$$\eta_t^r := \nabla_{(\Theta_t(X_t^{r,x}))^{-1}v_t^r} b_t^{(1)}(X_t^x,\mu_t) - \nabla_{(\Theta_t(X_t^x))^{-1}v_t^r} b_t^{(1)}(X_t^{r,x},\mu_t^r)$$
$$- \lambda\nabla_{(\Theta_t(X_t^{r,x}))^{-1}\eta_t^r} u_t(X_t^{r,x}) + \lambda\nabla_{(\Theta_t(X_t^x))^{-1}v_t^r} u_t(X_t^x)$$
$$+ \nabla_{\{(\nabla\Theta_t(X_t^x))^{-1} - (\nabla\Theta_t(X_t^{r,x}))^{-1}\}v_t^r} u_t(X_t^{r,x}),$$
$$\xi_t^r := \nabla_{(\Theta_t(X_t^x))^{-1}v_t^r}\left\{(\nabla\Theta_t)\sigma_t\right\}(X_t^x) - \nabla_{(\Theta_t(X_t^{r,x}))^{-1}v_t^r}\left\{(\nabla\Theta_t)\sigma_t\right\}(X_t^{r,x}).$$

By (4.4.17), $(A^{4.5})$ and Lemma 1.3.4, we find a constant $c_1 > 0$ such that

$$|\eta_t^r| + \|\xi_t^r\| \leq c_1|v_t^r|\left\{\|\nabla b_t^{(1)}(X_t^x,\mu_t) - \nabla b_t^{(1)}(X_t^{r,x},\mu_t^r)\|\right.$$
$$\left. + |X_t^{r,x} - X_t^x|\sum_{i=0}^{l}\left(1 + \mathcal{M}f_i(t,X_t^x) + \mathcal{M}f_i(t,X_t^{r,x})\right)\right\}.$$

By the boundedness of $\nabla b^{(1)}$ and (4.4.8), we have
$$\|\nabla b_t^{(1)}(X_t^x,\mu_t) - \nabla b_t^{(1)}(X_t^{r,x},\mu_t^r)\|$$
$$\leq n\left\{|X_t^x - X_t^{r,x}| + \mathbb{W}_k(\mu_t,\mu_t^r)\right\}^{\frac{1}{2j}} + s_n, \quad n \geq 1, \qquad (4.4.77)$$

where for $\varphi(r) := \sup_{|x-x'|+\mathbb{W}_k(\mu,\nu)\le r} \|\nabla b_t^{(1)}(x,\mu) - \nabla b_t^{(1)}(x',\nu)\|$,
$$s_n := \sup_{r\ge 0}\left\{\varphi(r) - nr^{\frac{1}{2j}}\right\} \downarrow 0 \text{ as } n\uparrow\infty.$$
Using the notation (4.4.72), by combining this with Lemma 4.4.8, (4.4.25) and (1.4.15) for the processes X_t^x and $X_t^{r,x}$, for any $j \ge 4$ we find positive function ε_1 with $\varepsilon_1(r) \downarrow 0$ as $r \downarrow 0$ such that for $\|\phi\|_{L^k(\mu)} \le 1$,
$$\mathbb{E}\left[\left(\int_0^T \{|\eta_s^r|^2 + \|\xi_s^r\|^2\}\mathrm{d}s\right)^j\right] \qquad (4.4.78)$$
$$\le \varepsilon_1(r)(1+|\phi(x)|^{2j}), \quad r\in[0,1], x\in\mathbb{R}^d.$$
Combining this with (4.4.76), ($A^{4.5}$) and BDG's inequality in Lemma 1.3.5, we find a constant $c_1 > 0$ such that
$$\gamma_t^r := \sup_{s\in[0,t]} |z_s^r|, \quad t\in[0,T]$$
satisfies
$$\mathbb{E}[\gamma_t^j] \le \varepsilon_2(r) + c_1\int_0^t \left\{\gamma_s^j + \gamma_s^{j-1}|\eta_s^r| + \gamma_s^{j-2}|\xi_s^r|^2\right\}\mathrm{d}s, \quad t\in[0,T], \quad (4.4.79)$$
where by (1.4.9), $\varepsilon_2(r) := \mathbb{E}[|z_0^r|^j] \to 0$ as $r \to 0$. Since $st \le s^{\frac{n}{n-1}} + t^n$ holds for $s,t \ge 0$ and $n \ge 1$, by taking $n = \frac{j}{2}$ and j for $j \ge 4$ respectively, we obtain
$$\int_0^t \left\{\gamma_s^{j-1}|\eta_s^r| + \gamma_s^{j-2}|\xi_s^r|^2\right\}\mathrm{d}s$$
$$\le \left(\int_0^t |z_s^r|^{2(j-1)}\mathrm{d}s\right)^{\frac{1}{2}}\left(\int_0^t |\eta_s^r|^2\mathrm{d}s\right)^{\frac{1}{2}}$$
$$+ \left(\int_0^t |z_s^r|^{2(j-2)}\mathrm{d}s\right)^{\frac{1}{2}}\left(\int_0^t |\xi_s^r|^2\mathrm{d}s\right)^{\frac{1}{2}}$$
$$\le \left(\int_0^t |z_s^r|^{2(j-1)}\mathrm{d}s\right)^{\frac{j}{2(j-1)}} + \left(\int_0^t |z_s^r|^{2(j-2)}\mathrm{d}s\right)^{\frac{j}{2(j-2)}} + \alpha^r,$$
where
$$\alpha^r := \left(\int_0^T |\eta_s^r|^2\mathrm{d}s\right)^{\frac{j}{2}} + \left(\int_0^T |\xi_s^r|^2\mathrm{d}s\right)^j.$$
So, there exists a constant $c_2 > 0$ such that
$$c_1\int_0^t \left\{\gamma_s^{j-1}|\eta_s^r| + \gamma_s^{j-2}|\xi_s^r|^2\right\}\mathrm{d}s \le c_1|\gamma_t^r|^{\frac{j(j-2)}{2(j-1)}}\left(\int_0^t |z_s^r|^j \mathrm{d}s\right)^{\frac{j}{2(j-1)}}$$
$$+ c_1|\gamma_t^r|^{\frac{j(j-4)}{2(j-2)}}\left(\int_0^t |z_s^r|^j\mathrm{d}s\right)^{\frac{j}{2(j-2)}} + c_1\alpha^r$$
$$\le \frac{1}{2}|\gamma_t^r|^j + c_2\int_0^t |\gamma_s^r|^j\mathrm{d}s + c_1\alpha^r.$$

Since (4.4.6) implies $\mathbb{E}[|\gamma_t^r|^j] < \infty$, combining this with (4.4.78), (4.4.79) and applying Gronwall's inequality, we derive (4.4.74) for some positive function ε with $\varepsilon(r) \downarrow 0$ as $r \downarrow 0$. $\square$

We are now ready to prove $\alpha_i(r) \to 0$ as $r \to 0$ for $i = 1, 2, 3, 4$ respectively and hence finish the proof of Theorem 4.4.3.

(a) $\alpha_1(r) \to 0$. As in (4.4.77), by $(A^{4.5})$ and (4.4.8) we find a sequence of positive numbers $s_n \downarrow 0$ as $n \uparrow \infty$ such that

$$\sup_{s \in [0,T]} \|\zeta_s(x) - \zeta_s(y)\|^2 \leq n|x-y|^{2(k-1)} + s_n, \quad n \geq 1, \tag{4.4.80}$$

$$\sup_{s \in [0,T]} \|D^L b_s^{(1)}(x,\mu)(y) - D^L b_s^{(1)}(x',\nu)(y')\|$$
$$\leq n\{|x-x'| + |y-y'| + \mathbb{W}_k(\mu,\nu)\}^{\frac{1}{k^*}} + s_n, \quad n \geq 1. \tag{4.4.81}$$

By (4.4.80), Lemma 4.4.8, Lemma 4.4.9 and (4.4.43), we find a constant $c_1 > 0$ such that for any $\phi \in T_{\mu,k}$ with $\|\phi\|_{L^k(\mu)} \leq 1$,

$\mathbb{E}[|F_\phi(r,x) - F_\phi(0,x)|]$

$$\leq \mathbb{E}\bigg(\int_0^t |\zeta_s(X_s^{\mu^r, x+r\phi(x)}) - \zeta_s(X_s^{\mu,x})|^2 \cdot |\nabla_{\phi(x)} X_s^{\mu^r, x+r\phi(x)}|^2 \mathrm{d}s\bigg)^{\frac{1}{2}}$$

$$+ \|\zeta\|_\infty \mathbb{E}\bigg(\int_0^t |\nabla_{\phi(x)} X_s^{\mu^r, x+r\phi(x)} - \nabla_{\phi(x)} X_s^{\mu,x}|^2 \mathrm{d}s\bigg)^{\frac{1}{2}}$$

$$\leq \bigg(\mathbb{E}\Big[\sup_{s\in[0,T]} |\nabla_{\phi(x)} X_s^{\mu^r, x+r\phi(x)}|^2\Big]$$

$$\int_0^t \mathbb{E}\big[n|X_s^{\mu^r, x+r\phi(x)} - X_s^{\mu,x}|^{2(k-1)} + s_n\big]\mathrm{d}s\bigg)^{\frac{1}{2}}$$

$$+ c_1|\phi(x)|\min\{1, \varepsilon(r)(1+|\phi(x)|)\}$$

$$\leq c_1|\phi(x)|\Big(\sqrt{n}(r+r|\phi(x)|)^{k-1} + \sqrt{s_n} + \min\{1, \varepsilon(r)(1+|\phi(x)|)\}\Big), n \geq 1.$$

Integrating with respect to $\mu(\mathrm{d}x)$ and letting first $r \to 0$ then $n \to \infty$, we derive $\alpha_1(r) \to 0$ as $r \to 0$.

(b) $\alpha_2(r) + \alpha_4(r) \to 0$. Let

$$R_\theta := \int_0^\theta \langle \zeta_s(X_s^{\mu,x})\nabla_{\phi(x)} X_s^{\mu,x}, \mathrm{d}W_s\rangle, \quad \theta \in [0,t].$$

By (4.4.6), we find a constant $c_1 > 0$ such that

$$\mathbb{E}[|R_t - R_\theta|] \leq c_1\sqrt{t-\theta}|\phi(x)|, \quad \theta \in [0,t], x \in \mathbb{R}^d. \tag{4.4.82}$$

On the other hand, as in (4.4.63) and (4.4.65), we find a constant $c_2 > 0$ such that for $\|\phi\|_{L^k(\mu)} \leq 1$,

$$\left|\mathbb{E}\left[f(X_t^{\mu^r, x+r\phi(x)}) - f(X_t^{\mu,x})\big|\mathcal{F}_\theta\right]\right| = \left|(P_{\theta,t}^{\mu^r}f)(X_\theta^{\mu^r, x+r\phi(x)}) - (P_{\theta,t}^{\mu}f)(X_\theta^{\mu,x})\right|$$

$$\leq \left|(P_{\theta,t}^{\mu^r}f)(X_\theta^{\mu^r, x+r\phi(x)}) - (P_{\theta,t}^{\mu^r}f)(X_\theta^{\mu,x})\right| + \left|(P_{\theta,t}^{\mu^r}f)(X_\theta^{\mu,x}) - (P_{\theta,t}^{\mu}f)(X_\theta^{\mu,x})\right|$$

$$\leq c_2\|f\|_\infty\left[1 \wedge \frac{|X_\theta^{\mu^r, x+r\phi(x)} - X_\theta^{\mu,x}|}{\sqrt{t-\theta}} + r\right], \quad \theta \in [0, t].$$

Combining this with (4.4.82) and Lemma 4.4.8, and using the Markov property, we find constants $c_3, c_4 > 0$ such that

$$\left|\mathbb{E}\left[\{f(X_t^{\mu^r, x+r\phi(x)}) - f(X_t^{\mu,x})\}F_\phi(0,x)\right]\right|$$

$$\leq 2\|f\|_\infty \mathbb{E}[|R_t - R_\theta|] + \left|\mathbb{E}\left[\mathbb{E}\big(f(X_t^{\mu^r, x+r\phi(x)}) - f(X_t^{\mu,x})\big|\mathcal{F}_\theta\big)R_\theta\right]\right|$$

$$\leq c_3\|f\|_\infty\left\{\sqrt{t-\theta}|\phi(x)| + \left(r + \frac{\min\{1, r(1+|\phi(x)|)\}}{\sqrt{t-\theta}}\right)(\mathbb{E}[|R_\theta|^2])^{\frac{1}{2}}\right\}$$

$$\leq c_4\|f\|_\infty\left\{\sqrt{t-\theta}|\phi(x)| + r|\phi(x)| + \frac{\{nr^{k-1}(1+|\phi(x)|)^{k-1} + s_n\}|\phi(x)|}{\sqrt{t-\theta}}\right\},$$

where

$$s_n := \sup_{s>0}\{s \wedge 1 - ns^{k-1}\} \downarrow 0 \text{ as } n \uparrow \infty.$$

Therefore, there exists a constant $c_5 > 0$ such that

$$\alpha_2(r) \leq \|f\|_\infty\left\{c_5\sqrt{t-\theta} + \frac{nr^{k-1} + \varepsilon_n}{\sqrt{t-\theta}} + r\right\}, \quad \theta \in (0, t).$$

By letting first $r \to 0$ then $n \to \infty$ and finally $\theta \to t$, we derive $\alpha_2(r) \to 0$ as $r \to 0$.

The proof of $\alpha_4(r) \to 0$ is completely similar.

(c) $\alpha_3(r) \to 0$. Write

$$\mathbb{E}\left[|G_\phi(r) - G_\phi(0)|\right] \leq \varepsilon_r(\phi) + \|\zeta\|_\infty \mathbb{E}\left[J_r(X_s^{\mu^r}, X_s^{\mu})\right],$$

where

$$\varepsilon_r(\phi) := \mathbb{E}\left[\left(\int_0^t |\zeta_s(X_s^r) - \zeta_s(X_s)|^2 (\mathbb{E}|\nabla_{\phi(X_0)}X_s^{\mu^r}|^k)^{\frac{2}{k}} ds\right)^{\frac{1}{2}}\right],$$

$$J_r(y, z) := \left(\int_0^t \left|\mathbb{E}[\langle D^L b_s^{(1)}(y_s, \mu_s^r)(X_s^r), \nabla_{\phi(X_0)}X_s^{\mu^r}\rangle\right.\right.$$

$$\left.\left. - \langle D^L b_s^{(1)}(z_s, \mu_s)(X_s), \nabla_{\phi(X_0)}X_s^{\mu}\rangle\right]\right|^2 ds\right)^{\frac{1}{2}}, \quad y, z \in C([0,t]; \mathbb{R}^d).$$

By (4.4.42) for $j = k$, we obtain
$$\sup_{r \in [0,1]} \mathbb{E}[|\nabla_{\phi(X_0)} X_s^{\mu^r}|^k] \leq c, \quad \|\phi\|_{L^k(\mu)} \leq 1 \tag{4.4.83}$$
for some constant $c > 0$, so that by (4.4.12) and (4.4.80), we find constants $c_1, c_2 > 0$ such that
$$\sup_{\|\phi\|_{L^k(\mu)} \leq 1} \varepsilon_r(\phi) \leq c_1 \mathbb{E}\left[\sup_{s \in [0,t]} n|X_s^r - X_s|^{k-1} + s_n\right] \leq c_2 n r^{k-1} + c_1 s_n, \quad n \geq 1.$$
Then $\sup_{\|\phi\|_{L^k(\mu)} \leq 1} \varepsilon_r(\phi) \to 0$ as $r \to 0$. It remains to prove
$$\lim_{r \downarrow 0} \sup_{\|\phi\|_{L^k(\mu)} \leq 1} \mathbb{E}[J_r(X^r, X)] = 0. \tag{4.4.84}$$
By $(A^{4.5})$, (4.4.12), (4.4.25), Lemma 4.4.9, (4.4.81) and (4.4.83), we find constants $c_3, c_4, c_5 > 0$ and positive function $\tilde{\varepsilon}(\cdot)$ on $[0,1]$ with $\tilde{\varepsilon}(r) \to 0$ as $r \to 0$, such that when $\|\phi\|_{L^k(\mu)} \leq 1$,
$$\mathbb{E}\big[\big|\langle D^L b_s^{(1)}(y_s, \mu_s^r)(X_s^r), \nabla_{\phi(X_0)} X_s^{\mu^r}\rangle - \langle D^L b_s^{(1)}(z_s, \mu_s)(X_s), \nabla_{\phi(X_0)} X_s^{\mu}\rangle\big|\big]$$
$$\leq c_3 \big(\mathbb{E}[|\nabla_{\phi(X_0)} X_s^{\mu^r} - \nabla_{\phi(X_0)} X_s^{\mu}|^k]\big)^{\frac{1}{k}}$$
$$+ \big(\mathbb{E}[|\nabla_{\phi(X_0)} X_s|^k]\big)^{\frac{1}{k}} \big(\mathbb{E}[|D^L b_s^{(1)}(z_s, \mu_s)(X_s) - D^L b_s^{(1)}(y_s, \mu_s^r)(X_s^r)|^{k^*}]\big)^{\frac{1}{k^*}}$$
$$\leq \tilde{\varepsilon}(r) + c_4 \big(\mathbb{E}[n^{k^*}\{|z_s - y_s| + |X_s^r - X_s| + r\} + s_n^{k^*}]\big)^{\frac{1}{k^*}}$$
$$\leq \tilde{\varepsilon}(r) + c_5 \big\{n|z_s - y_s|^{\frac{1}{k^*}} + nr^{\frac{1}{k^*}} + s_n\big\}, \quad n \geq 1.$$
Combining this with (4.4.12) we find a constant $c_6 > 0$ such that
$$\sup_{\|\phi\|_{L^k(\mu)} \leq 1} \mathbb{E}[J_r(X^r, X)] \leq c_6\big\{\tilde{\varepsilon}(r) + nr^{\frac{1}{k^*}} + s_n\big\}, \quad n \geq 1.$$
By letting first $r \to 0$ then $n \to \infty$ we derive (4.4.84).

4.5 Notes and further results

The power/log-Harnack inequalities and Bismut formulas have been established in [Huang et al. (2019)] and [Bao et al. (2021)] respectively for the path-distribution dependent SDE (3.8.7) when the noise coefficient is path-distribution independent. Moreover, when the noise coefficient $\sigma_t(x, \mu)$ depends on both x and μ, then the log-Harnack inequality has been derived in recent papers [Ren and Wang (2023); Huang et al. (2023); Qian et al. (2023)]. See also [Huang and Song (2021)] for the study on singular distribution dependent SPDEs.

In the following, we introduce Bismut formula for degenerate DDSDEs, two results on derivative estimates for distribution dependent noise, log-Harnack inequality and Bismut formula for DDSDEs with drifts singular in distribution.

4.5.1 Bismut formula for degenerate DDSDEs

Consider the following distribution dependent stochastic Hamiltonian system for $X_t = (X_t^{(1)}, X_t^{(2)})$ on $\mathbb{R}^d = \mathbb{R}^{d_1} \times \mathbb{R}^{d_2}$:

$$\begin{cases} dX_t^{(1)} = b_t^{(1)}(X_t)dt, \\ dX_t^{(2)} = b_t^{(2)}(X_t, \mathcal{L}_{X_t})dt + \sigma_t dW_t, \end{cases} \quad (4.5.1)$$

where $(W_t)_{t \geq 0}$ is a d_2-dimensional Brownian motion as before, and for each $t \geq 0$, σ_t is an invertible $d_2 \times d_2$-matrix,

$$b_t = (b_t^{(1)}, b_t^{(2)}) : \mathbb{R}^d \times \mathcal{P}_2 \to \mathbb{R}^d,$$

which is measurable with $b_t^{(1)}(x, \mu) = b_t^{(1)}(x)$ independent of the distribution μ. Let $\nabla = (\nabla^{(1)}, \nabla^{(2)})$ be the gradient operator on $\mathbb{R}^d = \mathbb{R}^{d_1} \times \mathbb{R}^{d_2}$, where $\nabla^{(i)}$ is the gradient in the i-th component, $i = 1, 2$. Let $\nabla^2 = \nabla\nabla$ denote the Hessian operator on $\mathbb{R}^d$. We assume

$(A^{4.6})$ *For every* $t \in [0,T]$, $b_t^{(1)} \in C_b^2(\mathbb{R}^d \to \mathbb{R}^{d_1})$, $b_t^{(2)} \in C^{1,1}(\mathbb{R}^d \times \mathcal{P}_2 \to \mathbb{R}^{d_2})$. *Moreover:*

(1) *There exists a constant* $K > 0$ *such that*

$$\|\nabla b_t(\cdot, \mu)(x)\| + \|D^L b_t^{(2)}(x, \cdot)(\mu)\| + \|\nabla^2 b_t^{(1)}(x)\| \leq K,$$
$$t \in [0,T], (x, \mu) \in \mathbb{R}^d \times \mathcal{P}_2.$$

(2) *There exist* $B \in \mathcal{B}_b([0,T] \to \mathbb{R}^{d_1} \otimes \mathbb{R}^{d_2})$, *an increasing function* $\theta \in C([0,T]; (0,\infty))$ *with* $\theta_t > 0$ *for* $t \in (0,T]$, *and* $\varepsilon \in (0,1)$ *such that*

$$\langle (\nabla^{(2)} b_t^{(1)} - B_t) B_t^* v, v \rangle \geq -\varepsilon |B_t^* v|^2, \quad v \in \mathbb{R}^{d_1},$$

$$\int_0^t s(T-s) K_{T,s} B_s B_s^* K_{T,s}^* ds \geq \theta_t I_{d_1 \times d_1}, \quad t \in (0,T],$$

where for any $s \geq 0$, $\{K_{t,s}\}_{t \geq s}$ *is the unique solution of the following linear random ODE on* $\mathbb{R}^{d_1} \otimes \mathbb{R}^{d_1}$:

$$\frac{d}{dt} K_{t,s} = (\nabla^{(1)} b_t^{(1)})(X_t) K_{t,s}, \quad t \geq s, K_{s,s} = I_{d_1 \times d_1}.$$

According to the proof of Theorem 1.1 in [Wang and Zhang (2013)], $(A^{4.6})$ implies that the matrices

$$Q_t := \int_0^t s(T-s) K_{T,s} \nabla^{(2)} b_s^{(1)}(X_s) B_s^* K_{T,s}^* \mathrm{d}s, \quad t \in (0,T]$$

are invertible with

$$\|Q_t^{-1}\| \le \frac{1}{(1-\varepsilon)\theta_t}, \quad t \in (0,T]. \tag{4.5.2}$$

For $(X_t)_{t \in [0,T]}$ solving (4.5.1) with $\mathcal{L}_{X_0} = \mu \in \mathcal{P}_2$ and $\phi = (\phi^{(1)}, \phi^{(2)}) \in L^2(\mathbb{R}^d \to \mathbb{R}^d, \mu)$, let

$$\alpha_t^{(2)} = \frac{T-t}{T} \phi^{(2)}(X_0) - \frac{t(T-t) B_t^* K_{T,t}^*}{\int_0^T \theta_s^2 \mathrm{d}s} \int_t^T \theta_s^2 Q_s^{-1} K_{T,0} \phi^{(1)}(X_0) \mathrm{d}s$$

$$- t(T-t) B_t^* K_{T,t}^* Q_T^{-1} \int_0^T \frac{T-s}{T} K_{T,s} \nabla^{(2)}_{\phi^{(2)}(X_0)} b_s^{(1)}(X_s) \mathrm{d}s,$$

$$\alpha_t^{(1)} = K_{t,0} \phi^{(1)}(X_0) + \int_0^t K_{t,s} \nabla^{(2)}_{\alpha_s^{(2)}} b_s^{(1)}(X_s(x)) \, \mathrm{d}s, \quad t \in [0,T],$$

and define

$$h_t^\alpha := \int_0^t \sigma_s^{-1} \Big\{ \big(\mathbb{E}\langle D^L b_s^{(2)}(y, \cdot)(\mathcal{L}_{X_s})(X_s), \alpha_s\rangle\big)\big|_{y=X_s} \\ + \nabla_{\alpha_s} b_s^{(2)}(\cdot, \mathcal{L}_{X_s})(X_s) - (\alpha_s^{(2)})' \Big\} \mathrm{d}s, \quad t \in [0,T]. \tag{4.5.3}$$

Let $(D^*, \mathcal{D}(D^*))$ be the Malliavin divergence operator associated with the Brownian motion $(W_t)_{t \in [0,T]}$, see Theorem 1.4.1. The following result is due to Theorem 2.3 in [Ren and Wang (2019)].

Theorem 4.5.1. *Assume $(A^{4.6})$. Then $h^\alpha \in \mathcal{D}(D^*)$ with $\mathbb{E}|D^*(h^\alpha)|^p < \infty$ for all $p \in [1, \infty)$. Moreover, for any $f \in \mathcal{B}_b(\mathbb{R}^d)$ and $T > 0$, $P_T f$ is L-differentiable such that*

$$D_\phi^L (P_T f)(\mu) = \mathbb{E}\big[f(X_T) D^*(h^\alpha)\big]$$

holds for $\mu \in \mathcal{P}_2, \phi \in L^2(\mathbb{R}^d \to \mathbb{R}^d, \mu)$ and h^α in (4.5.3). Consequently:

(1) *Let $\psi \in L^2(\mathbb{R}^d \to \mathbb{R}, P_T^* \mu)$ be such that $\psi(X_T) = \mathbb{E}(D^*(h^\alpha)|X_T)$. Then*

$$D_\phi^L P_T^* \mu := \lim_{\varepsilon \downarrow 0} \frac{P_T^* \mu \circ (\mathrm{Id} + \varepsilon \phi)^{-1} - P_T^* \mu}{\varepsilon} = \psi P_T^* \mu \tag{4.5.4}$$

exists in the total variational norm.

(2) *There exists a constant $c \ge 0$ such that for any $T > 0$,*

$$\|D^L(P_T f)(\mu)\| \le c\sqrt{P_T|f|^2(\mu) - (P_T f)^2(\mu)} \frac{\sqrt{T}(T^2 + \theta_T)}{\int_0^T \theta_s^2 \mathrm{d}s}, f \in \mathcal{B}_b(\mathbb{R}^d),$$

$$\|P_T^* \mu - P_T^* \nu\|_{TV} \le c \mathbb{W}_2(\mu, \nu) \frac{\sqrt{T}(T^2 + \theta_T)}{\int_0^T \theta_s^2 \mathrm{d}s}, \quad \mu, \nu \in \mathcal{P}_2.$$

4.5.2 L-derivative estimate for distribution dependent noise

Consider the DDSDE (3.1.1) with coefficients satisfying the following assumption which, by Theorem 3.3.1, implies the well-posedness.

$(A^{4.7})$ For any $t \geq 0$, $b_t, \sigma_t \in C^{1,1}(\mathbb{R}^d \times \mathcal{P}_2)$, and there exists an increasing function $K : [0, \infty) \to [1, \infty)$ such that for any $t \geq 0, x, y \in \mathbb{R}^d$ and $\mu \in \mathcal{P}_2$,

$$K_t^{-1} I_{d \times d} \leq (\sigma_t \sigma_t^*)(x, \mu) \leq K_t I_{d \times d},$$

$$|b_t(x, \mu)| + \|\nabla b_t(\cdot, \mu)(x)\| + \|D^L\{b_t(x, \cdot)\}(\mu)\|$$
$$+ \|\nabla\{\sigma_t(\cdot, \mu)\}(x)\|^2 + \|D^L\{\sigma_t(x, \cdot)\}(\mu)\|^2 \leq K_t,$$

$$\|D^L\{b_t(x, \cdot)\}(\mu) - D^L\{b_t(y, \cdot)\}(\mu)\|$$
$$+ \|D^L\{\sigma_t(x, \cdot)\}(\mu) - D^L\{\sigma_t(y, \cdot)\}(\mu)\| \leq K_t |x - y|.$$

Let $P_{s,t} f(\mu) := \mathbb{E}[f(X_{s,t})]$ for $f \in \mathcal{B}_b(\mathbb{R}^d)$ and $(X_{s,t})_{t \geq s \geq 0}$ solving (3.1.1) with $\mathcal{L}_{X_{s,s}} = \mu \in \mathcal{P}_2$. The following result is due to Theorem 1.1 in [Huang and Wang (2021b)].

Theorem 4.5.2. Assume $(A^{1.7})$. Then for any $t > s \geq 0$ and $f \in \mathcal{B}_b(\mathbb{R}^d)$, $P_{s,t} f$ is L-differentiable, and there exists an increasing function $C : [0, \infty) \to (0, \infty)$ such that

$$\|D^L P_t f(\mu)\| \leq \frac{C_t \|f\|_\infty}{\sqrt{t-s}}, \quad t > s, f \in \mathcal{B}_b(\mathbb{R}^d).$$

Consequently, for any $t > 0$ and $\mu, \nu \in \mathcal{P}_2$,

$$\|P^*_{s,t} \mu - P^*_{s,t} \nu\|_{TV} := 2 \sup_{\|f\|_\infty \leq 1} |P_{s,t} f(\mu) - P_{o,t} f(\nu)| \leq \frac{2 C_t}{\sqrt{t-s}} \mathbb{W}_2(\mu, \nu).$$

4.5.3 Derivative estimates for the transition density

Consider the decoupled SDE associated with (3.1.1):

$$\begin{aligned} \mathrm{d} X_t^{\mu, x} &= b_t(X_t^{\mu, x}, P_t^* \mu) \mathrm{d}t + \sigma_t(X_t^{\mu, x}, P_t^* \mu) \mathrm{d} W_t, \\ X_0^{\mu, x} &= x, t \in [0, T]. \end{aligned} \quad (4.5.5)$$

We have

$$P_t^\mu f(x) := \mathbb{E}[f(X_t^\mu)] = \int_{\mathbb{R}^d} f(y) p_t^\mu(x, y) \mathrm{d}y, \quad t \in (0, T], x \in \mathbb{R}^d, f \in \mathcal{B}_b(\mathbb{R}^d).$$

Derivative estimates on $p_t^\mu(x,y)$ have been studied in [Crisan and E. McMurray (2018)] and [Chaudru de Raynal and Frikha (2022)] under $(A^{3.8})$ and the following assumption involving the linear functional derivatives ∂_μ and ∂_μ^2 introduced in (3.8.1) and (3.8.2).

$(A^{4.8})$ $b_t(x,\cdot)$ and $a_t(x,\cdot) := (\sigma_t \sigma_t^*)(x,\cdot)$ have second order linear functional derivatives, such that for constants $K > 0$ and $\alpha \in (0,1]$, the following conditions hold for any $t \in [0,T], \nu, \mu \in \mathcal{P}$ and $x, x', y, y', z, z' \in \mathbb{R}^d$:

$$\|\sigma_t(x,\mu) - \sigma_t(y,\mu)\| \le K|x-y|,$$
$$\left|\partial_\mu b_t(x,\mu)(y) - \partial_\mu b_t(x,\mu)(y')\right| + \left|\partial_\mu^2 b_t(x,\mu)(y,z) - \partial_\mu^2 b_t(x,\mu)(y,z')\right|$$
$$+ \left|b_t(x,\mu) - b_t(x',\mu)\right| + \left|\partial_\mu^2 a_t(x,\mu)(y,z) - \partial_\mu^2 a_t(x,\mu)(y,z')\right|$$
$$\le K\big(|x-x'|^\alpha + |y-y'|^\alpha + |z-z'|^\alpha\big).$$

The following result is included in Theorem 3.6 in [Chaudru de Raynal and Frikha (2022)].

Theorem 4.5.3. *Assume* $(A^{3.8})$.

(1) *For any $\mu \in \mathcal{P}_2$ and $t \in (0,T]$, the density $p_t^\mu(x,y)$ exists. Moreover, there exists a constant $c > 1$ such that for any $t \in (0,T], x, y \in \mathbb{R}^d$ and $\mu \in \mathcal{P}$,*

$$\left|\nabla_x^i p_t^\mu(x,y)\right| \le c t^{-\frac{i+d}{2}} e^{-\frac{|x-y|^2}{ct}}, \quad i = 0, 1, 2,$$

and for any $\beta \in [0,\alpha)$ there exists a constant $c(\beta) > 0$ such that for any $x' \in \mathbb{R}^d$,

$$\left|\nabla_x^2 p_t^\mu(x,y) - \nabla_{x'}^2 p_t^\mu(x',y)\right| \le c(\beta)|x-x'|^\beta t^{-1-\frac{\beta+d}{2}} e^{-\frac{|x-y|^2}{c(\beta)t}}.$$

(2) *If $(A^{4.8})$ holds, then there exists a constant $c > 0$ such that for any $\beta \in [0,1], t \in (0,T], x, x', y, z \in \mathbb{R}^d$ and $\mu \in \mathcal{P}_2$,*

$$\left|\nabla_z^i \{D^L p_t^\mu(x,y)\}(z)\right| \le c t^{-\frac{1+i+d-\alpha}{2}} e^{-\frac{|x-y|^2}{ct}}, \quad i = 0, 1,$$
$$\left|\{D^L p_t^\mu(x,y)\}(z) \{D^L p_t^\mu(x',y)\}(z)\right| \le c|x-x'|^\beta t^{-\frac{1+\beta+d-\alpha}{2}} e^{-\frac{|x-y|^2+|x'-y|^2}{ct}},$$
$$\left|\nabla_x^i p_t^\mu(x,y) - \nabla_x^i p_t^\nu(x,y)\right| \le c(\beta) \mathbb{W}_2(\mu,\nu)^\beta t^{-\frac{i+\beta+d}{2}} e^{-\frac{|x-y|^2}{ct}},$$

and for any $\beta \in [0, \alpha)$ there exists a constant $c(\beta) > 0$ such that for any $z' \in \mathbb{R}^d$ and $\nu \in \mathcal{P}_2$,

$$\left|\nabla_z\{D^L p_t^\mu(x,y)\}(z) - \nabla_{z'}\{D^L p_t^\mu(x,y)\}(z')\right|$$
$$\leq c(\beta)(|x-x'|^\beta + |z-z'|^\beta)t^{-\frac{1+\beta+d-\alpha}{2}}e^{-\frac{|x-y|^2}{ct}},$$
$$\left|\nabla_x^2 p_t^\mu(x,y) - \nabla_x^2 p_t^\nu(x,y)\right| \leq c(\beta)\mathbb{W}_2(\mu,\nu)^\beta t^{-1-\frac{\beta+d}{2}}e^{-\frac{|x-y|^2}{ct}},$$
$$\left|\nabla_z\{D^L p_t^\mu(x,y)\}(z) - \nabla_z\{D^L p_t^\nu(x,y)\}(z)\right|$$
$$\leq c(\beta)\mathbb{W}_2(\mu,\nu)^\beta t^{-1-\frac{\beta+d-\alpha}{2}}e^{-\frac{|x-y|^2}{ct}}.$$

We remark that the above derivative estimates imply the derivative formula for $D^L P_t f$. Indeed, since

$$P_t f(\mu) = \int_{\mathbb{R}^d \times \mathbb{R}^d} p_t^\mu(x,y) f(y) \mu(\mathrm{d}x) \mathrm{d}y,$$

for any $f \in \mathcal{B}_b(\mathbb{R}^d)$ and $\phi \in T_{\mu,2}$, we have

$$D_\phi P_t f(\mu) = \frac{\mathrm{d}}{\mathrm{d}\varepsilon}\bigg|_{\varepsilon=0} \int_{\mathbb{R}^d \times \mathbb{R}^d} p_t^{\mu \circ (\mathrm{id}+\varepsilon\phi)^{-1}}(x+\varepsilon\phi(x),y) f(y) \mu(\mathrm{d}x) \mathrm{d}y$$
$$= \int_{\mathbb{R}^d \times \mathbb{R}^d} \langle D^L p_t^\mu(x,y)(\cdot), \phi \rangle_{L^2(\mu)} f(y) \mu(\mathrm{d}x) \mathrm{d}y$$
$$+ \int_{\mathbb{R}^d \times \mathbb{R}^d} \nabla_{\phi(x)} p_t^\mu(\cdot,y)(x) f(y) \mu(\mathrm{d}x) \mathrm{d}y.$$

4.5.4 Log-Harnack inequality and Bismut formula for DDSDEs with drifts singular in distributions

This part is taken from [Huang and Wang (2022b)] where Theorem 4.1.1 for log-Harnack inequality and Theorem 4.4.1 for Bismut formula are extended to the case that $b_t(x,\mu)$ is only Lipschitz continuous in μ with respect to the distance induced by the square root of Dini functions, so that it may be discontinuous in the distance induced by Dini functions.

4.5.4.1 *Log-Harnack inequality*

Let α be in the following class

$$\mathcal{A} := \bigg\{\alpha : [0,\infty) \to [0,\infty) \text{ is increasing and concave,}$$
$$\alpha(0) = 0, \int_0^1 \frac{\alpha(r)^2}{r} \mathrm{d}r \in (0,\infty)\bigg\},$$

where $\int_0^1 \frac{\alpha(r)^2}{r}\mathrm{d}r < \infty$ is the Dini condition for α^2. For a (real or Banach valued) function f, let

$$[f]_\alpha := \sup_{x\neq y} \frac{|f(x)-f(y)|}{\alpha(|x-y|)}$$

be its continuity modulus in α. Define the Wasserstein distance induced by α:

$$\mathbb{W}_\alpha(\mu,\nu) := \sup_{[f]_\alpha \leq 1} |\mu(f)-\nu(f)|, \quad \mu,\nu \in \mathcal{P}_\alpha := \{\mu \in \mathcal{P} : \mu(\alpha(|\cdot|)) < \infty\},$$

where f are real functions and $\mu(f) := \int_{\mathbb{R}^d} f \mathrm{d}\mu$.

By the concavity of α, $\mathbb{W}_\alpha$ is a complete distance on $\mathcal{P}_\alpha$, and $\mathcal{P}_k \subset \mathcal{P}_\alpha$ for $k \geq 1$.

$(A^{4.9})$ There exist $\alpha \in \mathcal{A}$, $k \in (1,\infty), \kappa \in [0,\infty), K \in (0,\infty), l \in \mathbb{N}$, and

$$1 \leq f_i \in \tilde{L}_{q_i}^{p_i}(T), \quad (p_i,q_i) \in \mathcal{K}, p_i > 2, \quad 0 \leq i \leq l$$

such that the following conditions hold.

(1) $(\sigma_t \sigma_t^*)(x)$ is invertible and $\sigma_t(x)$ is weakly differentiable in x such that

$$\|\sigma\sigma^*\|_\infty + \|(\sigma\sigma^*)^{-1}\|_\infty < \infty, \quad |\nabla \sigma| \leq \sum_{i=1}^l f_i,$$

$$\lim_{\varepsilon \downarrow 0} \sup_{t\in[0,T], |x-x'|\leq \varepsilon} \|(\sigma_t\sigma_t^*)(x) - (\sigma_t\sigma_t^*)(x')\| = 0.$$

(2) $b_t(x,\mu) = b_t^{(0)}(x) + b_t^{(1)}(x,\mu)$, where for any $t \in [0,T], x, y \in \mathbb{R}^d, \mu,\nu \in \mathcal{P}_k$,

$$|b_t^{(0)}(x)| \leq f_0(t,x), \quad |b_t^{(1)}(x,\mu)| \leq K + \kappa|x| + \kappa\|\mu\|_k,$$

$$|b_t^{(1)}(x,\mu) - b_t^{(1)}(y,\nu)| \leq K\{|x-y| + \mathbb{W}_\alpha(\mu,\nu) + \mathbb{W}_k(\mu,\nu)\}.$$

By Theorem 3.5.1, $(A^{4.9})$ implies the well-posedness of (3.4.1) for distributions in $\mathcal{P}_k$, and for any $n \geq 1$ there exists a constant $c_n > 0$ such that

$$\mathbb{E}\left[\sup_{t\in[0,T]} |X_t|^n \Big| \mathcal{F}_0\right] \leq c_n(1+|X_0|^n).$$

The following result due to [Huang and Wang (2022b)] extends Theorem 4.1.1.

Theorem 4.5.4. *Assume* $(A^{4.9})$ *with* $k = 2$, *let*

$$\tilde{\alpha}(r) := \left(\int_0^r \frac{\alpha(t)^2}{t} dt\right)^{\frac{1}{2}}, \quad r \geq 0.$$

Then there exists a constant $c > 0$ *such that for any* $t \in (0, T]$ *and* $\gamma, \tilde{\gamma} \in \mathcal{P}_2$,

$$\text{Ent}(P_t^*\gamma | P_t^*\tilde{\gamma})$$
$$\leq \mathbb{W}_2(\gamma, \tilde{\gamma})^2 \left\{\frac{c}{t} + \tilde{\alpha}(1 + \kappa\|\gamma\|_2 + \kappa\|\tilde{\gamma}\|_2)^2 e^{c\alpha(1+\kappa\|\gamma\|_2+\kappa\|\tilde{\gamma}\|_2)^2}\right\}.$$

If in particular $\kappa = 0$ (*i.e.* $b^{(1)}$ *is bounded*), *there exists a constant* $c > 0$ *such that*

$$\text{Ent}(P_t^*\gamma | P_t^*\tilde{\gamma}) \leq \frac{c}{t} \mathbb{W}_2(\gamma, \tilde{\gamma})^2, \quad t \in (0, T], \gamma, \tilde{\gamma} \in \mathcal{P}_2.$$

4.5.4.2 Bismut formula

In this part, we establish the Bismut formula for the intrinsic derivative of $P_t f(\gamma)$ for $\gamma \in \mathcal{P}_k$. To this end, we assume

$$b_t(x, \mu) = h_t^{(0)}(x) + B_t(x, \mu, \mu(V)), \quad t \in [0, T], x \in \mathbb{R}^d, \mu \in \mathcal{P}, \quad (4.5.6)$$

where for a Banach space $(\mathbb{B}, \|\cdot\|_\mathbb{B})$,

$$V : \mathbb{R}^d \to \mathbb{B}, \quad B : [0, T] \times \mathbb{R}^d \times \mathcal{P}_k \times \mathbb{B} \to \mathbb{R}^d$$

are measurable such that $[V]_\alpha \leq 1$ for some $\alpha \in \mathcal{A}$, i.e. V is only square root Dini continuous and hence $\mu \mapsto B_t(x, \mu, \mu(V))$ may be not intrinsically differentiable.

Recall that a real function f on a Banach space $\mathbb{B}$ is called Gateaux differentiable, if for any $z \in \mathbb{B}$,

$$v \mapsto \nabla_v^\mathbb{B} f(z) := \lim_{\varepsilon \downarrow 0} \frac{f(z + \varepsilon v) - f(z)}{\varepsilon}$$

is a well-defined bounded linear functional. In this case we denote

$$\|\nabla^\mathbb{B} f(z)\|_{\mathbb{B}^*} := \sup_{\|v\|_\mathbb{B} \leq 1} |\nabla_v^\mathbb{B} f(z)|.$$

Moreover, f is called Fréchet differentiable if it is Gateaux differentiable and

$$\lim_{\|v\|_\mathbb{B} \downarrow 0} \frac{|f(z + v) - f(z) - \nabla_v^\mathbb{B} f(z)|}{\|v\|_\mathbb{B}} = 0, \quad z \in \mathbb{B}.$$

It is well-known that a Gateaux differentiable function f is Fréchet differentiable provided $\nabla_v^\mathbb{B} f(z)$ is continuous in $(v, z) \in \mathbb{B} \times \mathbb{B}$. When $\mathbb{B} = \mathbb{R}^l$ for some $l \geq 1$, we simply denote $\nabla^\mathbb{B} = \nabla$.

$(A^{4.10})$ Let $k \in (1, \infty)$ and let b in (4.5.6).
  (1) $b^{(0)}$ and σ satisfy the corresponding conditions in $(A^{4.9})$.
  (2) There exists $\alpha \in \mathcal{A}$ such that $\alpha_k(s) := \alpha(s^{\frac{1}{k-1}})$ is concave in $s \geq 0$, and
$$[V]_\alpha := \sup_{x \neq y} \frac{\|V(x) - V(y)\|_{\mathbb{B}}}{\alpha(|x-y|)} \leq 1.$$
  (3) For any $t \in [0, T]$, $B_t \in C(\mathbb{R}^d \times \mathcal{P}_k \times \mathbb{B})$, $B_t(x, \mu, z)$ is differentiable in x, L-differentiable in $\mu \in \mathcal{P}_k$, and Fréchet differentiable in $z \in \mathbb{B}$, such that $\nabla_v^{\mathbb{B}} B_t(x, \mu, z)$ is continuous in $(v, z) \in \mathbb{B} \times \mathbb{B}$. Next, there exist constants $K > 0$ and $\kappa \geq 0$ such that
$$|B_t(x, \mu, z)| \leq K + \kappa\big(|x| + \|\mu\|_k + \|z\|_{\mathbb{B}}\big),$$
$$|\nabla B_t(\cdot, \mu, z)(x)| + \|D^L B_t(x, \cdot, z)(\mu)\|_{L^{k^*}(\mu)} + \|\nabla^{\mathbb{B}} B_t(x, \mu, \cdot)(z)\|_{\mathbb{B}^*}$$
$$\leq K, \quad (t, x, \mu, z) \in [0, T] \times \mathbb{R}^d \times \mathcal{P}_k \times \mathbb{B}.$$
Moreover, for any $(t, x, \mu) \in [0, T] \times \mathbb{R}^d \times \mathcal{P}_k$, there exists a constant $c(t, \mu, z) > 0$ such that
$$|\{D^L B_t(x, \cdot, z)(\mu)(y)| \leq c(t, x, \mu, z)(1 + |y|^{k-1}), \quad y \in \mathbb{R}^d.$$

Since α is concave, the concavity of α_k holds for $k \geq 2$. When $\alpha(s) = s^\varepsilon$ for some $\varepsilon \in (0, 1)$, α_k is concave for $k \geq 1 + \varepsilon$. Since $(A^{4.10})$ implies $(A^{4.9})$, as explained above that under this assumption (3.4.1) is well-posed for distributions in $\mathcal{P}_k$.

For $\mu \in \mathcal{P}_k$, consider the decoupled SDE
$$dX_t^{x,\mu} = \big\{b_t^{(0)}(X_t^{x,\mu}) + B_t(X_t^{x,\mu}, P_t^*\mu, P_t V(\mu))\big\}dt + \sigma_t(X_t^{x,\mu})dW_t,$$
$$X_0^{x,\mu} = x, t \in [0, T].$$
According to Theorem 1.3.1, $(A^{4.10})$ implies that this SDE is well-posed,
$$\nabla_v X_t^{x,\mu} := \lim_{\varepsilon \downarrow 0} \frac{X_t^{x+\varepsilon v, \mu} - X_t^{x, \mu}}{\varepsilon}, \quad t \in [0, T]$$
exists in $L^p(\Omega \to C([0, T]; \mathbb{R}^d); \mathbb{P})$ for any $p \geq 1$, and there exists a constant $c_p > 0$ such that
$$\mathbb{E}\left[\sup_{t \in [0,T]} |\nabla_v X_t^{x,\mu}|^p\right] \leq c_p |v|^p, \quad v \in \mathbb{R}^d, \mu \in \mathcal{P}_k, x \in \mathbb{R}^d.$$

To state the Bismut formula for $P_t f$, we introduce the following I_t^f which comes from the Bismut formula Theorem 1.4.2: for any $t \in (0, T]$, $\beta \in C^1([0, t])$ with $\beta_0 = 0$ and $\beta_t = 1$, let
$$I_t^f(\mu, \phi) := \int_{\mathbb{R}^d} \mathbb{E}\left[f(X_t^{x,\mu}) \int_0^t \langle \beta_s' \zeta_s(X_s^{x,\mu}) \nabla_{\phi(x)} X_s^{x,\mu}, dW_s \rangle\right] \mu(dx),$$
$$\zeta_s := \sigma_s^*(\sigma_s \sigma_s^*)^{-1}, \quad s \in [0, t], \mu \in \mathcal{P}_k, \phi \in T_{\mu,k}.$$

Then there exists a constant $c > 0$ such that
$$|I_t^f(\mu,\phi)| \le c\|\beta'\|_\infty \sqrt{t}\bigl(P_t|f|^{k^*}(\mu)\bigr)^{\frac{1}{k^*}}\|\phi\|_{L^k(\mu)}, \quad \mu \in \mathcal{P}_k, \phi \in T_{\mu,k}.$$

Next, let X_0^μ be $\mathcal{F}_0$-measurable such that $\mathcal{L}_{X_0^\mu} = \mu$, and let X_t^μ solve (3.4.1) with initial value X_0^μ. For any $\varepsilon \ge 0$, denote
$$\mu_\varepsilon := \mu \circ (id + \varepsilon\phi)^{-1}, \quad X_0^{\mu_\varepsilon} := X_0^\mu + \varepsilon\phi(X_0^\mu).$$
Let $X_t^{\mu_\varepsilon}$ solve (3.4.1) with initial value $X_0^{\mu_\varepsilon}$. So,
$$X_t^\mu = X_t^{\mu_0}, \quad P_t^*\mu_\varepsilon = \mathcal{L}_{X_t^{\mu_\varepsilon}}, \quad t \in [0,T], \varepsilon \ge 0.$$
Recall that
$$\nabla_\phi X_t^\mu := \lim_{\varepsilon \downarrow 0} \frac{X_t^{\mu_\varepsilon} - X_t^\mu}{\varepsilon}, \quad t \in [0,T]. \tag{4.5.7}$$

Theorem 4.5.5. *Assume* $(A^{4.10})$. *Then the following assertions hold.*

(1) *For any $t \in (0,T]$, $P_t V$ is intrinsically differentiable on $\mathcal{P}_k$, and there exists a constant $c > 0$ such that*
$$\|D^I P_t V(\mu)\|_{L^{k^*}(\mu)} \le \frac{c\alpha((1+\kappa\|\mu\|_k)t^{\frac{1}{2}})}{\sqrt{t}} e^{c\tilde{\alpha}(1+\kappa\|\mu\|)^2}, t \in (0,T], \mu \in \mathcal{P}_k.$$

(2) *The limit in (4.5.7) exists in $L^k(\Omega \to C([0,T],\mathbb{R}^d), \mathbb{P})$, and there exists a constant $c > 0$ such that*
$$\mathbb{E}\Bigl[\sup_{t \in [0,T]} |\nabla_\phi X_t^\mu|^k\Bigr] \le c\|\phi\|_{L^k(\mu)}^k, \quad t \in (0,T], \mu \in \mathcal{P}_k.$$

(3) *For any $t \in (0,T]$ and $f \in \mathcal{B}_b(\mathbb{R}^d)$, $P_t f$ is intrinsically differentiable on $\mathcal{P}_\mu$. Moreover, for any $\mu \in \mathcal{P}_k$ and $\phi \in T_{\mu,k}$,*
$$D_\phi^I P_t f(\mu) = I_t^f(\mu,\phi) + \mathbb{E}\Bigl[f(X_t^\mu)\int_0^t \bigl\langle \zeta_s(X_s^\mu)\{N_s + \tilde{N}_s\}, dW_s\bigr\rangle\Bigr],$$
$$N_s := \bigl\{\nabla^B_{D_\phi^I P_s V(\mu)} B_s(X_s^\mu, \mu, \cdot)\bigr\}(P_s V(\mu)),$$
$$\tilde{N}_s := \bigl\langle \mathbb{E}[\{D^L B_s^{(1)}(y,\cdot, P_s V(\mu))\}(P_s^*\mu)(X_s^\mu)]_{y=X_s^\mu}, \nabla_\phi X_s^\mu\bigr\rangle,$$
where X_t^μ solves (3.4.1) with initial distribution $\mathcal{L}_{X_0} = \mu$, and $\zeta_s := \sigma_s^*(\sigma_s\sigma_s^*)^{-1}$.

By taking $\beta_s = \frac{s}{t}$, we find a constant $c > 0$ such that for any $t \in (0,T]$, $f \in \mathcal{B}_b(\mathbb{R}^d)$, $\mu \in \mathcal{P}_k$,
$$\|D^I P_t f(\mu)\|_{L^{k^*}(\mu)} \le \frac{\{P_t|f|^{k^*}(\mu)\}^{\frac{1}{k^*}}}{\sqrt{t}} e^{c\tilde{\alpha}(1+\kappa\|\mu\|_k)^2}.$$

Chapter 5

DDSDEs: Long Time Behaviors

In this chapter, we study the exponential ergodicity for the time-homogeneous DDSDE:
$$dX_t = \sigma(X_t, \mathcal{L}_{X_t})dW_t + b(X_t, \mathcal{L}_{X_t})dt, \tag{5.0.1}$$
where W_t is the m-dimensional Brownian motion and
$$\sigma : \mathbb{R}^d \times \mathcal{P} \to \mathbb{R}^d \otimes \mathbb{R}^m, \quad b : \mathbb{R}^d \times \mathcal{P} \to \mathbb{R}^d$$
are measurable.

We first present a general result with application to the $\mathbb{W}_k$-exponential ergodicity, then consider the exponential ergodicity in variation distance for singular DDSDEs, investigate the exponential ergodicity in entropy and $\mathbb{W}_2$ for the dissipative case, and finally derive the exponential ergodicity in weighted Wasserstein distance for the partially dissipative and non-dissipative cases.

5.1 A general result with application to $\mathbb{W}_k$-exponential ergodicity

Let $\hat{\mathcal{P}} \subset \mathcal{P}$ be equipped with a complete metric $\mathbb{W}$. When (5.0.1) is well-posed for distributions in $\hat{\mathcal{P}}$, for any $t \geq 0$, let $P_t^*\mu := \mathcal{L}_{X_t}$ for the solution X_t with initial distribution $\mathcal{L}_{X_0} = \mu \in \hat{\mathcal{P}}$. The well-posedness implies the semigroup property
$$P_t^* P_s^* = P_{t+s}^*, \quad t, s \geq 0. \tag{5.1.1}$$
A point $\mu \in \hat{\mathcal{P}}$ is called P_t^*-invariant, if $P_t^*\mu = \mu$ for all $t \geq 0$.

Theorem 5.1.1. *If there exist constants $c \geq 1, \lambda > 0, t_0 > \frac{\log c}{\lambda}$ and a point $\mu_0 \in \hat{\mathcal{P}}$ such that*
$$\sup_{t \in [0, t_0]} \mathbb{W}(P_t^*\mu_0, \mu_0) < \infty, \tag{5.1.2}$$

$$\mathbb{W}(P_t^*\mu, P_t^*\nu) \leq ce^{-\lambda t}\mathbb{W}(\mu,\nu), \quad \mu,\nu \in \hat{\mathcal{P}}, \tag{5.1.3}$$

then there exists a unique P_t^*-invariant point $\bar{\mu} \in \hat{\mathcal{P}}$, and

$$\mathbb{W}(P_t^*\mu, \bar{\mu}) \leq ce^{-\lambda t}\mathbb{W}(\mu,\bar{\mu}), \quad t \geq 0, \mu \in \hat{\mathcal{P}}. \tag{5.1.4}$$

Proof. Since (5.1.4) follows from (5.1.3) if $\bar{\mu}$ is P_t^*-invariant, it suffices to prove that P_t^* has an invariant probability measure.

Simply denote $\mu_t = P_t^*\mu_0, t \geq 0$. By (5.1.2), we have

$$c_0 := \sup_{t \in [0, t_0]} \mathbb{W}(\mu_t, \mu_0) < \infty.$$

Moreover, by (5.1.1), (5.1.3) and $\varepsilon := ce^{-\lambda t_0} \in (0,1)$, we obtain

$$\sup_{s \in [0, t_0]} \mathbb{W}(\mu_{nt_0+s}, \mu_{nt_0}) \leq c_0 \varepsilon^n, \quad n \geq 1.$$

By this and the triangle inequality, we obtain

$$\sup_{s \geq 0} \mathbb{W}(\mu_s, \mu_0) \leq c_0 + \sum_{n=0}^{\infty} \mathbb{W}(\mu_{(n+1)t_0}, \mu_{nt_0})$$

$$\leq c_0 + \sum_{n=0}^{\infty} \varepsilon^n =: c_1 < \infty.$$

Thus,

$$\lim_{t \to \infty} \sup_{s \geq 0} \mathbb{W}(\mu_t, \mu_{t+s}) \leq \lim_{t \to \infty} ce^{-\lambda t} \sup_{s \geq 0} \mathbb{W}(\mu_0, \mu_s) = 0.$$

Therefore, $\{\mu_t\}_{t \geq 0}$ is a $\mathbb{W}$-Cauchy family as $t \to \infty$, hence there exists a unique $\bar{\mu} \in \hat{\mathcal{P}}$ such that

$$\lim_{t \to \infty} \mathbb{W}(\mu_t, \bar{\mu}) = 0.$$

This together with (5.1.1) yields that $\bar{\mu}$ is P_t^*-invariant. $\square$

Next, we consider the exponential ergodicity in $\mathbb{W}_k$ for $k \in [1, \infty)$ under the condition

$$\frac{k}{2}|x-y|^{k-2}\big(\|\sigma(x,\mu) - \sigma(y,\nu)\|_{HS}^2 + 2\langle b(x,\mu) - b(y,\nu), x-y\rangle$$
$$+ (k-2)\|\sigma(x,\mu) - \sigma(y,\nu)\|^2\big) \leq K_2 \mathbb{W}_k(\mu,\nu)^k - K_1|x-y|^k, \tag{5.1.5}$$
$$x, y \in \mathbb{R}^d, \ \mu, \nu \in \mathcal{P}_k,$$

for some constants $K_1 > K_2 > 0$, where for $k \in [1, 2)$, we assume that $\sigma(x,\mu) = \sigma(x)$ does not depend on μ.

Theorem 5.1.2. *Assume that (σ, b) satisfies $(A^{3.1})$. If (5.1.5) holds for some constants $K_1 > K_2 \geq 0$, then P_t^* has a unique invariant probability measure $\bar{\mu}$ in $\mathcal{P}_k$, and*

$$\mathbb{W}_k(P_t^*\mu, \bar{\mu})^k \leq e^{-(K_1-K_2)t}\mathbb{W}_k(\mu,\bar{\mu})^k, \quad t \geq 0, \mu \in \mathcal{P}_k.$$

Proof. By Theorem 3.3.1, (3.1.1) is well-posed for distributions in $\mathcal{P}_k$, and (5.1.2) holds for $\mu_0 = \delta_0$ and any $t_0 > 0$. Next, for any $\mu, \nu \in \mathcal{P}_k$, let X_t and Y_t solve (3.1.1) with

$$\mathcal{L}_{X_0} = \mu, \quad \mathcal{L}_{Y_0} = \nu, \quad \mathbb{W}_k(\mu,\nu)^k = \mathbb{E}[|X_0 - Y_0|^k]. \qquad (5.1.6)$$

We have

$$\begin{aligned}\mathrm{d}|X_t - Y_t|^2 =& 2\langle X_t - Y_t, (\sigma(X_t, P_t^*\mu) - \sigma(Y_t, P_t^*\nu))\mathrm{d}W_t\rangle \\ &+ \|\sigma(X_t, P_t^*\mu) - \sigma(Y_t, P_t^*\nu)\|_{HS}^2 \mathrm{d}t \\ &+ 2\langle b(X_t, P_t^*\mu) - b(Y_t, P_t^*\nu), X_t - Y_t\rangle \mathrm{d}t.\end{aligned}$$

Combining this with Itô's formula and applying (5.1.5), we derive

$$\mathrm{d}|X_t - Y_t|^k \leq \left\{K_2 \mathbb{W}_k(P_t^*\mu, P_t^*\nu)^k - K_1|X_t - Y_t|^k\right\}\mathrm{d}t + \mathrm{d}M_t$$

for some martingale M_t. This and (5.1.6) imply

$$\mathbb{W}_k(P_t^*\mu, P_t^*\nu)^k \leq \mathbb{E}[|X_t - Y_t|^k] \leq \mathbb{W}_k(\mu,\nu)^k \mathrm{e}^{-(K_1-K_2)t}, \quad t \geq 0.$$

Then the proof is finished by Theorem 5.1.1. $\square$

5.2 Ergodicity in variation distance: Singular case

The following result extends Theorem 1.6.1 to the distribution dependent setting. See [Wang (2023c)] for a result with weaker integral condition replacing (5.2.1).

Theorem 5.2.1. *Let $\sigma(x,\mu) = \sigma(x)$ not depend on μ. Assume that for any $\nu \in \mathcal{P}$, $(\sigma, b(\cdot,\nu))$ satisfies $(A^{1.4})$, and that*

$$|b(x,\mu_1) - b(x,\mu_2)| \leq \kappa\|\mu_1 - \mu_2\|_{var}, \quad x \in \mathbb{R}^d, \mu_1, \mu_2 \in \mathcal{P} \qquad (5.2.1)$$

holds for some constant $\kappa > 0$. Then (5.0.1) is well-posed, and when $\kappa > 0$ is small enough and Φ is convex with $\int_0^\infty \frac{\mathrm{d}s}{\Phi(s)} < \infty$, P_t^* has a unique invariant probability measure $\bar{\mu}$, $\bar{\mu}(\Phi(\varepsilon_0 V)) < \infty$ holds for some constant $\varepsilon_0 > 0$, and there exist constants $c, \lambda > 0$ such that

$$\|P_t^*\nu - \bar{\mu}\|_{var} \leq c\mathrm{e}^{-\lambda t}\|\bar{\mu} - \nu\|_{var}, \quad t \geq 0, \nu \in \mathcal{P}. \qquad (5.2.2)$$

5.2.1 Two lemmas

For any $\gamma \in \mathcal{P}$, consider the following SDE with fixed distribution parameter:
$$dX_t^\gamma = b(X_t^\gamma, \gamma) + \sigma(X_t^\gamma)dW_t. \tag{5.2.3}$$

The following result says that if (5.2.3) is exponentially ergodic with respect to $\|\cdot\|_{var}$ uniformly in γ, and if the dependence of $b(x, \mu)$ on μ is weak enough, then (5.0.1) is exponentially ergodic in $\|\cdot\|_{var}$.

Lemma 5.2.2. *Assume* (5.2.1) *and that for each* $\gamma \in \mathcal{P}$ *the SDE* (5.2.3) *is well-posed. Then* (5.0.1) *is well-posed. If the associated Markov semigroup* P_t^γ *of* (5.2.3) *satisfies*
$$\|(P_t^\gamma)^*\mu_1 - (P_t^\gamma)^*\mu_2\|_{var} \le ce^{-\lambda t}\|\mu_1 - \mu_2\|_{var}, \quad t \ge 0, \gamma, \mu_1, \mu_2 \in \mathcal{P} \tag{5.2.4}$$
for some constants $c, \lambda > 0$, *then* P_t^* *associated with* (5.0.1) *has a unique invariant probability measure* $\bar{\mu}$ *when* $\kappa < \frac{\sqrt{\lambda}}{2\sqrt{\log(2c)}}$. *If moreover* $\kappa \in (0, \hat{\kappa})$, *where*
$$\hat{\kappa} := \sup\left\{\kappa > 0 : \frac{(c\kappa)^2(2c)^{\frac{2\kappa^2}{\lambda}}}{\lambda + \kappa^2} < \frac{1}{2}\right\} > 0,$$
then there exists a constant $c' > 0$ *such that*
$$\|P_t^*\nu - \bar{\mu}\|_{var} \le c'e^{-\lambda' t}\|\nu - \bar{\mu}\|_{var}, \quad t \ge 0, \nu \in \mathcal{P} \tag{5.2.5}$$
holds for
$$\lambda' := -\frac{\lambda}{\log(2c)}\log\left(\frac{1}{2} + \frac{(c\kappa)^2(2c)^{\frac{2\kappa^2}{\lambda}}}{\lambda + \kappa^2}\right) > 0.$$

Proof. By Theorem 3.4.1, the well-posedness of (5.0.1) follows from that of (5.2.3) and (5.2.1).

(a) Existence and uniqueness of $\bar{\mu}$. For any $\gamma \in \mathcal{P}$, (5.2.4) implies that P_t^γ has a unique invariant probability measure μ_γ. It suffices to prove that the map $\gamma \mapsto \mu_\gamma$ has a unique fixed point $\bar{\mu}$, which is the unique invariant probability measure of P_t^*.

For $\gamma_1, \gamma_2 \in \mathcal{P}$, (5.2.3) implies
$$\|(P_t^{\gamma_1})^*\mu_{\gamma_2} - \mu_{\gamma_1}\|_{var} \le ce^{-\lambda t}\|\mu_{\gamma_2} - \mu_{\gamma_1}\|_{var}, \quad t \ge 0. \tag{5.2.6}$$

On the other hand, let (X_t^1, X_t^2) solve the SDEs
$$dX_t^i = b(X_t^i, \gamma_i) + \sigma(X_t^i)dW_t, \quad i = 1, 2$$

with $X_0^1 = X_0^2$ having distribution μ_{γ_2}. Since μ_{γ_2} is $(P_t^{\gamma_2})^*$-invariant, we have
$$\mathcal{L}_{X_t^2} = (P_t^{\gamma_2})^* \mu_{\gamma_2} = \mu_{\gamma_2}, \quad \mathcal{L}_{X_t^1} = (P_t^{\gamma_1})^* \mu_{\gamma_2}, \quad t \geq 0. \tag{5.2.7}$$
Let
$$\eta_t := \{\sigma^*(\sigma\sigma^*)^{-1}[b(\cdot, \gamma_2) - b(\cdot, \gamma_1)]\}(X_t^1),$$
$$R_t := e^{\int_0^t \langle \eta_s, \mathrm{d}W_s \rangle - \frac{1}{2}\int_0^t |\eta_s|^2 \mathrm{d}s}, \quad t \geq 0.$$
By (5.2.1), R_t is a martingale, and by Girsanov's theorem, for any $t > 0$,
$$\tilde{W}_r := W_r - \int_0^r \eta_s \mathrm{d}s, \quad r \in [0, t]$$
is a Brownian motion under $\mathbb{Q}_t := R_t \mathbb{P}$. Reformulating the SDE for X_r^1 as
$$\mathrm{d}X_r^1 = b(X_r^1, \gamma_2) \mathrm{d}r + \sigma(X_r^1) \mathrm{d}\tilde{W}_r, \quad r \in [0, t],$$
by $X_0^1 = X_0^2$ and the weak uniqueness, the law of X_t^1 under $\mathbb{Q}_t$ satisfies
$$\mathcal{L}_{X_t^1 | \mathbb{Q}_t} = \mathcal{L}_{X_t^2} = (P_t^{\gamma_2})^* \mu_{\gamma_2}.$$
Combining this with (5.2.7) and Pinsker's inequality (3.2.3), we obtain
$$\|(P_t^{\gamma_1})^* \mu_{\gamma_2} - \mu_{\gamma_2}\|_{var}^2 = \|(P_t^{\gamma_1})^* \mu_{\gamma_2} - (P_t^{\gamma_2})^* \mu_{\gamma_2}\|_{var}^2$$
$$= \sup_{|f| \leq 1} \left| \mathbb{E}[f(X_t^1)] - \mathbb{E}[f(X_t^1) R_t] \right|^2 \leq (\mathbb{E}|R_t - 1|)^2 \tag{5.2.8}$$
$$\leq 2\mathbb{E}[R_t \log R_t] = 2\mathbb{E}_{\mathbb{Q}_t}[\log R_t]$$
$$= \mathbb{E}_{\mathbb{Q}_t} \int_0^t \left| \{\sigma^*(\sigma\sigma^*)^{-1}[b(\cdot, \gamma_2) - b(\cdot, \gamma_1)]\}(X_s^1) \right|^2 \mathrm{d}s.$$
Thus, (5.2.1) implies
$$\|(P_t^{\gamma_1})^* \mu_{\gamma_2} - \mu_{\gamma_2}\|_{var}^2 \leq \kappa^2 \int_0^t \|\gamma_1 - \gamma_2\|_{var}^2 \mathrm{d}s = \kappa^2 t \|\gamma_1 - \gamma_2\|_{var}^2.$$
Combining this with (5.2.6) and taking $t = \frac{\log(2c)}{\lambda}$, we derive
$$\|\mu_{\gamma_1} - \mu_{\gamma_2}\|_{var} \leq \|(P_t^{\gamma_1})^* \mu_{\gamma_1} - \mu_{\gamma_1}\|_{var} + \|(P_t^{\gamma_1})^* - \mu_{\gamma_2}\|_{var}$$
$$\leq \left\{ \kappa\sqrt{t} + ce^{-\lambda t} \right\} \|\gamma_1 - \gamma_2\|_{var} = \left\{ \frac{1}{2} + \frac{\kappa\sqrt{\log(2c)}}{\sqrt{\lambda}} \right\} \|\gamma_1 - \gamma_2\|_{var}$$
$$=: \delta \|\gamma_1 - \gamma_2\|_{var}.$$
When $\kappa < \kappa_0 := \frac{\sqrt{\lambda}}{2\sqrt{\log(2c)}}$, we have $\delta < 1$ so that μ_γ is contractive in γ, hence it has a unique fixed point.

(b) Exponential ergodicity in variation distance. Let $\bar{\mu}$ be the unique invariant probability measure of P_t^*, and for any $\nu \in \mathcal{P}$, let $(\bar{X}_0, X_0)$ be $\mathcal{F}_0$-measurable such that

$$\mathbb{P}(\bar{X}_0 \neq X_0) = \frac{1}{2}\|\bar{\mu} - \nu\|_{var}, \quad \mathcal{L}_{\bar{X}_0} = \bar{\mu}, \quad \mathcal{L}_{X_0} = \nu.$$

Let $\bar{X}_t$ and X_t solve the following SDEs with initial values $\bar{X}_0$ and X_0 respectively:

$$d\bar{X}_t = b(\bar{X}_t, \bar{\mu})dt + \sigma(\bar{X}_t)dW_t,$$
$$dX_t = b(X_t, P_t^*\nu)dt + \sigma(X_t)dW_t.$$

Since $\bar{\mu}$ is P_t^*-invariant, we have

$$\mathcal{L}_{\bar{X}_t} = (P_t^{\bar{\mu}})^*\bar{\mu} = P_t^*\bar{\mu} = \bar{\mu}. \tag{5.2.9}$$

Moreover, $\mathcal{L}_{X_t} = P_t^*\nu$ by the definition of P_t^*. Let

$$\bar{\eta}_t := \{\sigma^*(\sigma\sigma^*)^{-1}[b(\cdot, \bar{\mu}) - b(\cdot, P_s^*\nu)]\}(X_t),$$
$$\bar{R}_t := e^{\int_0^t \langle \bar{\eta}_s, dW_s\rangle - \frac{1}{2}\int_0^t |\bar{\eta}_s|^2 ds}.$$

Similarly to (5.2.8), by (5.2.1), Girsanov's theorem and Pinsker's inequality (3.2.3), we obtain

$$\|(P_t^{\bar{\mu}})^*\nu - P_t^*\nu\|_{var}^2 = \sup_{|f|\leq 1}\left|\mathbb{E}[f(X_t)\bar{R}_t] - \mathbb{E}[f(X_t)]\right|^2$$
$$\leq \kappa^2 \int_0^t \|\bar{\mu} - P_s^*\nu\|_{var}^2 ds, \quad t \geq 0.$$

This together with (5.2.6) for $\gamma_1 = \mu$ and (5.2.9) gives

$$\|P_t^*\nu - \bar{\mu}\|_{var}^2 \leq 2\|P_t^*\nu - (P_t^{\bar{\mu}})^*\nu\|_{var}^2 + 2\|(P_t^{\bar{\mu}})^*\nu - \bar{\mu}\|_{var}^2$$
$$\leq 2\kappa^2 \int_0^t \|\bar{\mu} - P_s^*\nu\|_{var}^2 ds + 2c^2 e^{-2\lambda t}\|\nu - \bar{\mu}\|_{var}^2, \quad t \geq 0.$$

By Gronwall's inequality we obtain

$$\|P_t^*\nu - \bar{\mu}\|_{var}^2 \leq \|\bar{\mu} - \nu\|_{var}^2 \left(2c^2 e^{-2\lambda t} + 2\kappa^2 c^2 \int_0^t e^{-2\lambda s + 2\kappa^2(t-s)}ds\right)$$
$$\leq \left\{2c^2 e^{-2\lambda t} + \frac{(c\kappa)^2 e^{2\kappa^2 t}}{\lambda + \kappa^2}\right\}\|\bar{\mu} - \nu\|_{var}^2, \quad t \geq 0.$$

Taking $t = \hat{t} := \frac{\log(2c)}{\lambda}$, we arrive at

$$\|P_{\hat{t}}^*\nu - \bar{\mu}\|_{var}^2 \leq \delta_\kappa \|\bar{\mu} - \nu\|_{var}^2, \quad \nu \in \mathcal{P}$$

for

$$\delta_\kappa := \left(\frac{1}{2} + \frac{(c\kappa)^2 (2c)^{\frac{2\kappa^2}{\lambda}}}{\lambda + \kappa^2}\right) < 1, \quad \kappa < \hat{\kappa}.$$

So, (5.2.5) holds for some constant $c' > 0$ due to the semigroup property (5.1.1). □

To verify condition (5.2.4), we present below a Harris type theorem on the exponential ergodicity in variation distance for a family of Markov processes.

Lemma 5.2.3. *Let (E,ρ) be a metric space and let $\{(P_t^i)_{t\geq 0} : i \in I\}$ be a family of Markov semigroups on $\mathcal{B}_b(E)$. If there exist $t_0 > 0$ and measurable set $B \subset E$ such that*

$$\alpha := \inf_{i \in I, x \in E} P_{t_0}^i 1_B(x) > 0, \tag{5.2.10}$$

$$\beta := \sup_{i \in I, x, y \in B} \|(P_{t_1}^i)^* \delta_x - (P_{t_1}^i)^* \delta_y\|_{var} < 2, \tag{5.2.11}$$

then there exists $c > 0$ such that

$$\sup_{i \in I, x, y \in E} \|(P_t^i)^* \delta_x - (P_t^i)^* \delta_y\|_{var} \leq ce^{-\lambda t}, \quad t \geq 0 \tag{5.2.12}$$

holds for $\lambda := \frac{1}{t_0+t_1} \log \frac{2}{2-\alpha^2(2-\beta)} > 0$.

Proof. By the semigroup property, we have

$$\|(P_{t_0+t_1}^i)^* \delta_x - (P_{t_0+t_1}^i)^* \delta_y\|_{var}$$

$$= \sup_{|f|\leq 1} \left| \int_{E \times E} \left(P_{t_1}^i f(x') - P_{t_1}^i f(y') \right) \{(P_{t_0}^i)^* \delta_x\}(\mathrm{d}x') \{(P_{t_0}^i)^* \delta_y\}(\mathrm{d}y') \right|$$

$$\leq \int_{B \times B} \|(P_{t_1}^i)^* \delta_{x'} - (P_{t_1}^i)^* \delta_{y'}\|_{var} \{(P_{t_0}^i)^* \delta_x\}(\mathrm{d}x') \{(P_{t_0}^i)^* \delta_y\}(\mathrm{d}y')$$

$$+ 2 \int_{(B \times B)^c} \{(P_{t_0}^i)^* \delta_x\}(\mathrm{d}x') \{(P_{t_0}^i)^* \delta_y\}(\mathrm{d}y')$$

$$\leq \beta \{P_{t_0}^i 1_B(x)\} P_{t_0}^i 1_B(y) + 2\left[1 - \{P_{t_0}^i 1_B(x)\} P_{t_0}^i 1_B(y)\right] \leq 2 - \alpha^2(2-\beta).$$

Noting that $\|\delta_x - \delta_y\|_{var} = 2$ for $x \neq y$ and $\delta := \frac{2-\alpha^2(2-\beta)}{2} < 1$, we derive

$$\|(P_{t_0+t_1}^i)^* \delta_x - (P_{t_0+t_1}^i)^* \delta_y\|_{var} \leq \delta \|\delta_x - \delta_y\|_{var}, \quad x, y \in E.$$

Combining this with the semigroup property, we find constants $c > 0$ such that (5.2.12) holds for the claimed $\lambda > 0$. □

5.2.2 Proof of Theorem 5.2.1

According to Theorems 1.6.1 and Lemma 3.7.7, it suffices to verify (5.2.4). By Lemma 5.2.3, we only need to prove (5.2.10) and (5.2.11) for the family $\{(P_t^\gamma)_{t\geq 0} : \gamma \in \mathcal{P}\}$.

(a) Proof of (5.2.11). Let us fix $\gamma \in \mathcal{P}$, and let $X_t^{x,\gamma}$ solve (5.2.3) with $X_0^\gamma = x$. For any $\nu \in \mathcal{P}$, by Girsanov's theorem we have

$$P_t^\nu f(x) = \mathbb{E}[f(X_t^{x,\gamma})R_t^{x,\gamma,\nu}], \quad t \geq 0,$$

where

$$R_t^{x,\gamma,\nu} := e^{\int_0^t \langle \eta_s^{x,\gamma,\nu}, \mathrm{d}W_s \rangle - \frac{1}{2}|\eta_s^{x,\gamma,\nu}|^2 \mathrm{d}s},$$

$$\eta_s^{x,\gamma,\nu} := \{\sigma^*(\sigma\sigma^*)^{-1}[b(\cdot,\nu) - b(\cdot,\gamma)]\}(X_s^{x,\gamma}).$$

So, (5.2.1) and Pinsker's inequality (3.2.3) imply

$$\|(P_t^\gamma)^*\delta_z - (P_t^\nu)^*\delta_z\|_{var}^2 \leq (\mathbb{E}|R_t^{\gamma,\nu} - 1|)^2 \leq \kappa^2 t \|\gamma - \nu\|_{var}^2$$
$$\leq 4\kappa^2 t, \quad t \geq 0, z \in \mathbb{R}^d, \nu \in \mathcal{P}.$$

Taking $t_1 = \frac{1}{16\kappa^2}$, we obtain

$$\sup_{\nu \in \mathcal{P}} \|(P_{t_1}^\gamma)^*\delta_z - (P_{t_1}^\nu)^*\delta_z\|_{var} \leq \frac{1}{2}, \quad z \in \mathbb{R}^d, \nu \in \mathcal{P}. \tag{5.2.13}$$

On the other hand, by (1.3.4), there exists $x_0 \in D$ and a constant $\varepsilon > 0$ such that $B(x_0, \varepsilon) \subset D$ and

$$\|(P_{t_1}^\gamma)^*\delta_x - (P_{t_1}^\gamma)^*\delta_y\|_{var} \leq \frac{1}{4}, \quad x, y \in B(x_0, \varepsilon).$$

Combining this with (5.2.13) we derive

$$\sup_{\nu \in \mathcal{P}} \|(P_{t_1}^\nu)^*\delta_x - (P_{t_1}^\nu)^*\delta_y\|_{var} \leq \frac{3}{2} < 2, \quad x, y \in B(x_0, \varepsilon).$$

So, (5.2.11) holds for $B = B(x_0, \varepsilon)$.

(b) Let u solve (5.5.16) for $f = -b^{(0)}$ and large $\lambda > 0$ such that (1.3.16) holds, and let $\Theta(x) = x + u(x)$. Since $(\sigma, b(\cdot, \gamma))$ satisfies $(A^{1.4})$, (1.6.20) holds for $Y_t^{x,\nu} := \Theta(X_t^{x,\nu})$ replacing Y_t for all $\nu \in \mathcal{P}$. So, by $H(\infty) < \infty$ and the argument leading to (1.6.23), we obtain

$$\sup_{\nu \in \mathcal{P}, x \in \mathbb{R}^d} \mathbb{E}[V(Y_t^{x,\nu})] \leq \theta^{-1}k, \quad t \geq kH(\infty) =: t_2.$$

This together with (1.6.18) implies

$$\sup_{\nu \in \mathcal{P}, x \in \mathbb{R}^d} \mathbb{E}[V(X_t^{x,\nu})] \leq \theta^{-2}k, \quad t \geq t_2.$$

Letting $\mathbf{K} := \{V \leq 2\theta^{-2}k\}$, we derive

$$\inf_{\nu \in \mathcal{P}, x \in \mathbb{R}^d} P_{t_2}^\nu \mathbf{1}_{\mathbf{K}}(x) \geq \frac{1}{2}. \tag{5.2.14}$$

On the other hand, by Girsanov's theorem and Schwarz's inequality, we find a constant $c_0 > 0$ such that

$$P_1^\nu 1_{B(x_0,\varepsilon)}(x) = \mathbb{E}\big[1_{B(x_0,\varepsilon)}(X_1^{x,\gamma})R_1^{x,\gamma,\nu}\big]$$
$$\geq \frac{\{\mathbb{E}1_{B(x_0,\varepsilon)}(X_1^{x,\gamma})\}^2}{\mathbb{E}R_1^{x,\gamma,\nu}} \geq c_0(P_1^\gamma 1_{B(x_0,\varepsilon)}(x))^2.$$

Since $\mathbf{K}$ is bounded, combining this with Lemma 3.7.6 for P_t^γ, we find a constant $c_1 > 0$ such that

$$\inf_{\nu \in \mathcal{P}, x \in \mathbf{K}} P_1^\nu 1_{B(x_0,\varepsilon)}(x) \geq c_1.$$

This together with (5.2.14) and the semigroup property yields

$$P_{t_2+1}^\nu 1_{B(x_0,\varepsilon)}(x) \geq P_{t_2}^\nu\big\{1_{\mathbf{K}} P_1^\nu 1_{B(x_0,\varepsilon)}\big\}(x) \geq c_1 P_{t_2}^\nu 1_{\mathbf{K}}(x) \geq \frac{c_1}{2} > 0$$

for all $x \in \mathbb{R}^d, \nu \in \mathcal{P}$. Therefore, (5.2.10) holds for $t_0 = t_2 + 1$.

5.3 Exponential ergodicity in relative entropy and $\mathbb{W}_2$: Dissipative case

We first present a criterion on the exponential convergence in entropy by using the log-Harnack and Talagrand inequalities, then apply to (5.0.1) with non-degenerate and degenerate noises respectively.

5.3.1 *A criterion with application to Granular media type equations*

Theorem 5.3.1. *Assume that (5.0.1) is well-posed for distributions in $\mathcal{P}_2$, P_t^* has a unique invariant probability measure $\bar{\mu} \in \mathcal{P}_2$ such that for some constants $t_0, c_0, C > 0$ we have the log-Harnack inequality*

$$P_{t_0}(\log f)(\nu) \leq \log P_{t_0} f(\mu) + c_0 \mathbb{W}_2(\mu,\nu)^2, \quad \mu, \nu \in \mathcal{P}_2, \tag{5.3.1}$$

and the Talagrand inequality

$$\mathbb{W}_2(\mu,\bar{\mu})^2 \leq C\mathrm{Ent}(\mu|\bar{\mu}), \quad \mu \in \mathcal{P}_2. \tag{5.3.2}$$

(1) *If there exist constants $c_1, \lambda, t_1 \geq 0$ such that*

$$\mathbb{W}_2(P_t^*\mu,\bar{\mu})^2 \leq c_1 e^{-\lambda t}\mathbb{W}_2(\mu,\bar{\mu})^2, \quad t \geq t_1, \mu \in \mathcal{P}_2, \tag{5.3.3}$$

then for any $t \geq t_0 + t_1$,

$$\max\big\{c_0^{-1}\mathrm{Ent}(P_t^*\mu|\bar{\mu}), \mathbb{W}_2(P_t^*\mu,\bar{\mu})^2\big\}$$
$$\leq c_1 e^{-\lambda(t-t_0)} \min\big\{\mathbb{W}_2(\mu,\bar{\mu})^2, C\mathrm{Ent}(\mu|\bar{\mu})\big\}, \quad \mu \in \mathcal{P}_2. \tag{5.3.4}$$

(2) *If for some constants* $\lambda, c_2, t_2 > 0$,
$$\operatorname{Ent}(P_t^*\mu|\bar{\mu}) \leq c_2 \mathrm{e}^{-\lambda t}\operatorname{Ent}(\mu|\bar{\mu}), \quad t \geq t_2, \nu \in \mathcal{P}_2, \tag{5.3.5}$$
then for any $t \geq t_0 + t_2$ *and* $\mu \in \mathcal{P}_2$,
$$\begin{aligned}&\max\left\{\operatorname{Ent}(P_t^*\mu,\bar{\mu}), C^{-1}\mathbb{W}_2(P_t^*\mu,\bar{\mu})^2\right\} \\ &\leq c_2 \mathrm{e}^{-\lambda(t-t_0)}\min\left\{c_0\mathbb{W}_2(\mu,\bar{\mu})^2, \operatorname{Ent}(\mu|\bar{\mu})\right\}.\end{aligned} \tag{5.3.6}$$

Proof. (1) Since
$$\operatorname{Ent}(P_{t_0}^*\nu|P_{t_0}^*\mu) = \sup_{f \geq 0, (P_{t_0}f)(\mu)=1} P_{t_0}(\log f)(\nu),$$
(5.3.1) implies
$$\operatorname{Ent}(P_{t_0}^*\nu|P_{t_0}^*\mu) \leq c_0 \mathbb{W}_2(\mu, \nu)^2.$$
This together with $P_{t_0}^*\bar{\mu} = \bar{\mu}$ gives
$$\operatorname{Ent}(P_{t_0}^*\mu|\bar{\mu}) \leq c_0 \mathbb{W}_2(\mu, \bar{\mu})^2, \quad \mu \in \mathcal{P}_2. \tag{5.3.7}$$
Combining (5.3.3) with (5.3.2) and (5.3.7), we obtain
$$\begin{aligned}\mathbb{W}_2(P_t^*\mu,\bar{\mu})^2 &\leq c_1 \mathrm{e}^{-\lambda t}\mathbb{W}_2(\mu,\bar{\mu})^2 \\ &\leq c_1 \mathrm{e}^{-\lambda t}\min\left\{\mathbb{W}_2(\mu,\bar{\mu})^2, C\operatorname{Ent}(\mu|\bar{\mu})\right\}, \quad t \geq t_1,\end{aligned}$$
and for any $t \geq t_0 + t_1$,
$$\begin{aligned}\operatorname{Ent}(P_t^*\mu|\bar{\mu}) &= \operatorname{Ent}(P_{t_0}^* P_{t-t_0}^*\mu|\bar{\mu}) \\ &\leq c_0 \mathbb{W}_2(P_{t-t_0}^*\mu,\bar{\mu})^2 \leq c_0 c_1 \mathrm{e}^{-\lambda(t-t_0)}\mathbb{W}_2(\mu,\bar{\mu})^2 \\ &= \{c_0 c_1 \mathrm{e}^{\lambda t_0}\}\mathrm{e}^{-\lambda t}\min\left\{\mathbb{W}_2(\mu,\bar{\mu})^2, C\operatorname{Ent}(\mu|\bar{\mu})\right\}.\end{aligned}$$
Therefore, (5.3.4) holds.

(2) Similarly, if (5.3.5) holds, then (5.3.2) and (5.3.7) imply
$$\begin{aligned}\operatorname{Ent}(P_t^*\mu|\bar{\mu}) &\leq c_2 \mathrm{e}^{-\lambda(t-t_0)}\min\left\{\operatorname{Ent}(P_{t_0}^*\mu|\bar{\mu}), \operatorname{Ent}(\mu|\bar{\mu})\right\} \\ &\leq c_2 \mathrm{e}^{-\lambda(t-t_0)}\min\left\{c_0 \mathbb{W}_2(\mu,\bar{\mu})^2, \operatorname{Ent}(\mu|\bar{\mu})\right\}, \quad t \geq t_0 + t_2,\end{aligned}$$
and
$$\begin{aligned}C^{-1}\mathbb{W}_2(P_t^*\mu,\bar{\mu})^2 &\leq \operatorname{Ent}(P_{t-t_0}^* P_{t_0}^*\mu|\bar{\mu}) \\ &\leq c_2 \min\left\{\mathrm{e}^{-\lambda t}\operatorname{Ent}(\mu|\bar{\mu}), \mathrm{e}^{-\lambda(t-t_0)}\operatorname{Ent}(P_{t_0}^*\mu|\bar{\mu})\right\} \\ &\leq c_2 \mathrm{e}^{-\lambda(t-t_0)}\min\left\{\operatorname{Ent}(\mu|\bar{\mu}), c_0 \mathbb{W}_2(\mu,\bar{\mu})^2\right\}, \quad t \geq t_0 + t_2.\end{aligned}$$
Then (5.3.6) holds, and the proof is finished. $\square$

When $\sigma\sigma^*$ is invertible and does not depend on the distribution, the log-Harnack inequality (5.3.1) has been established in Theorem 4.1.1. The Talagrand inequality was first found in [Talagrand (1996)] for $\bar{\mu}$ being the Gaussian measure, and extended in [Bobkov et al. (2001)] to $\bar{\mu}$ satisfying the log-Sobolev inequality

$$\bar{\mu}(f^2 \log f^2) \leq C\bar{\mu}(|\nabla f|^2), \quad f \in C_b^1(\mathbb{R}^d), \bar{\mu}(f^2) = 1, \qquad (5.3.8)$$

see [Otto and Villani (2000)] for an earlier result under a curvature condition, and see [Wang (2004)] for further extensions.

To illustrate this result, we consider the granular media type equation for probability density functions $(\rho_t)_{t\geq 0}$ on $\mathbb{R}^d$:

$$\partial_t \rho_t = \mathrm{div}\big\{a\nabla\rho_t + \rho_t a\nabla(V + W \circledast \rho_t)\big\}, \qquad (5.3.9)$$

where

$$W \circledast \rho_t := \int_{\mathbb{R}^d} W(\cdot, y)\rho_t(y)\mathrm{d}y, \qquad (5.3.10)$$

and the functions

$$a : \mathbb{R}^d \to \mathbb{R}^d \otimes \mathbb{R}^d, \quad V : \mathbb{R}^d \to \mathbb{R}, \quad W : \mathbb{R}^d \times \mathbb{R}^d \to \mathbb{R}$$

satisfy the following assumptions.

($A^{5.1}$)
(1) $a := (a_{ij})_{1\leq i,j \leq d} \in C_b^2(\mathbb{R}^d \to \mathbb{R}^d \otimes \mathbb{R}^d)$, and $a \geq \lambda_a I_d$ for some constant $\lambda_a > 0$.
(2) $V \in C^2(\mathbb{R}^d), W \in C^2(\mathbb{R}^d \times \mathbb{R}^d)$ with $W(x,y) = W(y,x)$, and there exist constants $\kappa_0 \in \mathbb{R}$ and $\kappa_1, \kappa_2, \kappa_0' > 0$ such that

$$\mathrm{Hess}_V \geq \kappa_0 I_d, \quad \kappa_0' I_{2d} \geq \mathrm{Hess}_W \geq \kappa_0 I_{2d}, \qquad (5.3.11)$$

$$\langle x, \nabla V(x)\rangle \geq \kappa_1 |x|^2 - \kappa_2, \quad x \in \mathbb{R}^d. \qquad (5.3.12)$$

Moreover, for any $\lambda > 0$,

$$\int_{\mathbb{R}^d \times \mathbb{R}^d} \mathrm{e}^{-V(x) - V(y) - \lambda W(x,y)} \mathrm{d}x\mathrm{d}y < \infty. \qquad (5.3.13)$$

(3) There exists a function $b_0 \in L_{loc}^1([0,\infty))$ with

$$r_0 := \frac{\|\mathrm{Hess}_W\|_\infty}{4} \int_0^\infty \mathrm{e}^{\frac{1}{4}\int_0^t b_0(s)\mathrm{d}s} \mathrm{d}t < 1$$

such that

$$\langle y - x, \nabla V(x) - \nabla V(y) + \nabla W(\cdot, z)(x) - \nabla W(\cdot, z)(y)\rangle$$
$$\leq |x - y|b_0(|x-y|), \quad x, y, z \in \mathbb{R}^d.$$

For any $N \geq 2$, consider the Hamiltonian for the system of N particles:

$$H_N(x_1,\ldots,x_N) = \sum_{i=1}^N V(x_i) + \frac{1}{N-1}\sum_{1\leq i<j\leq N} W(x_i,x_j),$$

and the corresponding finite-dimensional Gibbs measure

$$\mu^{(N)}(\mathrm{d}x_1,\ldots,\mathrm{d}x_N) = \frac{1}{Z_N}\mathrm{e}^{-H_N(x_1,\ldots,x_N)}\mathrm{d}x_1\ldots\mathrm{d}x_N,$$

where $Z_N := \int_{\mathbb{R}^{dN}} \mathrm{e}^{-H_N(x)}\mathrm{d}x < \infty$ due to (5.3.13) in ($A^{5.5}$). For any $1 \leq i \leq N$, the conditional marginal of $\mu^{(N)}$ given $z \in \mathbb{R}^{d(N-1)}$ is

$$\mu_z^{(N)}(\mathrm{d}x) := \frac{1}{Z_N(z)}\mathrm{e}^{-H_N(x|z)}\mathrm{d}x,$$

where

$$Z_N(z) := \int_{\mathbb{R}^d} \mathrm{e}^{-H_N(x|z)}\mathrm{d}x,$$

$$H_N(x|z) := V(x) - \log\int_{\mathbb{R}^{d(N-1)}} \mathrm{e}^{-\sum_{i=1}^{N-1}\{V(z_i)+\frac{1}{N-1}W(x,z_i)\}}\mathrm{d}z_1\ldots\mathrm{d}z_{N-1}.$$

The following result allows V and W to be non-convex. For instance, let $V = V_1 + V_2 \in C^2(\mathbb{R}^d)$ such that $\|V_1\|_\infty \wedge \|\nabla V_1\|_\infty < \infty$, $\mathrm{Hess}_{V_2} \geq \lambda I_d$ for some $\lambda > 0$ (recall that I_d is the $d \times d$ identity matrix), and $W \in C^2(\mathbb{R}^d \times \mathbb{R}^d)$ with $\|W\|_\infty \wedge \|\nabla W\|_\infty < \infty$. Then the uniform log-Sobolev inequality (5.3.14) holds for some constant $\beta > 0$. Indeed, by the Bakry-Emery criterion, $\mu_2(\mathrm{d}x) := \frac{1}{\int_{\mathbb{R}^d} \mathrm{e}^{-V(x)}\mathrm{d}x}\mathrm{e}^{-V_2(x)}\mathrm{d}x$ satisfies the log-Sobolev inequality,

$$\mu_2(f^2 \log f^2) \leq \frac{2}{\lambda}\mu_2(|\nabla f|^2), \quad f \in C_b^1(\mathbb{R}^d), \mu_2(f^2) = 1.$$

Then (5.3.14) with some constant $\beta > 0$ follows by the stability of the log-Sobolev inequality under bounded perturbations (see [Chen and Wang (1997)]), as well as Lipschitz perturbations (see [Aida (1998)]) for the potential V_2. Moreover, assumptions ($A^{5.1}$) holds provided $\|\mathrm{Hess}_W\|_\infty$ is small enough such that $r_0 < 1$. So, the following Theorem 5.3.2 applies. See [Guillin et al. (2022)] for more concrete examples satisfying ($A^{5.1}$) and (5.3.14).

Theorem 5.3.2. *Assume* ($A^{5.1}$). *If there is a constant $a > 0$ such that the uniform log-Sobolev inequality*

$$\mu_z^{(N)}(f^2 \log f^2) \leq \frac{1}{\beta}\mu_z^{(N)}(|\nabla f|^2),$$

$$f \in C_b^1(\mathbb{R}^d), \mu_z^{(N)}(f^2) = 1, N \geq 2, z \in \mathbb{R}^{d(N-1)}$$

(5.3.14)

holds, then there exists a unique $\bar{\mu} \in \mathcal{P}_2$ and a constant $c > 0$ such that
$$\mathbb{W}_2(\mu_t, \bar{\mu})^2 + \mathrm{Ent}(\mu_t|\bar{\mu})$$
$$\leq c e^{-\lambda_a \beta(1-r_0)^2 t} \min\{\mathbb{W}_2(\mu_0, \bar{\mu})^2 + \mathrm{Ent}(\mu_0|\bar{\mu})\}, \quad t \geq 1 \tag{5.3.15}$$
holds for any probability density functions $(\rho_t)_{t \geq 0}$ solving (5.3.9), where $\mu_t(\mathrm{d}x) := \rho_t(x)\mathrm{d}x, t \geq 0$.

Proof. By Theorem 10 in [Guillin et al. (2022)], there exists a unique $\bar{\mu} \in \mathcal{P}_2$ such that
$$\mathrm{Ent}^{V,W}(\bar{\mu}) = 0. \tag{5.3.16}$$
Let $\mu_0 = \rho_0 \mathrm{d}x \in \mathcal{P}_2$. As in (5.3.10), let
$$W \circledast \mu := \int_{\mathbb{R}^d} W(\cdot, y) \mu(\mathrm{d}y).$$
We first note that $\mu_t = P_t^* \mu_0 := \mathcal{L}_{X_t}$ for X_t solving the distribution dependent SDE (5.0.1) with
$$b(x, \mu) = \sum_{j=1}^d \partial_j a_{\cdot, j}(x) - a\nabla\{V + W \circledast \mu\}(x),$$
$$\sigma(x, \mu) = \sqrt{2a(x)}, \quad x \in \mathbb{R}^d, \mu \in \mathcal{P}_2. \tag{5.3.17}$$
Obviously, for this choice of (σ, b), conditions $(A^{5.1})(1)$–(2) imply condition $(A^{3.1})$ for any $k > 1$, so that (5.0.1) SDE is well posed for distributions in $\mathcal{P}_k, k \geq 1$. Below we only consider $k = 2$. For any $N \geq 2$, let $\mu_t^{(N)} = \mathcal{L}_{X_t^{(N)}}$ for the mean field particle system $X_t^{(N)} = (X_t^{N,k})_{1 \leq i \leq N}$:
$$\mathrm{d}X_t^{N,k} = \sqrt{2}\sigma(X_t^{N,k})\mathrm{d}B_t^k$$
$$+ \left\{\sum_{j=1}^d \partial_j a_{\cdot,j}(X_t^{N,k}) - a(X_t^{N,k})\nabla_k H_N(X_t^{(N)})\right\}\mathrm{d}t, \quad t \geq 0, \tag{5.3.18}$$
where ∇_k denotes the gradient in the k th component, and $\{X_0^{N,k}\}_{1 \leq i \leq N}$ are i.i.d. with distribution $\mu_0 \in \mathcal{P}_2$. According to the propagation of chaos, see [Sznitman (1991)], $(A^{5.1})$ implies
$$\lim_{N \to \infty} \mathbb{W}_2(\mathcal{L}_{X_t^{N,1}}, P_t^* \mu_0) = 0. \tag{5.3.19}$$

Next, our conditions imply conditions (25) and (26) in [Guillin et al. (2022)] for $\rho_{LS} = \beta(1-r_0)^2$. So, by Theorem 8(2) in [Guillin et al. (2022)], we have the log-Sobolev inequality
$$\mu^{(N)}(f^2 \log f^2) \leq \frac{2}{\beta(1-r_0)^2} \mu^{(N)}(|\nabla f|^2),$$
$$f \in C_b^1(\mathbb{R}^{dN}), \mu^{(N)}(f^2) = 1. \tag{5.3.20}$$

By [Bobkov et al. (2001)], this implies the Talagrand inequality

$$\mathbb{W}_2(\nu^{(N)}, \mu^{(N)})^2 \leq \frac{2}{\beta(1-r_0)^2} \text{Ent}(\nu^{(N)}|\mu^{(N)}), \qquad (5.3.21)$$

$$t \geq 0, N \geq 2, \nu^{(N)} \in \mathcal{P}(\mathbb{R}^{dN}).$$

On the other hand, by Itô's formula we see that the generator of the diffusion process $X_t^{(N)}$ is

$$L^{(N)}(x^{(N)}) := \sum_{i,j,k=1}^d \left\{ a_{ij}(x^{N,k}) \partial_{x_i^{N,k}} \partial_{x_j^{N,k}} + \partial_j a_{ij}(x^{N,k}) \partial_{x_i^{N,k}} \right.$$

$$\left. - a_{ij}(x^{N,k}) [\partial_{x_i^{N,k}} H_N(x^{(N)})] \partial_{x_i^{N,k}} \right\},$$

for $x^{(N)} = (x^{N,1}, \ldots, x^{N,N}) \in \mathbb{R}^{dN}$, where $x_i^{N,k}$ is the i-th component of $x^{N,k} \in \mathbb{R}^d$. Using the integration by parts formula, we see that this operator is symmetric in $L^2(\mu^{(N)})$, i.e. for any $f, g \in C_0^\infty(\mathbb{R}^{dN})$,

$$\mathcal{E}^{(N)}(f,g) := \int_{\mathbb{R}^{dN}} \langle a^{(N)} \nabla f, \nabla g \rangle \mathrm{d}\mu^{(N)} = -\int_{\mathbb{R}^{dN}} (fL^{(N)}g) \mathrm{d}\mu^{(N)},$$

where

$$a^{(N)}(x^{(N)}) := \text{diag}\{a(x^{N,1}), \ldots, a(x^{N,N})\}.$$

So, the closure of the pre-Dirichlet form $(\mathcal{E}^{(N)}, C_0^\infty(\mathbb{R}^{dN}))$ in $L^2(\mu^{(N)})$ is the Dirichlet form for the Markov semigroup $P_t^{(N)}$ of $X_t^{(N)}$. By $(A^{5.1})$ we have $a^{(N)} \geq \lambda_a I_{dN}$, so that (5.3.20) implies

$$\mu^{(N)}(f^2 \log f^2) \leq \frac{2}{\beta \lambda_a (1-r_0)^2} \mathcal{E}^{(N)}(f,f), \quad f \in C_b^1(\mathbb{R}^{dN}), \mu^{(N)}(f^2) = 1.$$

It is well known that this log-Sobolev inequality implies the exponential convergence

$$\text{Ent}(\mu_t^{(N)}|\mu^{(N)}) \leq e^{-\lambda_a \beta(1-r_0)^2 t} \text{Ent}(\mu_0^{(N)}|\mu^{(N)})$$

$$= e^{-\lambda_a \beta(1-r_0)^2 t} \text{Ent}(\mu^{\otimes N}|\mu^{(N)}), \quad t \geq 0, N \geq 2, \qquad (5.3.22)$$

see for instance Theorem 5.2.1 in [Bakry et al. (2014)]. Moreover, since Hess$_V$ and Hess$_W$ are bounded from below, $(A^{5.1})$ implies that the Bakry-Emery curvature of the generator of $X_t^{(N)}$ is bounded by a constant. Then according to [Wang (2010)], there exists a constant $K \geq 0$ such that the Markov semigroup $P_t^{(N)}$ of $X_t^{(N)}$ satisfies the log-Harnack inequality

$$P_t^{(N)} \log f(x) \leq \log P_t^{(N)} f(y) + \frac{K\rho^{(N)}(x,y)^2}{2(1-e^{-2Kt})}, \qquad (5.3.23)$$

$$0 < f \in \mathcal{B}_b(\mathbb{R}^{dN}), t > 0, x, y \in \mathbb{R}^{dN},$$

where $\rho^{(N)}$ is the intrinsic distance induced by the Dirichlet form $\mathcal{E}^{(N)}$. Since $a^{(N)} \geq \lambda_a I_{dN}$, we have $\rho^{(N)}(x,y)^2 \leq \lambda_a^{-1}|x-y|^2$. So, (5.3.23) implies (5.3.1) for $P_t^{(N)}$ replacing P_{t_0} and $c_0 = \frac{K}{2\lambda_a(1-\mathrm{e}^{-2Kt})}$:

$$P_t^{(N)}(\log f)(\nu) \leq \log P_t^{(N)} f(\mu) + \frac{K\mathbb{W}_2(\mu,\nu)^2}{2\lambda_a(1-\mathrm{e}^{-2Kt})},$$
$$0 < f \in \mathcal{B}_b(\mathbb{R}^{dN}), t > 0, \mu,\nu \in \mathcal{P}_2(\mathbb{R}^{dN}).$$

Thus, by Theorem 5.3.1, (5.3.22) implies

$$\mathbb{W}_2(\mu_t^{(N)}|\mu^{(N)})^2 \leq \frac{c_1 \mathrm{e}^{-\lambda_a \beta (1-r_0)^2 t}}{1 \wedge t}\mathbb{W}_2(\mu^{\otimes N}, \mu^{(N)})^2, \qquad (5.3.24)$$
$$t > 0, N \geq 2$$

for some constant $c_1 > 0$. Moreover, (5.3.21), (5.3.16) and Lemma 17 in [Guillin et al. (2022)] yield

$$\lim_{N\to\infty} \frac{1}{N}\mathbb{W}_2(\bar{\mu}^{\otimes N}, \mu^{(N)})^2 \leq \limsup_{N\to\infty} \frac{2}{\beta(1-r_0)^2 N}\mathrm{Ent}(\bar{\mu}^{\otimes N}|\mu^{(N)})^2$$
$$= \frac{2}{\beta(1-r_0)^2}\mathrm{Ent}^{V,W}(\bar{\mu}) = 0. \qquad (5.3.25)$$

Combining this with (5.3.24) we derive

$$\limsup_{N\to\infty} \frac{1}{N}\mathbb{W}_2(\mu_t^{(N)}, \bar{\mu}^{\otimes N})^2 = \limsup_{N\to\infty} \frac{1}{N}\mathbb{W}_2(\mu_t^{(N)}, \mu^{(N)})^2$$
$$\leq \frac{c_1 \mathrm{e}^{-\lambda_a \beta(1-r_0)^2 t}}{1 \wedge t}\limsup_{N\to\infty}\frac{1}{N}\mathbb{W}_2(\mu_0^{\otimes N}, \mu^{(N)})^2$$
$$= \frac{c_1 \mathrm{e}^{-\lambda_a \beta(1-r_0)^2 t}}{1 \wedge t}\limsup_{N\to\infty}\frac{1}{N}\mathbb{W}_2(\mu_0^{\otimes N}, \bar{\mu}^{\otimes N})^2 \qquad (5.3.26)$$
$$= \frac{c_1 \mathrm{e}^{-\lambda_a \beta(1-r_0)^2 t}}{1 \wedge t}\mathbb{W}_2(\mu_0, \bar{\mu})^2, \quad t > 0.$$

Now, let $\xi = (\xi_i)_{1\leq i \leq N}$ and $\eta = (\eta_i)_{1\leq i\leq N}$ be random variables on $\mathbb{R}^{dN}$ such that $\mathcal{L}_\xi = \mu_t^{(N)}, \mathcal{L}_\eta = \bar{\mu}^{\otimes N}$ and

$$\sum_{i=1}^N \mathbb{E}|\xi_i - \eta_i|^2 = \mathbb{E}|\xi - \eta|^2 = \mathbb{W}_2(\mu_t^{(N)}, \bar{\mu}^{\otimes N})^2.$$

We have $\mathcal{L}_{\xi_i} = \mathcal{L}_{X_t^{N,1}}, \mathcal{L}_{\eta_i} = \bar{\mu}$ for any $1 \leq i \leq N$, so that

$$N\mathbb{W}_2(\mathcal{L}_{X_t^{N,1}}, \bar{\mu})^2 \leq \sum_{i=1}^N \mathbb{E}|\xi_i - \eta_i|^2 = \mathbb{W}_2(\mu_t^{(N)}, \bar{\mu}^{\otimes N})^2. \qquad (5.3.27)$$

Substituting this into (5.3.26), we arrive at
$$\limsup_{N\to\infty} \mathbb{W}_2(\mathcal{L}_{X_t^{N,1}}, \bar{\mu})^2 \le \frac{c_1 e^{-\lambda_a \beta(1-r_0)^2 t}}{1 \wedge t} \mathbb{W}_2(\mu, \bar{\mu})^2 \ , \quad t > 0.$$
This and (5.3.19) imply
$$\mathbb{W}_2(P_t^* \mu, \bar{\mu})^2 \le \frac{c_1 e^{-\lambda_a \beta(1-r_0)^2 t}}{1 \wedge t} \mathbb{W}_2(\mu, \bar{\mu})^2 \ , \quad t > 0. \tag{5.3.28}$$
Since $(A^{5.1})(1)$–(2) imply $(A^{4.1})$, by Theorem 4.1.1 we have the log-Harnack inequality
$$P_t(\log f)(\nu) \le \log P_t f(\mu) + \frac{c_2}{1 \wedge t} \mathbb{W}_2(\mu, \nu)^2, \quad \mu, \nu \in \mathcal{P}_2, t > 0 \tag{5.3.29}$$
for some constant $c_2 > 0$. Similarly to the proof of (5.3.27) we have
$$N \mathbb{W}_2(\bar{\mu}, \mu^{(N,1)})^2 \le \mathbb{W}_2(\bar{\mu}^{\otimes N}, \mu^{(N)})^2,$$
where $\mu^{(N,1)} := \mu^{(N)}(\cdot \times \mathbb{R}^{d(N-1)})$ is the first marginal distribution of $\mu^{(N)}$. This together with (5.3.25) implies
$$\lim_{N\to\infty} \mathbb{W}_2(\mu^{(N,1)}, \bar{\mu})^2 = 0.$$
Therefore, applying (5.3.20) to $f(x)$ depending only on the first component x_1, and letting $N \to \infty$, we derive the log-Sobolev inequality
$$\bar{\mu}(f^2 \log f^2) \le \frac{2}{\beta(1-r_0)^2} \bar{\mu}(|\nabla f|^2), \quad f \in C_b^1(\mathbb{R}^d), \bar{\mu}(f^2) = 1.$$
By [Bobkov et al. (2001)], this implies (5.3.2) for $C = \frac{2}{\beta(1-r_0)^2}$. Combining this with the log-Harnack inequality and (5.3.28), by Theorem 5.3.1 we prove (5.3.15) for some constant $c > 0$ and $\mu_t = \mathcal{L}_{X_t} = P_t^* \mu_0$ for solutions to (5.0.1) with b, σ in (5.3.17).

According to Example 3.1.3, for any probability density functions ρ_t solving (5.3.9), we have $\rho_t dx = P_t^* \mu_0$ for $\mu_0 = \rho_0 dx \in \mathcal{P}_2$. So, we have proved (5.3.15) for ρ_t solving (5.3.9) with $\mu_0 \in \mathcal{P}_2$. When $\mu_0 \notin \mathcal{P}_2$, it is easy to see that $\text{Ent}(\mu_0, \bar{\mu}) = \mathbb{W}_2(\mu, \bar{\mu}) = \infty$, so that (5.3.15) holds. □

5.3.2 The non-degenerate case

$(A^{5.2})$

(1) b is continuous on $\mathbb{R}^d \times \mathcal{P}_2$ and there exists a constant $K > 0$ such that
$$\langle b(x, \mu) - b(y, \nu), x - y \rangle^+ + \|\sigma(x, \mu) - \sigma(y, \nu)\|^2 \le K\{|x-y|^2 + \mathbb{W}_2(\mu, \nu)^2\},$$
$$|b(0, \mu)| \le c\left(1 + \sqrt{\mu(|\cdot|^2)}\right), \quad x, y \in \mathbb{R}^d, \mu, \nu \in \mathcal{P}_2.$$

(2) $\sigma\sigma^*$ is invertible with $\lambda := \|(\sigma\sigma^*)^{-1}\|_\infty < \infty$, and there exist constants $K_2 > K_1 \geq 0$ such that for any $x, y \in \mathbb{R}^d$ and $\mu, \nu \in \mathcal{P}_2$,

$$\|\sigma(x)-\sigma(y)\|_{HS}^2 + 2\langle b(x,\mu)-b(y,\nu), x-y\rangle \leq K_1 \mathbb{W}_2(\mu,\nu)^2 - K_2|x-y|^2.$$

According to Theorem 3.3.1, if $(A^{5.2})(1)$ holds and $b(x,\mu)$ is continuous on $\mathbb{R}^d \times \mathcal{P}_2$, then for any initial value $X_0 \in L^2(\Omega \to \mathbb{R}^d, \mathcal{F}_0, \mathbb{P})$, (5.0.1) has a unique solution which satisfies

$$\mathbb{E}\left[\sup_{t\in[0,T]} |X_t|^2\right] < \infty, \quad T \in (0, \infty). \tag{5.3.30}$$

Let $P_t^*\mu = \mathcal{L}_{X_t}$ for the solution with $\mathcal{L}_{X_0} = \mu$. We have the following result.

Theorem 5.3.3. *Assume* $(A^{5.2})$. *Then* P_t^* *has a unique invariant probability measure* $\bar{\mu}$ *such that*

$$\begin{aligned}&\max\left\{\mathbb{W}_2(P_t^*\mu, \bar{\mu})^2, \mathrm{Ent}(P_t^*\mu|\bar{\mu})\right\} \\ &\leq \frac{c_1}{t \wedge 1} e^{-(K_2-K_1)t} \mathbb{W}_2(\mu, \bar{\mu})^2, \quad t > 0, \mu \in \mathcal{P}_2\end{aligned} \tag{5.3.31}$$

holds for some constant $c_1 > 0$. *If moreover* $\sigma \in C_b^2(\mathbb{R}^d \to \mathbb{R}^d \otimes \mathbb{R}^m)$, *then there exists a constant* $c_2 > 0$ *such that for any* $\mu \in \mathcal{P}_2, t \geq 1$,

$$\begin{aligned}&\max\left\{\mathbb{W}_2(P_t^*\mu, \bar{\mu})^2, \mathrm{Ent}(P_t^*\mu|\bar{\mu})\right\} \\ &\leq c_2 e^{-(K_2-K_1)t} \min\left\{\mathbb{W}_2(\mu, \mu)^2, \mathrm{Ent}(\mu|\bar{\mu})\right\}.\end{aligned} \tag{5.3.32}$$

Proof. For any $\mu, \nu \in \mathcal{P}_2$, let X_t, Y_t solve (5.0.1) for initial values satisfying

$$\mathcal{L}_{X_0} = \mu, \quad \mathcal{L}_{Y_0} = \nu, \quad \mathbb{E}[|X_0 - Y_0|^2] = \mathbb{W}_2(\mu,\nu)^2. \tag{5.3.33}$$

Then $\mu_t := \mathcal{L}_{X_t} = P_t^*\mu$ and $\nu_t := \mathcal{L}_{Y_t} = P_t^*\nu$. By $(A^{5.2})$ and Itô's formula, we obtain

$$\mathrm{d}|X_t - Y_t|^2 \leq \left\{K_1 \mathbb{W}_2(\mu_t, \nu_t)^2 - K_2|X_t - Y_t|^2\right\}\mathrm{d}t + \mathrm{d}M_t$$

for some martingale M_t. Combining this with $\mathbb{W}_2(\mu_t, \nu_t)^2 \leq \mathbb{E}[|X_t - Y_t|^2]$, we obtain

$$\begin{aligned}\mathbb{W}_2(\mu_t, \nu_t)^2 &\leq \mathbb{E}[|X_t - Y_t|^2] \leq e^{-(K_2-K_1)t} \mathbb{E}[|X_0 - Y_0|^2] \\ &= e^{-(K_2-K_1)t} \mathbb{W}_2(\mu,\nu)^2, \quad t \geq 0.\end{aligned}$$

By Theorem 5.1.1, this and (5.3.30) imply that P_t^* has a unique invariant probability measure $\bar{\mu}$ and

$$\mathbb{W}_2(P_t^*\mu, \bar{\mu}) \leq e^{-\frac{1}{2}(K_2-K_1)t} \mathbb{W}_2(\mu, \bar{\mu}), \quad t \geq 0, \mu \subset \mathcal{P}_2. \tag{5.3.34}$$

Next, since $(A^{5.2})$ implies $(A^{4.1})$, by Theorem 4.1.1, there exists a constant $c_0 > 0$ such that

$$\text{Ent}(P_t^* \mu | \bar{\mu}) \leq \frac{c_0}{1 \wedge t} \mathbb{W}_2(\mu, \bar{\mu})^2, \quad t > 0, \mu \in \mathcal{P}_2. \tag{5.3.35}$$

Then for any $p > 1$, combining these with $P_t^* = P_{1 \wedge t}^* P_{(t-1)^+}^*$, we obtain

$$\text{Ent}(P_t^* \mu | \bar{\mu}) = \text{Ent}(P_{1 \wedge t}^* P_{(t-1)^+}^* \mu | \bar{\mu}) \leq \frac{c_0}{1 \wedge t} \mathbb{W}_2(P_{(t-1)^+}^* \mu, \bar{\mu})^2$$

$$\leq \frac{c_0 e^{-(K_2 - K_1)(t-1)^+}}{1 \wedge t} \mathbb{W}_2(\mu, \bar{\mu})^2 = \frac{c_0 e^{K_2 - K_1}}{1 \wedge t} e^{-(K_2 - K_1)t} \mathbb{W}_2(\mu, \bar{\mu})^2.$$

This together with (5.3.34) implies (5.3.31) for some constant $c_1 > 0$.

Now, let $\sigma \in C_b^2(\mathbb{R}^d \to \mathbb{R}^d \otimes \mathbb{R}^m)$. To deduce (5.3.32) from (5.3.31), it remains to find a constant $c > 0$ such that the following Talagrand inequality holds:

$$\mathbb{W}_2(\mu, \bar{\mu})^2 \leq c \, \text{Ent}(\mu | \bar{\mu}), \quad \mu \in \mathcal{P}_2.$$

According to [Bobkov et al. (2001)], this inequality follows from the log-Sobolev inequality

$$\bar{\mu}(f^2 \log f^2) \leq c \bar{\mu}(|\nabla f|^2), \quad f \in C_b^1(\mathbb{R}^d), \bar{\mu}(f^2) = 1. \tag{5.3.36}$$

To prove this inequality, we consider the diffusion process $\bar{X}_t$ on $\mathbb{R}^d$ generated by

$$\bar{L} := \frac{1}{2} \sum_{i,j=1}^d (\sigma \sigma^*)_{ij} \partial_i \partial_j + \sum_{i=1}^\infty b_i(\cdot, \bar{\mu}) \partial_i,$$

which can be constructed by solving the SDE

$$d\bar{X}_t = \sigma(\bar{X}_t) dW_t + b(\bar{X}_t, \bar{\mu}) dt. \tag{5.3.37}$$

Let $\bar{P}_t$ be the associated Markov semigroup. Since $P_t^* \bar{\mu} = \bar{\mu}$, when $\mathcal{L}_{X_0} = \bar{\mu}$, the SDE (5.3.37) coincides with (5.0.1) so that by the uniqueness, we see that $\bar{\mu}$ is an invariant probability measure of $\bar{P}_t$. Combining this with $(A^{5.2})(2)$ and Itô's formula, we obtain

$$\mathbb{W}_2(\mathcal{L}_{\bar{X}_t}, \bar{\mu})^2 \leq e^{-K_2 t} \mathbb{W}_2(\mathcal{L}_{\bar{X}_0}, \bar{\mu})^2, \quad t > 0. \tag{5.3.38}$$

To prove the log-Sobolev inequality (5.3.36), we first verify the hyperboundedness of $\bar{P}_t$, i.e. for large $t > 0$ we have

$$\|\bar{P}_t\|_{L^2(\bar{\mu}) \to L^4(\bar{\mu})} < \infty. \tag{5.3.39}$$

Since $(A^{5.2})$ implies that σ and $b(\cdot, \bar{\mu})$ satisfy conditions $(A1)$-$(A3)$ in [Wang (2011)] for $K = -(K_2 - K_1), \lambda_t^2 = \lambda$ and $\delta_t = \|\sigma\|_\infty$, by Theorem 1.1(3)

in [Wang (2011)], we find a constant $C > 0$ such that the following Harnack inequality holds:
$$(\bar{P}_t f(x))^2 \leq \bar{P}_t f^2(y) \exp\left[\frac{C|x-y|^2}{e^{(K_2-K_1)t}-1}\right], \quad t > 0.$$
Then for any f with $\bar{\mu}(f^2) \leq 1$, we have
$$(\bar{P}_t f(x))^2 \int_{\mathbb{R}^d} \exp\left[-\frac{C|x-y|^2}{e^{(K_2-K_1)t}-1}\right] \bar{\mu}(\mathrm{d}y)$$
$$\leq \bar{\mu}(\bar{P}_t f^2) = \bar{\mu}(f^2) \leq 1.$$
So,
$$\sup_{\bar{\mu}(f^2)\leq 1} |\bar{P}_t f(x)|^4 \leq \frac{1}{\left(\int_{\mathbb{R}^d} e^{-\frac{C|x-y|^2}{\exp[(K_2-K_1)t]-1}} \bar{\mu}(\mathrm{d}y)\right)^2}$$
$$\leq \frac{1}{\left(\int_{B(0,1)} e^{-\frac{C|x-y|^2}{\exp[(K_2-K_1)t]-1}} \bar{\mu}(\mathrm{d}y)\right)^2} \quad (5.3.40)$$
$$\leq C_1 \exp\left[C_1 e^{-(K_2-K_1)t}|x|^2\right], \quad t \geq 1, x \in \mathbb{R}^d.$$
Next, by $\|\sigma\|_\infty < \infty$, $(A^{5.2})(2)$ and Itô's formula, for any $k \in (0, K_2)$ there exists a constant $c_k > 0$ such that
$$\mathrm{d}|\bar{X}_t|^2 \leq 2\langle \bar{X}_t, \sigma(\bar{X}_t)\mathrm{d}W_t\rangle + \{c_k - k|\bar{X}_t|^2\}\mathrm{d}t.$$
Then for any $\varepsilon > 0$,
$$\mathrm{d}e^{\varepsilon|\bar{X}_t|^2} \leq 2\varepsilon e^{\varepsilon|\bar{X}_t|^2}\langle \bar{X}_t, \sigma(\bar{X}_t)\mathrm{d}W_t\rangle$$
$$+ \varepsilon e^{\varepsilon|\bar{X}_t|^2}\{c_k + 2\varepsilon\|\sigma\|_\infty^2|\bar{X}_t|^2 - k|\bar{X}_t|^2\}\mathrm{d}t. \quad (5.3.41)$$
When $\varepsilon > 0$ is small enough such that $2\varepsilon\|\sigma\|_\infty^2 < K_2$, there exist constants $c_1(\varepsilon), c_2(\varepsilon) > 0$ such that
$$\varepsilon e^{\varepsilon|X_t|^2}\{c_k + 2\varepsilon\|\sigma\|_\infty^2|\bar{X}_t|^2 - k|\bar{X}_t|^2\} \leq c_1(\varepsilon) - c_2(\varepsilon)e^{\varepsilon|\bar{X}_t|^2}.$$
Combining this with (5.3.41) we obtain
$$\mathrm{d}e^{\varepsilon|\bar{X}_t|^2} \leq c_1(\varepsilon) - c_2(\varepsilon)e^{\varepsilon|X_t|^2}\mathrm{d}t + 2\varepsilon e^{\varepsilon|X_t|^2}\langle \bar{X}_t, \sigma(\bar{X}_t)\mathrm{d}W_t\rangle.$$
Taking for instance $\bar{X}_0 = 0$, we get
$$\frac{c_2(\varepsilon)}{t}\int_0^t \mathbb{E}e^{\varepsilon|\bar{X}_s|^2}\mathrm{d}s \leq \frac{1 + c_1(\varepsilon)t}{t}, \quad t > 0.$$
This together with (5.3.38) yields
$$\bar{\mu}(e^{\varepsilon(|\cdot|^2 \wedge N)}) = \lim_{t\to\infty} \frac{1}{t}\int_0^t \mathbb{E}e^{\varepsilon(|\bar{X}_s|^2 \wedge N)}\mathrm{d}s \leq \frac{c_1(\varepsilon)}{c_2(\varepsilon)}, \quad N > 0.$$

By letting $N \to \infty$ we derive $\bar{\mu}(e^{\varepsilon|\cdot|^2}) < \infty$. Obviously, this and (5.3.40) imply (5.4.4) for large $t > 0$. Moreover, since $\|(\sigma\sigma^*)^{-1}\|_\infty < \infty$, $\sigma \in C_b^2(\mathbb{R}^d \to \mathbb{R}^d \otimes \mathbb{R}^m)$ and noting that $(A^{5.2})(2)$ for $\mu = \nu = \bar{\mu}$ implies

$$\langle v, \nabla_v b(\cdot, \bar{\mu})\rangle \leq -K_2|v|^2, \quad v \in \mathbb{R}^d,$$

we find a constant $K_0 \in \mathbb{R}$ such that for any $f \in C^\infty(\mathbb{R}^d)$,

$$\Gamma_2(f) := \frac{1}{2}\bar{L}|\sigma^*\nabla f|^2 - \langle \sigma^*\nabla f, \sigma^*\nabla \bar{L}f\rangle \geq K_0|\sigma^*\nabla f|^2,$$

i.e. the Bakry-Emery curvature of $\bar{L}$ is bounded below by a constant K_0. According to Theorem 2.1 in [Röckner and Wang (2003)], this and the hyperboundedness (5.4.4) imply the defective log-Sobolev inequality

$$\begin{aligned}\bar{\mu}(f^2 \log f^2) &\leq C_1\bar{\mu}(|\sigma^*\nabla f|^2) + C_2 \\ &\leq C_1\|\sigma\|_\infty^2 \bar{\mu}(|\nabla f|^2) + C_2, \quad f \in C_b^1(\mathbb{R}^d), \bar{\mu}(f^2) = 1\end{aligned} \quad (5.3.42)$$

for some constants $c_1, c_2 > 0$. Since $\bar{L}$ is elliptic, the invariant probability measure $\bar{\mu}$ is equivalent to the Lebesgue measure, see for instance Theorem 1.1(ii) in [Bogachev et al. (2001b)], so that the Dirichlet form

$$\mathcal{E}(f,g) := \bar{\mu}(\langle \nabla f, \nabla g\rangle), \quad f,g \in W^{1,2}(\mu)$$

is irreducible, i.e. $f \in W^{1,2}(\mu)$ and $\mathcal{E}(f,f) = 0$ imply that f is constant. Therefore, by Corollary 1.3 in [Wang (2014a)], the defective log-Sobolev inequality (5.3.42) implies the desired log-Sobolev inequality (5.3.36) for some constant $c > 0$. Hence, the proof is finished. $\square$

To illustrate this result, we consider the following example which is not included by Theorem 5.3.2 since the function W may be non-symmetric.

Example 5.2.1 (Granular media equation). Let $a = I_d$, let $V \in C^2(\mathbb{R}^d)$ and $W \in C^2(\mathbb{R}^d \times \mathbb{R}^d)$ such that

$$\text{Hess}_V \geq \lambda I_d, \quad \text{Hess}_W \geq \delta_1 I_d, \quad \|\text{Hess}_W\| \leq \delta_2 \quad (5.3.43)$$

holds for some constants $\lambda_1, \delta_2 > 0$ and $\delta_1 \in \mathbb{R}$. If $\lambda + \delta_1 - \delta_2 > 0$, then there exists a unique $\bar{\mu} \in \mathcal{P}_2$ and a constant $c > 0$ such that for any probability density functions $(\rho_t)_{t\geq 0}$ solving (5.3.9), $\mu_t(\mathrm{d}x) := \rho_t(x)\mathrm{d}x$ satisfies

$$\begin{aligned}&\max\{\mathbb{W}_2(\mu_t, \bar{\mu}), \text{Ent}(\mu_t|\bar{\mu})\} \\ &\leq ce^{-(\lambda+\delta_1-\delta_2)t} \min\{\mathbb{W}_2(\mu_0, \bar{\mu}), \text{Ent}(\mu_0|\bar{\mu})\}, \quad t \geq 1.\end{aligned} \quad (5.3.44)$$

Proof. Let $b(x, \mu) := -\nabla\{V + W \circledast \mu\}(x)$. It is easy to see that (5.5.20) implies $(A^{5.2})(1)$ and

$$\langle b(x,\mu) - b(y,\nu), x - y\rangle \leq -(\lambda_1 + \delta_1)|x-y|^2 + \delta_2|x-y|\mathbb{W}_1(\mu,\nu),$$

where we have used the formula
$$\mathbb{W}_1(\mu,\nu) = \sup\{\mu(f) - \nu(f) : \|\nabla f\|_\infty \le 1\}.$$
So, by taking $\alpha = \frac{\delta_2}{2}$ and noting that $\mathbb{W}_1 \le \mathbb{W}_2$, we obtain
$$\langle b(x,\mu) - b(y,\nu), x-y \rangle \le -(\lambda + \delta_1 - \alpha)|x-y|^2 + \frac{\delta_2^2}{4\alpha}\mathbb{W}_1(\mu,\nu)^2$$
$$\le -\left(\lambda + \delta_1 - \frac{\delta_2}{2}\right)|x-y|^2 + \frac{\delta_2}{2}\mathbb{W}_2(\mu,\nu)^2, \quad x,y \in \mathbb{R}^d, \mu,\nu \in \mathcal{P}_2.$$
Therefore, if (5.5.20) holds for $\lambda + \delta_1 - \delta_2 > 0$, Theorem 5.3.3 implies that P_t^* has a unique invariant probability measure $\bar\mu \in \mathcal{P}_2$, such that (5.3.44) holds for $\mu_0 \in \mathcal{P}_2$. When $\mu_0 \notin \mathcal{P}_2$, we have $\mathbb{W}_2(\mu_0,\bar\mu)^2 = \infty$ since $\bar\mu \in \mathcal{P}_2$. Combining this with the Talagrand inequality
$$\mathbb{W}_2(\mu_0,\bar\mu)^2 \le C\mathrm{Ent}(\mu_0|\bar\mu)$$
for some constant $C > 0$, see the proof of Theorem 5.3.3, we have $\mathrm{Ent}(\mu_0|\bar\mu) = \infty$ for $\mu_0 \notin \mathcal{P}_2$, so that (5.3.44) holds for all $\mu_0 \in \mathcal{P}$. □

5.3.3 The degenerate case

When $\mathbb{R}^k$ with some $k \in \mathbb{N}$ is considered, to emphasize the space, we use $\mathcal{P}(\mathbb{R}^k)$ ($\mathcal{P}_2(\mathbb{R}^k)$) to denote the class of probability measures (with finite second moment) on $\mathbb{R}^k$. Consider the following McKean-Vlasov stochastic Hamiltonian system for $(X_t, Y_t) \in \mathbb{R}^{d_1 + d_2} := \mathbb{R}^{d_1} \times \mathbb{R}^{d_2}$:
$$\begin{aligned} \mathrm{d}X_t &= BY_t\mathrm{d}t, \\ \mathrm{d}Y_t &= \sqrt{2}\mathrm{d}W_t - Y_t\mathrm{d}t \\ &\quad - \left\{B^*\nabla V(\cdot, \mathcal{L}_{(X_t,Y_t)})(X_t) + \beta B^*(BB^*)^{-1}X_t\right\}\mathrm{d}t, \end{aligned} \quad (5.3.45)$$
where $\beta > 0$ is a constant, B is a $d_1 \times d_2$-matrix such that BB^* is invertible, and
$$V : \mathbb{R}^{d_1} \times \mathcal{P}_2(\mathbb{R}^{d_1 + d_2}) \to \mathbb{R}^{d_2}$$
is measurable. Let
$$\psi_B((x,y),(\bar x,\bar y)) := \sqrt{|x-\bar x|^2 + |B(y-\bar y)|^2}, \quad (x,y),(\bar x,\bar y) \in \mathbb{R}^{d_1+d_2},$$
$$\mathbb{W}_2^{\psi_B}(\mu,\nu) := \inf_{\pi \in \mathcal{C}(\mu,\nu)} \left\{\int_{\mathbb{R}^{d_1+d_2} \times \mathbb{R}^{d_1+d_2}} \psi_B^2 \mathrm{d}\pi\right\}^{\frac{1}{2}}, \quad \mu,\nu \in \mathcal{P}_2(\mathbb{R}^{d_1+d_2}).$$
We assume

$(A^{5.3})$ $V(x,\mu)$ is differentiable in x such that $\nabla V(\cdot,\mu)(x)$ is Lipschitz continuous in $(x,\mu) \in \mathbb{R}^{d_1} \times \mathcal{P}_2(\mathbb{R}^{d_1+d_2})$. Moreover, there exist constants $\theta_1, \theta_2 \in \mathbb{R}$ with
$$\theta_1 + \theta_2 < \beta, \tag{5.3.46}$$
such that for any $(x,y), (x',y') \in \mathbb{R}^{d_1+d_2}$ and $\mu, \mu' \in \mathcal{P}_2(\mathbb{R}^{d_1+d_2})$,
$$\langle BB^*\{\nabla V(\cdot,\mu)(x) - \nabla V(\cdot,\mu')(x')\}, x - x' + (1+\beta)B(y-y')\rangle$$
$$\geq -\theta_1 \psi_B((x,y),(x',y'))^2 - \theta_2 \mathbb{W}_2^{\psi_B}(\mu,\mu')^2. \tag{5.3.47}$$

Obviously, $(A^{5.3})$ implies $(A^{3.1})(1)$ with $k = 2$ for $d = m = d_1 + d_2$, $\sigma = \mathrm{diag}\{0, \sqrt{2}I_{d_2}\}$, and
$$b((x,y),\mu) = \big(By, -B^*\nabla V(\cdot,\mu)(x) - \beta B^*(BB^*)^{-1}x - y\big).$$
So, by Theorem 3.3.1, (5.3.45) is well-posed for distributions in $\mathcal{P}_2(\mathbb{R}^{d_1+d_2})$ and
$$\sup_{t\in[0,T]} (P_t^*\mu)(|\cdot|^2) < \infty, \quad \mu \in \mathcal{P}_2(\mathbb{R}^{d_1+d_2}), T \in (0,\infty). \tag{5.3.48}$$
Let $P_t^*\mu = \mathcal{L}_{(X_t,Y_t)}$ for the solution with initial distribution $\mu \in \mathcal{P}_2(\mathbb{R}^{d_1+d_2})$. In this case, (5.3.45) becomes
$$\begin{cases} \mathrm{d}X_t = BY_t\mathrm{d}t, \\ \mathrm{d}Y_t = \sqrt{2}\mathrm{d}W_t + Z_t(X_t,Y_t)\mathrm{d}t, \end{cases}$$
where $Z_t(x,y) := -B^*\{\nabla V(\cdot, P_t^*\mu)\}(x) + \beta B^*(BB^*)^{-1}x + y$. According to Theorems 2.4 and 3.1 in [Wang (2014b)], when $\mathrm{Hess}_V(\cdot, P_t^*\mu)$ is bounded,
$$\rho_t(z) := \frac{(P_t^*\mu)(\mathrm{d}z)}{\mathrm{d}z} = \frac{\mathrm{d}(\mathcal{L}_{(X_t,Y_t)})(\mathrm{d}z)}{\mathrm{d}z}$$
exists and is differentiable in $z \in \mathbb{R}^{d_1+d_2}$. Moreover, since $(A^{5.3})$ implies that the class
$$\{\partial_{y_j}, [\partial_{y_j}, (Dy)_i \partial_{x_i}] : 1 \leq i \leq d_1, 1 \leq j \leq d_2\}$$
spans the tangent space at any point (i.e. the Hörmander condition of rank 1 holds), according to the Hörmander theorem, $\rho_t \in C^\infty(\mathbb{R}^{d_1+d_2})$ for $t > 0$ provided $Z_t \in C^\infty(\mathbb{R}^{d_1+d_2})$ for $t \geq 0$.

Theorem 5.3.4. *Assume $(A^{5.3})$. Then P_t^* has a unique invariant probability measure $\bar{\mu}$ such that for any $t > 0$ and $\mu \in \mathcal{P}_2(\mathbb{R}^{d_1+d_2})$,*
$$\max\big\{\mathbb{W}_2(P_t^*\mu,\bar{\mu})^2, \mathrm{Ent}(P_t^*\mu|\bar{\mu})\big\}$$
$$\leq \frac{ce^{-2\kappa t}}{(1\wedge t)^3} \min\big\{\mathrm{Ent}(\mu|\bar{\mu}), \mathbb{W}_2(\mu,\bar{\mu})^2\big\} \tag{5.3.49}$$
holds for some constant $c > 0$ and
$$\kappa := \frac{2(\beta - \theta_1 - \theta_2)}{2 + 2\beta + \beta^2 + \sqrt{\beta^4 + 4}} > 0. \tag{5.3.50}$$

Proof. (a) We first prove the exponential convergence of P_t^* in $\mathbb{W}_2$: there exists a constant $c_1 > 0$ such that

$$\mathbb{W}_2(P_t^*\mu, P_t^*\nu)^2 \leq c_1 e^{-\kappa t}\mathbb{W}_2(\mu,\nu)^2, \quad t \geq 0, \mu, \nu \in \mathcal{P}_2(\mathbb{R}^{d_1+d_2}). \quad (5.3.51)$$

By Theorem 5.1.1, this and (5.3.48) imply that P_t^* has a unique invariant probability measure $\mu \in \mathcal{P}_2(\mathbb{R}^{d_1+d_2})$.

Let

$$a := \Big(\frac{1+\beta+\beta^2}{1+\beta}\Big)^{\frac{1}{2}}, \quad r := \frac{1}{\sqrt{(1+\beta)(1+\beta+\beta^2)}} \in (0,1), \quad (5.3.52)$$

and consider the distance

$$\begin{aligned}\bar{\psi}_B((x,y),(\bar{x},\bar{y})) \\ := \sqrt{a^2|x-\bar{x}|^2 + |B(y-\bar{y})|^2 + 2ra\langle x-\bar{x}, B(y-\bar{y})\rangle}\end{aligned} \quad (5.3.53)$$

for $(x,y),(\bar{x},\bar{y}) \in \mathbb{R}^{d_1+d_2}$. Then there exists a constant $C > 1$ such that

$$C^{-1}|(x-\bar{x},y-\bar{y})| \leq \bar{\psi}_B((x,y),(\bar{x},\bar{y})) \leq C|(x-\bar{x},y-\bar{y})|. \quad (5.3.54)$$

Moreover, we claim that

$$\bar{\psi}_B((x,y),(\bar{x},\bar{y}))^2 \leq \frac{2+2\beta+\beta^2+\sqrt{\beta^4+4}}{2(1+\beta)}\bar{\psi}_B((x,y),(\bar{x},\bar{y}))^2. \quad (5.3.55)$$

Indeed, by (5.3.52) and (5.3.53), for any $\varepsilon > 0$ we have

$$\begin{aligned}\bar{\psi}_B((x,y),(\bar{x},\bar{y}))^2 \\ \leq a^2(1+\varepsilon)|x-\bar{x}|^2 + \Big(1+\frac{1}{\varepsilon(1+\beta)(1+\beta+\beta^2)}\Big)|B(y-\bar{y})|^2.\end{aligned} \quad (5.3.56)$$

Obviously, by (5.3.52),

$$\varepsilon := \frac{1-a^2+\sqrt{(a^2-1)^2+4a^2(1+\beta)^{-1}(1+\beta+\beta^2)^{-1}}}{2a^2} = \frac{\sqrt{\beta^4+4}-\beta^2}{2(1+\beta+\beta^2)}$$

satisfies

$$a^2(1+\varepsilon) = 1 + \frac{1}{\varepsilon(1+\beta)(1+\beta+\beta^2)} = \frac{2+2\beta+\beta^2+\sqrt{\beta^4+4}}{2(1+\beta)}.$$

Thus, (5.3.55) follows from (5.3.56).

Now, let (X_t, Y_t) and $(\bar{X}_t, \bar{Y}_t)$ solve (5.3.45) with $\mathcal{L}_{(X_0,Y_0)} = \mu$, $\mathcal{L}_{(\bar{X}_0,\bar{Y}_0)} = \nu$ such that

$$\mathbb{W}_2(\mu,\nu)^2 = \mathbb{E}|(X_0-\bar{X}_0, Y_0-\bar{Y}_0)|^2. \quad (5.3.57)$$

Simply denote $\mu_t = \mathcal{L}_{(X_t,Y_t)}, \bar{\mu}_t = \mathcal{L}_{(\bar{X}_t,\bar{Y}_t)}$. By $(A^{5.3})$ and Itô's formula, and noting that (5.3.52) implies

$$a^2 - \beta - ra = 0, \quad 1 - ra = ra\beta = \frac{\beta}{1+\beta},$$

we obtain

$$\frac{1}{2}\frac{\mathrm{d}}{\mathrm{d}t}\{\bar{\psi}_B((X_t,Y_t),(\bar{X}_t,\bar{Y}_t))^2\}$$
$$= \langle a^2(X_t - \bar{X}_t) + raB(Y_t - \bar{Y}_t), B(Y_t - \bar{Y}_t)\rangle$$
$$+ \langle B^*B(Y_t - \bar{Y}_t) + raB^*(X_t - \bar{X}_t),\ \beta B^*(BB^*)^{-1}(\bar{X}_t - X_t) + \bar{Y}_t - Y_t\rangle$$
$$+ \langle B^*B(Y_t - \bar{Y}_t) + raB^*(X_t - \bar{X}_t),\ B^*\{\nabla V(\bar{X}_t,\bar{\mu}_t) - \nabla V(X_t,\mu_t)\}\rangle$$
$$\leq -(1-ra)|B(Y_t - \bar{Y}_t)|^2$$
$$+ (a^2 - \beta - ra)\langle X_t - \bar{X}_t, B(Y_t - \bar{Y}_t)\rangle - ra\beta|X_t - \bar{X}_t|^2$$
$$+ \left\langle B^*B(Y_t - \bar{Y}_t) + \frac{B^*(X_t - \bar{X}_t)}{1+\beta},\ B^*\{\nabla V(\bar{X}_t,\bar{\mu}_t) - \nabla V(X_t,\mu_t)\}\right\rangle$$
$$\leq \frac{\theta_2}{1+\beta}\mathbb{W}_2^{\psi_B}(\mu_t,\bar{\mu}_t)^2 - \frac{\beta - \theta_1}{1+\beta}\psi_B((X_t,Y_t),(\bar{X}_t,\bar{Y}_t))^2.$$

By (5.3.55) and the fact that

$$\mathbb{W}_2^{\psi_B}(\mu_t,\bar{\mu}_t)^2 \leq \mathbb{E}[\psi_B((X_t,Y_t),(\bar{X}_t,\bar{Y}_t))^2],$$

for $\kappa > 0$ in (5.3.50), we obtain

$$\frac{1}{2}\{\mathbb{E}[\bar{\psi}_B((X_t,Y_t),(\bar{X}_t,\bar{Y}_t))^2] - \mathbb{E}[\bar{\psi}_B((X_s,Y_s),(\bar{X}_s,\bar{Y}_s))^2]\}$$
$$\leq -\frac{\beta - \theta_1 - \theta_2}{1+\beta}\int_s^t \mathbb{E}[\psi_B((X_r,Y_r),(\bar{X}_r,\bar{Y}_r))^2]\mathrm{d}r$$
$$\leq -\kappa \int_s^t \mathbb{E}[\psi_B((X_r,Y_r),(\bar{X}_r,\bar{Y}_r))^2]\mathrm{d}r,\quad t \geq s \geq 0.$$

Therefore, Gronwall's inequality implies

$$\mathbb{E}[\bar{\psi}_B((X_t,Y_t),(\bar{X}_t,\bar{Y}_t))^2] \leq \mathrm{e}^{-2\kappa t}\mathbb{E}[\bar{\psi}_B((X_0,Y_0),(\bar{X}_0,\bar{Y}_0))^2],\quad t \geq 0.$$

Combining this with (5.3.54) and (5.3.57), we derive (5.3.51) for some constant $c > 0$.

(b) By Theorem 5.3.1, (a) and the log-Harnack inequality in Theorem 4.1.2, we only need to verify the Talagrand inequality. As shown in the beginning of Section 3 in [Grothause and Wang (2019)] that $\bar{\mu}$ has the representation

$$\bar{\mu}(\mathrm{d}x,\mathrm{d}y) = Z^{-1}\mathrm{e}^{\bar{V}(x,y)}\mathrm{d}x\mathrm{d}y,\quad \bar{V}(x,y) := V(x,\bar{\mu}) + \frac{\beta}{2}|(BB^*)^{-\frac{1}{2}}x|^2 + \frac{1}{2}|y|^2,$$

where $Z := \int_{\mathbb{R}^{d_1+d_2}} \mathrm{e}^{-\bar{V}(x,y)}\mathrm{d}x\mathrm{d}y$ is the normalization constant. Since (5.3.47) implies

$$BB^*\mathrm{Hess}_{V(\cdot,\bar{\mu})} \geq -\theta_1 I_{d_1},$$

we deduce from (5.3.46) that
$$\text{Hess}_{\bar{V}} \geq \gamma I_{d_1+d_2}, \quad \gamma := 1 \wedge \frac{\beta - \theta_1}{\|B\|^2} > 0.$$
So, by the Bakry-Emery criterion [Bakry and Emery (1984)], we have the log-Sobolev inequality
$$\bar{\mu}(f^2 \log f^2) \leq \frac{2}{\gamma} \bar{\mu}(|\nabla f|^2), \quad f \in C_b^1(\mathbb{R}^{d_1+d_2}), \bar{\mu}(f^2) = 1.$$
According to [Bobkov et al. (2001)], this implies the Talagrand inequality
$$\mathbb{W}_2(\mu, \bar{\mu})^2 \leq \frac{2}{\gamma} \text{Ent}(\mu|\bar{\mu}).$$
Then the proof is finished. □

Example 5.2.2 (Degenerate granular media equation). Let $m \in \mathbb{N}$ and $W \in C^\infty(\mathbb{R}^m \times \mathbb{R}^{2m})$. Consider the following PDE for probability density functions $(\rho_t)_{t \geq 0}$ on $\mathbb{R}^{2m}$:
$$\begin{aligned}\partial_t \rho_t(x,y) &= \Delta_y \rho_t(x,y) - \langle \nabla_x \rho_t(x,y), y \rangle \\ &\quad + \langle \nabla_y \rho_t(x,y), \nabla_x(W \circledast \rho_t)(x) + \beta x + y \rangle,\end{aligned} \quad (5.3.58)$$
where $\beta > 0$ is a constant, $\Delta_y, \nabla_x, \nabla_y$ stand for the Laplacian in y and the gradient operators in x, y respectively, and $W \circledast \rho_t$ is in (5.3.10). If there exists a constant $\theta \in \left(0, \frac{2\beta}{1+3\sqrt{2+2\beta+\beta^2}}\right)$ such that
$$\begin{aligned}&|\nabla W(\cdot, z)(x) - \nabla W(\cdot, \bar{z})(\bar{x})| \\ &\leq \theta(|x - \bar{x}| + |z - \bar{z}|), \quad x, \bar{x} \in \mathbb{R}^m, z, \bar{z} \in \mathbb{R}^{2m},\end{aligned} \quad (5.3.59)$$
then there exists a unique probability measure $\bar{\mu} \in \mathcal{P}_2(\mathbb{R}^{2m})$ and a constant $c > 0$ such that for any probability density functions $(\rho_t)_{t \geq 0}$ solving (5.3.58), $\mu_t(dx) := \rho_t(x)dx$ satisfies
$$\begin{aligned}&\max\left\{\mathbb{W}_2(\mu_t, \bar{\mu})^2, \text{Ent}(\mu_t|\bar{\mu})\right\} \\ &\leq c e^{-\kappa t} \min\left\{\mathbb{W}_2(\mu_0, \bar{\mu})^2, \text{Ent}(\mu_0|\bar{\mu})\right\}, \quad t \geq 1,\end{aligned} \quad (5.3.60)$$
which holds for $\kappa = \frac{2\beta - \theta(1+3\sqrt{2+2\beta+\beta^2})}{2+2\beta+\beta^2+\sqrt{\beta^4+4}} > 0$.

Proof. Let $d_1 = d_2 = m$ and (X_t, Y_t) solve (5.3.45) for
$$B := I_m, \quad V(x, \mu) := \int_{\mathbb{R}^{2m}} W(x, z)\mu(dz). \quad (5.3.61)$$
We first observe that ρ_t solves (5.3.58) if and only if $\rho_t(z) = \frac{d(P_t^*\mu)(dz)}{dz}$ for $\mu(dz) = \rho_0(z)dz$, where $P_t^*\mu := \mathcal{L}_{(X_t, Y_t)}$.

Firstly, let $\rho_t(z) = \frac{\mathcal{L}_{(X_t,Y_t)}(\mathrm{d}z)}{\mathrm{d}z}$ which exists and is smooth as explained before Theorem 5.6.3. By Itô's formula and the integration by parts formula, for any $f \in C_0^2(\mathbb{R}^{2m})$ we have

$$\frac{\mathrm{d}}{\mathrm{d}t} \int_{\mathbb{R}^{2m}} (\rho_t f)(z) \mathrm{d}z = \frac{\mathrm{d}}{\mathrm{d}t} \mathbb{E}[f(X_t, Y_t)]$$
$$= \int_{\mathbb{R}^{2m}} \rho_t(x,y) \Big\{ \Delta_y f(x,y) + \langle \nabla_x f(x,y), y \rangle$$
$$\quad - \langle \nabla_y f(x,y), \nabla_x V(x, \rho_t(z)\mathrm{d}z) + \beta x + y \rangle \Big\} \mathrm{d}x\mathrm{d}y$$
$$= \int_{\mathbb{R}^{2m}} f(x,y) \Big\{ \Delta_y \rho_t(x,y) - \langle \nabla_x \rho_t(x,y), y \rangle$$
$$\quad + \langle \nabla_y \rho_t(x,y), \nabla_x \mu_t(W(x,\cdot)) + \beta x + y \rangle \Big\} \mathrm{d}x\mathrm{d}y.$$

Then ρ_t solves (5.3.58).

On the other hand, let ρ_t solve (5.3.58) with $\mu_0(\mathrm{d}z) := \rho_0(z)\mathrm{d}z \in \mathcal{P}_2(\mathbb{R}^{2m})$. By the integration by parts formula, $\mu_t(\mathrm{d}z) := \rho_t(z)\mathrm{d}z$ solves the nonlinear Fokker-Planck equation

$$\partial_t \mu_t = L_{\mu_t}^* \mu_t$$

in the sense that for any $f \in C_0^\infty(\mathbb{R}^{d_1+d_2})$ we have

$$\mu_t(f) = \mu_0(f) + \int_0^t \mu_s(L_{\mu_s} f) \mathrm{d}s, \quad t \geq 0,$$

where $L_\mu := \Delta_y + y \cdot \nabla_x - \{\nabla_x \mu(W(x,\cdot)) + \beta x - y\} \cdot \nabla_y$. By the superposition principle, see Section 2 in [Barbu and Röckner (2020)], we have $\mu_t = P_t^* \mu$.

Now, as explained in the proof of Example 5.2.1, by Theorem 5.6.3 we only need to verify $(A^{5.3})$ for B, V in (5.3.61) and

$$\theta_1 = \theta\Big(\frac{1}{2} + \sqrt{2 + 2\beta + \beta^2}\Big), \quad \theta_2 = \frac{\theta}{2}\sqrt{2 + 2\beta + \beta^2}, \tag{5.3.62}$$

so that the desired assertion holds for

$$\kappa := \frac{2(\beta - \theta_1 - \theta_2)}{2 + 2\beta + \beta^2 + \sqrt{\beta^4 + 4}} = \frac{2\beta - \theta(1 + 3\sqrt{2 + 2\beta + \beta^2})}{2 + 2\beta + \beta^2 + \sqrt{\beta^4 + 4}}.$$

By (5.3.59) and $V(x, \mu) := \mu(W(x, \cdot))$, for any constants $\alpha_1, \alpha_2, \alpha_3 > 0$ we have

$$I := \langle \nabla V(\cdot, \mu)(x) - \nabla V(\cdot, \bar{\mu})(\bar{x}), x - \bar{x} + (1+\beta)(y - \bar{y}) \rangle$$
$$= \int_{\mathbb{R}^{2m}} \langle \nabla W(\cdot, z)(x) - \nabla W(\cdot, z)(\bar{x}), x - \bar{x} + (1+\beta)(y - \bar{y}) \rangle \mu(\mathrm{d}z)$$
$$+ \langle \mu(\nabla_{\bar{x}} W(\bar{x}, \cdot)) - \bar{\mu}(\nabla_{\bar{x}} W(x, \cdot)), x - \bar{x} + (1+\beta)(y - \bar{y}) \rangle$$
$$\geq -\theta\{|x - \bar{x}| + \mathbb{W}_1(\mu, \bar{\mu})\} \cdot (|x - \bar{x}| + (1+\beta)|y - \bar{y}|)$$
$$\geq -\theta(\alpha_2 + \alpha_3)\mathbb{W}_2(\mu, \bar{\mu})^2$$
$$- \theta\left\{\left(1 + \alpha_1 + \frac{1}{4\alpha_2}\right)|x - \bar{x}|^2 + (1+\beta)^2\left(\frac{1}{4\alpha_1} + \frac{1}{4\alpha_3}\right)|y - \bar{y}|^2\right\}.$$

Take
$$\alpha_1 = \frac{\sqrt{2 + 2\beta + \beta^2} - 1}{2}, \quad \alpha_2 = \frac{1}{2\sqrt{2 + 2\beta + \beta^2}}, \quad \alpha_3 = \frac{(1+\beta)^2}{2\sqrt{2 + 2\beta + \beta^2}}.$$

We have
$$1 + \alpha_1 + \frac{1}{4\alpha_2} = \frac{1}{2} + \sqrt{2 + 2\beta + \beta^2},$$
$$(1+\beta)^2 \left(\frac{1}{4\alpha_1} + \frac{1}{4\alpha_3}\right) = \frac{1}{2} + \sqrt{2 + 2\beta + \beta^2},$$
$$\alpha_2 + \alpha_3 = \frac{1}{2}\sqrt{2 + 2\beta + \beta^2}.$$

Therefore,
$$I \geq -\frac{\theta}{2}\sqrt{2 + 2\beta + \beta^2}\mathbb{W}_2(\mu, \bar{\mu})^2 - \theta\left(\frac{1}{2} + \sqrt{2 + 2\beta + \beta^2}\right)|(x, y) - (\bar{x}, \bar{y})|^2,$$

i.e. ($A^{5.3}$) holds for B and V in (5.3.61) where $B = I_m$ implies that ψ_B is the Euclidean distance on $\mathbb{R}^{2m}$, and for θ_1, θ_2 in (5.3.62). □

5.4 Exponential ergodicity: Non-dissipative case

For any $t \geq 0$ and $\mu \in \mathcal{P}$, consider the second order differential operator

$$L_{t,\mu} := \frac{1}{2}\mathrm{tr}\{\sigma_t \sigma_t^* \nabla^2\} + b_t(\cdot, \mu) \cdot \nabla. \tag{5.4.1}$$

For any positive measurable function V on $\mathbb{R}^d$, let
$$\mathcal{P}_V := \{\mu \in \mathcal{P} : \mu(V) < \infty\}.$$
We assume the following Lyapunov condition.

$(A^{5.4})$ **(Lyapunov)** *There exists a function* $0 \leq V \in C^2(V)$ *with* $\lim_{|x|\to\infty} V(x) = \infty$ *and*

$$\sup_{t\geq 0,\, x\in\mathbb{R}^d} \frac{|\sigma_t(x)\nabla V(x)|}{1+V(x)} < \infty, \qquad (5.4.2)$$

such that for some $K_0, K_1 \in L^1_{loc}([0,\infty);\mathbb{R})$,

$$L_{t,\mu}V \leq K_0(t) - K_1(t)V, \quad t \geq 0, \mu \in \mathcal{P}_V. \qquad (5.4.3)$$

For any $l > 0$, consider the class

$$\Psi_l := \{\psi \in C^2([0,l];[0,\infty)) : \psi(0) = \psi'(l) = 0, \psi'|_{[0,l)} > 0\}.$$

For each $\psi \in \Psi_l$, we extend it to the half line by setting $\psi(r) = \psi(r \wedge l)$, so that ψ' is nonnegative and Lipschitz continuous with compact support. Then

$$c_\psi := \sup_{r>0} \frac{r\psi'(r)}{\psi(r)} < \infty. \qquad (5.4.4)$$

When $\psi'' \leq 0$, we have $\|\psi'\|_\infty := \sup |\psi'| = \psi'(0)$ and $c_\psi = \lim_{r\downarrow 0} \frac{r\psi'(r)}{\psi(r)} = 1$.

For any constant $\beta > 0$, the weighted Wasserstein distance is given by

$$\mathbb{W}_{\psi,\beta V}(\mu,\nu) := \inf_{\pi\in\mathcal{C}(\mu,\nu)} \int_{\mathbb{R}^d\times\mathbb{R}^d} \psi(|x-y|)\big(1 + \beta V(x) + \beta V(y)\big)\pi(\mathrm{d}x,\mathrm{d}y)$$

for $\mu,\nu \in \mathcal{P}_V$. In general, $\mathbb{W}_{\psi,\beta V}$ is only a quasi-distance on $\mathcal{P}_V$ as the triangle inequality may not hold. But it is complete in the sense that any $\mathbb{W}_{\psi,\beta V}$-Cauchy sequence in $\mathcal{P}_V$ is convergent. For any $\mu,\nu \in \mathcal{P}_V$, we introduce

$$\hat{\mathbb{W}}_{\psi,\beta V}(\mu,\nu) := \inf_{\pi\in\mathcal{C}(\mu,\nu)} \frac{\int_{\mathbb{R}^d\times\mathbb{R}^d} \psi(|x-y|)(1+\beta V(x)+\beta V(y))\pi(\mathrm{d}x,\mathrm{d}y)}{\int_{\mathbb{R}^d\times\mathbb{R}^d} \psi'(|x-y|)(1+\beta V(x)+\beta V(y))\pi(\mathrm{d}x,\mathrm{d}y)}, \qquad (5.4.5)$$

which will come naturally from Itô's formula for the process

$$\psi(|X_t - Y_t|)(1 + \beta V(X_t) + \beta V(Y_t))$$

for a coupling (X_t, Y_t) of the SDE. We observe that

$$\sup_{\pi\in\mathcal{C}(\mu,\nu)} \int_{\mathbb{R}^d\times\mathbb{R}^d} \psi'(|x-y|)(1+\beta V(x)+\beta V(y))\pi(\mathrm{d}x,\mathrm{d}y) \leq 1+\beta\mu(V)+\beta\nu(V),$$

so that $\hat{\mathbb{W}}_{\psi,\beta V} \geq \frac{\mathbb{W}_{\psi,\beta V}(\mu,\nu)}{1+\beta\mu(V)+\beta\nu(V)}$.

Moreover, let $\|\nabla f\|_\infty$ be the Lipschitz constant of a real function f on $\mathbb{R}^d$. We need the following non-degenerate and monotone conditions.

$(A^{5.5})$ **(Non-degeneracy)** *There exist* $\alpha \in L^1_{loc}([0,\infty);(0,\infty))$ *and measurable*

$$\hat{\sigma}:[0,\infty)\times\mathbb{R}^d\to\mathbb{R}^d\otimes\mathbb{R}^d$$

with $\int_0^T \|\nabla\hat{\sigma}_t\|_\infty dt < \infty$ *for* $T \in (0,\infty)$, *such that*

$$a_t(x) := (\sigma_t\sigma_t^*)(x) = \alpha_t I_d + (\hat{\sigma}_t\hat{\sigma}_t^*)(x), \quad t \geq 0, x \in \mathbb{R}^d. \quad (5.4.6)$$

$(A^{5.6})$ **(Monotonicity)** b *is bounded on bounded set in* $[0,\infty)\times\mathbb{R}^d\times\mathcal{P}_V$. *Moreover, there exist* $l > 0$, $K, \theta, q_l \in L^1_{loc}([0,\infty);[0,\infty))$ *and* $\psi \in \Psi_l$, *such that*

$$2\alpha_t\psi''(r) + K_t\psi'(r) \leq -q_l(t)\psi(r), \quad r \in [0,l], t \geq 0, \quad (5.4.7)$$

$$\langle b_t(x,\mu) - b_t(y,\nu), x-y\rangle + \frac{1}{2}\|\hat{\sigma}_t(x) - \hat{\sigma}_t(y)\|^2_{HS}$$
$$\leq K_t|x-y|^2 + \theta_t|x-y|\tilde{\mathbb{W}}_{\psi,\beta V}(\mu,\nu), \quad x,y \in \mathbb{R}^d, \mu,\nu \in \mathcal{P}_V, t \geq 0. \quad (5.4.8)$$

Remark 5.3.1. (1) Since $V \geq 0$ with $V(x) \to \infty$ as $|x| \to \infty$, we have

$$\kappa_{l,\beta}(t) := \inf_{|x-y|>l}\frac{K_1(t)V(x) + K_1(t)V(y) - 2K_0(t)}{\beta^{-1} + V(x) + V(y)} \in \mathbb{R}, \quad l > 0, \quad (5.4.9)$$

and $\kappa_{l,\beta}(t) > 0$ for large enough $l > 0$ and $K_1(t) > 0$.

(2) Consider the one-dimensional differential operator $L = 2\lambda\frac{d^2}{dr^2} + K\frac{d}{dr}$ on $[0,l]$. In (5.4.7) one may take ψ to be the first eigenfunction of L with Dirichlet boundary at 0 and Neumann boundary at l. In this case, $q_l > 0$ is the first mixed eigenvalue.

(3) (5.4.2) and $(A^{5.5})$ imply that

$$\alpha_{l,\beta}(t): c_\psi \sup_{|x-y|\in(0,l)}\left\{\alpha_t\frac{|\nabla V(x) - \nabla V(y)|}{|x-y|\{\beta^{-1} + V(x) + V(y)\}}\right.$$
$$\left. + \frac{|\{\hat{\sigma}_t(x) - \hat{\sigma}_t(y)\}[(\hat{\sigma}_t(\cdot)^*\nabla V)(x) + (\hat{\sigma}_t(\cdot)^*\nabla V)(y)]|}{|x-y|\{\beta^{-1} + V(x) + V(y)\}}\right\} < \infty \quad (5.4.10)$$

for any $\beta, l > 0$. In many cases, we have $\alpha_{l,\beta} \downarrow 0$ as $\beta \downarrow 0$. For instance, it is the case when $V(x) = e^{|x|^p}$ for $p \in (0,1)$ and large $|x|$, and $\hat{\sigma}$ is Lipschitz continuous with $\|\hat{\sigma}(x)\| \leq c(1 + |x|^q)$ for some constants $c > 0$ and $q \in (0, 1-p)$, or $V(x) = |x|^k$ for some $k > 0$ and large $|x|$.

For K_0, q_l, $\kappa_{l,\beta}$ and $\alpha_{l,\beta}$ given in $(A^{5.4})$, $(A^{5.6})$, (5.4.9) and (5.4.10) respectively, let

$$\lambda_{l,\beta}(t) := \min\{\kappa_{l,\beta}(t),\ q_l(t) - 2K_0(t)\beta - \alpha_{l,\beta}(t)\}. \tag{5.4.11}$$

Since $\alpha_{l,\beta}(t) \to 0$ as $\beta \to 0$, and since $\kappa_{l,\beta}(t) > 0$ for $K_1(t) > 0$ and large $l > 0$, when $K_1(t) > 0$ we may take large $l > 0$ and small $\beta > 0$ such that $\lambda_{l,\beta}(t) > 0$. The main result in this part is the following.

Theorem 5.4.1. *Assume* $(A^{5.4})$–$(A^{5.6})$, *with* $\psi'' \leq 0$ *when* $\hat{\sigma}_t(\cdot)$ *is non-constant for some* $t \geq 0$. *Then the SDE* (5.0.1) *is well-posed in* $\mathcal{P}_V$, *and* $P_t^* := P_{0,t}^*$ *satisfies*

$$\mathbb{W}_{\psi,\beta V}(P_t^*\mu, P_t^*\nu) \leq \mathrm{e}^{-\int_0^t \{\lambda_{l,\beta}(s) - \theta_s\}\mathrm{d}s} \mathbb{W}_{\psi,\beta V}(\mu,\nu), \tag{5.4.12}$$
$$t \geq 0, \mu, \nu \in \mathcal{P}_V.$$

Consequently, if (a,b) *does not depend on* t *and* $\lambda_{l,\beta} > \theta$, *then* P_t^* *has a unique invariant probability measure* $\bar{\mu} \in \mathcal{P}_V$ *such that*

$$\mathbb{W}_{\psi,\beta V}(P_t^*\mu,\bar{\mu}) \leq \mathrm{e}^{-(\lambda_{l,\beta}-\theta)t}\mathbb{W}_{\psi,\beta V}(\mu,\bar{\mu}),\quad t \geq 0, \mu \in \mathcal{P}_V. \tag{5.4.13}$$

Example 5.3.1. Let $a = I_d + \hat{\sigma}\hat{\sigma}^*$ for some Lipschitz continuous matrix valued function $\hat{\sigma}$, $V(x) = \mathrm{e}^{(1+|x|^2)^{p/2}}$ for some $p \in (0,1]$, and

$$b(x,\mu) := b_0(x) + \varepsilon \Phi(x, \log \mu(V))$$

for some $\varepsilon \in [0,1)$, $b_0 \in C^1(\mathbb{R}^d)$ with $b_0(x) = -|x|^{-p}x$ for $|x| \geq 1$, and $\Phi \in C_b^1(\mathbb{R}^d \times [0,\infty); \mathbb{R}^d)$. Let

$$\tilde{\mathbb{W}}_V(\mu,\nu) := \inf_{\pi \in \mathcal{C}(\mu,\nu)} \int_{\mathbb{R}^d \times \mathbb{R}^d} \{1 \wedge |x-y|\} \cdot \{1 + V(x) + V(y)\}\pi(\mathrm{d}x,\mathrm{d}y).$$

Then when $\varepsilon > 0$ is small enough, P_t^* has a unique invariant probability measure $\bar{\mu} \in \mathcal{P}_V$, and there exist constants $c, q > 0$ such that

$$\tilde{\mathbb{W}}_V(P_t^*\mu,\bar{\mu}) \leq c\mathrm{e}^{-qt}\tilde{\mathbb{W}}_V(\mu,\bar{\mu}),\quad t \geq 0, \mu \in \mathcal{P}_V.$$

Proof. It is easy to see that $(A^{5.4})$ holds for some constants $K_0, K_1 > 0$, $(A^{5.5})$ holds for $\alpha = 1$. Since $V(x) \to \infty$ as $|x| \to \infty$, we take $l > 0$ such that

$$\inf_{|x-y|\geq l}\{K_1 V(x) + K_1 V(y) - 2K_0\} \geq 1.$$

So, in (5.4.9) the constant $\kappa_{l,\beta} > 0$ for all $\beta > 0$. Next, take $\psi \in \Psi_l$ such that (5.4.7) holds for some $q_l > 0$, for instance ψ is the first mixed eigenfunction of $2\frac{d^2}{dr^2} + K\frac{d}{dr}$ on $[0, l]$ with Dirichlet condition at 0 and Neumann condition at l. Then there exists a constant $c_0 > 0$ such that

$$|V(x) - V(y)| \le c_0\psi(|x - y|)(V(x) + V(y)), \quad x, y \in \mathbb{R}^d. \quad (5.4.14)$$

Next, since for any $\pi \in \mathcal{C}(\mu, \nu)$ we have

$$\int_{\mathbb{R}^d \times \mathbb{R}^d} \psi'(|x - y|)(1 + \beta V(x) + \beta V(y))\pi(dx, dy)$$

$$\le \|\psi'\|_\infty \int_{\{|x-y|\le l\}} \{1 + (1 + e)\beta[V(x) \wedge V(y)]\}\pi(dx, dy)$$

$$\le (2 + e)[\mu(V) \wedge \nu(V)], \quad \beta \in (0, 1],$$

(5.4.5) implies

$$\hat{W}_{\psi,\beta V}(\mu, \nu) \ge \frac{\mathbb{W}_{\psi,\beta V}(\mu, \nu)}{(2 + e)[\mu(V) \wedge \nu(V)]}, \quad \beta \in (0, 1].$$

Combining this with $\Phi \in C_b^1$ and noting that (5.4.14) implies

$$|\mu(V) - \nu(V)| \le \inf_{\pi \in \mathcal{C}(\mu,\nu)} \int_{\mathbb{R}^d \times \mathbb{R}^d} |V(x) - V(y)|\pi(dx, dy) \le c_0 \beta^{-1} \mathbb{W}_{\psi,\beta V}(\mu, \nu)$$

for some constant $c_0 > 0$, we find a constant $c_1 > 0$ such that

$$|b(x, \mu) - b(x, \nu)| \le \varepsilon \|\nabla\Phi(x, \cdot)\|_\infty |\log\mu(V) - \log\nu(V)|$$

$$\le \frac{\varepsilon \|\nabla\Phi(x, \cdot)\|_\infty |\mu(V) - \nu(V)|}{\mu(V) \wedge \nu(V)} \le c_1 \varepsilon \beta^{-1} \hat{\mathbb{W}}_{\psi,\beta V}(\mu, \nu), \quad \beta \in (0, 1].$$

Noting that $\|\nabla b_0\|_\infty + \|\nabla \Phi\|_\infty + \|\nabla \hat{\sigma}\|_\infty < \infty$, this implies $(A^{5.6})$ holds for some constant $K > 0$ and $\theta = c_1 \varepsilon \beta^{-1}, \beta \in (0, 1]$.

Finally, as observed in Remark 5.3.1(3) that for the present V we have $\alpha_{l,\beta} \downarrow 0$ as $\beta \downarrow 0$. Then in (5.4.11), $\lambda_{l,\beta} > 0$ for small $\beta \in (0, 1]$. Therefore, by Theorem 5.4.1, when $\varepsilon > 0$ is small enough, P_t^* has a unique invariant probability measure $\bar{\mu} \in \mathcal{P}_V$, such that

$$\mathbb{W}_{\psi,\beta V}(P_t^*\mu, \bar{\mu}) \le e^{-qt}\mathbb{W}_{\psi,\beta V}(\mu, \bar{\mu}), \quad t \ge 0$$

holds for some constant $q > 0$. This completes the proof since

$$C^{-1}\tilde{\mathbb{W}}_V \le \mathbb{W}_{\psi,\beta V} \le C\tilde{\mathbb{W}}_V$$

holds for some constant $C > 1$. □

5.4.1 Proof of Theorem 5.4.1

Since $\psi(r) := \psi(r \wedge l)$ for $\psi \in \Psi_l$ is not second order differentiable at l, we introduce the following lemma ensuring Itô's formula for ψ of a semi-martingale which will be used frequently in the sequence.

Lemma 5.4.2. *Let ξ_t be a nonnegative continuous semi-martingale satisfying*

$$d\xi_t \leq A_t dt + dM_t$$

for a local martingale M_t and an integrable adapted process A_t. Then for any $\psi \in C^1([0,\infty))$ with ψ' nonnegative and Lipschitz continuous, we have

$$d\psi(\xi_t) \leq \psi'(\xi_t) A_t dt + \frac{1}{2}\psi''(\xi_t) d\langle M \rangle_t + \psi'(\xi_t) dM_t,$$

where

$$\psi''(r) := \limsup_{s \downarrow r} \limsup_{\varepsilon \downarrow 0} \frac{\psi'(s+\varepsilon) - \psi'(s)}{\varepsilon}, \quad r \geq 0$$

is a bounded measurable function on $[0, \infty)$.

Proof. By choosing stopping times τ such that $\xi_{t \wedge \tau}$, $\int_0^{\wedge \tau} tA_s ds$ and $M_{t \wedge \tau}$ are bounded, we may and do assume that these processes with t replacing $t \wedge \tau$ are bounded.

For any $n \geq 1$, let

$$\psi_n(r) = n \int_0^\infty \psi(r+s) e^{-ns} ds, \quad r \geq 0.$$

Then each ψ_n is C^∞-smooth, with $\psi'_n \geq 0$, $(\psi_n, \psi'_n) \to (\psi, \psi')$ locally uniformly, $\{\|\psi''_n\|_\infty\}_{n \geq 1}$ uniformly bounded, and by Fatou's lemma,

$$\limsup_{n \to \infty} \psi''_n(r) \leq \limsup_{n \to \infty} \int_0^\infty \limsup_{\varepsilon \downarrow 0} \frac{\psi'(r+s+\varepsilon) - \psi'(r+s)}{\varepsilon} n e^{-ns} ds$$
$$\leq \limsup_{s \downarrow 0} \limsup_{\varepsilon \downarrow 0} \frac{\psi'(r+s+\varepsilon) - \psi'(r+s)}{\varepsilon} = \psi''(r), \quad r \geq 0.$$

Therefore, by applying Itô's formula to $\psi_n(\xi_t)$ and letting $n \to \infty$, we finish the proof. □

A. The well-posedness. For any $T > 0$ and a subspace $\hat{\mathcal{P}} \subset \mathcal{P}$, let $C_w([0,T]; \hat{\mathcal{P}})$ be the class of all continuous maps from $[0,T]$ to $\hat{\mathcal{P}}$ under the weak topology.

Lemma 5.4.3. *Let $L_{t,\mu}$ be in (5.4.1). Assume that for some $K \in L^1_{loc}([0, \infty); (0, \infty))$,*

$$L_{t,\mu} V(x) \leq K_t(1 + \mu(V) + V(x)), \quad t \geq 0, x \in \mathbb{R}^d, \mu \in \mathcal{P}_V, \quad (5.4.15)$$

$$\|\sigma_t \nabla V(x)\| \leq K_t(1 + V(x)), \quad t \geq 0, x \in \mathbb{R}^d, \quad (5.4.16)$$

$$2\langle b_t(x,\mu) - b_t(y,\nu), x - y \rangle^+ + \|\sigma_t(x) - \sigma_t(y)\|_{HS}^2$$
$$\leq K_t|x-y|\{|x-y| + \mathbb{W}_{\psi,V}(\mu,\nu)\}, \quad (5.4.17)$$
$$t \geq 0, x, y \in \mathbb{R}^d, \mu, \nu \in \mathcal{P}_V.$$

Then (5.0.1) is well-posed for distributions in $\mathcal{P}_V$ with

$$\mathbb{E}[V(X_t)] < e^{2\int_0^T K_s \, ds} \mathbb{E}\left[V(X_0) \int_0^T K_s e^{2\int_s^T K_r \, dr} \, ds\right]. \quad (5.4.18)$$

Proof. It is easy to see that (5.4.18) follows from (5.4.15) and Itô's formula. To prove the well-posedness for distributions in $\mathcal{P}_V$, we adopt a fixed point theorem in distributions. For any $T > 0$, $\gamma := \mathcal{L}_{X_0} \subset \mathcal{P}_V$, and

$$\mu \in \mathcal{P}_{T,V}^\gamma := \{\mu \in C_w([0,T]; \mathcal{P}_V) : \mu_0 = \gamma\},$$

consider the following SDE

$$dX_t^\mu = b_t(X_t^\mu, \mu_t) + \sigma_t(X_t^\mu) dW_t, \quad X_0^\mu = X_0, t \in [0, T]. \quad (5.4.19)$$

It is well known that the monotone condition (5.4.8) in $(A^{5.6})$ implies the well-posedness of this SDE up to life time, while the Lyapunov condition (5.4.15) implies

$$\sup_{t \in [0,T]} \mathbb{E}[V(X_t^\mu)] < \infty.$$

Then by the continuity of X_t^μ in t we conclude that

$$\Phi^\gamma \mu := \mathcal{L}_{X^\mu} \in C_w([0,T]; \mathcal{P}_V).$$

It remains to prove that Φ^γ has a unique fixed point $\bar{\mu} \in \mathcal{P}_{V,T}$, so that $X_t^{\bar{\mu}}$ is the unique solution of (5.0.1) up to time T, and by the modified Yamada-Watanabe principle Theorem 3.2.3, this also implies the weak well-posedness of (5.0.1) up to time T.

To prove the existence and uniqueness of the fixed point of H, for any $N \geq 1$ let
$$\mathcal{P}_{V,T}^{\gamma,N} := \Big\{\mu \in C_w([0,T]; \mathcal{P}_V) : \mu_0 = \gamma,\ \sup_{t\in[0,T]} e^{-Nt}\mu_t(V) \leq N(1+\gamma(V))\Big\}.$$
Then as $N \uparrow \infty$, we have $\mathcal{P}_{V,T}^{\gamma,N} \uparrow \mathcal{P}_{V,T}^{\gamma}$ as $N \uparrow \infty$. So, it suffices to find $N_0 \geq 1$ such that for any $N \geq N_0$, $\Phi^\gamma \mathcal{P}_{T,V}^{\gamma,N} \subset \mathcal{P}_{T,V}^{\gamma,N}$ and Φ^γ has a unique fixed point in $\mathcal{P}_{T,V}^{\gamma,N}$. We prove this in the following two steps.

(a) Construction of N_0. Let
$$c := e^{\int_0^T K_s ds},\quad N_0 := 3c.$$
By Itô's formula and (5.4.15), for any $N \geq N_0$ and $\mu \in \mathcal{P}_{T,V}^{\gamma,N}$, we have
$$e^{-Nt}\mathbb{E}[V(X_t^\mu)] \leq \gamma(V)e^{\int_0^t K_s ds - Nt}$$
$$+ \int_0^t K_s\big\{1 + N(1+\gamma(V))\big\}e^{\int_s^t K_r dr - N(t-s)}ds$$
$$\leq c\gamma(V) + 2cN(1+\gamma(V))\sup_{t\in[0,T]}\int_0^t e^{-N(t-s)}ds$$
$$\leq c\gamma(V) + 2c(1+\gamma(V)) \leq N(1+\gamma(V)).$$
So, $\Phi^\gamma \mathcal{P}_{T,V}^{\gamma,N} \subset \mathcal{P}_{T,V}^{\gamma,N}$ for $N \geq N_0$.

(b) Let $N \geq N_0$. It remains to prove that H is contractive in $\mathcal{P}_{T,V}^{\gamma,N}$ under
$$\mathbb{W}_{\psi,V,\lambda}(\mu,\nu) := \sup_{t\in[0,T]} e^{-\lambda t}\mathbb{W}_{\psi,V}(\mu_t,\nu_t),\quad \mu,\nu \in \mathcal{P}_{T,V}^{\gamma,N}$$
for large $\lambda > 0$.

For $\mu, \nu \in \mathcal{P}_{T,V}^{\gamma,N}$, by (5.4.17) and the Itô-Tanaka formula, we find $C_0 \in L_{loc}^1([0,\infty); (0,\infty))$ such that
$$d|X_t^\mu - X_t^\nu| \leq C_0(t)(\mathbb{W}_{\psi,\beta V}(\mu_t,\nu_t) + |X_t^\mu - X_t^\nu|)dt$$
$$+ \Big\langle \frac{X_t^\mu - X_t^\nu}{|X_t^\mu - X_t^\nu|}, \{\sigma_t(X_t^\mu) - \sigma_t(X_t^\nu)\}dW_t\Big\rangle.$$
Since $\psi \in \Psi_l$, by extending to the half-line with $\psi(r) := \psi(r \wedge l)$, we see that ψ' is nonnegative and Lipschitz continuous. By Lemma 5.4.2, $\mu,\nu \in \mathcal{P}_{V,T}^{\gamma,N}$, and noting that $\psi'' \leq 0$ when σ_t is non-constant for some $t \geq 0$, we find $C_1 \in L^1([0,T]; (0,\infty))$ such that
$$d\psi(|X_t^\mu - X_t^\nu|) \leq C_1(t)\big\{\psi(|X_t^\mu - X_t^\nu|) + \mathbb{W}_{\psi,\beta V}(\mu_t,\nu_t)\big\}dt$$
$$+ \psi'(|X_t^\mu - X_t^\nu|)\Big\langle \frac{X_t^\mu - X_t^\nu}{|X_t^\mu - X_t^\nu|}, \{\sigma_t(X_t^\mu) - \sigma_t(X_t^\nu)\}dW_t\Big\rangle \quad (5.4.20)$$
holds for $t \in [0,T]$.

On the other hand, by (5.4.15) and $\mu, \nu \in \mathcal{P}_{V,T}^{\gamma,N}$, we find a constant $K(N) > 1$ such that

$$\begin{aligned}\mathrm{d}\{V(X_t^\mu) + V(X_t^\nu)\} &\leq K_t\{1 + \mu_t(V) + \nu_t(V) + V(X_t^\mu) + V(X_t^\nu)\}\mathrm{d}t \\ &\quad + \langle \sigma_t(X_t^\mu)\nabla V(X_t^\mu) + \sigma_t(X_t^\nu)\nabla V(X_t^\nu), \mathrm{d}W_t\rangle \\ &\leq K(N)K_t\{1 + V(X_t^\mu) + V(X_t^\nu)\}\mathrm{d}t \\ &\quad + \langle \sigma_t(X_t^\mu)\nabla V(X_t^\mu) + \sigma_t(X_t^\nu)\nabla V(X_t^\nu), \mathrm{d}W_t\rangle.\end{aligned}$$

Combining this with (5.4.4), (5.4.16), and (5.4.20), we find $C_2 \in L^1([0,T]; (0,\infty))$ such that

$$\xi_t := \psi(|X_t^\mu - X_t^\nu|)\bigl(1 + V(X_t^\mu) + V(X_t^\nu)\bigr)$$

satisfies

$$\mathrm{d}\xi_t \leq C_2(t)\bigl[\xi_t + (1 + V(X_t^\mu) + V(X_t^\nu))\mathbb{W}_{\psi,V}(\mu_t, \nu_t)\bigr]\mathrm{d}t + \mathrm{d}M_t$$

for some local martingale M_t, $t \in [0,T]$. Since $\Phi^\gamma \mu, \Phi^\gamma \nu \in \mathcal{P}_{V,T}^{\gamma,N}$, we have

$$\mathbb{E}V(X_t^\mu) + \mathbb{E}V(X_t^\nu) < N(1 + \gamma(V))\mathrm{e}^{NT} =: D(N) < \infty, \quad t \in [0,T],$$

which together with $\xi_0 = 0$ yields that for any $t \in [0,T], \lambda > 0$,

$$\mathrm{e}^{-\lambda t}\mathbb{E}\xi_t \leq \bigl(1 + D(N)\bigr)\mathbb{W}_{\psi,V,\lambda}(\mu, \nu)\int_0^t C_2(s)\mathrm{e}^{\int_s^t (C_2(r) - \lambda)\mathrm{d}r}\mathrm{d}s.$$

Noting that $\lim_{\lambda \to \infty} \sup_{t \in [0,T]} \int_0^t C_2(s)\mathrm{e}^{\int_s^t (C_2(r) - \lambda)\mathrm{d}r}\mathrm{d}s = 0$, we conclude that when $\lambda > 0$ is large enough,

$$\mathrm{e}^{-\lambda t}\mathbb{W}_{\psi,V}\bigl(\Phi_t^\gamma \mu, \Phi_t^\gamma \nu\bigr) \leq \mathrm{e}^{-\lambda t}\mathbb{E}\xi_t \leq \frac{1}{2}\mathbb{W}_{\psi,V,\lambda}(\mu, \nu), \quad t \in [0,T].$$

Therefore, $\Phi^\gamma : \mathcal{P}_{V,T}^{\gamma,N} \to \mathcal{P}_{V,T}^{\gamma,N}$ is contractive in $\mathbb{W}_{\psi,T,\lambda}$ for large enough $\lambda > 0$. □

B. Construction of coupling. Simply denote

$$\psi_{\beta V}(x, y) := \psi(|x - y|)(1 + \beta V(x) + \beta V(y)), \quad x, y \in \mathbb{R}^d.$$

For $s \geq 0$ and $\mu, \nu \in \mathcal{P}_V$, let X_s and Y_s be $\mathcal{F}_s$-measurable random variables such that

$$\mathcal{L}_{X_s} = P_s^*\mu, \quad \mathcal{L}_{Y_s} = P_s^*\nu, \quad \mathbb{E}\psi_{\beta V}(X_s, Y_s) = \mathbb{W}_{\psi,\beta V}(P_s^*\mu, P_s^*\nu). \quad (5.4.21)$$

Let $W_t^{(1)}$ and $W_t^{(2)}$ be two independent d-dimensional Brownian motions and consider the following SDE:

$$\mathrm{d}X_t = b_t(X_t, P_t^*\mu)\mathrm{d}t + \sqrt{\alpha_t}\mathrm{d}W_t^{(1)} + \hat{\sigma}_t(X_t)\mathrm{d}W_t^{(2)}, \quad t \geq s. \quad (5.4.22)$$

By ($A^{5.6}$), this SDE is well-posed. Indeed, since b is locally bounded, by Girsanov's transform to the regular SDE
$$\mathrm{d}X_t = \sqrt{\alpha_t}\mathrm{d}W_t^{(1)} + \hat{\sigma}_t(X_t)\mathrm{d}W_t^{(2)}, \quad t \geq s$$
up to the exit time of a large ball, we construct a weak solution to (5.4.22) up to the same stopping time. On the other hand, the monotone condition in ($A^{5.6}$) implies the pathwise uniqueness of (5.4.22), then the well-posedness is implied by the Yamada-Watanabe principle. Moreover, the Lyapunov condition in ($A^{5.4}$) ensures the non-explosion. By ($A^{5.5}$), we have
$$\sigma_t^*(\sigma_t\sigma_t^*)^{-1}\{\alpha_t + \hat{\sigma}_t\hat{\sigma}_t^*\}(\sigma_t\sigma_t^*)^{-1}\sigma_t + \{I_m - \sigma_t^*(\sigma_t\sigma_t^*)^{-1}\sigma_t\}^2$$
$$= \sigma_t^*(\sigma_t\sigma_t^*)^{-1}\sigma_t + I_m - \sigma_t^*(\sigma_t\sigma_t^*)^{-1}\sigma_t = I_m.$$
So, for an m-dimensional Brownian motion $W^{(3)}$ independent of ($W^{(1)}$, $W^{(2)}$),
$$W_t := \int_0^t \{\sigma_s^*(\sigma_s\sigma_s^*)^{-1}\}(X_s)\{\sqrt{\alpha_s}\mathrm{d}W_s^{(1)}$$
$$+ \hat{\sigma}_s(X_s)\mathrm{d}W_s^{(2)}\} + \int_0^t \{I_m - \sigma_s^*(\sigma_s\sigma_s^*)^{-1}\sigma_s\}(X_s)\mathrm{d}W_s^{(3)}$$
is an m-dimensional Brownian motion such that
$$\sigma_t(X_t)\mathrm{d}W_t = \sqrt{\alpha_t}\mathrm{d}W_t^{(1)} + \hat{\sigma}_t(X_t)\mathrm{d}W_t^{(2)}.$$
So, the solution to (5.4.22) is a weak solution to (5.0.1), the weak uniqueness of (5.4.22) implies that $\mathcal{L}_{X_t} = P_t^*\mu, t \geq s$.

To construct the coupling with reflection, let
$$u(x,y) = \frac{x-y}{|x-y|}, \quad x \neq y \in \mathbb{R}^d.$$
We consider the SDE:
$$\begin{aligned}\mathrm{d}Y_t = &\, b_t(Y_t, P_t^*\nu)\mathrm{d}t + \hat{\sigma}_t(Y_t)\mathrm{d}W_t^{(2)} \\ &+ \sqrt{\alpha_t}\{I_d - 2u(X_t,Y_t) \otimes u(X_t,Y_t)1_{\{t<\tau\}}\}\mathrm{d}W_t^{(1)}, \quad t \geq s,\end{aligned} \quad (5.4.23)$$
where
$$\tau := \inf\{t \geq s : Y_t = X_t\}$$
is the coupling time. Since the coefficients in noises are Lipschitz continuous in $Y_t \neq X_t$, by the same argument leading to the well-posedness of (5.4.22), we conclude that (5.4.23) has a unique solution up to the coupling time τ. When $t \geq \tau$, the equation of Y_t becomes
$$\begin{aligned}\mathrm{d}Y_t = &\, b_t(Y_t, P_t^*\nu)\mathrm{d}t + \sqrt{\alpha_t}\mathrm{d}W_t^{(1)} \\ &+ \hat{\sigma}_t(Y_t)\mathrm{d}W_t^{(2)},\end{aligned} \quad (5.4.24)$$
which is well-posed as explained above. Therefore, (5.4.23) has a unique solution up to life time. On the other hand, the Lyapunov condition in ($A^{5.4}$) implies that the solution is non-explosive, and by the same reason leading to $\mathcal{L}_{X_t} = P_t^*\mu$, we have $\mathcal{L}_{Y_t} = P_t^*\nu$.

Remark 5.3.2 The operator $I_d - 2u(X_t, Y_t) \otimes u(X_t, Y_t)$ in (5.4.23) is the reflection operator with respect to the vertical mirror at the middle of the line from X_t to Y_t, so that the term for $W_t^{(1)}$ in (5.4.23) together with that in (5.4.22) is called the coupling by reflection, while that for $W_t^{(2)}$ without change is called the coupling by parallel displacement. The construction of coupling by reflection is due to Lindvall and Rogers [Lindvall and Rogers (1986)], and has been extended by Chen and Li [Chen and Li (1989)] to diffusion processes on $\mathbb{R}^d$. The construction of (5.4.23) with split couplings by reflection for $W_t^{(1)}$ and by parallel displacement for $W_t^{(2)}$ is due to [Priola and Wang (2006)].

C. Proof of (5.4.12). By $(A^{5.6})$ and the Itô-Tanaka formula for (5.4.22) and (5.4.23), we obtain

$$d|X_t - Y_t| \leq \{\theta_t \hat{\mathbb{W}}_{\psi,\beta V}(P_t^*\mu, P_t^*\nu) + K_t|X_t - Y_t|\}dt$$
$$+ 2\sqrt{\alpha_t}\langle u(X_t, Y_t), dW_t^{(1)}\rangle + \langle u(X_t, Y_t), (\hat{\sigma}_t(X_t) - \hat{\sigma}_t(Y_t))dW_t^{(2)}\rangle, \quad t < \tau.$$

By Lemma 5.4.2 and noting that $\psi'' \leq 0$ when $\hat{\sigma}$ is non-constant, we derive

$$d\psi(|X_t - Y_t|)$$
$$\leq \theta_t \psi'(|X_t - Y_t|)\hat{\mathbb{W}}_{\psi,\beta V}(P_t^*\mu, P_t^*\nu)dt$$
$$+ \{K_t|X_t - Y_t|\psi'(|X_t - Y_t|) + 2\alpha_t\psi''(|X_t - Y_t|)\}dt$$
$$+ \psi'(|X_t - Y_t|)2\sqrt{\alpha_t}\langle u(X_t, Y_t), dW_t^{(1)}\rangle$$
$$+ \psi'(|X_t - Y_t|)\langle u(X_t, Y_t), (\hat{\sigma}_t(X_t) - \hat{\sigma}_t(Y_t))dW_t^{(2)}\rangle, \quad t \in [0, \tau).$$

Therefore, (5.4.7) yields

$$d\psi(|X_t - Y_t|) \leq \theta_t \psi'(|X_t - Y_t|)\mathbb{W}_{\psi,\beta V}(P_t^*\mu, P_t^*\nu)dt$$
$$- q_l(t)\psi(|X_t - Y_t|)1_{\{|X_t - Y_t| < l\}}dt$$
$$+ \psi'(|X_t - Y_t|)\Big[2\sqrt{\alpha_t}\langle u(X_t, Y_t), dW_t^{(1)}\rangle \quad (5.4.25)$$
$$+ \langle u(X_t, Y_t), (\hat{\sigma}_t(X_t) - \hat{\sigma}_t(Y_t))dW_t^{(2)}\rangle\Big], \quad t < \tau.$$

By $(A^{5.4})$ and Itô's formula, we obtain

$$d\{V(X_t) + V(Y_t)\} \leq \{2K_0(t) - K_1(t)V(X_t) - K_1(t)V(Y_t)\}dt$$
$$+ \sqrt{\alpha_t}\langle \nabla V(X_t) + \nabla V(Y_t) - 2\langle u(X_t, Y_t), \nabla V(Y_t)\rangle u(X_t, Y_t), dW_t^{(1)}\rangle$$
$$+ \langle \hat{\sigma}_t(X_t)^*\nabla V(X_t) + \hat{\sigma}_t(Y_t)^*\nabla V(Y_t), dW_t^{(2)}\rangle. \quad (5.4.26)$$

This together with (5.4.25) yields that
$$\phi_t := \psi_{\beta V}(X_t, Y_t) = \psi(|X_t - Y_t|)\{1 + \beta V(X_t) + \beta V(Y_t)\}$$
satisfies
$$\begin{aligned}\mathrm{d}\phi_t \leq &\Big\{\theta_t \psi'(|X_t - Y_t|)\hat{\mathbb{W}}_{\psi,\beta V}(P_t^*\mu, P_t^*\nu)\big[1 + \beta V(X_t) + \beta V(Y_t)\big]\\ &+ \beta\psi'(|X_t - Y_t|)\Big(\alpha_t|\nabla V(X_t) - \nabla V(Y_t)| - q_l(t)\phi_t \mathbf{1}_{\{|X_t - Y_t| < l\}}\\ &+ \beta\psi(|X_t - Y_t|)\big[2K_0(t) - K_1(t)V(X_t) - K_1(t)V(Y_t)\big]\\ &+ \big|\{\hat{\sigma}_t(X_t) - \hat{\sigma}_t(Y_t)\}[\hat{\sigma}_t(X_t)^*\nabla V(X_t) + \hat{\sigma}_t(Y_t)^*\nabla V(Y_t)]\big|\Big)\Big\}\mathrm{d}t\\ &+ \mathrm{d}M_t, \quad t < \tau\end{aligned} \quad (5.4.27)$$
for some martingale M_t. Combining (5.4.4), (5.4.9) and (5.4.10), we derive
$$\beta\psi(|X_t - Y_t|)\{2K_0(t) - K_1(t)V(X_t) - K_1(t)V(Y_t)\}$$
$$\leq 2K_0(t)\beta\phi_t \mathbf{1}_{\{|X_t - Y_t| < l\}} - \kappa_{l,\beta}(t)\phi_t \mathbf{1}_{\{|X_t - Y_t| \geq l\}},$$
$$\beta\psi'(|X_t - Y_t|)\Big\{\alpha_t|\nabla V(X_t) - \nabla V(Y_t)|$$
$$+ \big|\{\hat{\sigma}_t(X_t) - \hat{\sigma}_t(Y_t)\}[\hat{\sigma}_t(X_t)^*\nabla V(X_t) + \hat{\sigma}_t(Y_t)^*\nabla V(Y_t)]\big|\Big\}$$
$$\leq \alpha_{l,\beta}(t)\phi_t \mathbf{1}_{\{|X_t - Y_t| < l\}}.$$
Hence, it follows from (5.4.27) that for $t \in [0, \tau)$,
$$\begin{aligned}\mathrm{d}\phi_t - \mathrm{d}M_t &\leq \theta_t \psi'(|X_t - Y_t|)\hat{\mathbb{W}}_{\psi,\beta V}(P_t^*\mu, P_t^*\nu)\{1 + \beta V(X_t) + \beta V(Y_t)\}\mathrm{d}t\\ &- \big\{[q_l(t) - \alpha_{l,\beta}(t) - 2K_0(t)\beta]\phi_t \mathbf{1}_{\{|X_t - Y_t| < l\}} + \kappa_{l,\beta}(t)\phi_t \mathbf{1}_{\{|X_t - Y_t| \geq l\}}\big\}\mathrm{d}t\\ &\leq \big\{\theta\psi'(|X_t - Y_t|)\hat{\mathbb{W}}_{\psi,\beta V}(P_t^*\mu, P_t^*\nu)\{1 + \beta V(X_t) + \beta V(Y_t)\} - \lambda_{l,\beta}(t)\phi_t\big\}\mathrm{d}t.\end{aligned}$$
Since $\phi_{t \wedge \tau} = 0$ for $t \geq \tau$, this implies
$$\mathrm{e}^{\int_0^t \lambda_{l,\beta}(s)\mathrm{d}s}\mathbb{E}\phi_{t \wedge \tau} = \mathbb{E}[\phi_{t \wedge \tau}\mathrm{e}^{\int_0^{t \wedge \tau}\lambda_{l,\beta}(s)\mathrm{d}s}] \leq \mathrm{e}^{\int_0^s \lambda_{l,\beta}(r)\mathrm{d}r}\mathbb{E}\phi_s$$
$$+ \mathbb{E}\int_s^{t \wedge \tau}\mathrm{e}^{\int_0^r \lambda_{l,\beta}(p)\mathrm{d}p}\theta_t\psi'(|X_t - Y_t|)\hat{\mathbb{W}}_{\psi,\beta V}(P_r^*\mu, P_r^*\nu)$$
$$\times \big\{1 + \beta V(X_r) + \beta V(Y_r)\big\}\mathrm{d}r, \quad t \geq s.$$
Therefore, for any $t \geq s$, we have
$$\mathbb{E}\phi_{t \wedge \tau} \leq \mathrm{e}^{-\int_s^t \lambda_{l,\beta}(r)\mathrm{d}r}\mathbb{E}\phi_s$$
$$+ \mathrm{e}^{\int_s^t |\lambda_{l,\beta}|(r)\mathrm{d}r}\mathbb{E}\int_s^{t \wedge \tau}\theta_r \hat{\mathbb{W}}_{\psi,\beta V}(P_r^*\mu, P_r^*\nu)\psi'(|X_t - Y_t|) \quad (5.4.28)$$
$$\times \big\{1 + \beta V(X_r) + \beta V(Y_r)\big\}\mathrm{d}r.$$

On the other hand, for $t \geq \tau$, by Itô's formula for (5.4.22) and (5.4.24), and applying (5.4.8), we find $C_1 \in L^1_{loc}([0,\infty);(0,\infty))$ such that

$$\mathrm{d}\psi(|X_t - Y_t|) \leq \{C_1(t)\psi(|X_t - Y_t|) + \theta_t \psi'(|X_t - Y_t|)\hat{\mathbb{W}}_{\psi,\beta V}(P_t^*\mu, P_t^*\nu)\}\mathrm{d}t$$
$$+ \psi'(|X_t - Y_t|)\langle\{\hat{\sigma}_t(X_t) - \hat{\sigma}_t(Y_t)\}^* u(X_t, Y_t), \mathrm{d}W_t^{(2)}\rangle.$$

Combining this with (5.4.26), we find $C_2 \in L^1_{loc}([0,\infty);(0,\infty))$ such that for some martingale M_t and $t \in [0, \tau)$,

$$\mathrm{d}\phi_t - \mathrm{d}M_t$$
$$\leq \Big\{C_2(t)\phi_t + \theta_t \hat{\mathbb{W}}_{\psi,\beta V}(P_t^*\mu, P_t^*\nu)\psi'(|X_t - Y_t|)\big[1 + \beta V(X_t) + \beta V(Y_t)\big]\Big\}\mathrm{d}t.$$

Therefore, for any $t \geq s$, we have $t \wedge \tau \geq s$ so that

$$\mathbb{E}\big[1_{\{t>\tau\}}(\phi_t - \phi_{t\wedge\tau})\big]$$
$$\leq \mathbb{E}\int_{t\wedge\tau}^t \mathrm{e}^{\int_r^t C_2(p)\mathrm{d}p}\theta_r \hat{\mathbb{W}}_{\psi,\beta V}(P_r^*\mu, P_r^*\nu)\psi'(|X_r - Y_r|)$$
$$\times \{1 + \beta V(X_r) + \beta V(Y_r)\}\mathrm{d}r$$
$$\leq \mathrm{e}^{\int_s^t C_2(p)\mathrm{d}p}\mathbb{E}\int_{t\wedge\tau}^t \theta_r \hat{\mathbb{W}}_{\psi,\beta V}(P_r^*\mu, P_r^*\nu)\psi'(|X_r - Y_r|)$$
$$\times \{1 + \beta V(X_r) + \beta V(Y_r)\}\mathrm{d}r.$$

This together with (5.4.28), (5.4.21) and (5.4.5) yields

$$\mathbb{E}\psi_t = \mathbb{E}\phi_{t\wedge\tau} + \mathbb{E}\big[1_{\{t>\tau\}}(\phi_t - \phi_{t\wedge\tau})\big] \leq \mathrm{e}^{-\int_0^t \lambda_{l,\beta}(r)\mathrm{d}r}\mathbb{E}\phi_s$$
$$+ \mathrm{e}^{\int_s^t(|\lambda_{l,\beta}|+C_2)(r)\mathrm{d}r}\mathbb{E}\int_s^t \theta_r \hat{\mathbb{W}}_{\psi,\beta V}(P_r^*\mu, P_r^*\nu)\psi'(|X_r - Y_r|)$$
$$\times \{1 + \beta V(X_r) + \beta V(Y_r)\}\mathrm{d}r$$
$$\leq \mathrm{e}^{-\int_s^t \lambda_{l,\beta}(r)\mathrm{d}r}\mathbb{W}_{\psi,\beta V}(P_s^*\mu, P_s^*\nu)$$
$$+ \mathrm{e}^{\int_s^t(2|\lambda_{l,\beta}|+C_2(r))\mathrm{d}r}\int_s^t \theta_r \mathrm{e}^{\int_s^r \lambda_{l,\beta}(p)\mathrm{d}p}\mathbb{E}\phi_r \mathrm{d}r, \quad t \geq s,$$

where the last step follows from the definition of $\hat{\mathbb{W}}_{\psi,\beta V}$ which implies

$$\hat{\mathbb{W}}_{\psi,\beta V}(P_r^*\mu, P_r^*\nu) \leq \frac{\mathbb{E}\phi_r}{\mathbb{E}[\psi'(|X_r - Y_r|)\{1 + \beta V(X_r) + \beta V(Y_r)\}]}.$$

By Gronwall's lemma, we obtain

$$\mathrm{e}^{\int_s^t \lambda_{l,\beta}(r)\mathrm{d}r}\mathbb{E}\phi_t$$
$$\leq \mathbb{W}_{\psi,\beta V}(P_s^*\mu, P_s^*\nu)\exp\left[\mathrm{e}^{\int_s^t\{2|\lambda_{l,\beta}(r)|+C_2(r)\}\mathrm{d}r}\int_s^t \theta_r\,\mathrm{d}r\right],$$

for $t \geq s$. Thus, for a.e. $s \geq 0$,

$$\frac{\mathrm{d}^+}{\mathrm{d}s} \mathbb{W}_{\psi,\beta V}(P_s^*\mu, P_s^*\nu)$$
$$:= \limsup_{t \downarrow s} \frac{\mathbb{W}_{\psi,\beta V}(P_t^*\mu, P_t^*\nu) - \mathbb{W}_{\psi,\beta V}(P_s^*\mu, P_s^*\nu)}{t-s}$$
$$\leq \limsup_{t \downarrow s} \frac{\mathbb{E}\phi_t - \mathbb{W}_{\psi,\beta V}(P_s^*\mu, P_s^*\nu)}{t-s}$$
$$\leq -(\lambda_{l,\beta}(s) - \theta_s) \mathbb{W}_{\psi,\beta V}(P_s^*\mu, P_s^*\nu).$$

This implies (5.4.12).

D. Proof of (5.4.13). Let a, b be independent of the time parameter and

$$\kappa := \lambda_{l,\beta} - \theta > 0.$$

According to Theorem 5.1.1, (5.4.13) implies that P_t^* has a unique invariant probability measure $\bar{\mu}$ in $\mathcal{P}_V$, and (5.4.13) holds.

5.5 Exponential ergodicity in $\mathbb{W}_\psi$: Partially dissipative case

In this part, we do not assume the Lyapunov condition in $(A^{5.4})$ but use the following $(A^{5.7})$ to replace $(A^{5.6})$.

For any $\psi \in \Psi$, where

$$\Psi := \{\psi \in C^2([0,\infty)) : \psi(0) = 0, \psi' > 0, r\psi'(r) + r^2(\psi'')^+(r) \leq cr$$
$$\text{for some constant } c > 0\},$$

the quasi-distance

$$\mathbb{W}_\psi(\mu, \nu) := \inf_{\pi \in \mathcal{C}(\mu,\nu)} \int_{\mathbb{R}^d \times \mathbb{R}^d} \psi(|x-y|) \pi(\mathrm{d}x, \mathrm{d}y) \qquad (5.5.1)$$

on the space

$$\mathcal{P}_\psi := \{\mu \in \mathcal{P} : \|\mu\|_\psi := \mu(\psi(|\cdot|)) < \infty\}$$

is complete, i.e. a $\mathbb{W}_\psi$-Cauchy sequence in $\mathcal{P}_\psi$ converges with respect to $\mathbb{W}_\psi$. When ψ is concave, $\mathbb{W}_\psi$ satisfies the triangle inequality and is hence a metric on $\mathcal{P}_\psi$.

$(A^{5.7})$ (ψ-**Monotonicity**) Let $\psi \in \Psi$, $g \in C([0,\infty))$ with $g(r) \leq Kr$ for some constant $K > 0$ and all $r \geq 0$, such that

$$2\alpha_t \psi''(r) + (g\psi')(r) \leq -q_t \psi(r), \quad r \geq 0 \qquad (5.5.2)$$

holds for some $q \in L^1_{loc}([0,\infty);(0,\infty))$. Moreover, b is locally bounded on $[0,\infty) \times \mathbb{R}^d \times \mathcal{P}_\psi$, and there exists $\theta \in L^1_{loc}([0,\infty);(0,\infty))$ such that

$$\langle b_t(x,\mu) - b_t(y,\nu), x-y \rangle + \frac{1}{2}\|\hat{\sigma}_t(x) - \hat{\sigma}_t(y)\|^2_{HS}$$
$$\leq |x-y|\{\theta_t \mathbb{W}_\psi(\mu,\nu) + g(|x-y|)\}, \qquad (5.5.3)$$
$$t \geq 0, x, y \in \mathbb{R}^d, \mu, \nu \in \mathcal{P}_\psi.$$

When $a = I_d$ and

$$b(x,\mu) = b_0(x) + \int_{\mathbb{R}^d} Z(x,y)\mu(\mathrm{d}y)$$

for a drift b_0 and a Lipschitz continuous map $Z : \mathbb{R}^d \times \mathbb{R}^d \to \mathbb{R}^d$, the exponential convergence of (5.0.1) is presented in Theorems 2.3 and 2.4 in [Eberle et al. (2019)] under the condition that

$$\langle b_0(x) - b_0(y), x-y \rangle \leq \kappa(|x-y|)|x-y|^2, \quad x, y \in \mathbb{R}^d$$

for some function $\kappa \in C((0,\infty))$ with $\int_0^1 r\kappa^+(r)\mathrm{d}r < \infty$ and $\limsup_{r\to\infty} \kappa(r) < 0$, and that the Lipschitz constant of Z is small enough. It is clear that in this case (5.5.3) holds for $g(r) := r\kappa(r)$ and $\psi(r)$ comparable with r, for which we may choose $\psi \in \Psi$ as in (5.5.15) below such that (5.5.2) holds for $\alpha = 1$ and some $q > 0$. Therefore, this situation is included in Theorem 5.5.1 below.

5.5.1 Main results and example

Theorem 5.5.1. *Assume* $(A^{5.5})$ *and* $(A^{5.7})$, *with* $\psi'' \leq 0$ *if* $\hat{\sigma}_t(\cdot)$ *is nonconstant for some* $t \geq 0$. *Then* (5.0.1) *is well-posed in* $\mathcal{P}_\psi$, *and* P_t^* *satisfies*

$$\mathbb{W}_\psi(P_t^*\mu, P_t^*\nu) \leq \mathrm{e}^{\int_0^t [q_s - \theta_s\|\psi'\|_\infty]\mathrm{d}s} \mathbb{W}_\psi(\mu,\nu), \quad t \geq 0, \mu, \nu \in \mathcal{P}_\psi. \qquad (5.5.4)$$

Consequently, $(b_t, \sigma_t) = (b, \sigma)$ *does not depend on t and $q > \theta\|\psi'\|_\infty$, then P_t^* has a unique invariant probability measure $\bar{\mu} \in \mathcal{P}_\psi$ such that*

$$\mathbb{W}_\psi(P_t^*\mu, \bar{\mu}) \leq \mathrm{e}^{-(q-\theta\|\psi'\|_\infty)t} \mathbb{W}_\psi(\mu,\bar{\mu}), \quad t \geq 0, \mu \in \mathcal{P}_\psi. \qquad (5.5.5)$$

Proof. By $(A^{5.5})$ and $(A^{5.7})$, the well-posedness follows from the proof of Lemma 3.6.2 with $\mathbb{W}_\psi$ replacing $\mathbb{W}_{\psi,V}$, and the solution satisfies

$$\sup_{t \in [0,T]} \|P_t^*\mu\|_\psi < \infty, \quad \mu \in \mathcal{P}_\psi, T > 0. \qquad (5.5.6)$$

It remains to prove (5.5.4) and the existence of the invariant probability measure $\bar{\mu}$ in the time homogeneous case.

(1) Proof of (5.5.4). Let $s \geq 0$ and $\mu, \nu \in \mathcal{P}_\psi$. We make use of the coupling constructed by (5.4.22) and (5.4.23) for initial values (X_s, Y_s) satisfying
$$\mathcal{L}_{X_s} = P_s^*\mu, \quad \mathcal{L}_{Y_s} = P_s^*\nu, \quad \mathbb{W}_\psi(P_s^*\mu, P_s^*\nu) = \mathbb{E}\psi(X_s, Y_s). \tag{5.5.7}$$
By the same reason leading to (5.4.25), by $(A^{5.7})$ for $\psi \in \Psi$ with $\psi'' \leq 0$ when $\hat{\sigma}$ is non-constant, we derive
$$\begin{aligned}\mathrm{d}\psi(|X_t - Y_t|) &\leq \{\theta\psi'(|X_t - Y_t|)\mathbb{W}_\psi(P_t^*\mu, P_t^*\nu) - q\psi(|X_t - Y_t|)\}\mathrm{d}t \\ &+ \psi'(|X_t - Y_t|)\Big[2\sqrt{\lambda}\big\langle u(X_t, Y_t), \mathrm{d}W_t^{(1)}\big\rangle \\ &+ \big\langle u(X_t, Y_t), (\hat{\sigma}_t(X_t) - \hat{\sigma}_t(Y_t))\mathrm{d}W_t^{(2)}\big\rangle\Big], \quad t < \tau.\end{aligned} \tag{5.5.8}$$
By the same argument leading to (5.4.28), this implies
$$\begin{aligned}\mathbb{E}\psi(|X_{t\wedge\tau} - Y_{t\wedge\tau}|) &\leq \mathrm{e}^{-q(t-s)}\mathbb{E}\psi(|X_s - Y_s|) \\ &+ \theta\|\psi'\|_\infty \int_s^{t\wedge\tau} \mathbb{W}_\psi(P_r^*\mu, P_r^*\nu)\mathrm{d}r, \quad t \geq s.\end{aligned} \tag{5.5.9}$$
On the other hand, when $t \geq \tau$, by $(A^{5.7})$ and applying Itô's formula for (5.4.22) and (5.4.24), we find a constant $C > 0$ such that
$$\begin{aligned}\mathrm{d}\psi(|X_t - Y_t|) &\leq \{C\psi(|X_t - Y_t|)\mathrm{d}t + \theta\|\psi'\|_\infty \mathbb{W}_\psi(P_t^*\mu, P_t^*\nu)\}\mathrm{d}t \\ &+ \psi'(|X_t - Y_t|)\langle\{\hat{\sigma}_t(X_t) - \hat{\sigma}_t(Y_t)\}^*u(X_t, Y_t), \mathrm{d}W_t^{(2)}\rangle.\end{aligned}$$
Thus,
$$\mathbb{E}\big[1_{\{t>\tau\}}\psi(|X_t - Y_t|)\big] \leq \theta\|\psi'\|_\infty \mathrm{e}^{C(t-s)} \mathbb{E}\int_{t\wedge\tau}^t \mathbb{W}_\psi(P_r^*\mu, P_r^*\nu)\mathrm{d}r, \quad t \geq s.$$
Combining this with (5.5.9) and (5.5.7), we derive
$$\begin{aligned}\mathbb{W}_\psi(P_t^*\mu, P_t^*\nu) &\leq \mathbb{E}\psi(|X_t - Y_t|) = \mathbb{E}\psi(|X_{t\wedge\tau} - Y_{t\wedge\tau}|) + \mathbb{E}\big[1_{\{t>\tau\}}\psi(|X_t - Y_t|)\big] \\ &\leq \mathrm{e}^{-q(t-s)}\mathbb{E}\psi(|X_s - Y_s|) + \theta\|\psi'\|_\infty \mathrm{e}^{C(t-s)} \int_s^t \mathbb{W}_\psi(P_r^*\mu, P_r^*\nu)\mathrm{d}r \\ &= \mathrm{e}^{-q(t-s)}\mathbb{W}_\psi(P_s^*\mu, P_s^*\nu) + \theta\|\psi'\|_\infty \mathrm{e}^{C(t-s)} \int_s^t \mathbb{W}_\psi(P_r^*\mu, P_r^*\nu)\mathrm{d}r, \quad t \geq s.\end{aligned}$$
Therefore,
$$\frac{\mathrm{d}^+}{\mathrm{d}s}\mathbb{W}_\psi(P_s^*\mu, P_s^*\nu) := \limsup_{t\downarrow s} \frac{\mathbb{W}_\psi(P_t^*\mu, P_t^*\nu) - \mathbb{W}_\psi(P_s^*\mu, P_s^*\nu)}{t - s}$$
$$\leq -(q - \theta\|\psi'\|_\infty)\mathbb{W}_\psi(P_s^*\mu, P_s^*\nu), \quad s \geq 0.$$
This implies (5.5.4).

(2) Existence of $\bar{\mu} \in \mathcal{P}_\psi$. According to Theorem 5.1.1, (5.5.4) and (5.5.6) imply that P_t^* has a unique invariant probability measure $\bar{\mu} \in \mathcal{P}_\psi$, so that (5.5.5) follows from (5.5.4). $\square$

As a consequence of Theorem 5.5.1, we consider the non-dissipative case where $\nabla b_t(\cdot, \mu)(x)$ is positive definite in a possibly unbounded set but with bounded "one-dimensional puncture mass" in the sense of (5.5.12) below. Let $\mathcal{P}_1 = \{\mu \in \mathcal{P} : \mu(|\cdot|) < \infty\}$ and
$$S_b(x) := \sup \{\langle \nabla_v b_t(\cdot, \mu)(x), v \rangle : t \geq 0, |v| \leq 1, \mu \in \mathcal{P}_1\}, \quad x \subset \mathbb{R}^d.$$

$(A^{5.8})$ *There exist constants $\theta_0, \theta_1, \theta_2, \beta \geq 0$ such that*
$$\frac{1}{2}\|\hat{\sigma}_t(x) - \hat{\sigma}_t(y)\|_{HS}^2 \leq \theta_0 |x-y|^2, \quad t \geq 0, x, y \in \mathbb{R}^d; \tag{5.5.10}$$
$$S_b(x) \leq \theta_1, \quad |b_t(x,\mu) - b_t(x,\nu)| \leq \beta \mathbb{W}_1(\mu,\nu),$$
$$t \geq 0, x \in \mathbb{R}^d, \mu, \nu \in \mathcal{P}_1; \tag{5.5.11}$$
$$\kappa := \sup_{x, v \in \mathbb{R}^d, |v|=1} \int_{\mathbb{R}} 1_{\{S_b(x+sv) > -\theta_2\}} \mathrm{d}s < \infty. \tag{5.5.12}$$

Let $\mathbb{W}_1 = \mathbb{W}_\psi$ and $\mathcal{P}_1 = \mathcal{P}_\psi$ for $\psi(r) = r$.

Corollary 5.5.2. *Assume $(A^{5.5})$ and $(A^{5.8})$. Let*
$$g(r) := (\theta_1 + \theta_2)\{(\kappa r^{-1}) \wedge r\} - (\theta_2 - \theta_0)r, \quad r \geq 0,$$
$$k := \frac{2\lambda}{\int_0^\infty t \mathrm{e}^{\frac{1}{2\lambda}\int_0^t g(u)\mathrm{d}u} \mathrm{d}t} - \frac{\beta(\theta_2 - \theta_0)}{2\lambda} \int_0^\infty t \mathrm{e}^{\frac{1}{2\lambda}\int_0^t g(u)\mathrm{d}u} \mathrm{d}t. \tag{5.5.13}$$
Then there exists a constant $c > 0$ such that
$$\mathbb{W}_1(P_t^* \mu, P_t^* \nu) \leq c\mathrm{e}^{-kt} \mathbb{W}_1(\mu,\nu), \quad t \geq 0, \mu, \nu \in \mathcal{P}_1.$$
If $\theta_2 > \theta_0$ and
$$\beta < \frac{4\lambda^2}{(\theta_2 - \theta_2)(\int_0^\infty t \mathrm{e}^{\frac{1}{2\lambda}\int_0^t g(u)\mathrm{d}u} \mathrm{d}t)^2}, \tag{5.5.14}$$
then $\kappa > 0$ and P_t^* has a unique invariant probability measure $\bar{\mu} \in \mathcal{P}_1$ satisfying
$$\mathbb{W}_1(P_t^* \mu, \bar{\mu}) \leq c\mathrm{e}^{-kt} \mathbb{W}_1(\mu, \bar{\mu}), \quad t \geq 0, \mu \in \mathcal{P}_1.$$

Proof. For g in (5.5.13), let
$$q := \frac{2\lambda}{\int_0^\infty t\mathrm{e}^{\frac{1}{2\lambda}\int_0^t g(u)\mathrm{d}u}\mathrm{d}t}, \quad \theta := \frac{\beta(\theta_2 - \theta_0)}{2\lambda} \int_0^\infty t\mathrm{e}^{\frac{1}{2\lambda}\int_0^t g(u)\mathrm{d}u}\mathrm{d}t,$$
and take
$$\psi(r) := \int_0^r \mathrm{e}^{-\frac{1}{2\lambda}\int_0^s g(u)\mathrm{d}u} \int_s^\infty t \mathrm{e}^{\frac{1}{2\lambda}\int_0^t g(u)\mathrm{d}u}\mathrm{d}t, \quad r \geq 0. \tag{5.5.15}$$
By Theorem 5.5.1, it suffices to verify

(a) $\psi \in \Psi$ and $\psi'' \leq 0$;
(b) there exists a constant $C > 1$ such that $C^{-1}\mathbb{W}_\psi \leq \mathbb{W}_1 \leq C\mathbb{W}_\psi$;
(c) (5.5.2) and (5.5.3) hold.

(a) We have $\psi(0) = 0, \psi'(r) > 0$ and

$$\psi''(r) = -\frac{g(r)}{2\lambda} e^{-\frac{1}{2\lambda}\int_0^r g(u)du} \int_r^\infty t e^{\frac{1}{2\lambda}\int_0^t g(u)du} dt - r, \quad r \geq 0. \quad (5.5.16)$$

To prove $\psi \in \Psi$, it suffices to show $\psi'' \leq 0$. To this end, take

$$r_0 := \frac{\sqrt{\kappa(\theta_1 + \theta_2)}}{\sqrt{\theta_2 - \theta_0}}.$$

It is easy to see that g in (5.5.13) satisfies

$$g|_{[0,r_0]} \geq 0, \quad g|_{(r_0,\infty)} < 0. \quad (5.5.17)$$

Combining this with (5.5.16), we have $\psi''(r) \leq 0$ for $r \leq r_0$. On the other hand, for $r > r_0$, we have $g(r) < 0$ and

$$\frac{r}{-g(r)} = \frac{1}{(\theta_2 - \theta_0)r^{1-p} - (\theta_1 + \theta_2)\kappa r^{-(1+p)}}$$

is decreasing in $r > r_0$, so that

$$\int_r^\infty t e^{\frac{1}{2\lambda}\int_0^t g(u)du} dt = \int_r^\infty \frac{2\lambda t}{g(t)} \left(\frac{d}{dt} e^{\frac{1}{2\lambda}\int_0^t g(u)du}\right) dt$$

$$= -\frac{2\lambda r}{g(r)} e^{\frac{1}{2\lambda}\int_0^r g(u)du} + 2\lambda \int_r^\infty \left(\frac{d}{dt} \frac{2\lambda t}{-g(t)}\right) e^{\frac{1}{2\lambda}\int_0^t g(u)du} dt$$

$$\leq -\frac{2\lambda r}{g(r)} e^{\frac{1}{2\lambda}\int_0^r g(u)du}, \quad r > r_0.$$

This together with (5.5.16) yields $\psi''(r) \leq 0$ for $r > r_0$. In conclusion, $\psi \in \Psi$.

(b) Since $\psi \in \Psi$ with $\psi'' \leq 0$ implies that $\psi(r) \leq \psi'(0)r$ and $\frac{\psi(r)}{r}$ is decreasing in $r > 0$, we have $\mathbb{W}_\psi \leq \psi'(0)\mathbb{W}_1$ and

$$\inf_{r>0} \frac{\psi(r)}{r} = \lim_{r\to\infty} \frac{\psi(r)}{r} = \lim_{r\to\infty} \psi'(r)$$

$$= \lim_{r\to\infty} \frac{\int_r^\infty t \exp[\frac{1}{2\lambda}\int_0^t g(u)du]dt}{\exp[\frac{1}{2\lambda}\int_0^r g(u)du]} \quad (5.5.18)$$

$$= \lim_{r\to\infty} \frac{2\lambda r}{-g(r)} = \frac{2\lambda}{\theta_2 - \theta_0} \in (0,\infty).$$

Thus,

$$\frac{1}{\psi'(0)}\mathbb{W}_\psi \leq \mathbb{W}_1(\mu,\nu) \leq \frac{\theta_2 - \theta_0}{2\lambda}\mathbb{W}_\psi.$$

(c) By (5.5.13) we have
$$2\lambda\psi''(r) + g(r)\psi'(r) = -2\lambda r, \quad r \geq 0.$$
Since $\psi(r) \leq \psi'(0)r$, this implies
$$2\lambda\psi''(r) + g(r)\psi'(r) \leq -\frac{2\lambda r}{\psi'(0)r}\psi(r) =: -q\psi(r), \quad r \geq 0.$$
Therefore, (5.5.2) holds.

Next, for $x \neq y$, let $v = \frac{x-y}{|x-y|}$. Then (5.5.11) implies
$$\langle b_t(x,\mu) - b_t(y,\nu), x - y \rangle$$
$$= |x-y|\langle b_t(x,\mu) - b_t(y,\mu), v \rangle + |x-y|\langle b_t(y,\mu) - b_t(y,\nu), v \rangle$$
$$\leq \beta|x-y|\mathbb{W}_1(\mu,\nu) + |x-y|\int_0^{|x-y|} S_b(y + s(x-y))\mathrm{d}s \quad (5.5.19)$$
$$= \beta|x-y|\mathbb{W}_1(\mu,\nu) + \int_0^{|x-y|^2} S_b(y + sv)\mathrm{d}s, \quad \mu,\nu \in \mathcal{P}_1.$$

On the other hand, by (5.5.11) and (5.5.12) we obtain
$$\int_0^{|x-y|^2} S_b(y+sv)\mathrm{d}s$$
$$\leq \theta_1 \int_0^{|x-y|^2} 1_{\{S_b(x+sv) > -\theta_2\}}\mathrm{d}s - \theta_2 \int_0^{|x-y|^2} 1_{\{S_b(x+sv) \leq -\theta_2\}}\mathrm{d}s$$
$$= (\theta_1 + \theta_2) \int_0^{|x-y|^2} 1_{\{S_b(x+sv) > -\theta_2\}}\mathrm{d}s - \theta_2|x-y|^2$$
$$\leq (\theta_1 + \theta_2)(\kappa \wedge |x-y|^2) - \theta_2|x-y|^2.$$

Combining this with (5.5.10) and (5.5.19), we derive (5.5.3). □

Example 5.4.1. Let a satisfy $(A^{5.5})$ with $\hat{\sigma}$ satisfying (5.5.10). Consider (5.3.9) with $G \in C^2(\mathbb{R}^d)$ and $W \in C^2(\mathbb{R}^d \times \mathbb{R}^d)$ such that
$$\nabla^2\{G + W(\cdot, z)\} \geq \theta_2 1_{\{|\cdot| \geq \lambda_0\}} - \theta_1 1_{\{|\cdot| < \lambda_0\}}, \quad z \in \mathbb{R}^d,$$
$$\|\nabla_x \nabla_y W(x,y)\| \leq \tilde{\theta}, \quad x, y \in \mathbb{R}^d \quad (5.5.20)$$
holds for some constants $\lambda_0, \theta_1, \theta_2 > 0$. Then the assertion in Corollary 5.5.2 holds for $\kappa = 4\lambda_0$ and $(P_t^*\mu)(\mathrm{d}x) := \rho_t(x)\mathrm{d}x$, where ρ_t solves (5.3.9) with $\rho_0(x)\mathrm{d}x \in \mathcal{P}_1$.

Proof. It is easy to see that (5.5.20) implies (5.5.11). So, it remains to verify that κ in (5.5.12) satisfies $\kappa \leq 4\lambda_0$. By the second inequality in (5.5.20) we have
$$S_b(x) \leq -\theta_2 1_{\{|x| \geq \lambda_0\}} + \theta_1 1_{\{|x| < \lambda_0\}}, \quad x \in \mathbb{R}^d.$$

For $x, v \in \mathbb{R}^d$ with $|v| = 1$, if there exists $s_0 \in \mathbb{R}^d$ such that $|x + s_0 v| < \lambda_0$, then
$$|x + sv| \geq |s - s_0| - |x + s_0 v| > |s - s_0| - \lambda_0,$$
so that
$$\{s \in \mathbb{R} : |x + sv| < \lambda_0\} \subset (s_0 - 2\lambda_0, s_0 + 2\lambda_0),$$
which implies
$$\kappa := \sup_{x, v \in \mathbb{R}^d, |v|=1} \int_{\{S_b(x+sv) > -\theta_2\}} \mathrm{d}s \leq \sup_{x, v \in \mathbb{R}^d, |v|=1} \int_{\{|x+sv|<\lambda_0\}} \mathrm{d}s \leq 4\lambda_0. \quad \Box$$

5.6 Donsker-Varadhan large deviations

The LDP (large deviation principle) is a fundamental tool characterizing asymptotic behaviours of probability measures $\{\mu_\varepsilon\}_{\varepsilon>0}$ on a topological space E, see [Dembo and Zeitouni (1998)] and references within. Recall that μ_ε for small $\varepsilon > 0$ is said to satisfy the LDP with speed $\lambda(\varepsilon) \to +\infty$ (as $\varepsilon \to 0$) and rate function $I : E \to [0, +\infty]$, if I has compact level sets (i.e. $\{I \leq r\}$ is compact for $r \in \mathbb{R}^+$), and for any Borel subset A of E,
$$-\inf_{A^o} I \leq \liminf_{\varepsilon \to 0} \frac{1}{\lambda(\varepsilon)} \log \mu_\varepsilon(A) \leq \limsup_{\varepsilon \to 0} \frac{1}{\lambda(\varepsilon)} \log \mu_\varepsilon(A) \leq -\inf_{\bar{A}} I,$$
where A^o and $\bar{A}$ stand for the interior and the closure of A in E respectively.

In this part, we consider the Donsker-Varadhan type long time LDP [Donsker and Varadhan (1975)] for $\mu_\varepsilon := \mathcal{L}_{L_{\varepsilon^{-1}}}$, where
$$L_t := \frac{1}{t} \int_0^t \delta_{X_s} \mathrm{d}s, \quad t > 0$$
is the empirical measure for a path-distribution dependent SPDE.

We first introduce the main results and illustrate them by concrete examples, then recall some facts on LDP for Markov processes due to Liming Wu [Wu (2000)] and [Wu (2000b)], and finally present the proof of the main result.

5.6.1 *Main result and examples*

Consider
$$\mathrm{d}X_t = b(X_t, \mathcal{L}_{X_t})\mathrm{d}t + \sigma(\mathcal{L}_{X_t})\mathrm{d}W_t, \tag{5.6.1}$$
where $b : \mathbb{R}^d \times \mathcal{P}_2 \to \mathbb{R}^d$, $\sigma : \mathcal{P}_2 \to \mathbb{R}^d \otimes \mathbb{R}^m$ and W_t is the m-dimensional Brownian motion. We assume

($A^{5.9}$) b is continuous, σ is bounded and continuous such that
$$2\langle b(x,\mu) - b(y,\nu), x-y\rangle + \|\sigma(\mu) - \sigma(\nu)\|_{HS}^2$$
$$\leq -\kappa_1 |x-y|^2 + \kappa_2 \mathbb{W}_2(\mu,\nu)^2$$
holds for some constants $\kappa_1 > \kappa_2 \geq 0$ and all $x, y \subset \mathbb{R}^d, \mu, \nu \in \mathcal{P}_2$.

According to Theorem 5.1.2 for $k=2$, ($A^{5.9}$) implies that (5.6.1) is well-posed for distributions in $\mathcal{P}_2$, and P_t^* has a unique invariant probability measure $\bar{\mu} \in \mathcal{P}_2$ such that
$$\mathbb{W}_2(P_t^*\nu, \bar{\mu})^2 \leq e^{-(\kappa_1-\kappa_2)t} \mathbb{W}_2(\nu,\bar{\mu})^2, \quad t \geq 0, \nu \in \mathcal{P}_2. \tag{5.6.2}$$
Let X_t^ν be the solution of (5.6.1) with initial distribution ν. We study the long time LDP for the empirical measure
$$L_t^\nu := \frac{1}{t}\int_0^t \delta_{X_s^\nu} ds, \quad t > 0.$$

Definition 5.6.1. Let $\mathcal{P}$ be equipped with the weak topology, let $\mathcal{A} \subset \mathcal{P}$, and let $J : \mathcal{P} \to [0, \infty]$ have compact level sets, i.e. $\{J \leq r\}$ is compact in $\mathcal{P}$ for any $r > 0$.

(1) $\{L_t^\nu\}_{\nu \in \mathcal{A}}$ is said to satisfy the upper bound uniform LDP with rate function J, denoted by $\{L_t^\nu\}_{\nu \in \mathcal{A}} \in LDP_u(J)$, if for any closed $A \subset \mathcal{P}$,
$$\limsup_{t \to \infty} \frac{1}{t} \sup_{\nu \in \mathcal{A}} \log \mathbb{P}(L_t^\nu \in A) \leq -\inf_A J.$$

(2) $\{L_t^\nu\}_{\nu \in \mathcal{A}}$ is said to satisfy the lower bound uniform LDP with rate function J, denoted by $\{L_t^\nu\}_{\nu \in \mathcal{A}} \in LDP_l(J)$, if for any open $A \subset \mathcal{P}$,
$$\liminf_{t \to \infty} \frac{1}{t} \inf_{\nu \in \mathcal{A}} \log \mathbb{P}(L_t^\nu \in A) \geq -\inf_A J.$$

(3) $\{L_t^\nu\}_{\nu \in \mathcal{A}}$ is said to satisfy the uniform LDP with rate function J, denoted by $\{L_t^\nu\}_{\nu \in \mathcal{A}} \in LDP(J)$, if $\{L_t^\nu\}_{\nu \in \mathcal{A}} \in LDP_u(J)$ and $\{L_t^\nu\}_{\nu \in \mathcal{A}} \in LDP_l(J)$.

Let P be a sub-Markov operator on $\mathcal{B}_b(\mathbb{R}^d)$, i.e. it is a positivity-preserving linear operator with $P1 \leq 1$. P is called strong Feller if $P\mathcal{B}_b(\mathbb{R}^d) \subset C_b(\mathbb{R}^d)$, is called Feller if $PC_b(\mathbb{R}^d) \subset C_b(\mathbb{R}^d)$, and is called μ-irreducible for some $\mu \in \mathcal{P}(\mathbb{R}^d)$ if $\mu(1_A P 1_B) > 0$ holds for any $A, B \in \mathcal{B}(\mathbb{R}^d)$ with $\mu(A)\mu(B) > 0$.

Recall that the strong Feller property has been introduced in Definition 1.6.1, where the irreducibility is stronger than μ-irreducibility.

Consider the reference SDE

$$d\bar{X}_t = b(\bar{X}_t, \bar{\mu})dt + \sigma(\bar{\mu})dW_t. \tag{5.6.3}$$

It is standard that under $(A^{5.9})$ the equation (5.6.3) has a unique solution $\bar{X}_t^x$ for any starting point $x \in \mathbb{R}^d$, and $\bar{\mu}$ is the unique invariant probability measure of the associated Markov semigroup

$$\bar{P}_t f(x) := \mathbb{E}[f(\bar{X}_t^x)], \quad t \geq 0, x \in \mathbb{R}^d, f \in \mathcal{B}_b(\mathbb{R}^d).$$

Consequently, $\bar{P}_t$ uniquely extends to $L^\infty(\bar{\mu})$. If $f \in L^\infty(\bar{\mu})$ satisfies

$$\bar{P}_t f = f + \int_0^t \bar{P}_s g \, ds, \quad \bar{\mu}\text{-a.e.}$$

for some $g \in L^\infty(\bar{\mu})$ and all $t \geq 0$, we write $f \in \mathcal{D}(\bar{A})$ and denote $\bar{A}f = g$. Obviously, we have $\mathcal{D}(\bar{A}) \supset C_c^\infty(\mathbb{R}^d) := \{f \in C_b^\infty(\mathbb{R}^d) : \nabla f$ has compact support$\}$ and

$$\bar{A}f(x) = \frac{1}{2} \sum_{i,j=1}^d \{\sigma\sigma^*\}_{ij}(\bar{\mu}) \partial_i \partial_j f(x) + \sum_{i=1}^d b_i(x, \bar{\mu}) \partial_i f(x), \quad f \in C_c^\infty(\mathbb{R}^d).$$

The Donsker-Varadhan level 2 entropy function J for the diffusion process generated by $\bar{A}$ has compact level sets in $\mathcal{P}$ under the τ and weak topologies, and

$$J(\nu) = \begin{cases} \sup\left\{ \int_{\mathbb{R}^d} \frac{-\bar{A}f}{f} d\nu : 1 \leq f \in \mathcal{D}(\bar{A}) \right\}, & \text{if } \frac{d\nu}{d\bar{\mu}} \text{ exists,} \\ \infty, & \text{otherwise.} \end{cases}$$

Theorem 5.6.1. *Assume* $(A^{5.9})$. *For any* $r, R > 0$, *let* $\mathcal{B}_{r,R} = \{\nu \in \mathcal{P} : \nu(e^{|\cdot|^r}) \leq R\}$.

(1) *We have* $\{L_t^\nu\}_{\nu \in \mathcal{B}_{r,R}} \in LDP_u(J)$ *for all* $r, R > 0$. *If* $\bar{P}_t$ *is strong Feller and* $\bar{\mu}$-*irreducible for some* $t > 0$, *then* $\{L_t^\nu\}_{\nu \in \mathcal{B}_{r,R}} \in LDP(J)$ *for all* $r, R > 0$.

(2) *If there exist constants* $\varepsilon, c_1, c_2 > 0$ *such that*

$$\langle x, b(x, \nu) \rangle \leq c_1 - c_2 |x|^{2+\varepsilon}, \quad x \in \mathbb{R}^d, \nu \in \mathcal{P}_2, \tag{5.6.4}$$

then $\{L_t^\nu\}_{\nu \in \mathcal{P}_2} \in LDP_u(J)$. *If moreover* $\bar{P}_t$ *is strong Feller and* $\bar{\mu}$-*irreducible for some* $t > 0$, *then* $\{L_t^\nu\}_{\nu \in \mathcal{P}_2} \in LDP(J)$.

To apply this result, we first recall some facts on the strong Feller property and the $\bar{\mu}$-irreducibility of diffusion semigroups.

Remark 5.6.1. (1) Let $\bar{P}_t$ be the (sub-)Markov semigroup generated by the second order differential operator

$$\bar{\mathcal{A}} := \sum_{i=1}^{m} U_i^2 + U_0,$$

where $\{U_i\}_{i=1}^{m}$ are C^1-vector fields and U_0 is a continuous vector field. According to Theorem 5.1 in [Lanconelli and Polidoro (1994)], if $\{U_i : 1 \leq i \leq m\}$ together with their Lie brackets with U_0 span $\mathbb{R}^d$ at any point (i.e. the Hörmander condition holds), then the Harnack inequality

$$P_t f(x) \leq \psi(t,s,x,y) P_{t+s} f(y), \quad t,s > 0, x,y \in \mathbb{R}^d, f \in \mathcal{B}^+(\mathbb{R}^d)$$

holds for some map $\psi : (0,\infty)^2 \times (\mathbb{R}^d)^2 \to (0,\infty)$. Consequently, if moreover $\bar{P}_t$ has an invariant probability measure $\bar{\mu}$, then $\bar{P}_t$ is $\bar{\mu}$-irreducible for any $t > 0$. Finally, if $\{U_i\}_{0 \leq i \leq m}$ are smooth with bounded derivatives of all orders, then the above Hörmander condition implies that $\bar{P}_t$ has smooth heat kernel with respect to the Lebesgue measure, in particular it is strong Feller for any $t > 0$.

(2) Let $\bar{P}_t$ be the Markov semigroup generated by

$$\bar{\mathcal{A}} := \sum_{i,j=1}^{d} \bar{a}_{ij} \partial_i \partial_j + \sum_{i=1}^{d} \bar{b}_i \partial_j,$$

where $(\bar{a}_{ij}(x))$ is strictly positive definite for any x, $\bar{a}_{ij} \in H_{loc}^{p,1}(\mathrm{d}x)$ and $\bar{b}_i \in L_{loc}^p(\mathrm{d}x)$ for some $p > d$ and all $1 \leq i,j \leq d$. Moreover, let $\bar{\mu}$ be an invariant probability measure of $\bar{P}_t$. Then by Theorem 4.1 in [Bogachev et al (2001)], $\bar{P}_t$ is strong Feller for all $t > 0$. Moreover, as indicated in (1) that Theorem 5.1 in [Lanconelli and Polidoro (1994)] ensures the $\bar{\mu}$-irreducibility of $\bar{P}_t$ for $t > 0$.

We present below two examples to illustrate this result, where the first is a distribution dependent perturbation of the Ornstein-Ulenbeck process, and the second is the distribution dependent stochastic Hamiltonian system.

Example 5.6.1. Let $\sigma(\nu) = I + \varepsilon \sigma_0(\nu)$ and $b(x,\nu) = -\frac{1}{2}(\sigma \sigma^*)(\nu)x$, where I is the identity matrix, $\varepsilon > 0$ and σ_0 is a bounded Lipschitz continuous map from $\mathcal{P}_2$ to $\mathbb{R}^d \otimes \mathbb{R}^d$. When $\varepsilon > 0$ is small enough, assumption $(A^{5.4})$ holds and $\bar{P}_t$ satisfies conditions in Remark 5.6.1(2). So, Theorem 5.6.1(1) implies $\{L_t^\nu\}_{\nu \in \mathcal{B}_{r,R}} \in LDP(J)$ for all $r, R > 0$.

If we take $b(x,\nu) = -x - c|x|^\theta x$ for some constants $c, \theta > 0$, then when $\varepsilon > 0$ is small enough such that $(A^{5.4})$ and (5.6.4) are satisfied, Theorem 5.6.1(2) and Remark 5.6.1(2) imply $\{L_t^\nu\}_{\nu \in \mathcal{P}_2} \in LDP(J)$.

Example 5.6.2. Let $d = 2m$ and consider the following distribution dependent SDE for $X_t = (X_t^{(1)}, X_t^{(2)})$ on $\mathbb{R}^m \times \mathbb{R}^m$:

$$\begin{cases} dX_t^{(1)} = \{X_t^{(2)} - \lambda X_t^{(1)}\}dt \\ dX_t^{(2)} = \{Z(X_t, \mathcal{L}_{X_t}) - \lambda X_t^{(2)}\}dt + \sigma dW_t, \end{cases}$$

where $\lambda > 0$ is a constant, σ is an invertible $m \times m$-matrix, W_t is the m-dimensional Brownian motion, and $Z : \mathbb{R}^{2m} \times \mathcal{P}_2(\mathbb{R}^{2m}) \to \mathbb{R}^m$ satisfies

$$|Z(x_1, \nu_1) - Z(x_2, \nu_2)| \le \alpha_1 |x_1^{(1)} - x_2^{(1)}| + \alpha_2 |x_1^{(2)} - x_2^{(2)}| + \alpha_3 \mathbb{W}_2(\nu_1, \nu_2)$$

for some constants $\alpha_1, \alpha_2, \alpha_3 \ge 0$ and all $x_i = (x_i^{(1)}, x_i^{(2)}) \in \mathbb{R}^{2m}$, $\nu_i \in \mathcal{P}_2(\mathbb{R}^{2m}), 1 \le i \le 2$. If

$$4\lambda > \inf_{s > 0} \left\{ 2\alpha_3 s + \alpha_3 s^{-1} + 2\alpha_2 + \sqrt{4(1+\alpha_1)^2 + (2\alpha_2 + \alpha_3 s^{-1})^2} \right\}, \quad (5.6.5)$$

then $\{L_t^\nu\}_{\nu \in \mathcal{B}_{r,R}} \in LDP(J)$ for all $r, R > 0$.

Indeed, $b(x, \nu) := (x^{(2)} - \lambda x^{(1)}, Z(x, \nu) - \lambda x^{(2)})$ satisfies

$$2\langle b(x_1, \nu_1) - b(x_2, \nu_2), x_1 - x_2 \rangle$$
$$\le -2\lambda |x_1^{(1)} - x_2^{(1)}|^2 - 2(\lambda - \alpha_2)|x_1^{(2)} - x_2^{(2)}|^2$$
$$+ 2|x_1^{(2)} - x_2^{(2)}|\{(1+\alpha_1)|x_1^{(1)} - x_2^{(1)}| + \alpha_3 \mathbb{W}_2(\nu_1, \nu_2)\}$$
$$\le \alpha_3 s \mathbb{W}_2(\nu_1, \nu_2)^2 - \{2\lambda - \delta(1+\alpha_1)\}|x_1^{(1)} - x_2^{(1)}|^2$$
$$- \{2\lambda - 2\alpha_2 - \delta^{-1}(1+\alpha_1) - \alpha_3 s^{-1}\}|x_1^{(2)} - x_2^{(2)}|^2, \quad s, \delta > 0$$

for all $x_1, x_2 \in \mathbb{R}^{2m}$ and $\nu_1, \nu_2 \in \mathcal{P}_2(\mathbb{R}^{2m})$. Taking

$$\delta = \frac{2\alpha_2 + \alpha_3 s^{-1} + \sqrt{4(1+\alpha_1)^2 + (2\alpha_2 + \alpha_3 r^{-1})^2}}{2(1+\alpha_1)}$$

such that $\delta(1+\alpha_1) = 2\alpha_2 + \delta^{-1}(1+\alpha_1) + \alpha_3 s^{-1}$, we see that $(A^{5.9})$ holds for some $\kappa_1 > \kappa_2$ provided $2\lambda - \delta(1+\alpha_1) > \alpha_3 s$ for some $s > 0$, i.e. (5.6.5) implies $(A^{5.9})$. Moreover, it is easy to see that conditions in Remark 5.6.1(1) hold, see also [Guillin and Wang (2012); Wang and Zhang (2013)] for Harnack inequalities and gradient estimates on stochastic Hamiltonian systems which also imply the strong Feller and $\bar{\mu}$-irreducibility of $\bar{P}_t$. Therefore, the claimed assertion follows from Theorem 5.6.1(1).

5.6.2 LDP for Markov processes

We first introduce the rate function, i.e. the Donsker-Varadhan level 2 entropy function for continuous Markov processes on a Polish space E.

Consider the path space
$$\mathbf{C}_E := C([0,\infty) \to E) = \{w : [0,\infty) \ni t \mapsto w_t \in E \text{ is continuous}\}.$$
Let $\mathcal{P}(\mathbf{C}_E)$ be the set of all probability measures on $\mathbf{C}_E$, and $\mathcal{P}^s(\mathbf{C}_E)$ the set of all stationary (i.e. time-shift-invariant) elements in $\mathcal{P}(\mathbf{C}_E)$. For any $Q \in \mathcal{P}^s(\mathbf{C}_E)$, let $\bar{Q}$ be the unique stationary probability measure on $\bar{\mathbf{C}}_E := C(\mathbb{R} \to E)$ such that
$$\bar{Q}(\{w \in \bar{\mathbf{C}}_E : w(t_i) \in A_i, 1 \le i \le n\})$$
$$= Q(\{w \in \mathbf{C}_E : w(t_i + s) \in A_i, 1 \le i \le n\})$$
holds for any $n \ge 1, -\infty < t_1 < t_2 < \ldots < t_n < \infty, s \ge -t_1$, and $\{A_i\}_{1 \le i \le n} \subset \mathcal{B}(E)$. We call $\bar{Q}$ the stationary extension of Q to $\bar{\mathbf{C}}_E$. For any $s \le t$, let $\mathcal{F}_t^s := \sigma(\bar{\mathbf{C}}_E \ni w \mapsto w(u) : s \le u \le t)$. For a probability measure $\bar{Q}$ on $\bar{\mathbf{C}}_E$, let $\bar{Q}_{w-}$ be the regular conditional distribution of $\bar{Q}$ given $\mathcal{F}_0^{-\infty}$. Moreover, let $\operatorname{Ent}_{\mathcal{F}_1^0}$ be the Kullback-Leibler divergence (i.e. relative entropy) on the σ-field $\mathcal{F}_1^0$; that is, for any two probability measures μ_1, μ_2 on $\mathbf{C}_E$,
$$\operatorname{Ent}_{\mathcal{F}_1^0}(\mu_1|\mu_2) := \begin{cases} \int_{\mathbf{C}_E} (h \log h) \mathrm{d}\mu_2, & \text{if } \mathrm{d}\mu_1|_{\mathcal{F}_1^0} = h \mathrm{d}\mu_2|_{\mathcal{F}_1^0}, \\ \infty, & \text{otherwise}. \end{cases}$$

Now, for a standard Markov process on E with $\{P^x : x \in E\} \subset \mathcal{P}(\mathbf{C}_E)$, where P^x stands for the distribution of the process starting at x, the process level entropy function of Donsker-Varadhan is given by
$$H(Q) := \begin{cases} \int_{\bar{\mathbf{C}}_E} \operatorname{Ent}_{\mathcal{F}_1^0}(\bar{Q}_{w-}|P^{w(0)}) \bar{Q}(\mathrm{d}w), & \text{if } Q \subset \mathcal{P}^s(\mathbf{C}_E), \\ \infty, & \text{otherwise}. \end{cases}$$
Then the Donsker-Varadhan level 2 entropy function is defined as
$$J(\nu) := \inf\{H(Q) : Q \in \mathcal{P}^s(\mathbf{C}_E), Q(w(0) \in \cdot) = \nu\}, \quad \nu \in \mathcal{P}(E). \quad (5.6.6)$$
This function has compact level sets in $\mathcal{P}(E)$ under the τ- (hence the weak) topology, see for instance [Wu (2000)] and [Wu (2000b)]. For any $\nu \in \mathcal{P}(E)$, let $(X_t^\nu)_{t \ge 0}$ be the Markov process with initial distribution ν. Consider its empirical measure
$$L_t^\nu := \frac{1}{t} \int_0^t \delta_{X_s^\nu} \mathrm{d}s, \quad t > 0.$$
When $\nu = \delta_x$, we denote $X_t^\nu = X_t^x$ and $L_t^\nu = L_t^x$. Let μ be an invariant probability measure of P_t, where P_t is the Markov semigroup given by
$$P_t f(x) = \mathbb{E}[f(X_t^x)], \quad x \in E, t \ge 0, f \in \mathcal{B}_b(E).$$
We write $f \in \mathcal{D}_\mu(\mathcal{A})$ if $f \in L^\infty(\mu)$ and there exists $g \in L^\infty(\mu)$ such that $P_t f - f = \int_0^t P_s g \mathrm{d}s$ holds μ-a.e. for all $t \ge 0$. In this case, we denote $\mathcal{A}f = g$. We have the following formula for J.

Theorem 5.6.2 ([Wu (2000b)], Proposition B.10, Corollary B.11).
Assume that P_t has a unique invariant probability measure μ. Then

$$J(\nu) = \begin{cases} \sup\{\int_E \frac{-\mathcal{A}f}{f}d\nu : 1 \leq f \in \mathcal{D}_\mu(\mathcal{A})\}, & \text{if } \nu \lambda \mu, \\ \infty, & \text{otherwise.} \end{cases} \quad (5.6.7)$$

In particular, if the Markov process is associated with a symmetric Dirichlet form $(\mathcal{E}, \mathcal{D}(\mathcal{E}))$ in $L^2(\mu)$, then

$$J(\nu) = \begin{cases} \mathcal{E}(h^{\frac{1}{2}}, h^{\frac{1}{2}}), & \text{if } \nu = h\mu, h^{\frac{1}{2}} \in \mathcal{D}(\mathcal{E}), \\ \infty, & \text{otherwise.} \end{cases} \quad (5.6.8)$$

We now recall another result due to [Wu (2000b)] on the LDP for uniformly integrable Markov semigroups. Let $p \geq 1$ and let P be a bounded linear operator on $L^p(\mu)$. We call P uniformly integrable in $L^p(\mu)$ if

$$\lim_{R\to\infty} \sup_{\mu(|f|^p)\leq 1} \mu(|Pf|^p 1_{\{|Pf|>R\}}) = 0.$$

This LDP is established under the τ-topology induced by $f \in \mathcal{B}_b(E)$, and hence also holds under the weak topology. Let $\nu \in I_{q,L} := \{\nu = h\mu : \|h\|_{L^q(\mu)} \leq L\}$ for $q, L \in (1, \infty)$.

Theorem 5.6.3 ([Wu (2000b)], Theorem 5.1). Let μ be the unique invariant probability measure of P_t. If there exists $T \in (1, \infty)$ and $p \in (1, \infty)$ such that P_T is μ-irreducible and uniformly integrable in $L^p(\mu)$, then $\{L_t^\nu\}_{\nu \in I_{q,L}} \in LDP(J)$ under the τ-topology for all $q, L \in (1, \infty)$.

The next result due to [Wu (2000)] provides criteria on the LDP using the hitting time to compact sets. For any set $K \subset E$ and any $x \in E$, let

$$\tau_K^x := \inf\{t \geq 0 : X_t^x \in K\},$$

where X_t^x is the Markov process starting at x. We will use the following conditions where **(D1)** is weaker than **(D2)**:

(D1) For any $\lambda > 0$ there exist a constant $s > 0$ and a compact set $K \subset E$ such that for any compact set $K' \subset E$,

$$\sup_{x\in K} \mathbb{E}[e^{\lambda \tau_K^{X_s^x}}] < \infty, \quad \sup_{x\in K'} \mathbb{E}[e^{\lambda \tau_K^x}] < \infty. \quad (5.6.9)$$

(D2) For any $\lambda > 0$ there exists a compact set $K \subset E$ such that

$$\sup_{x\in E} \mathbb{E}[e^{\lambda \tau_K^x}] < \infty. \quad (5.6.10)$$

Theorem 5.6.4 ([Wu (2000)], Theorems 1.1,1.2). *Assume that P_t is a Feller Markov semigroup.*

(1) **(D1)** *implies $\{L_t^x\}_{x \in D} \in LDP_u(J)$ for any compact set $D \subset E$, and the inverse holds provided E is locally compact. If P_t is strong Feller and μ-irreducible for some $t > 0$, then $\{L_t^x\}_{x \in D} \in LDP(J)$ for compact $D \subset E$ if and only if* **(D1)** *holds.*

(2) **(D2)** *implies $\{L_t^\nu\}_{\nu \in \mathcal{P}(E)} \in LDP_u(J)$, and the inverse holds when E is locally compact. If moreover P_t is strong Feller and μ-irreducible for some $t > 0$, then $\{L_t^\nu\}_{\nu \in \mathcal{P}(E)} \in LDP(J)$ if and only if* **(D2)** *holds.*

Moreover, we introduce the following approximation lemma which is easy to prove and useful in applications, see for instance Theorem 3.2 in [Röckner et al. (2006)] for a stronger version called generalized contraction principle.

Lemma 5.6.5 (Approximation Lemma for LDP). *Let $\{(L_t^\nu)_{t>0}, (\bar{L}_t^\nu)_{t>0} : \nu \in \mathcal{I}\}$ be two families of stochastic processes on a Polish space (E, ρ) for an index set $\mathcal{I}$. If $(L_t^\nu)_{\nu \in \mathcal{I}} \in LDP_u(J)$ (respectively $LDP_l(J)$) and*

$$\lim_{t \to \infty} \frac{1}{\lambda(t)} \sup_{\nu \in \mathcal{I}} \log \mathbb{P}(\rho(L_t^\nu, \bar{L}_t^\nu) > \delta) = -\infty, \quad \delta > 0,$$

then $(\bar{L}_t^\nu)_{\nu \in \mathcal{I}} \in LDP_u(J)$ (respectively $LDP_l(J)$).

5.6.3 Proof of Theorem 5.6.1

Proof of Theorem 5.6.1(1). Let $\bar{X}_t^x$ denote the solution of (5.6.3) starting at x. According to Theorem 5.6.4 and Lemma 5.6.5, we only need to prove the following assertions:

(1_a) For any $\lambda > 0$, there exist a constant $s > 0$ and compact set $K \subset \mathbb{R}^d$, such that (5.6.9) holds for any compact set $K' \subset \mathbb{R}^d$ and

$$\tau_K^\perp := \inf\{t \geq 0 : X_t^x \subset K\}, \quad x \in \mathbb{R}^d.$$

(1_b) For any $N \geq 1$,

$$\sup_{\nu \in \mathcal{B}_{r,R}} \mathbb{E} e^{N \int_0^\infty \{1 \wedge |X_s^\nu - \bar{X}_s^0|^2\} ds} < \infty.$$

Indeed, by Theorem 5.6.4(1), (1_a) implies the upper LDP (LDP if $\bar{P}_t$ is strong Feller and $\bar{\mu}$-irreducible) for $\bar{L}_t^x$ locally uniformly in x, in particular, $\bar{L}_t^0$ satisfies the upper LDP (LDP if $\bar{P}_t$ is strong Feller and $\bar{\mu}$-irreducible).

On the other hand, by Lemma 5.6.5, (1_b) implies the equivalence of $\bar{L}_t^\nu$ and L_t^ν in $\mathrm{LDP}_u(J)$ and $\mathrm{LDP}_l(J)$. Then we prove the desired assertion for L_t^ν with $\nu \in \mathcal{B}_{r,R}$.

Verify (1_a). By $(A^{5.9})$, there exist constants $\alpha, \beta > 0$ such that

$$\mathrm{d}|\bar{X}_t|^2 \leq 2\{\alpha - \beta|\bar{X}_t|^2\}\mathrm{d}t + 2\langle \bar{X}_t, \sigma(\bar{\mu})\mathrm{d}W_t\rangle. \tag{5.6.11}$$

Let $\theta = \|\sigma\|_\infty^2$. Then for any $\varepsilon \in (0, \beta/\theta)$, there exist constants $c_1, c_2 > 0$ such that

$$\mathrm{d}\mathrm{e}^{\varepsilon|\bar{X}_t|^2} \leq 2\varepsilon\{\alpha - (\beta - \varepsilon\theta)|\bar{X}_t|^2\}\mathrm{e}^{\varepsilon|\bar{X}_t|^2}\mathrm{d}t + \mathrm{d}M_t$$
$$\leq \{c_1 - c_2 \mathrm{e}^{\varepsilon|\bar{X}_t|^2}\}\mathrm{d}t + \mathrm{d}M_t$$

for some martingale M_t. So,

$$\mathbb{E}\mathrm{e}^{\varepsilon|\bar{X}_t^x|^2} \leq \mathrm{e}^{\varepsilon|x|^2} + \frac{c_1}{c_2}, \quad x \in \mathbb{R}^d. \tag{5.6.12}$$

To estimate τ_K^x for $K := B_0(N)$, we take $N \geq N_0 := (2\alpha/\beta)^{\frac{1}{2}}$. Then (5.6.11) implies

$$\mathrm{d}|\bar{X}_t|^2 \leq -\beta|\bar{X}_t|^2\mathrm{d}t + 2\langle \bar{X}_t^x, \sigma(\bar{\mu})\mathrm{d}W_t\rangle, \quad t \leq \tau_K^x.$$

For any $\delta > 0$, we obtain

$$\mathbb{E}\mathrm{e}^{\delta \int_0^{t\wedge\tau_K^x} |\bar{X}_s^x|^2 \mathrm{d}s} \leq \mathrm{e}^{\delta\beta^{-1}|x|^2} \mathbb{E}\mathrm{e}^{2\delta\beta^{-1} \int_0^{t\wedge\tau_K^x} \langle \bar{X}_s^x, \sigma(\bar{X}_s^x, \mu)\mathrm{d}W_s\rangle}$$
$$\leq \mathrm{e}^{\delta\beta^{-1}|x|^2} \left(\mathbb{E}\mathrm{e}^{8\delta^2\beta^{-2}\theta \int_0^{t\wedge\tau_K^x} |\bar{X}_s^x|^2 \mathrm{d}s}\right)^{\frac{1}{2}}.$$

Thus, taking $\delta \leq \frac{\beta^2}{8\theta}$, we arrive at

$$\mathbb{E}\mathrm{e}^{\delta N^2(t\wedge\tau_K^x)} \leq \mathbb{E}\mathrm{e}^{\delta \int_0^{t\wedge\tau_K^x} |\bar{X}_s^x|^2 \mathrm{d}s} \leq \mathrm{e}^{2\delta\beta^{-1}|x|^2}.$$

Letting $t \uparrow \infty$ implies

$$\mathbb{E}\mathrm{e}^{\delta N^2 \tau_K^x} \leq \mathrm{e}^{2\delta\beta^{-1}|x|^2}, \quad x \in \mathbb{R}^d, N \geq N_0. \tag{5.6.13}$$

Combining this with the Markov property and (5.6.12), when $\delta \leq \frac{\varepsilon\beta}{2}$, we have

$$\mathbb{E}\mathrm{e}^{\delta N^2 \tau_K^{\bar{X}_s^x}} \leq \mathbb{E}\mathrm{e}^{2\delta\beta^{-1}|\bar{X}_s^x|^2} \leq \mathrm{e}^{\varepsilon|x|^2} + \frac{c_1}{c_2}, \quad x \in \mathbb{R}^d, s \geq 0, N \geq N_0.$$

Therefore, for any $\lambda > 0$ there exists compact $K \subset \mathbb{R}^d$ such that (5.6.9) holds.

Verify (1_b). Simply denote $X_t = X_t^\nu, \bar{X}_t = \bar{X}_t^0$ and $\nu_t = \mathcal{L}_{X_t^\nu} = P_t^*\nu$ for $\nu \in \mathcal{B}_{r,R}$. By ($A^{5.9}$), (5.6.2) and Itô's formula, we obtain
$$d|X_t - \bar{X}_t|^2 \leq \{-\kappa_1|X_t - \bar{X}_t|^2 + \kappa_2 e^{-(\kappa_1-\kappa_2)t}\mathbb{W}_2(\bar{\mu},\nu)^2\}dt$$
$$+ 2\langle X_t - \bar{X}_t, \{\sigma(\nu_t) - \sigma(\bar{\mu})\}dW_t\rangle.$$

Letting $g_t = \frac{|X_t - \bar{X}_t|^2}{1+|X_t - \bar{X}_t|^2}$, we derive
$$d\log(1 + |X_t - \bar{X}_t|^2) \leq \{-\kappa_1 g_t + \kappa_2 e^{-(\kappa_1-\kappa_2)t}\mathbb{W}_2(\bar{\mu},\nu)^2\}dt$$
$$+ \frac{2}{1+|X_t - \bar{X}_t|^2}\langle X_t - \bar{X}_t, \{\sigma(\nu_t) - \sigma(\bar{\mu})\}dW_t\rangle.$$

We deduce from this and (5.6.2) that for any $\lambda > 0$,

$$e^{-\frac{\lambda\kappa_2}{\kappa_1-\kappa_2}\mathbb{W}_2(\bar{\mu},\nu)^2}\mathbb{E}\left[e^{\lambda\kappa_1\int_0^t g_s ds}\right]$$
$$\leq \mathbb{E}\left[(1+|X_0|^2)^\lambda e^{\lambda\int_0^t \frac{2\langle X_s - \bar{X}_s, \{\sigma(\nu_s) - \sigma(\mu)\}dW_s\rangle}{1+|X_s - \bar{X}_s|^2}}\right]$$
$$\leq \mathbb{E}\left[(1+|X_0|^2)^\lambda \left(\mathbb{E}\left[e^{8\kappa_2\lambda^2\int_0^t g_s \mathbb{W}_2(\nu_s,\bar{\mu})^2 ds}\big|\mathcal{F}_0\right]\right)^{\frac{1}{2}}\right] \qquad (5.6.14)$$
$$\leq \{\nu((1+|\cdot|^2)^{2\lambda})\}^{\frac{1}{2}}\left(\mathbb{E}\left[e^{8\kappa_2\lambda^2\mathbb{W}_2(\nu,\bar{\mu})^2\int_0^t g_s e^{-(\kappa_1-\kappa_2)s}ds}\right]\right)^{\frac{1}{2}}$$
$$\leq C(\lambda, R)\left(\mathbb{E}\left[e^{\lambda\kappa_1\int_0^t g_s ds}\right]\right)^{\frac{1}{2}}, \quad t > 0$$

holds for some constant $C(\lambda, R) > 0$, where the last step is due to $g_s \leq 1$ and $\nu \subset \mathcal{B}_{r,R}$. Therefore,
$$\sup_{\nu \in \mathcal{B}_{r,R}} \mathbb{E}\left[e^{\lambda\kappa_1\int_0^\infty \frac{|X_s^\nu - \bar{X}_s^0|^2}{1+|X_s^\nu - \bar{X}_s^0|^2}ds}\right] < \infty, \quad \lambda > 0,$$
which implies (1_b). □

Proof of Theorem 5.6.1(2). Assume (5.6.4). For any $\lambda > 0$, it suffices to find a compact set $K \subset \mathbb{R}^d$ such that (5.6.10) holds for $\bar{X}$, and
$$\sup_{\nu \in \mathcal{P}_2} \mathbb{E}e^{N\int_0^\infty \{1 \wedge |X_s^\nu - \bar{X}_s^\nu|^2\}ds} < \infty, \quad N \geq 1.$$

Indeed, by Theorem 5.6.4(2), (1_b) and Lemma 5.6.5, this implies the upper LDP (LDP if $\bar{P}_t$ is strong Feller and $\bar{\mu}$-irreducible) for L_t^ν uniformly in $\nu \in \mathcal{P}_2$.

By (5.6.4), there exist constants $c_1, c_2 > 0$ such that
$$de^{|\bar{X}_t|^2} \leq \{c_1 - c_2|\bar{X}_t|^{2+\varepsilon}e^{|\bar{X}_t|^2}\}dt$$
$$+ 2e^{|\bar{X}_t|^2}\langle \bar{X}_t, \sigma(\bar{\mu})dW_t\rangle. \qquad (5.6.15)$$

This implies

$$h_x(t) := \mathbb{E} e^{|\bar{X}_t^x|^2} \leq c_1 t + e^{|x|^2} < \infty, \quad t \geq 0, x \in \mathbb{R}^d.$$

Moreover, by Jensen's inequality and the convexity of $[1,\infty) \ni r \mapsto r \log^{1+\varepsilon/2} r$, we deduce from (5.6.15) that

$$h_x(t) \leq h_x(0) + c_1 t - c_2 \int_0^t h_x(s) \log^{1+\varepsilon/2} h_x(s) ds, \quad t \geq 0.$$

This and the comparison theorem imply $h_x(t) \leq \psi(t)$, where $\psi(t)$ solves the ODE

$$\psi'(t) = c_1 - c_2 \psi(t) \log^{1+\varepsilon/2} \psi(t), \quad \psi(0) = h_x(0) = e^{|x|^2}.$$

So,

$$\sup_{x \in \mathbb{R}^d} h_x(t) \leq \sup_{\psi(0) \geq 1} \psi(t) =: c(t) < \infty. \tag{5.6.16}$$

On the other hand, by (5.6.15), there exist constants $N_0, \beta > 0$ such that for any $N \geq N_0$ and $K = B_0(N)$, we have

$$\begin{aligned} de^{|\bar{X}_t^x|^2} &\leq 2e^{|\bar{X}_t^x|^2} \langle \bar{X}_t^x, \sigma(\bar{\mu}) dW_t \rangle \\ &\quad - \beta |\bar{X}_t^x|^{2+\varepsilon} e^{|\bar{X}_t^x|^2} dt, \quad t \leq \tau_K^x. \end{aligned} \tag{5.6.17}$$

Combining this with (5.6.13) and using the Markov property, when $2\delta \leq \beta^2$, we find a constant $c > 0$ such that

$$\begin{aligned} \mathbb{E}[e^{\delta N^2 \tau_K^x}] &\leq e^{\delta N^2} + \mathbb{E}\big[e^{\delta N^2 \tau_K^x} 1_{\{\tau_K^x \geq 1\}}\big] \\ &\leq e^{\delta N^2} + \mathbb{E}\big[e^{\delta N^2 (1+\tau_K^{\bar{X}_1^x})} 1_{\{\tau_K^x \geq 1\}}\big] \\ &\leq e^{\delta N^2} (1 + \mathbb{E} e^{|\bar{X}_1^x|^2}) \leq e^{\delta N^2}(1+c) < \infty, \quad x \in \mathbb{R}^d, N \geq N_0. \end{aligned}$$

Therefore, for any $\lambda > 0$, there exists compact set K such that (5.6.10) holds.

Finally, repeating the proof of (6.3.3) using X_t^ν replacing $\bar{X}_t^x$, we derive

$$\sup_{\nu \in \mathcal{P}_2} \mathbb{E}[e^{|X_1^\nu|^2}] < \infty.$$

This together with (6.3.3) yields

$$\sup_{\nu \in \mathcal{P}_2} \mathbb{E}\big[e^{|X_1^\nu|^2} + e^{|\bar{X}_1^\nu|^2}\big] < \infty. \tag{5.6.18}$$

On the other hand, as in (5.6.14) but integrating from time 1, we obtain

$$e^{-\frac{\lambda\kappa_2}{\kappa_1-\kappa_2}W_2(\bar{\mu},\nu)^2}\mathbb{E}\Big[e^{\lambda\kappa_1\int_1^t\frac{|X_s^\nu-\bar{X}_s^\nu|^2}{1+|X_s^\nu-\bar{X}_s^\nu|^2}ds}\Big]$$

$$\leq \mathbb{E}\Big[(1+|X_1^\nu-\bar{X}_1^\nu|^2)^\lambda e^{\lambda\int_1^t\frac{2\langle X_s^\nu-\bar{X}_s^\nu,\{\sigma(\nu_s)-\sigma(\bar{\mu})\}dW_s\rangle}{1+|X_s^\nu-\bar{X}_s^\nu|^2}}\Big]$$

$$\leq \Big\{\mathbb{E}\big[(1+|X_1^\nu-\bar{X}_1^\nu|^2)^{2\lambda}\big]\Big\}^{\frac{1}{2}}$$

$$\times \Big(\mathbb{E}\Big[e^{\lambda\kappa_1 W_2(P_1^*\nu,\bar{\mu})^2\int_1^t\frac{|X_s^\nu-\bar{X}_s^\nu|^2 e^{-(\kappa_1-\kappa_2)s}}{1+|X_s^\nu-\bar{X}_s^\nu|^2}ds}\Big]\Big)^{\frac{1}{2}}, \quad t > 1.$$

Combining this with (5.6.18), we derive

$$\sup_{\nu\in\mathcal{P}_2}\mathbb{E}e^{\lambda\kappa_1\int_1^\infty\frac{|X_s^\nu-\bar{X}_s^\nu|^2}{1+|X_s^\nu-\bar{X}_s^\nu|^2}ds} < \infty, \quad \lambda \geq 1.$$

Therefore, the desired assertion holds. □

5.7 Notes

The condition (5.1.2) in Theorem 5.1.1 is new, in references one uses the following stronger condition:

$$\sup_{t\geq 0}\mathbb{W}(P_t^*\mu_0,\mu_0) < \infty,$$

see for instance [Wang (2018)] for $\mathbb{W} = \mathbb{W}_2$.

Theorem 5.6.4 is taken from [Wang (2023c)] where the exponential ergodicity is derived for a weighted variation norm $\|\cdot\|_V$ replacing $\|\cdot\|_{var}$, see also Theorem 6.5.1 in Chapter 6 for the case with reflection.

Section 5.3 is organized from [Ren and Wang (2021b)], while Sections 5.4–5.5 are due to [Wang (2023a)], where a result for order-preserving McKean-Vlasov SDEs is also presented, see also [Ren et al. (2021)] for extensions to the time periodic setting.

There are a number of papers studying the Freidlin-Wentzell type large deviations for DDSDEs with small noise, see for instance [Fan et al. (2023)] and references therein, where fractional noise is considered.

Finally, Section 5.6 is taken from [Ren and Wang (2021a)], where the Donsker-Varadhan large deviations are derived for more general models, including path-distribution dependent SDEs in Hilbert space.

Chapter 6

DDSDEs with Reflecting Boundary

In this chapter, we study reflected DDSDEs, i.e. DDSDEs in a domain $D \subset \mathbb{R}^d$ with reflecting boundary. We first introduce the link of reflected DDSDE and nonlinear Neumann problem, then study this type of DDSDEs for the well-posedness, regularity estimates and exponential ergodicity respectively.

6.1 Reflected DDSDE for nonlinear Neumann problem

Let $\mathcal{P}(\bar{D})$ be the space of all probability measures on the closure $\bar{D}$ of D, equipped with the weak topology. We regard $\mathcal{P}(\bar{D}) \subset \mathcal{P}$ by letting $\mu(\mathbb{R}^d \setminus \bar{D}) = 0$ for $\mu \in \mathcal{P}(\bar{D})$. For any $k \geq 0$, let

$$\mathcal{P}_k(D) := \mathcal{P}(\bar{D}) \cap \mathcal{P}_k.$$

Consider the following reflected DDSDE on $\bar{D}$ for fixed $T > 0$:

$$\mathrm{d}X_t = b_t(X_t, \mathcal{L}_{X_t})\mathrm{d}t + \sigma_t(X_t, \mathcal{L}_{X_t})\mathrm{d}W_t + \mathbf{n}(X_t)\mathrm{d}l_t, \quad t \in [0,T], \quad (6.1.1)$$

where

$$b : [0,T] \times D \times \mathcal{P}(\bar{D}) \to \mathbb{R}^d, \quad \sigma : [0,T] \times D \times \mathcal{P}(\bar{D}) \to \mathbb{R}^d \otimes \mathbb{R}^m$$

are measurable, and $W_t, \mathbf{n}(X_t)$ and l_t are as in (2.0.1).

For a subspace $\hat{\mathcal{P}}$ of $\mathcal{P}(\bar{D})$ equipped with a complete metric $\hat{d}$, let $C^w([0,T]; \hat{\mathcal{P}})$ and $C_b^w([0,T]; \hat{\mathcal{P}})$ be in (3.1.2).

Definition 6.1.1. (1) A pair $(X_t, l_t)_{t \in [0,T]}$ is called a solution of (6.1.1), if X_t is an adapted continuous process on $\bar{D}$, l_t is an adapted continuous increasing process with $\mathrm{d}l_t$ supported on $\{t \geq 0 : X_t \in \partial D\}$, such that $\mathbb{P}$-a.s.

$$\int_0^t \{|b_r(X_r, \mathcal{L}_{X_r})| + \|\sigma_r(X_r, \mathcal{L}_{X_r})\|^2\}\mathrm{d}r < \infty, \quad t \in [0,T],$$

and for some measurable map $\partial D \ni x \mapsto \mathbf{n}(x) \in \mathcal{N}_x$, $\mathbb{P}$-a.s.
$$X_t = X_0 + \int_0^t b_r(X_r, \mathcal{L}_{X_r}) \mathrm{d}r + \int_0^t \sigma_r(X_r, \mathcal{L}_{X_r}) \mathrm{d}W_r + \int_0^t \mathbf{n}(X_r) \mathrm{d}l_r, \ t \in [0, T].$$
In this case, l_t is called the local time of X_t on ∂D. We call (6.1.1) strongly well-posed for distributions in a subspace $\hat{\mathcal{P}} \subset \mathcal{P}(\bar{D})$, if for any $\mathcal{F}_0$-measurable variable X_0 with $\mathcal{L}_{X_0} \in \hat{\mathcal{P}}$, the equation has a unique solution with $\mathcal{L}_{X.} \in C_b^w([0,T]; \hat{\mathcal{P}})$; if this is true for $\hat{\mathcal{P}} = \mathcal{P}(\bar{D})$, we call it strongly well-posed.

(2) A triple $(X_t, l_t, W_t)_{t \in [0,T]}$ is called a weak solution of (6.1.1), if W_t is an m-dimensional Brownian motion under a probability space and $(X_t, l_t)_{t \in [0,T]}$ solves (6.1.1). (6.1.1) is called weakly unique (resp. jointly weakly unique), if for any two weak solutions $(X_t, l_t, W_t)_{t \in [0,T]}$ under probability $\mathbb{P}$ and $(\tilde{X}_t, \tilde{l}_t, \tilde{W}_t)_{t \in [0,T]}$ under probability $\tilde{\mathbb{P}}$, $\mathcal{L}_{X_0|\mathbb{P}} = \mathcal{L}_{\tilde{X}_0|\tilde{\mathbb{P}}}$ implies $\mathcal{L}_{(X_t, l_t)_{t \geq 0}|\mathbb{P}} = \mathcal{L}_{(\tilde{X}_t, \tilde{l}_t)_{t \in [0,T]}|\tilde{\mathbb{P}}}$ (resp. $\mathcal{L}_{(X_t, l_t, W_t)_{t \in [0,T]}|\mathbb{P}} = \mathcal{L}_{(\tilde{X}_t, \tilde{l}_t, \tilde{W}_t)_{t \in [0,T]}|\tilde{\mathbb{P}}}$). We call (6.1.1) weakly well-posed for distributions in $\hat{\mathcal{P}} \subset \mathcal{P}(\bar{D})$, if it has a unique weak solution for initial distributions in $\hat{\mathcal{P}}$ with $\mathcal{L}_{X.} \in C_b^w([0,T]; \hat{\mathcal{P}})$; it is called weakly well-posed if moreover $\hat{\mathcal{P}} = \mathcal{P}(\bar{D})$.

(3) We call (6.1.1) well-posed (for distributions in $\hat{\mathcal{P}}$), if it is both strongly and weakly well-posed (for distributions in $\hat{\mathcal{P}}$).

To characterize the nonlinear Fokker-Planck equation associated with (6.1.1), consider the following time-distribution dependent second order differential operator:
$$L_{t,\mu} := \frac{1}{2} \mathrm{tr}\{(\sigma_t \sigma_t^*)(\cdot, \mu) \nabla^2\} + \nabla_{b_t(\cdot, \mu)}, \ t \in [0,T], \mu \in \mathcal{P}(\bar{D}). \quad (6.1.2)$$
Assume that for any $\mu \in C^w([0,\infty); \mathcal{P}(\bar{D}))$, see (3.1.2),
$$\sigma_t^\mu(x) := \sigma_t(x, \mu_t), \ b_t^\mu(x) := b_t(x, \mu_t) \quad (6.1.3)$$
satisfy $\|\sigma^\mu\|^2 + |b^\mu| \in L^1_{loc}([0,T] \times \bar{D}; \mathrm{d}t\, \mu_t(\mathrm{d}x))$.

Let $C_N^2(\bar{D})$ be the class of C^2-functions on $\bar{D}$ with compact support satisfying the Neumann boundary condition $\nabla_\mathbf{n} f|_{\partial D} = 0$. By Itô's formula, for any (weak) solution X_t to (6.1.1), $\mu_t := \mathcal{L}_{X_t}$ solves the nonlinear Fokker-Planck equation
$$\partial_t \mu_t = L_{t,\mu_t}^* \mu_t \ \text{ with respect to } C_N^2(\bar{D}), \ t \in [0,T] \quad (6.1.4)$$
for probability measures on $\bar{D}$, in the sense that $\mu_\cdot \in C^w([0,\infty); \mathcal{P}(\bar{D}))$ and
$$\mu_t(f) := \int_{\bar{D}} f \mathrm{d}\mu_t = \mu_0(f) + \int_0^t \mu_s(L_{s,\mu_s} f) \mathrm{d}s, \quad (6.1.5)$$
$$t \in [0,T], f \in C_N^2(\bar{D}).$$

To understand (6.1.4) as a nonlinear Neumann problem on D, let L^*_{t,μ_t} be the adjoint operator of L_{t,μ_t}: for any
$$g \in L^1_{loc}(D, (\|\sigma_t(x,\mu_t)\|^2 + |b_t(x,\mu_t)|)dx),$$
$L^*_{t,\mu_t} g$ is the linear functional on $C_0^2(D)$ (the class of C^2-functions on D with compact support) given by
$$C_0^2(D) \ni f \mapsto \int_D \{fL^*_{t,\mu_t}g\}(x)dx := \int_D \{gL_{t,\mu_t}f\}(x)dx. \tag{6.1.6}$$
Assume that $\mathcal{L}_{X_t}$ has a density function ρ_t, i.e. $\mu_t := \mathcal{L}_{X_t} = \rho_t(x)dx$. It is the case under a general non-degenerate or Hörmander condition (see for instance [Bogachev et al. (2015)]), and Krylov's estimate (2.2.3) or (2.2.34) implies that ρ_t exists for a.e. $t \in (0,T]$. When $\partial D \in C^2$, (6.1.4) implies that ρ_t solves the following nonlinear Neumann problem on $\bar{D}$:
$$\partial_t \rho_t = L^*_{t,\rho_t} \rho_t, \quad \nabla_{t,\mathbf{n}} \rho_t|_{\partial D} = 0, \ t \in [0,T] \tag{6.1.7}$$
in the weak sense, where $L_{t,\rho_t} := L_{t,\rho_t(x)dx}$, and for a function g on ∂D
$$\nabla_{t,\mathbf{n}} g := \nabla_{\sigma_t \sigma_t^* \mathbf{n}} g + \mathrm{div}_{\partial D}(g\pi \sigma_t \sigma_t^* \mathbf{n})$$
for the divergence $\mathrm{div}_{\partial D}$ on ∂D and the projection π to the tangent space of ∂D.
$$\pi_x v := v - \langle v, \mathbf{n}(x)\rangle \mathbf{n}(x), \quad v \in \mathbb{R}^d, x \in \partial D.$$
If in particular $\sigma\sigma^* \mathbf{n} = \lambda \mathbf{n}$ holds on $[0,\infty) \times \partial D$ for a function $\lambda \neq 0$ a.e., $\nabla_{t,\mathbf{n}} \rho_t|_{\partial D} = 0$ is equivalent to the standard Neumann boundary condition $\nabla_{\mathbf{n}} \rho_t|_{\partial D} = 0$.

We now deduce (6.1.7) from (6.1.5). Firstly, by (6.1.6), (6.1.5) implies
$$\int_D (f\rho_t)(x)dx = \int_D (f\rho_0)(x)dx + \int_0^t ds \int_D (fL^*_{s,\rho_s}\rho_s)(x)dx,$$
$$f \in C_0^2(D), t \in [0,T],$$
so that $\partial_t \rho_t = L^*_{t,\rho_t}\rho_t$. Next, by the integration by parts formula, (6.1.5) implies
$$I := \int_D (f\rho_t)(x)dx - \int_D (f\rho_0)(x)dx = \int_0^t ds \int_D (\rho_s L_{s,\rho_s}f)(x)dx$$
$$= \int_0^t \left(\int_D (fL^*_{s,\rho_s}\rho_s)(x)dx + \int_{\partial D} \{f\nabla_{\sigma_s\sigma_s^*\mathbf{n}}\rho_s - \rho_s \nabla_{\sigma_s\sigma_s^*\mathbf{n}}f\}(x)dx \right) ds$$
$$= \int_D f(x)dx \int_0^t (\partial_s \rho_s)(x)ds$$
$$+ \int_0^t ds \int_{\partial D} \{f \nabla_{\sigma_s\sigma_s^*\mathbf{n}} \rho_s + f\mathrm{div}_{\partial D}(\rho_s \pi \sigma_s \sigma_s^* \mathbf{n})\}(x)dx$$
$$= I + \int_0^t ds \int_{\partial D} \{f(\nabla_{s,\mathbf{n}}\rho_t)\}(x)dx, \quad f \in C_N^2(\bar{D}), t \in [0,T].$$
Thus, $\nabla_{t,\mathbf{n}} \rho_t|_{\partial D} = 0$.

6.2 Well-posedness: Singular case

By (3.5.1), $\|\mu\|_k := \mu(|\cdot|^k)^{\frac{1}{k}}$ for $k > 0$, and $\|\mu\|_0 := 1$. We make the following assumption.

$(A^{6.1})$ Let $k \geq 0$. $\sigma^\mu = \sigma$ does not depend on μ, and there exists $\hat{\mu} \in \mathcal{P}_k(\bar{D})$ such that at least one of the following two conditions holds.
(1) $(A^{2.3})$ holds for $\hat{b} := b(\cdot, \hat{\mu})$ replacing b, and there exists a constant $\alpha \geq 0$ such that for any $t \in [0, T]$, $x \in \bar{D}$, and $\mu, \nu \in \mathcal{P}_k(\bar{D})$,

$$|b_t(x, \mu) - \hat{b}_t^{(1)}(x)| \leq f_0(t, x) + \alpha \|\mu\|_k, \qquad (6.2.1)$$

$$|b_t(x, \mu) - b_t(x, \nu)| \leq \{\|\mu - \nu\|_{k,var} + \mathbb{W}_k(\mu, \nu)\} \sum_{i=0}^{l} f_i(t, x). \qquad (6.2.2)$$

(2) $(A^{2.2})$ holds for $\hat{b} := b(\cdot, \hat{\mu})$ replacing b, and (6.2.1)–(6.2.2) hold for $\sqrt{f_i}$ replacing f_i, $0 \leq i \leq l$.

Since $\hat{b}_t^{(1)}$ is regular, (6.2.1) gives a control for the singular term of b^μ. Moreover, (6.2.2) is a Lipschitz condition on $b_t(x, \cdot)$ in $\|\cdot\|_{k,var} + \mathbb{W}_k$ with Lipschitz coefficient singular in (t, x).

Theorem 6.2.1. *Assume* $(A^{6.1})$.

(1) (6.1.1) *is weakly well-posed for distributions in* $\mathcal{P}_k(\bar{D})$. *Moreover, for any* $\gamma \in \mathcal{P}_k(\bar{D})$, *and any* $n > 0$, *there exists a constant* $c > 0$, *such that*

$$\mathbb{E}\left[\sup_{t \in [0,T]} |X_t|^n \Big| X_0\right] \leq c(1 + |X_0|^n), \quad \mathbb{E}e^{nl_T} \leq c \qquad (6.2.3)$$

holds for the solution with $\mathcal{L}_{X_0} = \gamma$.
(2) (6.1.1) *is well-posed for distributions in* $\mathcal{P}_k(\bar{D})$ *in each of the following situations:*
  (i) $d = 1$ *and* $(A^{6.1})(2)$ *holds.*
  (ii) $(A^{6.1})(1)$ *holds with* $p_1 > 2$ *in* $(A^{2.3})$ *for* $\hat{b}$ *replacing* b.

To prove Theorem 6.2.1, we first present a general result on the well-posedness of the reflected DDSDE (6.1.1) by using that of the reflected SDE (3.1.5).

For any $k \geq 0, \gamma \in \mathcal{P}_k, N \geq 2$, let

$$\mathcal{P}_{k,\gamma}^{T,N}(\bar{D}) = \Big\{\mu \in C_b^w([0,T]; \mathcal{P}_k(\bar{D})) : \mu_0 = \gamma, \sup_{t \in [0,T]} e^{-Nt}(1 + \mu_t(|\cdot|^k)) \leq N\Big\}.$$

Then as $N \uparrow \infty$,
$$\mathcal{P}_{k,\gamma}^{T,N}(\bar{D}) \uparrow \mathcal{P}_{k,\gamma}^{T}(\bar{D}) := \{\mu \in C_b^w([0,T]; \mathcal{P}_k(\bar{D})) : \mu_0 = \gamma\}. \quad (6.2.4)$$
For any $\mu \in \mathcal{P}_{k,\gamma}^{T}(\bar{D})$, we will assume that the reflected SDE
$$\begin{aligned} &\mathrm{d}X_t^{\mu,\gamma} = b_t(X_t^{\mu,\gamma}, \mu_t)\mathrm{d}t + \sigma_t(X_t^{\mu,\gamma})\mathrm{d}W_t + \mathbf{n}(X_t^{\mu,\gamma})\mathrm{d}l_t^{\mu,\gamma}, \\ &t \in [0,T], \mathcal{L}_{X_0^{\mu,\gamma}} = \gamma \end{aligned} \quad (6.2.5)$$
has a unique weak solution with
$$\Phi_t^\gamma \mu := \mathcal{L}_{X_t^{\mu,\gamma}} \in \mathcal{P}_k(\bar{D}), \quad t \in [0,T].$$

$(A^{6.2})$ Let $k \geq 0$. For any $\gamma \in \mathcal{P}_k(\bar{D})$ and $\mu \in \mathcal{P}_{k,\gamma}^T(\bar{D})$, (6.2.5) has a unique weak solution, and there exist constants $l \in \mathbb{N}, \{p_i, q_i\}_{0 \leq i \leq l} \subset (1, \infty), N_0 \geq 2$ and increasing maps $C : [N_0, \infty) \to (0, \infty)$ and $F : [N_0, \infty) \times [0, \infty) \to (0, \infty)$, such that for any $N \geq N_0$ and $\mu \in \mathcal{P}_{k,\gamma}^{T,N}(\bar{D})$, the (weak) solution satisfies
$$\Psi_\cdot^\gamma := \mathcal{L}_{(X^{\mu,\gamma})} \in \mathcal{P}_{k,\gamma}^{T,N}, \quad (6.2.6)$$
$$\left(\mathbb{E}\left[(1+|X_t^{\mu,\gamma}|^k)^2 | X_0^{\mu,\gamma}\right]\right)^{\frac{1}{2}} \leq C(N)(1+|X_0^{\mu,\gamma}|^k), \quad t \in [0,T], \quad (6.2.7)$$
$$\begin{aligned} &\mathbb{E}\left(\int_0^t g_s(X_s^{\mu,\gamma})\mathrm{d}s\right)^2 \leq C(N)\|g\|_{\tilde{L}_{q_i}^{p_i}(t_0,t_1)}^2, \\ &\mathbb{E}\mathrm{e}^{\int_0^t g_s(X_s^{\mu,\gamma})\mathrm{d}s} \leq F(N, \|g\|_{\tilde{L}_{q_i}^{p_i}(T,D)}), \end{aligned} \quad (6.2.8)$$
$$t \in [0,T], \ g \in \tilde{L}_{q_i}^{p_i}(t,D), \ 0 \leq i \leq l.$$

Obviously, when $k = 0$, conditions (6.2.6) and (6.2.7) hold for $N_0 = 2$.

Theorem 6.2.2. *Assume* $(A^{6.2})$ *and let* $\sigma_t(x, \mu) = \sigma_t(x)$ *not depend on* μ. *Let* $1 \leq f_i$ *with* $|f_i|^2 \in \tilde{L}_{q_i}^{p_i}(T, D), 0 \leq i \leq l$. *Assume that there exist a measurable map* $\Gamma : [0, T] \times \bar{D} \times \mathcal{P}(\bar{D}) \to \mathbb{R}^m$ *such that*
$$\begin{aligned} &b_t(x, \nu) - b_t(x, \mu) = \sigma_t(x)\Gamma_t(x, \nu, \mu), \\ &x \in \bar{D}, t \in [0,T], \nu, \mu \in \mathcal{P}_k(\bar{D}). \end{aligned} \quad (6.2.9)$$

(1) *If for any* $x \in \bar{D}, t \in [0,T]$, *and* $\nu, \mu \in \mathcal{P}_k(\bar{D})$,
$$|\Gamma_t(x, \nu, \mu)| \leq \|\nu - \mu\|_{k,\mathrm{var}} \sum_{i=0}^{l} f_i(t,x), \quad (6.2.10)$$

then (6.1.1) *is weakly well-posed for distributions in* $\mathcal{P}_k(\bar{D})$. *If, furthermore, in* $(A^{6.2})$ *the SDE* (6.2.5) *is strongly well-posed for any* $\gamma \in \mathcal{P}_k(\bar{D})$ *and* $\mu \in \mathcal{P}_{k,\gamma}^T(\bar{D})$, *so is* (6.1.1) *for distributions in* $\mathcal{P}_k(\bar{D})$.

(2) Let $k > 1$. If for any $\mu, \nu \in \mathcal{P}_k(\bar{D})$ and $(t,x) \in [0,T] \times \bar{D}$,

$$|\Gamma_t(x,\nu,\mu)| \leq \{\|\nu - \mu\|_{k,var} + \mathbb{W}_k(\mu,\nu)\} \sum_{i=0}^{l} f_i(t,x), \qquad (6.2.11)$$

and for any $\gamma \in \mathcal{P}_k(\bar{D})$ and $N \geq N_0$, there exists a constant $C(N) > 0$ such that for any $\mu, \nu \in \mathcal{P}_{k,\gamma}^{T,N}(\bar{D})$,

$$\mathbb{W}_k(\Phi_t^\gamma \mu, \Phi_t^\gamma \nu)^{2k} \\ \leq C(N) \int_0^t \{\|\mu_s - \nu_s\|_{k,var}^{2k} + \mathbb{W}_k(\mu_s,\nu_s)^{2k}\} \mathrm{d}s, \quad t \in [0,T], \qquad (6.2.12)$$

then assertions in (1) holds.

Proof. Let $\gamma \in \mathcal{P}_k(\bar{D})$. Then the weak solution to (6.2.5) is a weak solution to (6.1.1) if and only if μ is a fixed point of the map Φ^γ in $\mathcal{P}_{k,\gamma}^T(\bar{D})$. So, if Φ^γ on $\mathcal{P}_{k,\gamma}^T(\bar{D})$ has a unique fixed point in $\mathcal{P}_{k,\gamma}^T(\bar{D})$, then the (weak) well-posedness of (6.2.5) implies that of (6.1.1). Thus, by (6.2.4), it suffices to show that for any $N \geq N_0$, Φ^γ has a unique fixed point in $\mathcal{P}_{k,\gamma}^{T,N}(\bar{D})$. By (6.2.6) and the fixed point theorem, we only need to prove that for any $N \geq N_0$, Φ^γ is contractive with respect to a complete metric on $\mathcal{P}_{k,\gamma}^{T,N}(\bar{D})$.

(1) For any $\lambda > 0$, consider the metric

$$\mathbb{W}_{k,\lambda,var}(\mu,\nu) := \sup_{t \in [0,T]} \mathrm{e}^{-\lambda t} \|\mu_t - \nu_t\|_{k,var}, \quad \mu,\nu \in \mathcal{P}_{k,\gamma}^{T,N}(\bar{D}).$$

Let $(X_t^{\mu,\gamma}, l_t^{\mu,\gamma})$ solve (6.2.5) for some Brownian motion W_t on a complete probability filtration space $(\Omega, \{\mathcal{F}_t\}, \mathbb{P})$. By (6.2.8), (6.2.10) or (6.2.11) with $|f|^2 \in \tilde{L}_{q'}^p(T,D)$, we find a constant $c_1 > 0$ depending on N such that

$$\sup_{\mu,\nu \in \mathcal{P}_{k,\gamma}^{T,N}(\bar{D})} \mathbb{E}\big(\mathrm{e}^{2\int_0^T |\Gamma_s(X_s^{\mu,\gamma},\nu_s,\mu_s)|^2 \mathrm{d}s} \big| \mathcal{F}_0\big) \leq c_1^2,$$

$$\sup_{\mu \in \mathcal{P}_{k,\gamma}^{T,N}(\bar{D})} \mathbb{E}\bigg(\bigg(\int_0^T g_s(X_s^{\mu,\gamma})\mathrm{d}s\bigg)^2 \bigg| \mathcal{F}_0\bigg) \leq c_1^2 \|g\|_{\tilde{L}_{q_i}^{p_i}(T,D)}^2, \qquad (6.2.13)$$

$$g \in \tilde{L}_{q_i}^{p_i}(T), 0 \leq i \leq l.$$

Then by Girsanov's theorem,

$$\tilde{W}_t := W_t - \int_0^t \Gamma_s(X_s^{\mu,\gamma},\nu_s,\mu_s)\mathrm{d}s, \quad t \in [0,T]$$

is a Brownian motion under the probability $\mathbb{Q} := R_T \mathbb{P}$, where

$$R_t := \mathrm{e}^{\int_0^t \langle \Gamma_s(X_s^{\mu,\gamma},\nu_s,\mu_s),\mathrm{d}W_s\rangle - \frac{1}{2}\int_0^t |\Gamma_s(X_s^{\mu,\gamma},\nu_s,\mu_s)|^2 \mathrm{d}s}, \quad t \in [0,T]$$

is a $\mathbb{P}$-martingale. By (6.2.9), we may formulate (6.2.5) as

$$dX_t^{\mu,\gamma} = b_t(X_t^{\mu,\gamma},\nu_t)dt + \sigma_t(X_t^{\mu,\gamma})d\tilde{W}_t + \mathbf{n}(X_t^{\mu,\gamma})dl_t^{\mu,\gamma}, \ t\in[0,T], \mathcal{L}_{X_0^{\mu,\gamma}} = \gamma.$$

By the weak uniqueness due to $(A^{6.2})$, the definition of $\|\cdot\|_{k,var}$, (6.2.7) and (6.2.9), we obtain

$$\begin{aligned}\|\Phi_t^\gamma \mu - \Phi_t^\gamma \nu\|_{k,var} &= \sup_{|\tilde{f}|\leq 1+|\cdot|^k} \left|\mathbb{E}\left[(R_t-1)\tilde{f}(X_t^{\mu,\gamma})\right]\right| \\ &\leq \mathbb{E}\left[(1+|X_t^{\mu,\gamma}|^k)|R_t-1|\right] \\ &\leq \mathbb{E}\left[\left\{\mathbb{E}\left((1+|X_t^{\mu,\gamma}|^k)^2|\mathcal{F}_0\right)\right\}^{\frac{1}{2}}\left\{\mathbb{E}(|R_t-1|^2|\mathcal{F}_0)\right\}^{\frac{1}{2}}\right] \\ &\leq C(N)\mathbb{E}\left[(1+|X_0^{\mu,\gamma}|^k)\left\{\mathbb{E}(e^{\int_0^t |\Gamma_s(X_s^{\mu,\gamma},\nu_s,\mu_s)|^2 ds} - 1|\mathcal{F}_0)\right\}^{\frac{1}{2}}\right].\end{aligned}$$ (6.2.14)

Moreover, (6.2.13) implies

$$\begin{aligned}&\mathbb{E}(e^{\int_0^t |\Gamma_s(X_s^{\mu,\gamma},\nu_s,\mu_s)|^2 ds} - 1|\mathcal{F}_0) \\ &< \mathbb{E}\left(e^{\int_0^t |\Gamma_s(X_s^{\mu,\gamma},\nu_s,\mu_s)|^2 ds} \int_0^t |\Gamma_s(X_s^{\mu,\gamma},\nu_s,\mu_s)|^2 ds \Big| \mathcal{F}_0\right) \\ &\leq c_1 \sum_{i=0}^l \left\{\mathbb{E}\left(\left(\int_0^t |f_i(s,X_s^{\mu,\gamma})|^2 \|\mu_s-\nu_s\|_{k,var}^2 ds\right)^2 \Big| \mathcal{F}_0\right)\right\}^{\frac{1}{2}} \\ &\leq c_1 \sum_{i=0}^l e^{2\lambda t} \mathbb{W}_{k,\lambda,var}(\mu,\nu)^2 \mathbb{E}\left(\left(\int_0^t |f_i(s,X_s^{\mu,\gamma})|^2 e^{-2\lambda(t-s)} ds\right)^2 \Big| \mathcal{F}_0\right)^{\frac{1}{2}} \\ &\leq c_1^2 \sum_{i=0}^l e^{2\lambda t} \|f_i^2 e^{-2\lambda(t-\cdot)}\|_{\tilde{L}_{q_i}^{p_i}(t,D)} \mathbb{W}_{k,\lambda,var}(\mu,\nu)^2, \ t\in[0,T].\end{aligned}$$

Combining this with (6.2.14) and the definition of $\mathbb{W}_{k,\lambda,var}$, we obtain

$$\begin{aligned}&\mathbb{W}_{k,\lambda,var}(\Phi^\gamma\mu,\Phi^\gamma\nu) \\ &\leq C(N)(1+\gamma(|\cdot|^k))c_1\sqrt{\varepsilon(\lambda)}\mathbb{W}_{k,\lambda,var}, \ \lambda > 0,\end{aligned}$$ (6.2.15)

where

$$\varepsilon(\lambda) := \sup_{t\in[0,T]} \sum_{i=0}^l \|f_i^2 e^{-2\lambda(t-\cdot)}\|_{\tilde{L}_{q_i}^{p_i}(t)} \downarrow 0 \ \text{ as } \ \lambda \uparrow \infty.$$

So, Φ^γ is contractive on $(\mathcal{P}_{k,\gamma}^{T,N}(\bar{D}), \mathbb{W}_{k,\lambda,var})$ for large enough $\lambda > 0$.

(2) Let $k > 1$. We consider the metric $\tilde{\mathbb{W}}_{k,\lambda,var} := \mathbb{W}_{k,\lambda,var} + \mathbb{W}_{k,\lambda}$, where

$$\mathbb{W}_{k,\lambda}(\mu,\nu) := \sup_{t\in[0,T]} e^{-\lambda t}\mathbb{W}_k(\mu_t,\nu_t), \ \mu,\nu \in \mathcal{P}_{k,\gamma}^{T,N}(\bar{D}).$$

By using (6.2.11) replacing (6.2.10), instead of (6.2.15) we find constants $\{C(N,\lambda) > 0\}_{\lambda>0}$ with $C(N,\lambda) \to 0$ as $\lambda \to \infty$ such that

$$\mathbb{W}_{k,\lambda,var}(\Phi^\gamma \mu, \Phi^\gamma \nu) \\ \leq C(N,\lambda)\tilde{\mathbb{W}}_{k,\lambda,var}(\mu,\nu), \quad \lambda > 0, \mu, \nu \in \mathcal{P}_{k,\gamma}^{T,N}(\bar{D}). \tag{6.2.16}$$

On the other hand, (6.2.12) yields

$$\mathbb{W}_{k,\lambda}(\Phi^\gamma \mu, \Phi^\gamma \nu) \\ \leq \sup_{t \in [0,T]} \left(C(N) e^{-\lambda k t} \int_0^t \{\|\mu_s - \nu_s\|_{k,var}^{2k} + \mathbb{W}_k(\mu_s, \nu_s)^{2k}\} ds \right)^{\frac{1}{2k}} \\ \leq \tilde{\mathbb{W}}_{k,\lambda,var}(\mu,\nu) \sup_{t \in [0,T]} \left(C(N) \int_0^t e^{-2\lambda k(t-s)} ds \right)^{\frac{1}{2k}} \\ \leq \frac{C(N)^{\frac{1}{2k}}}{(2\lambda k)^{\frac{1}{2k}}} \tilde{\mathbb{W}}_{k,\lambda,var}(\mu,\nu), \quad \lambda > 0.$$

Combining this with (6.2.16), we conclude that Φ^γ is contractive in $\mathcal{P}_{k,\gamma}^{T,N}(\bar{D})$ under the metric $\tilde{\mathbb{W}}_{k,\lambda,var}$ when λ is large enough, and hence finish the proof. □

Proof of Theorem 6.2.1. Let $\gamma \in \mathcal{P}_k(\bar{D})$ be fixed. By (6.2.1), for any $i = 2, 3$ and $\mu \in C_b^w([0,T]; \mathcal{P}_k(\bar{D}))$, condition $(A^{2.i})$ for $\hat{b}$ replacing b implies the same condition for μ. So, by Theorem 6.2.2, $(A^{6.1})$ implies the weak well-posedness of (6.2.5) for distributions in $\mathcal{P}_k(\bar{D})$ with

$$\Phi_t^\gamma \mu \in \mathcal{P}_k(\bar{D}), \quad \mathbb{E} e^{\lambda l_T^{\mu,\gamma}} < \infty, \\ \lambda > 0, \gamma \in \mathcal{P}_k(\bar{D}), \mu \subset C([0,\infty); \mathcal{P}_k(\bar{D})), \tag{6.2.17}$$

and also implies the strong well-posedness of (6.2.5) in each situation of Theorem 6.2.1(2). Moreover, by Lemma 2.2.1 and Lemma 2.2.3, $(A^{6.1})$ implies that (6.2.8) holds for any $(p,q) \in \mathcal{K}$, as well as for $(p,q) = (p_2/2, q_2/2)$ under $(A^{2.3})$ for $\hat{b}$ replacing b, (6.2.9) with (6.2.10) holds for $k \leq 1$ due to (3.7.19), and (6.2.9) with (6.2.11) holds for $k > 1$. Therefore, by Theorem 6.2.2, it remains to verify (6.2.3), (6.2.6), (6.2.7), and (6.2.12) for $k > 1$. Since (6.2.7) and (6.2.6) are trivial for $k = 0$, we only need to prove:

- (6.2.3);
- (6.2.7) and (6.2.6) for $k > 0$;
- (6.2.12) for $k > 1$ for case (i);
- (6.2.12) for $k > 1$ for case (ii).

In the following we simply denote
$$f(s,x) = \sum_{i=0}^{l} f_i(s,x).$$

(a) Let $X_t^{\mu,\gamma}$ solve (6.2.5). We first prove that under $(A^{6.1})$, there exist a constant $c > 0$ and an increasing function $c : [1,\infty) \to (0,\infty)$ such that for any $m \geq 1$, $\mu \in \mathcal{P}_{k,\gamma}^T(\bar{D})$, and $t \in [0,T]$,

$$\begin{aligned}\mathbb{E}\left(\int_0^t |f_s(X_s^{\mu,\gamma})|^2 \mathrm{d}s\right)^m &\leq c(m) + c(m)\left(\int_0^t \|\mu_s\|_k^2 \mathrm{d}s\right)^m, \\ \mathbb{E}\exp\left[m\int_0^t |f_s(X_s^{\mu,\gamma})|^2 \mathrm{d}s\right] &\leq c(m)\exp\left[c\int_0^t \|\mu_s\|_k^2 \mathrm{d}s\right].\end{aligned} \qquad (6.2.18)$$

We will prove these estimates by Lemmas 2.2.1 and 2.2.3 for the following reflected SDE:
$$\mathrm{d}\hat{X}_s = \hat{b}_s(\hat{X}_s)\mathrm{d}s + \sigma_s(\hat{X}_s)\mathrm{d}W_s + \mathbf{n}(\hat{X}_s)\mathrm{d}\hat{l}_s, \quad \hat{X}_0 = X_0^{\mu,\gamma}, s \in [0,t].$$

By (2.2.35) under $(A^{6.1})(1)$, and (2.2.4) under $(A^{6.1})(2)$, for any $m \geq 1$ we find a constant $c_1(m) > 0$ such that
$$\mathbb{E}e^{m\int_0^t (|\hat{b}_s^{(0)}|^2 + |f_s|^2)(X_s^{\mu,\gamma})\mathrm{d}s} \leq c_1(m), \quad t \in [0,T]. \qquad (6.2.19)$$

Let $\gamma_s = \{[\sigma_s^*(\sigma_s\sigma_s^*)^{-1}](b_s^\mu - \hat{b}_s)\}(\hat{X}_s)$, and
$$R_t := e^{\int_0^t \langle \gamma_s, \mathrm{d}W_s\rangle - \frac{1}{2}\int_0^t |\gamma_s|^2 \mathrm{d}s}, \quad \tilde{W}_s := W_s - \int_0^s \gamma_r \mathrm{d}r, \quad s \in [0,t].$$

By Girsanov theorem, $(\tilde{W}_s)_{s\in[0,t]}$ is a Brownian motion under $R_t\mathbb{P}$, and the SDE for $\hat{X}_s$ becomes
$$\mathrm{d}\hat{X}_s = b_s^\mu(\hat{X}_s)\mathrm{d}s + \sigma_s(\hat{X}_s)\mathrm{d}\tilde{W}_s + \mathbf{n}(\hat{X}_s)\mathrm{d}\hat{l}_s, \quad \hat{X}_0 = X_0^{\mu,\gamma}, s \in [0,t].$$

So, by (6.2.1), (6.2.19) and Hölder's inequality, we find constants c_1, c, $c(m) > 0$ such that
$$\mathbb{E}e^{m\int_0^t |f_s(X_s^{\mu,\gamma})|^2\mathrm{d}s} = \mathbb{E}\left[R_t e^{m\int_0^t |f_s(\hat{X}_s)|^2 \mathrm{d}s}\right] \leq \left(\mathbb{E}e^{2m\int_0^t |f_s(\hat{X}_s)|^2\mathrm{d}s}\right)^{\frac{1}{2}}\left(\mathbb{E}[R_t^2]\right)^{\frac{1}{2}}$$
$$\leq \sqrt{c_1(2m)}\left(\mathbb{E}e^{c_1\int_0^t\{|\hat{b}_s^{(0)}|^2 + (f_s + \alpha\|\mu_s\|_k)^2\}(\hat{X}_s)\mathrm{d}s}\right)^{\frac{1}{2}} \leq c(m)e^{c\int_0^t \|\mu_s\|_k^2\mathrm{d}s}.$$

Next, taking $c_2(m) > 0$ large enough such that the function $r \mapsto [\log(r + c_2(m))]^m$ is concave for $r \geq 0$, this and Jensen's inequality imply

$$\mathbb{E}\left(\int_0^t |f_s(X_s^{\mu,\gamma})|^2\mathrm{d}s\right)^m \leq \mathbb{E}\left(\left[\log(c_2(m) + e^{\int_0^t |f_s(X_s^{\mu,\gamma})|^2\mathrm{d}s})\right]^m\right)$$
$$\leq \left[\log(c_2(m) + \mathbb{E}e^{\int_0^t |f_s(X_s^{\mu,\gamma})|^2\mathrm{d}s})\right]^m \leq c(m) + c(m)\left(\int_0^t \|\mu_s\|_k^2 \mathrm{d}s\right)^m$$

holds for some constant $c(m) > 0$. Therefore, (6.2.18) holds.

(b) Proof of (6.2.6). Simply denote $X_t = X_t^{\mu,\gamma}$. By (6.2.1), the boundedness of σ and the condition on $\hat{b}^{(1)}$ in $(A^{2.1})$ which follows from $(A^{2.3})$ due to Lemma 2.2.2, we find a constant $c_1 > 0$ such that

$$L_{t,\mu} := \frac{1}{2}\mathrm{tr}\{\sigma_t \sigma_t^* \nabla^2\} + \nabla_{b_t^\mu}, \quad L^{\sigma,\hat{b}^{(1)}} := \frac{1}{2}\mathrm{tr}\{\sigma_t \sigma_t^* \nabla^2\} + \nabla_{\hat{b}_t^{(1)}}$$

satisfy

$$L_{t,\mu}\tilde{\rho} \geq L_t^{\sigma,\hat{b}^{(1)}}\tilde{\rho} - |b_t^\mu - \hat{b}_t^{(1)}| \cdot |\nabla \tilde{\rho}| \geq -c_1(f_t + \|\mu_t\|_k).$$

Since $\langle \mathbf{n}, \tilde{\rho} \rangle|_{\partial D} \geq 1$, by Itô's formula we obtain

$$d\tilde{\rho}(X_t) \geq -c_1\{f_t(X_t) + \|\mu_t\|_k\}dt + dM_t + dl_t \qquad (6.2.20)$$

for some martingale M_t with $\langle M \rangle_t \leq ct$ for some constant $c > 0$. This together with (6.2.18) yields that for some constant $k_0 > 0$,

$$\mathbb{E}l_t^k \leq k_0 + k_0 \mathbb{E}\left(\int_0^t \{f_s(X_s) + \|\mu_s\|_k\}ds\right)^k.$$

Combining this with (2.2.3), (6.2.2), (6.2.18) and $\|\sigma\|_\infty < \infty$, and using the formula

$$X_t = X_0 + \int_0^t b_s^\mu(X_s)ds + \int_0^t \sigma_s(X_s)dW_s + \mathbf{n}(X_t)dl_t, \quad \mathcal{L}_{X_0} = \gamma,$$

we find constants $k_1, k_2 > 0$ such that

$$\mathbb{E}(1+|X_t|^k)$$
$$\leq k_1(1+\|\gamma\|_k^k) + k_1 \mathbb{E}\left(\int_0^t \{|X_s| + |f_s(X_s)| + \|\mu_s\|_k\}ds\right)^k \qquad (6.2.21)$$
$$\leq k_2 + k_2 \mathbb{E}\left(\int_0^t \{|X_s|^2 + \|\mu_s\|_k^2\}ds\right)^{\frac{k}{2}}, \quad t \in [0, T].$$

(b1) When $k \geq 2$, by (6.2.21) we find a constant $k_3 > 0$ such that

$$\mathbb{E}(1+|X_t|^k) \leq k_2 + k_3 \int_0^t \{\mathbb{E}|X_s|^k + \|\mu_s\|_k^k\}ds, \quad t \in [0, T].$$

By Gronwall's lemma, and noting that $\mu \in \mathcal{P}_{k,\gamma}^{T,N}(\bar{D})$, we find constant $k_4 > 0$ such that

$$\mathbb{E}(1+|X_t|^k) \leq k_4 + k_4 \int_0^t (1+\|\mu_s\|_k^k)ds$$
$$\leq k_4 + k_4 N e^{NT} \int_0^t e^{-N(t-s)}ds \leq 2k_4 e^{Nt}, \quad t \in [0, T].$$

Taking $N_0 = 2k_4$ we derive
$$\sup_{t\in[0,T]} e^{-Nt}(1+\|\Phi_t^\gamma \mu\|_k^k)$$
$$= \sup_{t\in[0,T]} e^{-Nt}\mathbb{E}(1+|X_t|^k) \le N_0 \le N, \quad N \ge N_0, \mu \in \mathcal{P}_{k,\gamma}^{T,N}(\bar{D}),$$
so that (6.2.6) holds.

(b2) When $k \in (0,2)$, by BDG's inequality, and by the same reason leading to (6.2.21), we find constants $k_5, k_6, k_7 > 0$ such that
$$U_t := \mathbb{E}\Big[\sup_{s\in[0,t]}(1+|X_s|^k)\Big] \le k_5 + k_5 \mathbb{E}\Big(\int_0^t \{|X_s|^2 + \|\mu_s\|_k^2\}ds\Big)^{\frac{k}{2}}$$
$$\le k_6 + k_6\Big(\int_0^t \|\mu_s\|_k^2 ds\Big)^{\frac{k}{2}}$$
$$+ k_6\mathbb{E}\Big\{\Big[\sup_{s\in[0,t]}(|X_s|^k + \|\mu_s\|_k^k)\Big]^{1-\frac{k}{2}}\Big(\int_0^t |X_s|^k ds\Big)^{\frac{k}{2}}\Big\}$$
$$\le K_6 + \frac{1}{2}U_t + k_7\int_0^t U_s ds + k_6\Big(\int_0^t \|\mu_s\|_k^2 ds\Big)^{\frac{k}{2}}, \quad t \in [0,T].$$

By Gronwall's lemma, we find constants $k_8, k_9 > 0$ such that for any $\mu \in \mathcal{P}_{k,\gamma}^{T,N}(\bar{D})$,
$$\mathbb{E}(1+|X_t|^k) \le U_t \le k_8 + k_8\Big(\int_0^t \|\mu_s\|_k^2 ds\Big)^{\frac{k}{2}}$$
$$\le k_8 + k_8 N e^{Nt}\Big(\int_0^t e^{-2N(t-s)/k} ds\Big)^{\frac{k}{2}} \le k_8 + k_9 N^{1-\frac{k}{2}} e^{Nt}, \quad t \in [0,T].$$

Thus, there exists $N_0 > 0$ such that for any $N \ge N_0$,
$$\sup_{t\in[0,T]} e^{-Nt}(1+\|\Phi_t^\gamma \mu\|_k^k) = \sup_{t\in[0,T]} e^{-Nt}\mathbb{E}(1+|X_t|^k)$$
$$\le k_8 + k_9 N^{1-\frac{k}{2}} \le N, \quad \mu \in \mathcal{P}_{k,\gamma}^{T,N}(\bar{D}),$$
which implies (6.2.6).

(c) Proofs of (6.2.7) and (6.2.3). Simply denote $(\hat{X}_t, \hat{l}_t) = (X_t^{\mu,\gamma}, l_t^{\mu,\gamma})$ in (6.2.5) for $\mu_t = \hat{\mu}, t \in [0,T]$; that is,
$$d\hat{X}_t = \hat{b}_t(\hat{X}_t)dt + \sigma(\hat{X}_t)dW_t + \mathbf{n}(\hat{X}_t)d\hat{l}_t, \quad \mathcal{L}_{\hat{X}_0} = \gamma. \tag{6.2.22}$$

By $(A^{6.1})$ and Theorem 6.2.2, this SDE has a unique weak solution, and for any $n \ge 1$ there exists a constant $c > 0$ such that
$$\mathbb{E}\Big[\sup_{t\in[0,T]} |\hat{X}_t|^n \Big| \hat{X}_0\Big] \le c(1+|\hat{X}_0|^n), \quad \mathbb{E}e^{n\hat{l}_T} \le c. \tag{6.2.23}$$

So, by (6.2.2), Lemma 2.2.1, Lemma 2.2.3 under $(A^{2.3})$ for $\hat{b}$ replacing b, and Girsanov's theorem,

$$\tilde{W}_t := W_t - \int_0^t \{\sigma_s^*(\sigma_s\sigma_s^*)^{-1}\}(\hat{X}_s)\{b_s^\mu(\hat{X}_s) - \hat{b}_s(\hat{X}_s)\}\mathrm{d}s, \quad t \in [0,T]$$

is a $\mathbb{Q}$-Brownian motion for $\mathbb{Q} := R_T\mathbb{P}$, where

$$R_T := \mathrm{e}^{\int_0^T \langle \eta_s, \mathrm{d}W_s\rangle - \frac{1}{2}\int_0^T |\eta_s|^2\mathrm{d}s},$$

$$\eta_s := \{\sigma_s^*(\sigma_s\sigma_s^*)^{-1}\}(\hat{X}_s)\{b_s^\mu(\hat{X}_s) - \hat{b}_s(\hat{X}_s)\}.$$

By $(A^{6.1})$, (6.2.23), Lemma 2.2.1 when $|f|^2 \in \tilde{L}_q^p(T)$ for some $(p,q) \in \mathcal{K}$, and Lemma 2.2.3 when $(A^{2.3})$ holds for $\hat{b}$ replacing b, we find an increasing function F such that

$$\mathbb{E}(|R_T|^2|\mathcal{F}_0) \leq \mathbb{E}(\mathrm{e}^{\int_0^T |f_s(\hat{X}_s)|^2\{\|\mu_s-\hat{\mu}\|_{k,var}+\mathbb{W}_k(\mu_s,\hat{\mu})\}^2\mathrm{d}s}|\mathcal{F}_0) \leq F(\|\mu\|_{k,T}),$$

where $\|\mu\|_{k,T} := \sup_{t\in[0,T]}\mu_t(|\cdot|^k)$. Reformulating (6.2.22) as

$$\mathrm{d}\hat{X}_t = b_t^\mu(\hat{X}_t)\mathrm{d}t + \sigma_t(\hat{X}_t)\mathrm{d}\tilde{W}_t + \mathbf{n}(\hat{X}_t)\mathrm{d}\hat{l}_t, \quad \mathcal{L}_{\hat{X}_0} = \gamma,$$

by the weak uniqueness we have $\mathcal{L}_{\hat{X}|\mathbb{Q}} = \mathcal{L}_{X^{\mu,\gamma}}$, so that (6.2.23) with $2n$ replacing n implies

$$\mathbb{E}\Big[\sup_{t\in[0,T]}|X_t^{\mu,\gamma}|^n\Big|\mathcal{F}_0\Big] = \mathbb{E}_\mathbb{Q}\Big[\sup_{t\in[0,T]}|\hat{X}_t|^n\Big|\mathcal{F}_0\Big]$$

$$\leq \Big(\mathbb{E}\Big[\sup_{t\in[0,T]}|\hat{X}_t|^{2n}\Big|\mathcal{F}_0\Big]\Big)^{\frac{1}{2}}(\mathbb{E}R_T^2|\mathcal{F}_0)^{\frac{1}{2}} \leq c(1+|\hat{X}_0|^n)F(\|\mu\|_{k,T}).$$

Since $\sup_{\mu\in\mathcal{P}_{k,\gamma}^{T,N}(\bar{D})}\|\mu\|_{k,T}$ is a finite increasing function of N, this implies (6.2.7).

Finally, since $X_t := X_t^{\mu,\gamma}$ solves (6.1.1) with initial distribution γ and $\mu_t = \mathcal{L}_{X_t}$ (i.e. μ is the fixed point of Φ^γ), and since Φ^γ has a unique fixed point in $\mathcal{P}_{k,\gamma}^{T,N}(\bar{D})$ for some $N > 0$ depending on γ as proved in the proof of Theorem 6.2.2 using (6.2.8) and (6.2.6), we have $\mathcal{L}_X \in \mathcal{P}_{k,\gamma}^{T,N}(\bar{D})$, and hence (6.2.3) follows from (2.3.1).

(d) Proof of (6.2.12) for $k > 1$ in case (i). Let u_t^λ and Θ_t^λ be constructed for b^μ replacing b in the proof of Theorem 2.4.1 under $(A^{2.2})$ for $d = 1$. Let $X_0^{(1)} = X_0^{(2)}$ be $\mathcal{F}_0$-measurable with $\mathcal{L}_{X_0^{(i)}} = \gamma, i = 1,2$. As explained in the beginning of the present proof, the following reflected SDEs are well-posed:

$$\mathrm{d}X_t^{(1)} = b_t(X_t^{(1)},\mu_t)\mathrm{d}t + \sigma_t(X_t^{(1)})\mathrm{d}W_t + \mathbf{n}(X_t^{(1)})\mathrm{d}l_t^{(1)},$$
$$\mathrm{d}X_t^{(2)} = b_t(X_t^{(2)},\nu_t)\mathrm{d}t + \sigma_t(X_t^{(2)})\mathrm{d}W_t + \mathbf{n}(X_t^{(2)})\mathrm{d}l_t^{(2)}, \quad t \in [0,T].$$

Then instead of (2.4.15), the processes
$$Y_t^{(i)} := \Theta_t^\lambda(X_t^{(i)}), \quad i = 1, 2$$
satisfy
$$dY_t^{(1)} = B_t(Y_t^{(1)})dt + \Sigma_t(Y_t^{(1)})dW_t + \{1 + \nabla u_t^\lambda(X_t^{(1)})\}\mathbf{n}(X_t^{(1)})dl_t^{(1)},$$
$$dY_t^{(2)} = B_t(Y_t^{(2)})dt + \Sigma_t(Y_t^{(2)})dW_t + \{1 + \nabla u_t^\lambda(X_t^{(2)})\}\mathbf{n}(X_t^{(2)})dl_t^{(2)}$$
$$+ \{b_t(X_t^{(2)}, \nu_t) - b_t(X_t^{(2)}, \mu_t)\}dt.$$

By (6.2.2), $Y_0^{(1)} = Y_0^{(2)}$, Itô's formula to $|Y_t^{(1)} - Y_t^{(2)}|^{2k}$ with this formula replacing (2.4.15), the calculations in the proof of Theorem 2.4.1 under $(A^{2.2})$ for $d = 1$ yield that when λ is large enough,

$$|Y_t^{(1)} - Y_t^{(2)}|^{2k} + M_t \leq c_1 \int_0^t |Y_s^{(1)} - Y_s^{(2)}|^{2k} d\mathcal{L}_s$$
$$+ c_1 \int_0^t |Y_s^{(1)} - Y_s^{(2)}|^{2k-1} f_s(X_s^{(2)})\{\|\mu_s - \nu_s\|_{k,var} + \mathbb{W}_k(\mu_s, \nu_s)\}ds$$
$$\leq c_1 \int_0^t |Y_s^{(1)} - Y_s^{(2)}|^{2k} d\tilde{\mathcal{L}}_s + c_1 \int_0^t \{\|\mu_s - \nu_s\|_{k,var} + \mathbb{W}_k(\mu_s, \nu_s)\}^{2k} ds$$

holds for all $t \in [0, T]$ and some constant $c_1 > 0$ depending on N uniformly in $\mu \in \mathcal{P}_{k,\gamma}^{T,N}(\bar{D})$, some martingale M_t, $\mathcal{L}_t$ in (2.4.19), and

$$\tilde{\mathcal{L}}_t := \mathcal{L}_t + \int_0^t |f_s(X_s^{(2)})|^{\frac{2k}{2k-1}} ds \leq \mathcal{L}_t + \int_0^t |f_s(X_s^{(2)})|^2 ds.$$

By the stochastic Gronwall lemma, Lemma 2.2.1, we find a constant $c_2 > 0$ depending on N such that

$$\left(\mathbb{E}\left[\sup_{s \in [0,t]} |Y_s^{(1)} - Y_s^{(2)}|^k\right]\right)^2 \leq c_2 \int_0^t \{\|\mu_s - \nu_s\|_{k,var} + \mathbb{W}_k(\mu_s, \nu_s)\}^{2k} ds,$$

which implies (6.2.12), since by (2.4.14) and the definition of Φ^γ, there exists a constant $c \geq 0$ depending on N such that
$$(\mathbb{E}|Y_t^{(1)} - Y_t^{(2)}|^k)^2 \geq c(\mathbb{E}|X_t^{(1)} - X_t^{(2)}|^k)^2 \geq c\mathbb{W}_k(\Phi_t^\gamma \mu, \Phi_t^\gamma \nu)^{2k}.$$

(e) Proof of (6.2.12) for $k > 1$ in case (ii). Let $u_t^{\lambda,n}$ solve (2.4.20) for $l_t = l_{t,\nu}, b^{(0)} = b_t^{(0)}(\cdot, \nu_t)$ and the mollifying approximation $b^{0,n} = b_t^{0,n}(\cdot, \nu_t)$. Then in (2.4.24) the equation for ξ_t becomes

$$d\xi_t = \left\{\lambda u_t^{\lambda,n}(X_t^{(1)}) - \lambda u_t^{\lambda,n}(X_t^{(2)}) + (b_t^{(0)} - b_t^{0,n})(X_t^{(1)})\right.$$
$$\left. - (b_t^{(0)} - b_t^{0,n})(X_t^{(2)}) + b(X_t^{(2)}, \mu_t) - b_t(X_t^{(2)}, \nu_t)\right\}dt$$
$$+ \{[(\nabla \Theta_t^{\lambda,n})\sigma_t](X_t^{(1)}) - [(\nabla \Theta_t^{\lambda,n})\sigma_t](X_t^{(2)})\}dW_t$$
$$+ \mathbf{n}(X_t^{(1)})dl_t^{(1)} - \mathbf{n}(X_t^{(2)})dl_t^{(2)}.$$

So, as shown in step (d) by (6.2.2), instead of (2.4.33), we have

$$|X^{(1)}_{t\wedge\tau_m} - X^{(2)}_{t\wedge\tau_m}|^{2k}$$
$$\leq G_m(t) + c_2 \int_0^{t\wedge\tau_m} |X^{(1)}_{s\wedge\tau_m} - X^{2}_{s\wedge\tau_m}|^{2k} \mathrm{d}\tilde{\mathcal{L}}_s + \tilde{M}_t$$

for some local martingale $\tilde{M}_t$,

$$\tilde{\mathcal{L}}_t := \mathcal{L}_t + \int_0^t |f_s(X^{(2)}_s)|^2 \mathrm{d}s, \quad t \in [0,T]$$

for $\mathcal{L}_t$ in (2.4.32), and due to $X^{(1)}_0 = X^{(2)}_0 = X_0$ in the present setting,

$$G_m(t) := \int_0^t \Big\{ c_2 m^{2(k-1)} \sum_{i=1}^{2} |b_s^{(0)} - b_s^{0,n}|^2(X_s^{(i)}) + \big(\|\mu_s - \nu_s\|_{k,var} + \mathbb{W}_k(\mu_s, \nu_s)\big)^{2k} \Big\} \mathrm{d}s.$$

By the stochastic Gronwall inequality, Lemma 2.2.3 and (6.2.18), we find a constant $c > 0$ such that

$$\mathbb{W}_k(\Phi_t^\gamma \mu, \Phi_t^\gamma \nu)^{2k} \leq (\mathbb{E}|X_t^{(1)} - X_t^{(2)}|^k)^2$$
$$\leq c \liminf_{m\to\infty} \liminf_{n\to\infty} \mathbb{E} G_m(t) \qquad (6.2.24)$$
$$= c \int_0^t \big\{ \|\mu_s - \nu_s\|^{2k}_{k,var} + \mathbb{W}_k(\mu_s, \nu_s)^{2k} \big\} \mathrm{d}s.$$

Thus, (6.2.12) holds.

6.3 Well-posedness: Monotone case

For any $k \geq 0$, $\mathcal{P}_k(\bar{D})$ is a complete metric space under the L^k-Wasserstein distance $\mathbb{W}_k$, where $\mathbb{W}_0(\mu, \nu) := \frac{1}{2}\|\mu - \nu\|_{var}$ and

$$\mathbb{W}_k(\mu, \nu) := \inf_{\pi \in \mathcal{C}(\mu,\nu)} \left(\int_{\bar{D}\times\bar{D}} |x-y|^k \pi(\mathrm{d}x, \mathrm{d}y) \right)^{\frac{1}{1\vee k}}, \quad \mu, \nu \in \mathcal{P}_k(\bar{D}), \ k > 0.$$

In the following, we first study the well-posedness of (6.1.1) for distributions in $\mathcal{P}_k(\bar{D})$ with $k > 1$, then extend to a setting including $k = 1$.

($A^{6.3}$) Let $k > 1$. (**D**) holds, b and σ are bounded on bounded subsets of $[0,T] \times \bar{D} \times \mathcal{P}_k(\bar{D})$, and the following two conditions hold.

(1) There exists $0 < K \in L^1([0,T])$ such that
$$\|\sigma_t(x,\mu) - \sigma_t(y,\nu)\|_{HS}^2 + 2\langle x-y, b_t(x,\mu) - b_t(y,\nu)\rangle^+$$
$$\leq K_t\{|x-y|^2 + |x-y|\mathbb{W}_k(\mu,\nu) + 1_{\{k\geq 2\}}\mathbb{W}_k(\mu,\nu)^2\},$$
$$t \in [0,T], x,y \in \bar{D}, \mu,\nu \in \mathcal{P}_k(\bar{D}).$$

(2) There exists a subset $\tilde{\partial}D \subset \partial D$ such that
$$\langle y-x, \mathbf{n}(x)\rangle \geq 0, \quad x \in \partial D \setminus \tilde{\partial}D, \ y \in \bar{D}, \qquad (6.3.1)$$
and when $\tilde{\partial}D \neq \emptyset$, there exists $\tilde{\rho} \in C_b^2(\bar{D})$ such that $\tilde{\rho}|_{\partial D} = 0$, $\langle \nabla \tilde{\rho}, \mathbf{n}\rangle|_{\partial D} \geq 1_{\tilde{\partial}D}$ and
$$\sup_{(t,x)\in[0,T]\times\bar{D}} \{\|\{\sigma_t(\cdot,\mu_t)\}^*\nabla\tilde{\rho}\|^2 + \langle b_t(\cdot,\mu_t), \nabla\tilde{\rho}\rangle^-\}(x) \qquad (6.3.2)$$
$$< \infty, \quad \mu \in C_b^w([0,T]; \mathcal{P}_k(\bar{D})).$$

$(A^{6.3})(1)$ is a monotone condition, when $k \geq 2$ it allows $\sigma_t(x,\mu)$ depending on μ, but when $k \in [1,2)$ it implies that $\sigma_t(x,\mu) = \sigma_t(x)$ does not depend on μ.

$(A^{6.3})(2)$ holds for $\tilde{\partial}D = \emptyset$ when D is convex, and it holds for $\tilde{\partial}D = \partial D$ if $\partial D \in C_b^2$ and for some $r_0 > 0$
$$\sup_{(t,x)\in[0,T]\times\partial_{r_0}D} \{\|(\sigma_t^\mu)^*\nabla\rho\|^2(x) + \langle b_t^\mu, \nabla\rho\rangle^-(x)\} < \infty, \mu \in C_b^w([0,T]; \mathcal{P}_k(\bar{D})),$$
where in the second case we may take $\tilde{\rho} = h \circ \rho$ for $0 \leq h \in C^\infty([0,\infty))$ with $h(r) = r$ for $r \leq r_0/2$ and $h(r) = r_0$ for $r \geq r_0$. In general, $(A^{6.3})(2)$ includes the case where ∂D is partly convex and partly C_b^2.

Theorem 6.3.1. *Assume* $(A^{6.3})$. *Then* (6.1.1) *is well-posed for distributions in* $\mathcal{P}_k(\bar{D})$, *and there exist a constant* $C > 0$ *and a map* $c : [1,\infty) \to (0,\infty)$ *such that for any solution* (X_t, l_t) *of* (6.1.1) *with* $\mathcal{L}_{X_0} \in \mathcal{P}_k(\bar{D})$,
$$\mathbb{E}\Big[\sup_{t\in[0,T]}|X_t|^k\Big] \leq C(1 + \mathbb{E}|X_0|^k), \qquad (6.3.3)$$

$$\mathbb{E}e^{n\tilde{l}_T} \leq c(n), \quad n \geq 1, \tilde{l}_T := \int_0^T 1_{\tilde{\partial}D}(X_t)\mathrm{d}l_t. \qquad (6.3.4)$$

Proof. Let X_0 be $\mathcal{F}_0$-measurable with $\gamma := \mathcal{L}_{X_0} \in \mathcal{P}_k(\bar{D})$. Then
$$\mathcal{P}_{k,\gamma}^T(\bar{D}) := \{\mu \in C_b^w([0,T]; \mathcal{P}_k(\bar{D})) : \mu_0 = \gamma\}$$
is a complete space under the following metric for any $\lambda > 0$:
$$\mathbb{W}_k^{\lambda,T}(\mu,\nu) := \sup_{t\in[0,T]} e^{-\lambda t}\mathbb{W}_k(\mu_t,\nu_t), \quad \mu,\nu \in \mathcal{P}_{k,\gamma}^T(\bar{D}).$$

By Lemma 2.3.4, ($A^{6.3}$) implies the well-posedness of the following reflected SDE for any $\mu \in \mathcal{P}_{k,\gamma}^T(\bar{D})$:

$$dX_t^\mu = b_t(X_t^\mu, \mu_t)dt + \sigma_t(X_t^\mu, \mu_t)dW_t + \mathbf{n}(X_t^\mu)dl_t^\mu, \quad X_0^\mu = X_0, \quad (6.3.5)$$

and the solution satisfies

$$\mathbb{E}\left[\sup_{t \in [0,T]} |X_t^\mu|^k\right] < \infty. \quad (6.3.6)$$

So, as explained in the proof of Theorem 6.2.2, for the well-posedness of (6.1.1), it suffices to prove the contraction of the map

$$\mathcal{P}_{k,\gamma}^T(\bar{D}) \ni \mu \mapsto \Phi.\mu := \mathcal{L}_{X^\mu} \in \mathcal{P}_{k,\gamma}^T(\bar{D})$$

under the metric $\mathbb{W}_k^{\lambda,T}$ for large enough $\lambda > 0$.

Denote

$$\tilde{l}_t^\mu := \int_0^t 1_{\partial D}(X_s^\mu) dl_s^\mu, \quad \tilde{l}_t^\nu := \int_0^t 1_{\partial D}(X_s^\nu) dl_s^\nu, \quad t \geq 0.$$

By (2.1.2), ($A^{6.3}$) and Itô's formula, for any $k \geq 1$ we find a constant $c_1 > 0$ such that

$$d|X_t^\mu - X_t^\nu|^k \leq c_1 K_t\{|X_t^\mu - X_t^\nu|^k + \mathbb{W}_k(\mu_t, \nu_t)^k\}dt$$
$$+ \frac{k}{r_0}|X_t^\mu - X_t^\nu|^k(d\tilde{l}_t^\mu + d\tilde{l}_t^\nu) + dM_t \quad (6.3.7)$$

for some martingale M_t with

$$d\langle M\rangle_t \leq c_1 K_t\{|X_t^\mu - X_t^\nu|^{2k} + \mathbb{W}_k(\mu_t, \nu_t)^{2k}\}dt.$$

To estimate $\int_0^t |X_s^\mu - X_s^\nu|^k(d\tilde{l}_s^\mu + d\tilde{l}_s^\nu)$, we take

$$0 \leq h \in C_b^\infty([0,\infty)) \text{ such that } h' \leq 0,$$
$$h'(0) = -(1 + 2r_0^{-1}k), \quad h(0) = 1, \quad (6.3.8)$$

where $r_0 > 0$ is in (2.1.2). Let

$$F(x,y) := |x-y|^k\{(h \circ \tilde{\rho})(x) + (h \circ \tilde{\rho})(y)\}, \quad x, y \in \bar{D}.$$

By ($A^{6.3}$)(2), we have $\tilde{\rho}|_{\partial D} = 0$ and $\nabla_\mathbf{n}\tilde{\rho}|_{\partial D} \geq 1_{\partial D}$, so that (6.3.8) and (2.1.2) imply

$$\nabla_\mathbf{n} F(\cdot, X_t^\nu)(X_t^\mu)dl_t^\mu + \nabla_\mathbf{n} F(X_t^\mu, \cdot)(X_t^\nu)dl_t^\nu \leq -|X_t^\mu - X_t^\nu|^k(d\tilde{l}_t^\mu + d\tilde{l}_t^\nu).$$

Therefore, by ($A^{6.3}$) and applying Itô's formula, we find a constant $c_2 > 0$ such that

$$dF(X_t^\mu, X_t^\nu) \leq c_2\{|X_t^\mu - X_t^\nu|^k + \mathbb{W}_k(\mu_t, \nu_t)^k\}dt$$
$$- |X_t^\mu - X_t^\nu|^k(d\tilde{l}_t^\mu + d\tilde{l}_t^\nu) + d\tilde{M}_t$$

for some martingale $\tilde{M}_t$. This and $F(X_0^\mu, X_0^\nu) = F(X_0, X_0) = 0$ imply

$$\mathbb{E} \int_0^t |X_s^\mu - X_s^\nu|^k (\mathrm{d}\tilde{l}_s^\mu + \mathrm{d}\tilde{l}_s^\nu) \\ \leq c_2 \int_0^t K_s \{\mathbb{E}|X_s^\mu - X_s^\nu|^k + \mathbb{W}_k(\mu_s, \nu_s)^k\} \mathrm{d}s. \quad (6.3.9)$$

Substituting (6.3.9) into (6.3.7) and applying BDG's inequality, we find a constant $c_3 > 0$ such that

$$\zeta_t := \sup_{s \in [0,t]} |X_s^\mu - X_s^\nu|^k, \quad t \in [0, T]$$

satisfies

$$\mathbb{E}\zeta_t \leq c_3 \int_0^t K_s \{\mathbb{E}\zeta_s + \mathbb{W}_k(\mu_s, \nu_s)^k\} \mathrm{d}s, \quad t \in [0, T], \quad (6.3.10)$$

so that for any $\lambda > 0$,

$$\mathbb{E}\zeta_t \leq c_3 \int_0^t e^{c_3 \int_s^t K_r \mathrm{d}r} \mathbb{W}_k(\mu_s, \nu_s)^k \mathrm{d}s \\ \leq c_3 e^{k\lambda t} \mathbb{W}_k^{\lambda, T}(\mu, \nu)^k \int_0^t e^{\int_s^t \{c_r K_r - k\lambda\} \mathrm{d}r} \mathrm{d}s \quad (6.3.11) \\ \leq c_3 e^{k\lambda t} \delta(\lambda) \mathbb{W}_k^{\lambda, T}(\mu, \nu)^k, \quad t \in [0, T],$$

where

$$\delta(\lambda) := \sup_{t \in [0,T]} \int_0^t e^{\int_s^t \{c_r K_r - k\lambda\} \mathrm{d}r} \mathrm{d}s \downarrow 0 \text{ as } \lambda \uparrow \infty.$$

Therefore, Φ is contractive in $\mathbb{W}_k^{\lambda, T}$ for large $\lambda > 0$ as desired.

It remains to prove (6.3.3) and (6.3.4). Let X_t be the unique solution to (6.1.1). By $(A^{6.3})$, for any $k > 1$, we find a constant $c(k) > 0$ such that

$$\mathrm{d}|X_t|^k \leq c(k) K_t \{1 + |X_t|^k + \mathbb{E}|X_t|^k\} \mathrm{d}t + \\ k|X_t|^{k-2} \langle X_t, \sigma_t(X_t, \mathcal{L}_{X_t}) \mathrm{d}W_t \rangle + k|X_t|^{k-1} \mathrm{d}\tilde{l}_t, \quad (6.3.12)$$

where $\mathrm{d}\tilde{l}_t := 1_{\partial D}(X_t) \mathrm{d}l_t$. By applying Itô's formula to $(1+|X_t|^k)(h \circ \tilde{\rho})(X_t)$, similarly to (6.3.9) we obtain

$$\mathbb{E}\int_0^t (1 + |X_s|^k) \mathrm{d}\tilde{l}_s \leq \tilde{c}(k) \int_0^t K_s \mathbb{E}\{1 + |X_s|^k\} \mathrm{d}s \quad (6.3.13)$$

for some constant $\tilde{c}(k) > 0$. Combining (6.3.13) with (6.3.12) and using Gronwall's lemma, we derive

$$\mathbb{E}\Big[\sup_{t \in [0,T]} |X_t|^k\Big] \leq c'(1 + \mathbb{E}|X_0|^k)$$

for some constant $c' > 0$. Substituting this into (6.3.12) and using BDG's inequality, we derive (6.3.3) for some constant $c > 0$.

Finally, by $(A^{6.3})(2)$ and applying Itô's formula to $\tilde{\rho}(X_t)$, we derive (6.3.4). □

We now solve (6.1.1) for distributions in

$$\mathcal{P}_\psi(\bar{D}) := \{\mu \in \mathcal{P}(\bar{D}) : \|\mu\|_\psi := \mu(\psi(|\cdot|)) < \infty\},$$

where ψ belongs to the following class for some $\kappa > 0$:

$$\Psi_\kappa := \{\psi \in C^2((0,\infty)) \cap C^1([0,\infty)) : \psi(0) = 0, \ \psi'|_{(0,\infty)} > 0, \\ \|\psi'\|_\infty < \infty, \ r\psi'(r) + r^2 \{\psi''\}^+(r) \le \kappa\psi(r) \text{ for } r > 0\}. \quad (6.3.14)$$

Let

$$\mathbb{W}_\psi(\mu,\nu) := \inf_{\pi \in \mathcal{C}(\mu,\nu)} \int_{\bar{D}\times\bar{D}} \psi(|x-y|)\pi(\mathrm{d}x,\mathrm{d}y), \quad \mu,\nu \in \mathcal{P}_\psi(\bar{D}). \quad (6.3.15)$$

If $\psi'' \le 0$ then $\mathbb{W}_\psi$ is a complete metric on $\mathcal{P}_\psi$. In general, it is only a complete quasi-metric since the triangle inequality not necessarily holds.

$(A^{6.4})$ **(D)** holds, $\sigma_t(x,\mu) = \sigma_t(x)$ does not depend on μ, b and σ are bounded on bounded subsets of $[0,\infty) \times \bar{D} \times \mathcal{P}_\psi(\bar{D})$ for some $\psi \in \Psi_\kappa$ and $\kappa > 0$. Moreover, there exists $0 < K \in L^1([0,T])$ such that

$$\|\sigma_t(x) - \sigma_t(y)\|_{HS}^2 + 2\langle x-y, b_t(x,\mu) - b_t(y,\nu)\rangle^+ \\ \le K_t|x-y|\{|x-y| + \mathbb{W}_\psi(\mu,\nu)\}, t \in [0,T], x, y \in \bar{D}, \mu, \nu \in \mathcal{P}_k(\bar{D}).$$

Theorem 6.3.2. *Assume* $(A^{6.4})$ *and* $(A^{6.3})(2)$. *Then* (6.1.1) *is well-posed for distributions in* $\mathcal{P}_\psi(\bar{D})$, *and*

$$\mathbb{E}\left[\sup_{t\in[0,T]} \psi(|X_t|)\right] < \infty, \quad T > 0, \mathcal{L}_{X_0} \in \mathcal{P}_\psi(\bar{D}). \quad (6.3.16)$$

Proof. Let X_0 be $\mathcal{F}_0$-measurable with $\mathbb{E}\psi(|X_0|) < \infty$, and consider the path space

$$\mathcal{P}_\psi^T(\bar{D}) := \{\mu \in C_b^w([0,T]; \mathcal{P}_\psi(\bar{D})) : \mu_0 = \mathcal{L}_{X_0}\}.$$

For any $\lambda > 0$, the quasi-metric

$$\mathbb{W}_{\lambda,\psi}(\mu,\nu) := \sup_{t\in[0,T]} e^{-\lambda t}\mathbb{W}_\psi(\mu_t,\nu_t), \quad \mu,\nu \in \mathcal{P}_\psi^T(\bar{D})$$

is complete. By Lemma 2.3.4, $(A^{6.4})$ implies the well-posedness of the SDE (6.3.5) for any $\mu \in \mathcal{P}_\psi^T(\bar{D})$. By $(A^{6.3})(2)$ and Itô's formula for $g_t := \sqrt{1 + |X_t^\mu - X_0|^2}$, we find a constant $c_1 > 0$ such that

$$\mathrm{d}g_t \le c_1 K_t\{\|\mu_t\|_\psi + g_t\}\mathrm{d}t + g_t^{-1}\langle X_t^\mu - X_0, \sigma_t(X_t^\mu)\mathrm{d}W_t\rangle + \mathrm{d}\tilde{l}_t^\mu,$$

where $\mathrm{d}\tilde{l}_t^\mu := 1_{\tilde{\partial}D}(X_t^\mu)\mathrm{d}l_t^\mu$. Combining this with $\psi \in \Psi_\kappa$ and the linear growth of $\|\sigma_t\|$ implied by $(A^{6.4})$, we find a constant $c_2 > 0$ such that

$$\mathrm{d}\psi(g_t) \leq c_2 K_t\{\|\mu_t\|_\psi + \psi(g_t)\}\mathrm{d}t \\ + \psi'(g_t)g_t^{-1}\langle X_t^\mu - X_0, \sigma_t(X_t^\mu)\mathrm{d}W_t\rangle + \psi'(g_t)\mathrm{d}\tilde{l}_t^\mu. \quad (6.3.17)$$

Next, by $(A^{6.3})(2)$, $\psi \in \Psi_\kappa$ which implies $\psi'(g_t) \leq \kappa\psi(g_t)$ since $g_t \geq 1$, and applying Itô's formula to $\psi(g_t)\{\|\tilde{\rho}\|_\infty - \tilde{\rho}(X_t^\mu)\}$, we find a constant $c_3 > 0$ such that similarly to (6.3.9),

$$\mathbb{E}\int_0^t \psi'(g_s)\mathrm{d}\tilde{l}_s^\mu \leq \kappa\mathbb{E}\int_0^t \psi(g_s)\mathrm{d}\tilde{l}_s^\mu \\ \leq c_3\mathbb{E}\int_0^t K_s\{1 + \|\mu_s\|_\psi + \psi(|X_s^\mu|)\}\mathrm{d}s, \quad t \in [0,T]. \quad (6.3.18)$$

Combining this with (6.3.17), $r\psi'(r) \leq \kappa\psi(r)$, the linear growth of σ_t ensured by $(A^{6.4})$, and applying BDG's inequality, we obtain

$$\mathbb{E}\left[\sup_{t\in[0,T]} \psi(|X_t^\mu|)\right] < \infty.$$

Consequently, (6.3.16) holds for solutions of (6.1.1) with $\mathcal{L}_X \in \mathcal{P}_\psi^T(\bar{D})$. So, as explained in the proof of Theorem 6.2.2, it remains to prove the contraction of the map

$$\mathcal{P}_\psi^T(\bar{D}) \ni \mu \mapsto \Phi.\mu := \mathcal{L}_{X^\mu} \in \mathcal{P}_\psi^T(\bar{D})$$

under the metric $\mathbb{W}_{\lambda,\psi}$ for large enough $\lambda > 0$.

By (2.1.2), $(A^{6.3})(2)$, $\|\psi'\|_\infty < \infty$ and $r\psi'(r) \leq \kappa\psi(r)$, we obtain

$$\nabla_\mathbf{n}\{\psi(|\cdot - y|)\}(x) \leq \frac{\kappa}{2r_0}1_{\tilde{\partial}D}(x)\psi(|x-y|), \quad x \in \partial D, y \in \bar{D}. \quad (6.3.19)$$

Combining this with $(A^{6.4})$ and Itô's formula, we find a constant $c_4 > 0$ such that

$$\mathrm{d}\psi(|X_t^\mu - X_t^\nu|) \leq c_4 K_t\{\psi(|X_t^\mu - X_t^\nu|) + \mathbb{W}_\psi(\mu_t, \nu_t)\}\mathrm{d}t \\ + c_4\psi(|X_t^\mu - X_t^\nu|)(\mathrm{d}\tilde{l}_t^\mu + \mathrm{d}\tilde{l}_t^\nu) + \mathrm{d}M_t \quad (6.3.20)$$

for some martingale M_t.

On the other hand, let $\varepsilon = \frac{r_0}{2\kappa}$ and take $h \in C^\infty([0,\infty))$ with $h' \geq 0$, $h(r) = r$ for $r \leq \varepsilon/2$ and $h(r) = \varepsilon$ for $r \geq \varepsilon$. Consider

$$\eta_t := \psi(|X_t^\mu - X_t^\nu|)\{2\varepsilon - h \circ \tilde{\rho}(X_t^\mu) \quad h \circ \tilde{\rho}(X_t^\nu)\}.$$

By (6.3.19), $(A^{6.3})(2)$, $\varepsilon = \frac{r_0}{2\kappa}$ and Itô's formula, we find a constant $c_5 > 0$ such that
$$d\eta_t + d\tilde{M}_t$$
$$\leq c_5 K_t\{\psi(|X_t^\mu - X_t^\nu|) + \mathbb{W}_\psi(\mu_t, \nu_t)\}dt$$
$$+ \left(\frac{2\varepsilon\kappa}{2r_0} - 1\right)\psi(|X_t^\mu - X_t^\nu|)(d\tilde{l}_t^\mu + d\tilde{l}_t^\nu)$$
$$= c_5 K_t\{\psi(|X_t^\mu - X_t^\nu|) + \mathbb{W}_\psi(\mu_t, \nu_t)\}dt - \frac{1}{2}\psi(|X_t^\mu - X_t^\nu|)(d\tilde{l}_t^\mu + d\tilde{l}_t^\nu).$$
Since $X_0^\mu = X_0^\nu = X_0$, this implies
$$\mathbb{E}\int_0^t \psi(|X_s^\mu - X_s^\nu|)(d\tilde{l}_s^\mu + d\tilde{l}_t^\nu) \leq 2c_5 \int_0^t K_s\{\mathbb{E}\psi(|X_s^\mu - X_s^\nu|) + \mathbb{W}_\psi(\mu_s, \nu_s)\}ds.$$
Substituting this into (6.3.20), we find a constant $c_6 > 0$ such that
$$\mathbb{W}_\psi(\Phi_t\mu, \Phi_t\nu) \leq \mathbb{E}\psi(|X_t^\mu - X_t^\nu|) \leq c_6 \int_0^t K_s \mathbb{W}_\psi(\mu_s, \nu_s)ds, \quad t \in [0, T],$$
so that as in (6.3.11), we conclude that Φ^γ is contractive in $\mathbb{W}_{\lambda, \psi}$ for large $\lambda > 0$. Therefore, the proof is finished. □

6.4 Log-Harnack inequality and applications

6.4.1 *Singular case*

$(A^{6.5})$ Let $\partial D \in C_b^{2,L}$, let $\sigma_t(x, \mu) = \sigma_t(x)$ be distribution free. There exists $\hat{\mu} \in \mathcal{P}_2(\bar{D})$ such that $(A^{2.3})$ for $\hat{b}$ replacing b holds with $p_i > 2$, where $\hat{b} := b(\cdot, \hat{\mu})$ with regular term $\hat{b}^{(1)}$. Moreover, there exist a constant $\alpha \geq 0$ and a function $1 \leq f_0 \in \tilde{L}_{q_0}^{p_0}(T, D)$ for some $(p_0, q_0) \in \mathcal{K}, p_0 > 2$, such that for any $(t, x) \in [0, T] \times \bar{D}$,
$$|b_t(x, \mu) - \hat{b}_t^{(1)}(x)| \leq f_0(t, x) + \alpha\|\mu\|_2, \quad \mu \in \mathcal{P}_2(\bar{D}), \quad (6.4.1)$$
$$|b_t(x, \mu) - b_t(x, \nu)| \leq \mathbb{W}_2(\mu, \nu)\sum_{i=1}^l f_i(t, x), \quad \mu, \nu \in \mathcal{P}_2(\bar{D}). \quad (6.4.2)$$

According to Theorem 6.2.1, $(A^{6.5})$ implies the well-posedness of (6.1.1) for distributions in $\mathcal{P}_2(\bar{D})$. Let
$$P_t^*\mu = \mathcal{L}_{X_t} \text{ for } X_t \text{ solving } (6.1.1) \text{ with } \mathcal{L}_{X_0} = \mu \in \mathcal{P}_2(\bar{D}), \quad t \geq 0.$$
We consider
$$P_t f(\mu) := \int_{\bar{D}} f d(P_t^*\mu), \quad t \geq 0, \mu \in \mathcal{P}_2(\bar{D}), f \in \mathcal{B}_b(\bar{D}),$$
where $\mathcal{B}_b(\bar{D})$ is the class of all bounded measurable functions on $\bar{D}$.

Theorem 6.4.1. *Assume* $(A^{6.5})$. *For any* $N > 0$, *let* $\mathcal{P}_{2,N}(\bar{D}) := \{\mu \in \mathcal{P}_2(\bar{D}) : \|\mu\|_2 \leq N\}$.

(1) *For any* $N > 0$, *there exists a constant* $C(N) > 0$ *such that for any* $\nu \in \mathcal{P}_{2,N}(\bar{D})$ *and any* $t \in (0,T]$, *the following inequalities hold:*

$$\mathbb{W}_2(P_t^*\mu, P_t^*\nu)^2 \leq C(N)\mathbb{W}_2(\mu,\nu)^2, \quad \mu \in \mathcal{P}_2(\bar{D}), \tag{6.4.3}$$

$$P_t \log f(\nu) \leq \log P_t f(\mu) + \frac{C(N)}{t}\mathbb{W}_2(\mu,\nu)^2, \\ 0 < f \in \mathcal{B}_b(\bar{D}), \mu \in \mathcal{P}_{2,N}(\bar{D}), \tag{6.4.4}$$

$$\frac{1}{2}\|P_t^*\mu - P_t^*\nu\|_{var}^2 \leq \mathrm{Ent}(P_t^*\nu|P_t^*\mu) \\ \leq \frac{C(N)}{t}\mathbb{W}_2(\mu,\nu)^2, \quad \mu \in \mathcal{P}_{2,N}(\bar{D}), \tag{6.4.5}$$

$$\|\nabla P_t f(\nu)\|_{\mathbb{W}_2} := \limsup_{\mu \to \nu \text{ in } \mathbb{W}_2} \frac{|P_t f(\nu) - P_t f(\mu)|}{\mathbb{W}_2(\mu,\nu)} \\ \leq \frac{\sqrt{2C(N)}}{\sqrt{t}}\|f\|_\infty, \quad f \in \mathcal{B}_b(\bar{D}). \tag{6.4.6}$$

(2) *If* (6.4.2) *holds for* $\alpha = 0$, *then there exists a constant* $C > 0$ *such that*

$$\mathbb{W}_2(P_t^*\mu, P_t^*\nu)^2 \leq C\mathbb{W}_2(\mu,\nu)^2, \quad \mu,\nu \in \mathcal{P}_2(\bar{D}). \tag{6.4.7}$$

Moreover, if either $\sum_{i=1}^l \|f_i\|_\infty < \infty$ *or* D *is bounded, then* (6.4.4)–(6.4.6) *hold for some constant* C *replacing* $C(N)$ *and all* $\mu,\nu \in \mathcal{P}_2(\bar{D})$.

Proof. (1) Since the relative entropy of μ with respect to ν is given by

$$\mathrm{Ent}(\nu|\mu) = \sup_{g \in \mathcal{B}^+(\bar{D}), \mu(g)=1} \nu(\log g),$$

(6.4.4) is equivalent to

$$\mathrm{Ent}(P_t^*\nu|P_t^*\mu) \leq \frac{C(N)}{t}\mathbb{W}_2(\mu,\nu)^2, \quad t \in (0,T], \mu,\nu \in \mathcal{P}_{2,N}(\bar{D}). \tag{6.4.8}$$

By Pinsker's inequality (3.2.3), we conclude that (6.4.8) implies (6.4.5), which further yield (6.4.6). So, we only need to prove (6.4.3) and (6.4.8).

For any $\mu, \nu \in \mathcal{P}_2(\bar{D})$, let X_t solve (6.1.1) for $\mathcal{L}_{X_0} = \mu$, and denote

$$\mu_t := P_t^*\mu = \mathcal{L}_{X_t}, \quad \nu_t := P_t^*\nu, \quad \bar{\mu}_t := \mathcal{L}_{\bar{X}_t}, \quad t \in [0,T],$$

where $\bar{X}_t$ solves

$$\mathrm{d}\bar{X}_t = b_t(\bar{X}_t, \nu_t)\mathrm{d}t + \sigma_t(\bar{X}_t)\mathrm{d}W_t, \quad t \in [0,T], \bar{X}_0 = X_0.$$

Let σ and $\hat{b} := b(\cdot, \hat{\mu}) = \hat{b}^{(1)} + \hat{b}^{(0)}$ satisfy $(A^{2.3})$ for $\hat{b}$ replacing b. Consider the decomposition
$$b_t^\nu := b_t(\cdot, \nu_t) = \hat{b}_t^{(1)} + b_t^{\nu,0}, \quad b_t^{\nu,0} := b_t^\nu - \hat{b}_t^{(1)}.$$
Denote $f_t(x) := \sum_{i=1}^l f_i(t,x)$. By (6.2.3) and (6.4.2), there exists a constant $K(N) > 0$ such that
$$|b_t^{\nu,0}| \leq |\hat{b}_t^{(0)}| + K(N) f_t, \quad \|\nu\|_2 \leq N, \quad t \in [0,T]. \tag{6.4.9}$$
So, by Theorem 2.4.1, the estimate (2.4.1) and the log-Harnack inequality (2.4.5) hold for solutions of (3.1.5) with b^ν replacing b with a constant depending on N; that is, there exists a constant $c_1(N) > 0$ such that
$$\mathbb{W}_2(\bar{\mu}_t, \nu_t)^2 \leq c_1(N) \mathbb{W}_2(\mu, \nu)^2, \quad t \in [0,T], \mu \in \mathcal{P}_2(\bar{D}), \tag{6.4.10}$$
$$\begin{aligned}\mathrm{Ent}(\nu_t | \bar{\mu}_t) &= \sup_{f>0, \bar{\mu}(f)=1} (P_t f)(\nu) \\ &\leq \frac{c_1(N)}{t} \mathbb{W}_2(\mu, \nu)^2, \quad t \in (0,T], \mu \in \mathcal{P}_2(\bar{D}).\end{aligned} \tag{6.4.11}$$

Moreover, repeating step (e) in the proof of Theorem 2.4.1 for $k = 2$ and $(X_t, \bar{X}_t)$ replacing $(X_t^{(1)}, X_t^{(2)})$, and using (6.4.2) replacing (6.2.2), instead of (6.2.24) where $\|\mu_s - \nu_s\|_{k,var}^2$ disappears in the present case, we derive
$$\mathbb{W}_2(\mu_t, \bar{\mu}_t)^4 \leq (\mathbb{E}|X_t - \bar{X}_t|^2)^2 \leq c_2(N) \int_0^t \mathbb{W}_2(\mu_s, \nu_s)^4 ds, \quad t \in [0,T]$$
for some constant $c_2(N) > 0$. This together with (6.4.10) yields
$$\begin{aligned}\mathbb{W}_2(\mu_t, \nu_t)^4 &\leq 8 \mathbb{W}_2(\mu_t, \bar{\mu}_t)^4 + 8 \mathbb{W}_2(\bar{\mu}_t, \nu_t)^2 \\ &\leq 8 c_1(N)^2 \mathbb{W}_2(\mu, \nu)^4 + 8 c_2(N) \int_0^t \mathbb{W}_2(\mu_s, \mu_s)^4 ds, \quad t \in [0,T].\end{aligned}$$
Therefore, Gronwall's inequality implies (6.4.3) for some constant $C(N) > 0$.

On the other hand, let $\|\mu\|_2 \leq N$ and define
$$\begin{aligned}R_t &:= \exp\left[-\int_0^t \langle g_s, dW_s \rangle - \frac{1}{2} \int_0^t |g_s|^2 ds \right], \\ g_s &:= \{\sigma_s^* (\sigma_s \sigma_s^*)^{-1}\}(X_s) [b_s^\mu(X_s) - b_s^\nu(X_s)].\end{aligned}$$
By Girsanov's theorem, we obtain
$$\int_{\bar{D}} \left(\frac{d\bar{\mu}_t}{d\mu_t} \right)^2 d\mu_t = \mathbb{E}\left[\left(\frac{d\bar{\mu}_t}{d\mu_t}(X_t) \right)^2 \right] = \mathbb{E}\left[\left(\mathbb{E}[R_t | X_t] \right)^2 \right] \leq \mathbb{E} R_t^2.$$

By the same argument leading to (4.1.38), we derive
$$\text{Ent}(\nu_t|\mu_t) \leq 2\text{Ent}(\nu_t|\bar{\mu}_t) + \log \mathbb{E}R_t^2. \tag{6.4.12}$$
By (6.4.2), (6.4.3), $\|\sigma^*(\sigma\sigma^*)^{-1}\|_\infty < \infty$ and (2.2.35) due to $(A^{2.3})$ for b^μ replacing b, we find constants $c_3(N), c_4(N) > 0$ such that

$$\begin{aligned}
\mathbb{E}[R_t^2] &\leq \left(\mathbb{E}[R_t^2]\right)^2 \leq \mathbb{E} e^{c_3(N)\mathbb{W}_2(\mu,\nu)^2 \int_0^t f_s(X_s)^2 ds} \\
&\leq 1 + \mathbb{E}\left[c_3(N)\mathbb{W}_2(\mu,\nu)^2 \left(\int_0^t f_s(X_s)^2 ds\right) \right. \\
&\qquad \left. \times e^{c_3(N)\mathbb{W}_2(\mu,\nu)^2 \int_0^t f_s(X_s)^2 ds}\right] \\
&\leq 1 + c_3(N)\mathbb{W}_2(\mu,\nu)^2 \left[\mathbb{E}\left(\int_0^t f_s(X_s)^2 ds\right)^2\right]^{\frac{1}{2}} \\
&\qquad \times \left[\mathbb{E}e^{2c_3(N)\mathbb{W}_2(\mu,\nu)^2 \int_0^t f_s(X_s)^2 ds}\right]^{\frac{1}{2}} \\
&\leq 1 + c_4(N)\mathbb{W}_2(\mu,\nu)^2.
\end{aligned} \tag{6.4.13}$$

Combining this with (6.4.11) and (6.4.12), we derive (6.4.8) for some constant $C(N) > 0$.

(2) When $\alpha = 0$, (6.4.9) holds for $K(N) = K$ independent of N, so that (6.4.10) and (6.4.11) hold for some constant $C_1(N) = C_1 > 0$ independent of N and all $\mu, \nu \in \mathcal{P}_2(\bar{D})$, and in (6.4.13) the constant $C_3(N) = C_3$ is independent of N as well. Consequently, (6.4.7) holds and

$$\mathbb{E}[R_t^2] \leq \mathbb{E}e^{C_3\mathbb{W}_2(\mu,\nu)^2 \int_0^t f_s(X_s)^2 ds} \leq e^{C\mathbb{W}_2(\mu,\nu)^2}$$

if $\|f\|_\infty < \infty$, and when D is bounded we conclude that $C_4(N) = C_4$ in (6.4.13) is uniform in $N > 0$. Therefore, (6.4.4) and hence its consequent inequalities hold for some constant independent of N. □

6.4.2 Monotone case

$(A^{6.6})$ **(D)** and $(A^{6.3})(2)$ hold, $\sigma_t(x,\mu) = \sigma_t(x)$ does not depend on μ and is locally bounded on $[0,\infty) \times \bar{D}$, $\sigma\sigma^*$ is invertible, b is bounded on bounded subsets of $[0,\infty) \times \mathbb{R}^d \times \mathcal{P}_2(\bar{D})$, and there exists a constant $L > 0$ such that

$$\begin{aligned}
&\|\sigma_t(x) - \sigma_t(y)\|_{HS}^2 + 2\langle x-y, b_t(x,\mu) - b_t(y,\nu)\rangle^+ \\
&\leq L|x-y|^2 + L|x-y|\mathbb{W}_2(\mu,\nu), \\
&\|\sigma_t(x)(\sigma_t\sigma_t^*)^{-1}(x)\| \leq L, \quad t \in [0,T], x, y \in \bar{D}, \mu, \nu \in \mathcal{P}_2(\bar{D}).
\end{aligned}$$

By Theorem 6.3.1, $(A^{6.6})$ implies that (6.1.1) is well-posed for distributions in $\mathcal{P}_2(\bar{D})$.

Theorem 6.4.2. *Assume* $(A^{6.6})$. *Then there exists a constant* $C > 0$ *such that the following inequalities hold for all* $t \in (0, T]$ *and* $\nu \in \mathcal{P}_2(\bar{D})$:

$$\mathbb{W}_2(P_t^*\mu, P_t^*\nu)^2 \leq C\mathbb{W}_2(\mu, \nu)^2, \quad \mu \in \mathcal{P}_2(\bar{D}), \tag{6.4.14}$$

$$P_t \log f(\nu) \leq \log P_t f(\mu) + \frac{C}{t}\mathbb{W}_2(\mu,\nu)^2, \quad 0 < f \in \mathcal{B}_b(\bar{D}), \mu \in \mathcal{P}_2(\bar{D}), \tag{6.4.15}$$

$$\frac{1}{2}\|P_t^*\mu - P_t^*\nu\|_{var}^2 \leq \mathrm{Ent}(P_t^*\nu|P_t^*\mu) \leq \frac{C}{t}\mathbb{W}_2(\mu,\nu)^2, \quad \mu \in \mathcal{P}_2(\bar{D}), \tag{6.4.16}$$

$$\begin{aligned}\|\nabla P_t f(\nu)\|_{\mathbb{W}_2} &:= \limsup_{\mu \to \nu \text{ in } \mathbb{W}_2} \frac{|P_t f(\mu) - P_t f(\nu)|}{\mathbb{W}_2(\mu,\nu)} \\ &\leq \frac{\sqrt{2C}\|f\|_\infty}{\sqrt{t}}, \quad f \in \mathcal{B}_b(\bar{D}).\end{aligned} \tag{6.4.17}$$

Proof. As explained in the proof of Theorem 6.4.1, it suffices to prove (6.4.14) and (6.4.15).

Firstly, for $\mu_0, \nu_0 \in \mathcal{P}_2(\bar{D})$, let (X_0, Y_0) be $\mathcal{F}_0$-measurable such that

$$\mathcal{L}_{X_0} = \mu_0, \quad \mathcal{L}_{Y_0} = \nu_0, \quad \mathbb{E}|X_0 - Y_0|^2 = \mathbb{W}_2(\mu_0, \nu_0)^2. \tag{6.4.18}$$

Denote

$$\mu_t := P_t^*\mu_0, \quad \nu_t := P_t^*\nu_0, \quad t \geq 0.$$

Let X_t solve (6.1.1). We have

$$\mathrm{d}X_t = b_t(X_t, \mu_t)\mathrm{d}t + \sigma_t(X_t)\mathrm{d}W_t + \mathbf{n}(X_t)\mathrm{d}l_t^X, \quad t \in [0, T], \tag{6.4.19}$$

where l_t^X is the local time of X_t on ∂D. Next, for any $t_0 \in (0, T]$ consider the SDE

$$\begin{aligned}\mathrm{d}Y_t = &\left\{b_t(Y_t, \nu_t) + \frac{\sigma_t(Y_t)\{\sigma_t^*(\sigma_t\sigma_t^*)^{-1}\}(X_t)(X_t - Y_t)}{\xi_t}\right\}\mathrm{d}t \\ &+ \sigma_t(Y_t)\mathrm{d}W_t + \mathbf{n}(Y_t)\mathrm{d}l_t^Y, \quad t \in [0, t_0),\end{aligned} \tag{6.4.20}$$

where l_t^Y is the local time of Y_t on ∂D. For the constant $L > 0$ in $(A^{6.6})$, let

$$\xi_t := \frac{1}{L}\left(1 - e^{L(t-t_0)}\right), \quad t \in [0, t_0). \tag{6.4.21}$$

The construction of Y_t goes back to [Wang (2011)] for the classical SDEs, see also [Wang (2018)] for the extension to DDSDEs. According to Theorem 2.3.2, $(A^{6.6})$ implies that (6.4.20) has a unique solution up to times

$$\tau_{n,m} := \frac{T_0 n}{n+1} \wedge \inf\{t \in [0, t_0) : |Y_t| \geq m\}, \quad n, m \geq 1.$$

Let h be in (6.3.8) for $k = 2$. By (2.1.2) and $(A^{6.3})(2)$, we have

$$\langle \nabla\{(1 + h \circ \tilde{\rho})|\cdot - x_0|^2\}(Y_t), \mathbf{n}(Y_t)\rangle \mathrm{d}l_t^Y \leq 0, \quad x_0 \in \bar{D},$$

so that $(A^{6.6})$, for any $n \geq 1$ we find a constant $c(n) > 0$ such that

$$\mathrm{d}\{(1 + h \circ \tilde{\rho})(Y_t)|Y_t - x_0|^2\} \leq c(n)(1 + |Y_t|^2)\mathrm{d}t + \mathrm{d}M_t, \quad t \in [0, \tau_{n,m}], \, n, m \geq 1$$

holds for some martingale M_t. This implies $\lim_{m \to \infty} \tau_{n,m} = \frac{T_0 n}{n+1}$, and hence (6.4.20) has a unique solution up to time t_0.

Next, let $\tilde{Y}_t$ solve the SDE

$$\mathrm{d}\tilde{Y}_t = b_t(\tilde{Y}_t, \nu_t)\mathrm{d}t + \sigma_t(\tilde{Y}_t)\mathrm{d}W_t + \mathbf{n}(\tilde{Y}_t)\mathrm{d}l_t^{\tilde{Y}}, \quad \tilde{Y}_0 = Y_0, t \in [0, T], \quad (6.4.22)$$

where $l_t^{\tilde{Y}}$ is the local time of $\tilde{Y}_t$ on ∂D. By $(A^{6.6})$, (2.1.2) and Itô's formula, we find a constant $c_2 > 0$ such that

$$\begin{aligned}
&\mathbb{E}|X_t - \tilde{Y}_t|^2 - \mathbb{W}_2(\mu_0, \nu_0)^2 \\
&\leq c_2 \int_0^t \{\mathbb{E}|X_s - \tilde{Y}_s|^2 + \mathbb{W}_2(\mu_s, \nu_s)^2\}\mathrm{d}s \\
&\quad + \frac{2}{r_0}\mathbb{E}\int_0^t |X_s - \tilde{Y}_s|^2(\mathrm{d}\tilde{l}_s^X + \mathrm{d}\tilde{l}_s^{\tilde{Y}}), \quad t \in [0, T].
\end{aligned} \quad (6.4.23)$$

For h in (6.3.8) with $k = 2$, we deduce from $(A^{6.3})(2)$ that

$$\begin{aligned}
\langle \nabla\{|X_t - \cdot|^2(h \circ \rho(X_t) + h \circ \rho)\}(\tilde{Y}_t), \mathbf{n}(\tilde{Y}_t)\rangle \mathrm{d}\tilde{l}_t^{\tilde{Y}} \\
\leq -|X_t - \tilde{Y}_t|^2 \mathrm{d}l_t^Y, \\
\langle \nabla\{|\tilde{Y}_t - \cdot|^2(h \circ \rho(\tilde{Y}_t) + h \circ \rho)\}(X_t), \mathbf{n}(X_t)\rangle \mathrm{d}\tilde{l}_t^X \\
\leq |X_t - \tilde{Y}_t|^2 \mathrm{d}\tilde{l}_t^X.
\end{aligned} \quad (6.4.24)$$

So, applying Itô's formula to

$$\eta_t := |X_t - \tilde{Y}_t|^2 (h \circ \rho(X_t) + h \circ \rho(\tilde{Y}_t)),$$

and using $(A^{6.6})$ and (2.1.2), we find a constant $c_3 > 0$ such that

$$\begin{aligned}
&\mathrm{d}\eta_t + \mathrm{d}M_t \\
&\leq c_3\{|X_t - \tilde{Y}_t|^2 + \mathbb{W}_2(\mu_t, \nu_t)^2\}\mathrm{d}t + \mathrm{d}M_t - |X_t - \tilde{Y}_t|^2(\mathrm{d}\tilde{l}_t^X + \mathrm{d}\tilde{l}_t^{\tilde{Y}})
\end{aligned}$$

holds for some martingale M_t. This together with (6.4.23) yields

$$\mathbb{E}|X_t - \tilde{Y}_t|^2$$
$$\leq \mathbb{W}_2(\mu_0, \nu_0)^2 + \mathbb{E}\eta_0 + (c_2 + c_3)\int_0^t \{\mathbb{E}|X_s - \tilde{Y}_s|^2 + \mathbb{W}_2(\mu_s, \nu_s)^2\}\mathrm{d}s$$
$$\leq 3\mathbb{W}_2(\mu_0, \nu_0)^2 + 2(c_2 + c_3)\int_0^t \mathbb{E}|X_s - \tilde{Y}_s|^2 \mathrm{d}s, \quad t \in [0, T],$$

where we have used the fact that $\mathbb{W}_2(\mu_s, \nu_s)^2 \leq \mathbb{E}|X_s - \tilde{Y}_s|^2$ by definition. By Gronwall's lemma, this and $\mathbb{W}_2(\mu_t, \nu_t)^2 \leq \mathbb{E}|X_t - \tilde{Y}_t|^2$, we find a constant $c_4 > 0$ such that

$$\mathbb{W}_2(\mu_t, \nu_t)^2 \leq \mathbb{E}|X_t - \tilde{Y}_t|^2 \leq c_4 \mathbb{W}_2(\mu_0, \nu_0)^2, \quad t \in [0, T], \tag{6.4.25}$$

so that (6.4.14) holds.

Moreover, for any $n \geq 1$, let

$$\tau_n := \frac{t_0 n}{n+1} \wedge \inf\{t \in [0, t_0) : |X_t - Y_t| \geq n\},$$
$$\beta_s := \frac{1}{\xi_s}\{\sigma_s^*(\sigma_s \sigma_s^*)^{-1}\}(X_s)(X_s - Y_s), \quad s \in [0, \tau_n]. \tag{6.4.26}$$

By Girsanov's theorem,

$$\tilde{W}_t := W_t + \int_0^t \beta_s \mathrm{d}s, \quad t \in [0, \tau_n]$$

is an m-dimensional Brownian motion under the probability $\mathbb{Q}_n := R_n \mathbb{P}$, where

$$R_n := \mathrm{e}^{-\int_0^{\tau_n} \langle \beta_s, \mathrm{d}W_s \rangle - \frac{1}{2}\int_0^{\tau_n} |\beta_s|^2 \mathrm{d}s}. \tag{6.4.27}$$

Then (6.4.19) and (6.4.20) imply

$$\mathrm{d}X_t = \left\{b_t(X_t, \mu_t) - \frac{X_t - Y_t}{\xi_t}\right\}\mathrm{d}t + \sigma_t(X_t)\mathrm{d}\tilde{W}_t + \mathbf{n}(X_t)\mathrm{d}l_t^X,$$
$$\mathrm{d}Y_t = b_t(Y_t, \nu_t)\mathrm{d}t + \sigma_t(Y_t)\mathrm{d}\tilde{W}_t + \mathbf{n}(Y_t)\mathrm{d}l_t^Y, \quad t \in [0, \tau_n], n \geq 1. \tag{6.4.28}$$

Combining this with $(A^{6.6})$, (2.1.2), (6.4.25) and Itô's formula, we obtain

$$\begin{aligned}
\mathrm{d}&\frac{|X_t - Y_t|^2}{\xi_t} - \mathrm{d}M_t \\
&\leq \frac{L|X_t - Y_t|^2 + L|X_t - Y_t|\mathbb{W}_2(\mu_t, \nu_t)}{\xi_t}\mathrm{d}t \\
&\quad - \frac{|X_t - Y_t|^2(2 + \xi'_t)}{\xi_t^2}\mathrm{d}t + \frac{|X_t - Y_t|^2}{\xi_t^2}(\mathrm{d}\tilde{l}_t^X + \mathrm{d}\tilde{l}_t^Y) \\
&\leq \left\{\frac{L^2\mathbb{W}_2(\mu_t, \nu_t)^2}{2} - \frac{|X_t - Y_t|^2(2 + \xi'_t - L\xi_t - \frac{1}{2})}{\xi_t^2}\right\}\mathrm{d}t \qquad (6.4.29) \\
&\quad + \frac{|X_t - Y_t|^2}{\xi_t^2}(\mathrm{d}\tilde{l}_t^X + \mathrm{d}\tilde{l}_t^Y) \\
&\leq \left\{\frac{L^2 e^{2Lt}\mathbb{W}_2(\mu_0, \nu_0)^2}{2} - \frac{|X_t - Y_t|^2}{2\xi_t^2}\right\}\mathrm{d}t \\
&\quad + \frac{|X_t - Y_t|^2}{\xi_t^2}(\mathrm{d}\tilde{l}_t^X + \mathrm{d}\tilde{l}_t^Y), \quad t \in [0, \tau_n],
\end{aligned}$$

where $\mathrm{d}M_t := \frac{2}{\xi_t}\langle X_t - Y_t, \{\sigma_t(X_t) - \sigma_t(Y_t)\}\mathrm{d}\tilde{W}_t\rangle$ is a $\mathbb{Q}_n$-martingale. By (6.4.24) for $(Y_t, \tilde{l}_t^Y)$ replacing $(\tilde{Y}_t, \tilde{l}_t^{\tilde{Y}})$, and applying Itô's formula to $g_t := \frac{|X_t - Y_t|^2}{\xi_t}(h \circ \rho(X_t) + h \circ \rho(Y_t))$, we find a constant $c_5 > 0$ such that

$$\mathrm{d}g_t \leq c_5 g_t \mathrm{d}t + \mathrm{d}\tilde{M}_t - \frac{|X_t - Y_t|^2}{\xi_t}(\mathrm{d}\tilde{l}_t^X + \mathrm{d}\tilde{l}_t^Y), \quad t \in [0, \tau_n], n \geq 1$$

holds for some $\mathbb{Q}_n$-martingale $\tilde{M}_t$. This and (6.4.18) imply that for some constants $c_6, c_7 > 0$,

$$\mathbb{E}_{\mathbb{Q}_n} g_{t \wedge \tau_n} \leq e^{c_4 t_0} \mathbb{E} g_0 \leq \frac{c_6}{t_0}\mathbb{W}_2(\mu_0, \nu_0)^2, \quad t \geq 0,$$

$$\mathbb{E}_{\mathbb{Q}_n} \int_0^{\tau_n} \frac{|X_t - Y_t|^2}{\xi_t}(\mathrm{d}\tilde{l}_t^X + \mathrm{d}\tilde{l}_t^Y) \leq \frac{c_7}{t_0}\mathbb{W}_2(\mu_0, \nu_0)^2, \quad n \geq 1.$$

Combining this with (6.4.25), (6.4.29) and $(A^{6.6})$, we derive

$$\begin{aligned}
\mathbb{E}[R_n \log R_n] &= \mathbb{E}_{\mathbb{Q}_n}[\log R_n] \\
&= \frac{1}{2}\mathbb{E}_{\mathbb{Q}_n}\int_0^{\tau_n} \frac{|\{\sigma_s^*(\sigma_s \sigma_s^*)^{-1}\}(X_s)(X_s - Y_s)|^2}{|\xi_s|^2}\mathrm{d}s \qquad (6.4.30) \\
&\leq \frac{c}{t_0}\mathbb{W}_2(\mu_0, \nu_0)^2, \quad n \geq 1
\end{aligned}$$

for some constant $c > 0$ uniformly in $t_0 \in (0, T]$. Therefore, by the martingale convergence theorem, $R_\infty := \lim_{n \to \infty} R_n$ exists, and

$$N_t := e^{-\int_0^t \langle \beta_s, \mathrm{d}W_s\rangle - \frac{1}{2}\int_0^t |\beta_s|^2 \mathrm{d}s}, \quad t \in [0, t_0]$$

is a $\mathbb{P}$-martingale.

Finally, let $\mathbb{Q} := N_{t_0}\mathbb{P}$. By Girsanov's theorem, $(\tilde{W}_t)_{t\in[0,t_0]}$ is an m-dimensional Brownian motion under the probability $\mathbb{Q}$, and $(X_t)_{t\in[0,t_0]}$ solves the SDE

$$\mathrm{d}X_t = \left\{b_t(X_t,\mu_t) - \frac{X_t - Y_t}{\xi_t}\right\}\mathrm{d}t \\ + \sigma_t(X_t)\mathrm{d}\tilde{W}_t + \mathbf{n}(X_t)\mathrm{d}l_t^X, \quad t \in [0, t_0]. \tag{6.4.31}$$

Let $(Y_t)_{t\in[0,t_0]}$ solve

$$\mathrm{d}Y_t = b_t(Y_t, \nu_t)\mathrm{d}t + \sigma_t(Y_t)\mathrm{d}\tilde{W}_t + \mathbf{n}(Y_t)\mathrm{d}l_t^Y, \quad t \in [0, t_0]. \tag{6.4.32}$$

By the well-posedness of (6.1.1), this extends the second equation in (6.4.28) with $\mathcal{L}_{Y_{t_0}|\mathbb{Q}} = \nu_{t_0}$. Moreover, (6.4.30) and Fatou's lemma implies

$$\frac{1}{2}\mathbb{E}_{\mathbb{Q}}\int_0^{t_0} \frac{|\{\sigma_s^*(\sigma_s\sigma_s^*)^{-1}\}(X_s)(X_s - Y_s)|^2}{|\xi_s|^2}\mathrm{d}s \\ = \mathbb{E}[N_{t_0}\log N_{t_0}] \le \liminf_{n\to\infty}\mathbb{E}[R_n\log R_n] \le \frac{c}{t_0}W_2(\mu_0,\nu_0)^2, \tag{6.4.33}$$

which in particular implies $\mathbb{Q}(X_{t_0} = Y_{t_0}) = 1$. Indeed, by $(A^{6.6})$, if $X_{t_0}(\omega) \ne Y_{t_0}(\omega)$ then there exists a small constant $\varepsilon > 0$ such that

$$|\eta_s|^2(\omega) = |\{\sigma_s^*(\sigma_s\sigma_s^*)^{-1}\}(X_s)(X_s - Y_s)|^2(\omega) \ge \varepsilon, \quad s \in [t_0 - \varepsilon, t_0],$$

which implies $\int_0^{t_0}\frac{|\eta_s|^2}{|\xi_s|^2}(\omega)\mathrm{d}s = \infty$. So, (6.4.33) implies $\mathbb{Q}(X_{t_0} = Y_{t_0}) = 1$. Combining this with the Young's inequality, we arrive at

$$P_{t_0}\log f(\nu_0) = \mathbb{E}[N_{t_0}\log f(Y_{t_0})] = \mathbb{E}[N_{t_0}\log f(X_{t_0})] \\ \le \mathbb{E}[N_{t_0}\log N_{t_0}] + \log \mathbb{E}[f(X_{t_0})] \\ \le \log P_{t_0}f(\mu_0) + \frac{c}{t_0}W_2(\mu_0,\nu_0)^2, \quad t_0 \in (0,T].$$

Hence, (6.4.15) holds. $\square$

6.5 Exponential ergodicity

Let $(b_t, \sigma_t) = (b, \sigma)$ not depend on t. The SDE (6.1.1) becomes

$$\mathrm{d}X_t = b(X_t, \mathcal{L}_{X_t})\mathrm{d}t + \sigma(X_t, \mathcal{L}_{X_t})\mathrm{d}W_t + \mathbf{n}(X_t)\mathrm{d}l_t, \quad t \ge 0. \tag{6.5.1}$$

In this case, a probability measure $\bar{\mu}$ is called P_t^*-invariant, if $P_t^*\bar{\mu} = \bar{\mu}$ holds for all $t \ge 0$, where $P_t^*\mu := \mathcal{L}_{X_t}$ for the solution with $\mathcal{L}_{X_0} = \mu$.

6.5.1 Singular case

The following result can be proved by repeating the proof of Theorem 5.2.1, see [Wang (2023c)] for a result under a weaker integral condition replacing the following condition (6.5.2).

Theorem 6.5.1. *Assume* $(A^{2.3})$ *for* $b(\cdot,\nu)$ *replacing* b *for any* $\nu \in \mathcal{P}(\bar{D})$, *and that*

$$|b(x,\mu_1) - b(x,\mu_2)| \leq \kappa\|\mu_1 - \mu_2\|_{var}, \quad x \in \bar{D}, \mu_1,\mu_2 \in \mathcal{P} \quad (6.5.2)$$

holds for some constant $\kappa > 0$. *Then (6.1.1) is well-posed, and when* $\kappa > 0$ *is small enough and* Φ *is convex with* $\int_0^\infty \frac{ds}{\Phi(s)} < \infty$, P_t^* *has a unique invariant probability measure* $\bar{\mu}$ *such that for some constants* $\varepsilon_0, c, \lambda > 0$, $\bar{\mu}(\Phi(\varepsilon_0 V)) < \infty$ *and*

$$\|P_t^*\nu - \bar{\mu}\|_{var} \leq c e^{-\lambda t}\|\bar{\mu} - \nu\|_{var}, \quad t \geq 0, \nu \in \mathcal{P}.$$

6.5.2 Dissipative case

In this part, we study the exponential ergodicity of P_t^* in entropy and $\mathbb{W}_2$, such that Theorem 5.6.2 is extended to the reflected case.

Theorem 6.5.2. *Let* D *be convex and* (σ,b) *satisfy* $(A^{6.3})$ *with* $k = 2$. *Let* $K_1, K_2 \in L^1_{loc}([0,\infty);\mathbb{R})$ *such that*

$$2\langle b_t(x,\mu) - b_t(y,\nu), x - y\rangle + \|\sigma_t(x,\mu) - \sigma_t(y,\nu)\|_{HS}^2 \\ \leq K_1(t)|x-y|^2 + K_2(t)\mathbb{W}_2(\mu,\nu)^2, \quad t \geq 0. \quad (6.5.3)$$

Then (6.1.1) with $t \in [0,\infty)$ *is well-posed for distributions in* $\mathcal{P}_2(\bar{D})$, *and* P_t^* *satisfies*

$$\mathbb{W}_2(P_t^*\mu, P_t^*\nu)^2 \leq e^{\int_0^t (K_1+K_2)(r)dr}\mathbb{W}_2(\mu,\nu)^2, \quad \mu,\nu \in \mathcal{P}_2(\bar{D}), t \geq 0. \quad (6.5.4)$$

Consequently, the following assertions hold for $(b_t, \sigma_t) = (b, \sigma)$ *independent of* t *provided* $\lambda := -(K_1 + K_2) > 0$.

(1) P_t^* *has a unique invariant probability measure* $\bar{\mu}$ *such that*

$$\mathbb{W}_2(P_t^*\mu, \bar{\mu})^2 \leq e^{-\lambda t}\mathbb{W}_2(\mu,\bar{\mu})^2, \quad \mu \in \mathcal{P}_2(\bar{D}), t \geq 0. \quad (6.5.5)$$

If moreover $\sigma(x,\mu) = \sigma(x)$ *does not depend on* μ *and* $\sigma\sigma^*$ *is invertible with* $\|\sigma\|_\infty + \|(\sigma\sigma^*)^{-1}\|_\infty < \infty$, *then there exists a constant* $c > 0$ *such that*

$$\mathrm{Ent}(P_t^*\mu|\bar{\mu}) \leq c e^{-\lambda t}\mathbb{W}_2(\mu,\bar{\mu})^2, \quad t \geq 1, \mu \in \mathcal{P}_2(\bar{D}). \quad (6.5.6)$$

(2) If $\sigma(x,\mu) = \sigma(\mu)$ does not depend on x, then there exists a constant $c > 0$ such that $\bar{\mu}$ satisfies the following log-Sobolev inequality and Talagrand inequality:

$$\bar{\mu}(f^2 \log f^2) \leq c\bar{\mu}(|\nabla f|^2), \quad f \in C_b^1(\mathbb{R}^d), \bar{\mu}(f^2) = 1, \quad (6.5.7)$$

$$\mathbb{W}_2(\mu, \bar{\mu})^2 \leq c\mathrm{Ent}(\mu|\bar{\mu}), \quad \mu \in \mathcal{P}_2. \quad (6.5.8)$$

If furthermore $\sigma(x, \mu) = \sigma$ is constant with $\sigma\sigma^*$ invertible, then there exists a constant $c > 0$ such that

$$\begin{aligned}&\mathbb{W}_2(P_t^*\mu, \bar{\mu})^2 + \mathrm{Ent}(P_t^*\mu|\bar{\mu}) \\ &\leq ce^{-\lambda t} \min\{\mathbb{W}_2(\mu, \bar{\mu})^2, \mathrm{Ent}(\mu|\bar{\mu})\}, \quad t \geq 1, \mu \in \mathcal{P}_2(\bar{D}).\end{aligned} \quad (6.5.9)$$

Proof. The well-posedness is ensured by Theorem 6.3.1. Since D is convex, (2.1.3) holds. For any $\mu, \nu \in \mathcal{P}_2(\bar{D})$, let X_0^μ and X_0^ν be $\mathcal{F}_0$-measurable such that

$$\mathcal{L}_{X_0^\mu} = \mu, \quad \mathcal{L}_{X_0^\nu} = \nu, \quad \mathbb{E}|X_0^\mu - X_0^\nu|^2 = \mathbb{W}_2(\mu, \nu)^2. \quad (6.5.10)$$

By (6.5.3), (2.1.3), and applying Itô's formula to $|X_t^\mu - X_t^\nu|^2$, where $(X_t^\mu)_{t \geq 0}$ and $(X_t^\nu)_{t \geq 0}$ solve (6.1.1), we obtain

$$d|X_t^\mu - X_t^\nu|^2 \leq \{K_1(t)|X_t^\mu - X_t^\nu|^2 + K_2(t)\mathbb{W}_2(\mathbb{P}_t^*\mu, P_t^*\nu)^2\}dt + dM_t$$

for some martingale M_t. Combining this with (6.5.10), $\mathbb{W}_2(P_t^*\mu, P_t^*\nu)^2 \leq \mathbb{E}|X_t^\mu - X_t^\nu|^2$, and Gronwall's lemma, we derive (6.5.4).

(1) Let (b_t, σ_t) not depend on t and $\lambda := -(K_1 + K_2) > 0$. Then (6.5.4) implies the uniqueness of P_t^*-invariant probability measure $\bar{\mu} \in \mathcal{P}_2(\bar{D})$ and (6.5.5).

Next, by Theorem 5.1.1, the existence of $\bar{\mu}$ follows from a standard argument by showing that for $x_0 \in D$, $\{P_t^*\delta_{x_0}\}_{t \geq 0}$ is a $\mathbb{W}_2$-Cauchy family as $t \to \infty$. Since the term of local time does not make trouble due to (2.1.3), the proof is completely similar to that of Theorem 5.3.3.

Finally, when $\sigma_t(x, \mu) = \sigma_t(x)$ and $\sigma\sigma^*$ is invertible with $\|\sigma\|_\infty + \|(\sigma\sigma^*)^{-1}\|_\infty < \infty$, by Theorem 6.4.2, $(A^{6.3})$ with $k = 2$ implies the log-Harnack inequality

$$\mathrm{Ent}(P_1^*\mu|\bar{\mu}) \leq c\mathbb{W}_2(\mu, \bar{\mu})^2, \quad \mu \in \mathcal{P}_2(\bar{D})$$

for some constant $c > 0$. So, (6.5.6) follows from (6.5.5) and $P_t^* = P_1^* P_{t-1}^*$ for $t \geq 1$.

(2) Let $\sigma(x, \mu) = \sigma(\mu)$ be independent of x. Consider the SDE

$$d\bar{X}_t^x = b(\bar{X}_t^x, \bar{\mu})dt + \sigma(\bar{\mu})dW_t + \mathbf{n}(\bar{X}_t^x)dl_t, \quad t \geq s, \bar{X}_0^x = x \in \bar{D}. \quad (6.5.11)$$

The associated Markov semigroup $\{\bar{P}_t\}_{t\geq 0}$ is given by

$$\bar{P}_t f(x) := \mathbb{E} f(\bar{X}_t^x), \quad t \geq 0, f \in \mathcal{B}_b(\bar{D}), x \in \bar{D}.$$

Let $\bar{P}_t^*$ be given by

$$(\bar{P}_t^* \mu)(f) := \mu(\bar{P}_t f), \quad \mu \in \mathcal{P}(\bar{D}), t \geq 0, f \in \mathcal{B}_b(D).$$

Since (6.5.3) with $x = y$ implies $K_2 \geq 0$, we have

$$K_1 \leq -\lambda < 0. \tag{6.5.12}$$

As explained in the above proofs of (6.5.4) and (6.5.5), this implies that $\bar{P}_t^*$ has a unique invariant probability measure $\tilde{\mu}$ such that

$$\lim_{t\to\infty} \bar{P}_t f(x) = \tilde{\mu}(f), \quad f \in C_b(\bar{D}), x \in \bar{D}. \tag{6.5.13}$$

Since $\bar{\mu}$ is the unique invariant probability measure of P_t^*, and when the initial distribution is $\bar{\mu}$, the SDE (6.5.11) coincides with (6.1.1), we conclude that $\tilde{\mu} = \bar{\mu}$. Hence, (6.5.13) yields

$$\bar{\mu}(f) = \lim_{t\to\infty} P_t f(x_0), \quad f \in C_b(\bar{D}), x_0 \in D. \tag{6.5.14}$$

Now, by Itô's formula, (2.1.3) and (6.5.3) with (b_t, σ_t) independent of t, we obtain

$$|\bar{X}_t^x - \bar{X}_t^y|^2 \leq e^{K_1 t} |x - y|^2, \quad x, y \in \bar{D}, t \geq 0.$$

This and (6.5.12) imply

$$\begin{aligned}|\nabla \bar{P}_t f(x)| &:= \limsup_{y \to x} \frac{|\bar{P}_t f(x) - \bar{P}_t f(y)|}{|x-y|} \\ &\leq \limsup_{y \to x} \frac{\mathbb{E}|f(\bar{X}_t^x) - f(\bar{X}_t^y)|}{|x-y|} \\ &\leq e^{-\frac{\lambda t}{2}} \limsup_{y \to x} \mathbb{E} \frac{|f(\bar{X}_t^x) - f(\bar{X}_t^y)|}{|\bar{X}_t^x - \bar{X}_t^y|} \\ &= e^{-\lambda t/2} \bar{P}_t |\nabla f|(x), \quad t \geq 0, f \in C_b^1(\bar{D}).\end{aligned} \tag{6.5.15}$$

On the other hand, we have

$$\partial_t \bar{P}_t f = \bar{L} \bar{P}_t f, \quad \langle \mathbf{n}, \nabla \bar{P}_t f\rangle|_{\partial D} = 0, \quad t \geq 0, f \in C_N^2(\bar{D}),$$

where $C_N^2(\bar{D})$ is the set of $f \in C_b^2(\bar{D})$ satisfying $\langle \mathbf{n}, \nabla f\rangle|_{\partial D} = 0$, and

$$\bar{L} := \frac{1}{2} \mathrm{tr}\{(\bar{\sigma}\bar{\sigma}^*)\nabla^2\} + \nabla_{b(\cdot,\bar{\mu})}, \quad \bar{\sigma} := \sigma(\bar{\mu}), \quad s \geq 0.$$

So, by Itô's formula, for any $\varepsilon > 0$ and $f \in C_N^2(\bar{D})$,

$$\mathrm{d}\{(\bar{P}_{t-s}(\varepsilon + f^2))\log \bar{P}_{t-s}(\varepsilon + f^2)\}(\bar{X}_s)$$
$$= \left\{\frac{|\bar{\sigma}^* \nabla \bar{P}_{t-s} f^2|^2}{\varepsilon + \bar{P}_{t-s} f^2}\right\}\mathrm{d}t + \mathrm{d}M_s^\varepsilon, \quad s \in [0,t]$$

holds for some martingale $(M_s^\varepsilon)_{s\in[0,t]}$. Combining this with (6.5.15), we find a constant $c > 0$ such that for any $f \in C_N^2(\mathbb{R}^d)$,

$$\bar{P}_t\{(\varepsilon + f^2)\log(\varepsilon + f^2)\} - (\varepsilon + \bar{P}_t f^2)\log(\varepsilon + \bar{P}_t f^2)$$
$$= \int_0^t \bar{P}_s \frac{|\bar{\sigma}^* \nabla \bar{P}_{t-s} f^2|^2}{\varepsilon + \bar{P}_{t-s} f^2}\mathrm{d}s \leq 4(c_1 \|\bar{\sigma}\|_\infty)^2 \int_0^t \mathrm{e}^{-\lambda(t-s)} \bar{P}_s \bar{P}_{t-s}|\nabla f|^2 \mathrm{d}s$$
$$= 4(c_1 \|\bar{\sigma}\|_\infty)^2 (\bar{P}_t |\nabla f|^2) \int_0^t \mathrm{e}^{-\lambda(t-s)}\mathrm{d}s \leq c\bar{P}_t|\nabla f|^2, \quad t \geq 0, \varepsilon > 0.$$

By letting first $\varepsilon \downarrow 0$ then $t \to \infty$, we deduce from this and (6.5.14) that

$$\bar{\mu}(f^2 \log f^2) \leq c_2 \bar{\mu}(|\nabla f|^2), \quad f \in C_N^2(\bar{D}), \bar{\mu}(f^2) = 1$$

holds for some constant $c_2 > 0$. This implies (6.5.7) by an approximation argument. Indeed the inequality holds for $f \in H^{1,2}(\bar{\mu})$ with $\bar{\mu}(f^2) = 1$. According to Lemma 6.5.3 below, (6.5.8) holds.

Finally, let σ be constant with $\sigma\sigma^*$ invertible. Then (6.5.9) follows from (6.5.5), (6.5.6) and (6.5.8). $\square$

Note that under the log-Sobolev inequality, the Talagrand inequality has been derived in [Bobkov et al. (2001)] for a probability measure $\bar{\mu}$ of type $\mathrm{e}^{V(x)}\mathrm{d}x$ for some $V \in C(\mathbb{R}^d)$. Below we extend assertion to general probability measures and such that $\bar{\mu}$ supported in the domain D can be applied.

Lemma 6.5.3. *Let $c > 0$ be a constant and $\bar{\mu} \in \mathcal{P}_2(\mathbb{R}^d)$. Then the log-Sobolev inequality (6.5.7) implies (6.5.8).*

Proof. By an approximation argument, we only need to prove for $\mu = \varrho\bar{\mu}$ for some density $\varrho \in C_b(\mathbb{R}^d)$. Let $P_t^{(0)}$ be the Ornstein-Uhlenbeck semigroup generated by $\Delta - x \cdot \nabla$ on $\mathbb{R}^d$. We have

$$|\nabla P_t^{(0)} f| \leq P_t^{(0)}|\nabla f|, \quad P_t^{(0)}(f^2 \log f^2)$$
$$\leq t P_t^{(0)}|\nabla f|^2 + (P_t^{(0)} f^2)\log P_t^{(0)} f^2, \quad f \in C_b^1(\mathbb{R}^d).$$

Combining this with (6.5.7), we see that $\bar{\mu}_t := (P_t^{(0)})^*\bar{\mu}$ satisfies

$$\bar{\mu}_t(f^2 \log f^2) = \bar{\mu}(P_t^{(0)}(f^2 \log f^2))$$
$$\leq t\bar{\mu}_t(|\nabla f|^2) + \bar{\mu}((P_t^{(0)} f^2) \log P_t^{(0)} f^2)$$
$$\leq t\mu_t(|\nabla f|^2) + c\bar{\mu}\left(\left|\nabla\sqrt{P_t^{(0)} f^2}\right|^2\right) + \bar{\mu}_t(f^2)\log\bar{\mu}_t(f^2)$$
$$\leq (t+c)\bar{\mu}_t(|\nabla f|^2) + \bar{\mu}_t(f^2)\log\bar{\mu}_t(f^2), \quad f \in C_b^1(\mathbb{R}^d), \ t > 0,$$

where the last step follows from the gradient estimate $|\nabla P_t^{(0)} f| \leq P_t^{(0)}|\nabla f|$, which the Schwarz inequality imply

$$\left|\nabla\sqrt{P_t^{(0)} f^2}\right|^2 = \frac{|\nabla P_t^{(0)} f^2|^2}{4 P_t^{(0)} f^2} \leq \frac{\{P_t^{(0)}(|f\nabla f|)\}^2}{P_t^{(0)} f^2} \leq P_t^{(0)}|\nabla f|^2.$$

Therefore, $\bar{\mu}_t$ satisfies the log-Sobolev inequality with constant $t+c$ and has smooth strictly positive density. According to [Bobkov et al. (2001)], we have

$$\mathbb{W}_2(\mu, \bar{\mu}_t)^2 \leq (t+c)\mathrm{Ent}(\mu|\bar{\mu}_t), \quad \mu \in \mathcal{P}_2(\mathbb{R}^d).$$

Since $\mathbb{W}_2(\bar{\mu}_t, \bar{\mu}) \to 0$ as $t \to 0$, and $\mu = \varrho\bar{\mu}$ with $\varrho \in C_b(\mathbb{R}^d)$, this implies

$$\mathbb{W}_2(\mu,\bar{\mu})^2 = \lim_{t\downarrow 0}\mathbb{W}_2(\mu,\bar{\mu}_t)^2 \leq \lim_{t\downarrow 0}(t+c)\mathrm{Ent}(\mu|\bar{\mu}_t)$$
$$= \lim_{t\downarrow 0}(t+c)\bar{\mu}((P_t^{(0)}\varrho)\log P_t^{(0)}\varrho) = c\bar{\mu}(\varrho\log\varrho).$$

Therefore, (6.5.8) holds. □

6.5.3 Partially dissipative case

In this part, we consider the partially dissipative case such that Theorem 5.5.1 is extended to the reflected setting. Let $\psi \subset \Psi_\kappa$ and $\mathbb{W}_\psi$ be given in (5.5.1) for $\bar{D}$ replacing $\mathbb{R}^d$. Then $\mathbb{W}_\psi$ is a complete quasi-metric on the space

$$\mathcal{P}_\psi(\bar{D}) := \{\mu \in \mathcal{P}(\bar{D}) : \mu(\psi(|\cdot|)) < \infty\}.$$

$(A^{6.7})$ $\sigma_t(x,\mu) = \sigma_t(x)$ does not depend on μ.
(1) (*Ellipticity*) There exist $\alpha \in C([0,\infty);(0,\infty))$ and $\hat{\sigma} \in \mathcal{B}([0,\infty) \times \bar{D}; \mathbb{R}^d \otimes \mathbb{R}^d)$ such that

$$\sigma_t(x)\sigma_t(x)^* = \alpha_t I_d + \hat{\sigma}_t(x)\hat{\sigma}_t(x)^*, \quad t \geq 0, x \in \bar{D}.$$

(2) (*Partial dissipativity*) Let $\psi \in \Psi_\kappa$ in (6.3.14) for some $\kappa > 0$, $g \in C([0,\infty))$ with $g(r) \leq Kr$ for some constant $K > 0$ and all $r \geq 0$, such that

$$2\alpha_t \psi''(r) + (g\psi')(r) \leq -q_t \psi(r), \quad r \geq 0, t \geq 0 \qquad (6.5.16)$$

holds for some $q \in C([0,\infty); \mathbb{R})$. Moreover, $b \in C([0,\infty) \times \bar{D} \times \mathcal{P}_\psi(\bar{D}); \mathbb{R}^d)$, and there exists $\theta \in C([0,\infty); [0,\infty))$ such that

$$\begin{aligned}&\langle b_t(x,\mu) - b_t(y,\nu), x-y\rangle + \frac{1}{2}\|\hat{\sigma}_t(x) - \hat{\sigma}_t(y)\|_{HS}^2 \\ &\leq |x-y|\{\theta_t \mathbb{W}_\psi(\mu,\nu) + g(|x-y|)\}, \\ &t \geq 0, x,y \in \bar{D}, \mu,\nu \in \mathcal{P}_\psi(\bar{D}).\end{aligned} \qquad (6.5.17)$$

Theorem 6.5.4. *Let D be convex and assume $(A^{6.7})$, where $\psi'' \leq 0$ if $\hat{\sigma}$ is non-constant. Then (6.1.1) with $t \in [0,\infty)$ is well-posed for distributions in $\mathcal{P}_\psi(\bar{D})$, and P_t^* satisfies*

$$\begin{aligned}&\mathbb{W}_\psi(P_t^*\mu, P_t^*\nu) \leq e^{-\int_0^t \{q_s - \theta_s\|\psi'\|_\infty\}ds} \mathbb{W}_\psi(\mu,\nu), \\ &t \geq 0, \mu,\nu \in \mathcal{P}_\psi(\bar{D}).\end{aligned} \qquad (6.5.18)$$

Consequently, if $(b_t, \sigma_t, q_t, \theta_t)$ do not depend on t and $q > \theta\|\psi'\|_\infty$, then P_t^* has a unique invariant probability measure $\bar{\mu} \in \mathcal{P}_\psi(\bar{D})$ such that

$$\mathbb{W}_\psi(P_t^*\mu, \bar{\mu}) \leq e^{-(q-\theta\|\psi'\|_\infty)t} \mathbb{W}_\psi(\mu, \bar{\mu}), \quad t \geq 0, \mu \in \mathcal{P}_\psi(\bar{D}). \qquad (6.5.19)$$

Proof. Since D is convex, the proof is similar to that of Theorem 5.5.1. We outline it below for completeness.

By Theorem 6.3.2, the well-posedness follows from $(A^{6.7})(1)$ and $(A^{6.7})(2)$. Next, according to the proof of Theorem 6.5.2(2) with $\mathbb{W}_\psi$ replacing $\mathbb{W}_2$, the second assertion follows from the first. So, in the following we only prove (6.5.18).

For any $s \geq 0$, let (X_s, Y_s) be $\mathcal{F}_s$-measurable such that

$$\mathcal{L}_{X_s} = P_s^*\mu, \quad \mathcal{L}_{Y_s} = P_s^*\nu, \quad \mathbb{W}_\psi(P_s^*\mu, P_s^*\nu) = \mathbb{E}\psi(|X_s - Y_s|). \qquad (6.5.20)$$

Let $W_t^{(1)}$ and $W_t^{(2)}$ be two independent d-dimensional Brownian motions and consider the following SDE for $t \geq s$:

$$dX_t = b_t(X_t, P_t^*\mu)dt + \sqrt{\alpha_t}dW_t^{(1)} + \hat{\sigma}_t(X_t)dW_t^{(2)} + \mathbf{n}(X_t)dl_t^X, \qquad (6.5.21)$$

where l_t^X is the local time of X_t on ∂D. By Theorem 6.5.2, $(A^{6.7})(1)$ and $(A^{6.7})(2)$ imply that this SDE is well-posed and

$$\sqrt{\alpha_t}dW_t^{(1)} + \hat{\sigma}_t(X_t)dW_t^{(2)} = \sigma_t(X_t)dW_t$$

for the m-dimensional Brownian motion

$$W_t := \int_s^t \{\sigma_r^*(\sigma_r\sigma_r^*)^{-1}\}(X_r)\{\sqrt{\alpha_r}\mathrm{d}W_r^{(1)} + \hat{\sigma}_r(X_r)\mathrm{d}W_r^{(2)}\}, \quad t \geq s,$$

so that the weak uniqueness of (6.1.1) implies $\mathcal{L}_{X_t} = P_{s,t}^* P_s^* \mu = P_t^* \mu, t \geq s$, where for $\gamma \in \mathcal{P}_\psi$ we denote $P_{s,t}^* \gamma = \mathcal{L}_{X_t}$ for X_t solving (6.5.21) with $\mathcal{L}_{X_s} = \gamma$.

To construct the coupling with reflection, let

$$u(x,y) = \frac{x-y}{|x-y|}, \quad x \neq y \in \mathbb{R}^d.$$

We consider the SDE for $t \geq s$:

$$\mathrm{d}Y_t = b_t(Y_t, P_t^*\nu)\mathrm{d}t + \sqrt{\alpha_t}\{I_d - 2u(X_t, Y_t) \otimes u(X_t, Y_t)\mathbf{1}_{\{t<\tau\}}\}\mathrm{d}W_t^{(1)}$$
$$+ \hat{\sigma}_t(Y_t)\mathrm{d}W_t^{(2)} + \mathrm{d}l_t^Y, \qquad (6.5.22)$$

where

$$\tau := \inf\{t \geq s : Y_t = X_t\}$$

is the coupling time. Since the coefficients in noises are Lipschitz continuous outside a neighborhood of the diagonal, according to Theorem 2.3.2, (6.5.22) has a unique solution up to the coupling time τ. When $t \geq \tau$, the equation of Y_t becomes

$$\mathrm{d}Y_t = b_t(Y_t, P_t^*\nu)\mathrm{d}t + \sqrt{\alpha_t}\mathrm{d}W_t^{(1)} + \hat{\sigma}_t(Y_t)\mathrm{d}W_t^{(2)} + \mathrm{d}l_t^Y, \qquad (6.5.23)$$

which is well posed under $(A^{6.7})(1)$ and $(A^{6.7})(2)$ according to Theorem 6.3.2. So, (6.5.22) is well-posed and $\mathcal{L}_{Y_t} = P_t^*\nu$ by the same reason leading to $\mathcal{L}_{X_t} = P_t^*\mu$. Since D is convex, (2.1.3) holds. So, by $(A^{6.7})(1)$ and $(A^{6.7})(2)$ for $\psi \in \Psi$ with $\psi'' < 0$ when $\hat{\sigma}_t$ is non-constant, and applying Itô's formula, we obtain

$$\mathrm{d}\psi(|X_t - Y_t|)$$
$$\leq \{\theta_t \psi'(|X_t - Y_t|)\mathbb{W}_\psi(P_t^*\mu, P_t^*\nu) - q_t\psi(|X_t - Y_t|)\}\mathrm{d}t$$
$$+ \psi'(|X_t - Y_t|)\Big[2\sqrt{\alpha_t}\langle u(X_t, Y_t), \mathrm{d}W_t^{(1)}\rangle \qquad (6.5.24)$$
$$+ \langle u(X_t, Y_t), (\hat{\sigma}_t(X_t) - \hat{\sigma}_t(Y_t))\mathrm{d}W_t^{(2)}\rangle\Big], \quad s \leq t < \tau.$$

By a standard argument and noting that $\psi(|X_{t\wedge\tau}, Y_{t\wedge\tau}|)\mathbf{1}_{\{\tau\leq t\}} = 0$, this implies

$$e^{\int_s^t q_p \mathrm{d}p}\mathbb{E}\big[\psi(|X_{t\wedge\tau} - Y_{t\wedge\tau}|)\big] = \mathbb{E}\big[e^{\int_s^{t\wedge\tau} q_p \mathrm{d}p}\mathbb{E}\psi(|X_{t\wedge\tau} - Y_{t\wedge\tau}|)\big]$$
$$\leq \mathbb{E}\psi(|X_s - Y_s|) + \|\psi'\|_\infty \int_s^{t\wedge\tau} \theta_r e^{\int_s^r q_p \mathrm{d}p}\mathbb{W}_\psi(P_r^*\mu, P_r^*\nu)\mathrm{d}r, \quad t \geq s.$$

Consequently,

$$\mathbb{E}\psi(|X_{t\wedge\tau} - Y_{t\wedge\tau}|) \leq e^{-\int_s^t q_r dr}\mathbb{E}\psi(|X_s - Y_s|) \\ + \|\psi'\|_\infty \int_s^{t\wedge\tau} \theta_r e^{-\int_r^t q_p dp} \mathbb{W}_\psi(P_r^*\mu, P_r^*\nu) dr, \quad t \geq s. \tag{6.5.25}$$

On the other hand, when $t \geq \tau$, by $(A^{6.7})(2)$ and applying Itô's formula for (6.5.21) and (6.5.23), we find a constant $C > 0$ such that

$$d\psi(|X_t - Y_t|) \leq \{C\psi(|X_t - Y_t|)dt + \theta_t\|\psi'\|_\infty \mathbb{W}_\psi(P_t^*\mu, P_t^*\nu)\}dt \\ + \psi'(|X_t - Y_t|)\langle\{\hat\sigma_t(X_t) - \hat\sigma_t(Y_t)\}^* u(X_t, Y_t), dW_t^{(2)}\rangle.$$

Noting that $\psi(|X_\tau - Y_\tau|) = 0$, we obtain

$$\mathbb{E}\big[1_{\{t>\tau\}}\psi(|X_t - Y_t|)\big] \leq \|\psi'\|_\infty e^{C(t-s)}\mathbb{E}\int_{t\wedge\tau}^t \theta_r \mathbb{W}_\psi(P_r^*\mu, P_r^*\nu) dr, \quad t \geq s.$$

Combining this with (6.5.25) and (6.5.20), we derive

$$\mathbb{W}_\psi(P_t^*\mu, P_t^*\nu) \leq \mathbb{E}\psi(|X_t - Y_t|) \\ = \mathbb{E}\psi(|X_{t\wedge\tau} - Y_{t\wedge\tau}|) + \mathbb{E}\big[1_{\{t>\tau\}}\psi(|X_t - Y_t|)\big] \\ \leq e^{-\int_s^t q_r dr}\mathbb{E}\psi(|X_s - Y_s|) + \|\psi'\|_\infty e^{C(t-s)}\int_s^t \theta_r \mathbb{W}_\psi(P_r^*\mu, P_r^*\nu) dr \\ = e^{-\int_s^t q_r dr}\mathbb{W}_\psi(P_s^*\mu, P_s^*\nu) + \|\psi'\|_\infty e^{C(t-s)}\int_s^t \theta_r \mathbb{W}_\psi(P_r^*\mu, P_r^*\nu) dr, \quad t \geq s.$$

Therefore,

$$\frac{d^+}{ds}\mathbb{W}_\psi(P_s^*\mu, P_s^*\nu) := \limsup_{t\downarrow s} \frac{\mathbb{W}_\psi(P_t^*\mu, P_t^*\nu) - \mathbb{W}_\psi(P_s^*\mu, P_s^*\nu)}{t - s} \\ \leq -(q_s - \theta_s\|\psi'\|_\infty)\mathbb{W}_\psi(P_s^*\mu, P_s^*\nu), \quad s \geq 0.$$

This implies (6.5.18). □

As a consequence of Theorem 6.5.4, we consider the non-dissipative case where $\nabla b_t(\cdot, \mu)(x)$ is positive definite in a possibly unbounded set but with bounded "one-dimensional puncture mass" in the sense of (6.5.28) below.

Let $\mathbb{W}_1 = \mathbb{W}_\psi$ and $\mathcal{P}_1(\bar D) = \mathcal{P}_\psi(\bar D)$ for $\psi(r) = r$, and define

$$S_b(x) := \sup\{\langle \nabla_v b_t(\cdot, \mu)(x), v\rangle : t \geq 0, |v| \leq 1, \mu \in \mathcal{P}_1(\bar D)\}, \quad x \in \bar D.$$

$(A^{6.7})$ (3) There exist constants $\theta_0, \theta_1, \theta_2, \beta \geq 0$ such that

$$\frac{1}{2}\|\sigma_t(x) - \sigma_t(y)\|_{HS}^2 \leq \theta_0 |x-y|^2, \quad t \geq 0, x, y \in \bar{D}; \tag{6.5.26}$$

$$S_b(x) \leq \theta_1, \quad |b_t(x,\mu) - b_t(x,\nu)| \leq \beta \mathbb{W}_1(\mu,\nu),$$
$$t \geq 0, x \subset \bar{D}, \mu, \nu \in \mathcal{P}_1(\bar{D}); \tag{6.5.27}$$

$$\kappa := \sup_{x, v \in \bar{D}, |v|=1} \int_{\mathbb{R}} 1_{\{S_b(x+sv) > -\theta_2\}} ds < \infty. \tag{6.5.28}$$

According to the proof of Corollary 5.5.2, the following result follows from Theorem 6.5.4.

Corollary 6.5.5. *Let D be convex. Assume $(A^{6.7})(1)$ and $(A^{6.7})(3)$. Let*

$$g(r) := (\theta_1 + \theta_2)\{(\kappa r^{-1}) \wedge r\} - (\theta_2 - \theta_0)r, \quad r \geq 0,$$

$$k := \frac{2\alpha}{\int_0^\infty t e^{\frac{1}{2\beta} \int_0^t g(u)du} dt} - \frac{\beta(\theta_2 - \theta_0)}{2\alpha} \int_0^\infty t e^{\frac{1}{2\alpha} \int_0^t g(u)du} dt. \tag{6.5.29}$$

Then there exists a constant $c > 0$ such that

$$\mathbb{W}_1(P_t^*\mu, P_t^*\nu) \leq c e^{-kt} \mathbb{W}_1(\mu, \nu), \quad t \geq 0, \mu, \nu \in \mathcal{P}_1(\bar{D}).$$

If (b_t, σ_t) does not depend on t and $\theta_2 > \theta_0$ with

$$\beta < \frac{4\alpha^2}{(\theta_2 - \theta_0)(\int_0^\infty t e^{\frac{1}{2\alpha} \int_0^t g(u)du} dt)^2}, \tag{6.5.30}$$

then $k > 0$ and P_t^* has a unique invariant probability measure $\bar{\mu} \subset \mathcal{P}_1(\bar{D})$ satisfying

$$\mathbb{W}_1(P_t^*\mu, \bar{\mu}) \leq c e^{-kt} \mathbb{W}_1(\mu, \bar{\mu}), \quad t \geq 0, \mu \in \mathcal{P}_1(\bar{D}).$$

6.5.4 Non-dissipative case

Finally, we consider the fully non-dissipative case such that Theorem 5.4.1 is extended to the reflected setting. Let $L_{t,\mu}$ be in (6.1.2) for any $t \geq 0$ and $\mu \in \mathcal{P}(\bar{D})$. We assume the following Lyapunov condition.

$(A^{6.8})$ (*Lyapunov Condition*) There exists a function $0 \leq V \in C^2(\bar{D})$ with $\langle \nabla V, \mathbf{n} \rangle|_{\partial D} \leq 0$, $\lim_{|x| \to \infty} V(x) = \infty$ and

$$\sup_{t \geq 0; x, y \in \bar{D}} \left\{ \frac{|\nabla V(x) - \nabla V(y)|}{|x-y|\{1 + V(x) + V(y)\}} + \frac{\|\sigma_t(x)\|^2 \cdot |\nabla V(x)| + \|\sigma_t(y)\|^2 \cdot |\nabla V(y)|}{1 + V(x) + V(y)} \right\} < \infty, \tag{6.5.31}$$

such that for some $K_0, K_1 \in L^1_{loc}([0,\infty); \mathbb{R})$ and any

$$\mu \in \mathcal{P}_V(\bar{D}) := \{\mu \in \mathcal{P}(\bar{D}) : \mu(V) < \infty\}, \tag{6.5.32}$$

we have

$$L_{t,\mu} V \leq K_0(t) - K_1(t) V, \quad t \geq 0. \tag{6.5.33}$$

Next, we introduce the monotone condition with respect to a weighted Wasserstein distance induced by V and a function ψ in the following class for some $l > 0$:

$$\tilde{\Psi}_l := \{\psi \in C^2([0,l]; [0,\infty)) : \psi(0) = \psi'(l) = 0, \psi'|_{[0,l)} > 0\}.$$

For each $\psi \in \tilde{\Psi}_l$, we extend it to the half line by setting $\psi(r) = \psi(r \wedge l)$, so that ψ' is nonnegative, Lipschitz continuous with compact support, and satisfies

$$\|\psi'\|_\infty := \sup_{r>0} |\psi'(r)| = \sup_{r \in (0,l)} \psi'(r) \in (0,\infty).$$

For any constant $\beta > 0$, define the quasi-distance on $\mathcal{P}_V(\bar{D})$:

$$\mathbb{W}_{\psi,\beta V}(\mu,\nu) := \inf_{\pi \in \mathcal{C}(\mu,\nu)} \int_{\mathbb{R}^d \times \mathbb{R}^d} \psi(|x-y|)(1 + \beta V(x) + \beta V(y)) \pi(\mathrm{d}x, \mathrm{d}y),$$

$$\hat{\mathbb{W}}_{\psi,\beta V}(\mu,\nu) := \inf_{\pi \in \mathcal{C}(\mu,\nu)} \frac{\int_{\bar{D} \times \bar{D}} \psi(|x-y|)(1 + \beta V(x) + \beta V(y)) \pi(\mathrm{d}x, \mathrm{d}y)}{\int_{\bar{D} \times \bar{D}} \psi'(|x-y|)(1 + \beta V(x) + \beta V(y)) \pi(\mathrm{d}x, \mathrm{d}y)}.$$

Obviously, $\hat{\mathbb{W}}_{\psi,\beta V}(\mu,\nu) \geq \frac{\mathbb{W}_{\psi,\beta V}(\mu,\nu)}{\|\psi'\|_\infty (1+\beta\mu(V)+\beta\nu(V))}$.

$(A^{6.9})$ (*Local monotonicity*) σ satisfies $(A^{6.7})(1)$, b is bounded on bounded set in $[0,\infty) \times \bar{D} \times \mathcal{P}_V(\bar{D})$. Moreover, there exist $K, \theta, q \in L^1_{loc}([0,\infty); [0,\infty))$ and a function $\psi \in \tilde{\Psi}_l$ for some $l > 0$ satisfying

$$2\alpha_t \psi''(r) + K_t \psi'(r) \leq -q_t \psi(r), \quad r \in [0,l],$$

such that

$$\langle b_t(x,\mu) - b_t(y,\nu), x-y \rangle + \frac{1}{2} \|\hat{\sigma}_t(x) - \hat{\sigma}_t(y)\|^2_{HS}$$
$$\leq K_t |x-y|^2 + \theta_t |x-y| \hat{\mathbb{W}}_{\psi,\beta V}(\mu,\nu), \quad x,y \in \bar{D}, \mu,\nu \in \mathcal{P}_V(\bar{D}).$$

By $(A^{6.8})$, $V(x) \to \infty$ as $|x| \to \infty$, when $K_1(t) > 0$ and l is large enough, we have

$$\zeta_{l,\beta}(t) := \inf_{|x-y|>l} \frac{K_1(t)V(x) + K_1(t)V(y) - 2K_0(t)}{\beta^{-1} + V(x) + V(y)} > 0. \quad (6.5.34)$$

Moreover, $(A^{6.7})(1)$ and $(A^{6.8})$ imply

$$\begin{aligned}\alpha_{l,\beta}(t) := \|\psi'\|_\infty \sup_{|x-y|\in(0,l)} &\left\{ \frac{|\nabla V(x) - \nabla V(y)|}{|x-y|\{\beta^{-1} + V(x) + V(y)\}} \right.\\ &\left. + \frac{|\{\hat\sigma_t(x) - \hat\sigma_t(y)\}[(\hat\sigma_t^*\nabla V)(x) + (\hat\sigma_t^*\nabla V)(y)]|}{|x-y|\{\beta^{-1} + V(x) + V(y)\}} \right\} < \infty\end{aligned} \quad (6.5.35)$$

for any $\beta, l > 0$. For constants K_0, $\zeta_{l,\beta}$, $\alpha_{l,\beta}$ and q given in $(A^{6.8})$, $(A^{6.9})$, (6.5.34) and (6.5.35) respectively, let

$$\lambda_{l,\beta}(t) := \min\{\zeta_{l,\beta}(t),\ q_t - 2K_0(t)\beta - \alpha_{l,\beta}(t)\}.$$

Theorem 6.5.6. *Let D be convex. Assume $(A^{6.8})$ and $(A^{6.9})$, where $\psi'' \leq 0$ if $\hat\sigma_t(\cdot)$ is non-constant. Then (6.1.1) with $t \in [0,\infty)$ is well-posed for distributions in $\mathcal{P}_V(\bar D)$, and P_t^* satisfies*

$$\begin{aligned}&\mathbb{W}_{\psi,\beta V}(P_t^*\mu, P_t^*\nu) \leq e^{-\int_0^t \{\lambda_{l,\beta}(s) - \theta_s\}ds} \mathbb{W}_{\psi,\beta V}(\mu,\nu),\\ &t \geq 0, \mu, \nu \in \mathcal{P}_V(\bar D).\end{aligned} \quad (6.5.36)$$

Consequently, if (σ_t, b_t) does not depend on t and $\lambda_{l,\beta} > \theta$, then P_t^* has a unique invariant probability measure $\bar\mu \in \mathcal{P}_V(\bar D)$ such that

$$\mathbb{W}_{\psi,\beta V}(P_t^*\mu, \bar\mu) \leq e^{-(\lambda_{l,\beta} - \theta)t} \mathbb{W}_{\psi,\beta V}(\mu, \bar\mu), \quad t \geq 0, \mu \in \mathcal{P}_V(\bar D). \quad (6.5.37)$$

Proof. We first prove the well-posedness. Let X_0 be $\mathcal{F}_0$-measurable with $\mathcal{L}_{X_0} =: \gamma \in \mathcal{P}_V(\bar D)$. For any $T > 0$ and

$$\mu \in \mathcal{P}_{V,\gamma}^T(\bar D) := \{\mu \in C_b^w([0,T]; \mathcal{P}_V(D)) : \mu_0 = \gamma\},$$

consider the following reflected SDE on $\bar D$:

$$dX_t^\mu = b_t(X_t^\mu, \mu_t)dt + \sigma_t(X_t^\mu)dW_t + \mathbf{n}(X_t)dL_{t,\mu}, \quad X_0^\mu = X_0, t \in [0,T].$$

According to Theorem 2.3.2, $(A^{6.9})$ implies that this SDE is well-posed up to life time. By $\langle\nabla V, \mathbf{n}\rangle|_{\partial D} \leq 0$ and (6.5.26) in $(A^{6.8})$, and applying Itô's formula, we obtain

$$\begin{aligned}&dV(X_t^\mu)\\ &= L_{t,\mu_t}V(X_t^\mu)dt + \langle\nabla V(X_t^\mu), \sigma_t(X_t^\mu)dW_t\rangle + \langle\nabla V(X_t^\mu), \mathbf{n}(X_t^\mu)\rangle dL_{t,\mu}\\ &\leq \{K_0(t) - K_1(t)V(X_t^\mu)\}dt + \langle\nabla V(X_t^\mu), \sigma_t(X_t^\mu)dW_t\rangle.\end{aligned}$$

By Gronwall's lemma and $\lim_{|x|\to\infty} V(x) = \infty$, this implies the non-explosion of X_t^μ, and

$$\Phi.\mu := \mathcal{L}_{X^\mu} \in \mathcal{P}_{V,\gamma}^T(\bar{D}).$$

So, as shown in the proof of Theorem 6.2.2, it suffices to verify the contraction of Φ on $\mathcal{P}_{V,\gamma}^T(\bar{D})$ under the metric

$$\mathbb{W}_{\psi,V,\lambda}(\mu,\nu) := \sup_{t\in[0,T]} e^{-\lambda t} \mathbb{W}_{\psi,V}(\mu_t,\nu_t), \quad \mu,\nu \in \mathcal{P}_{V,\gamma}^T(\bar{D})$$

for large $\lambda > 0$. Let $\mu,\nu \in \mathcal{P}_{V,\gamma}^T(\bar{D})$. By $(A^{6.9})$, $\langle n, \nabla V\rangle|_{\partial D} \leq 0$, (2.1.3), and applying Itô-Tanaka formula, we find a constant $C_1 > 0$ such that

$$\begin{aligned}d|X_t^\mu - X_t^\nu| \leq {} & C_1(\hat{\mathbb{W}}_{\psi,\beta V}(\mu_t,\nu_t) + |X_t^\mu - X_t^\nu|)dt \\ & + \Big\langle \frac{X_t^\mu - X_t^\nu}{|X_t^\mu - X_t^\nu|}, \{\sigma_t(X_t^\mu) - \sigma_t(X_t^\nu)\}dW_t\Big\rangle.\end{aligned}$$

Then the remainder of the proof is the same as that of Lemma 5.4.3.

Next, we prove (6.5.36) which implies (6.5.37) in the time homogenous case. For any $\mu,\nu \in \mathcal{P}_V(\bar{D})$, let X_0, Y_0 be $\mathcal{F}_0$-measurable such that $\mathcal{L}_{X_0} = \mu$, $\mathcal{L}_{Y_0} = \nu$, and

$$\mathbb{E}\big[\psi(|X_0 - Y_0|)(1 + \beta V(X_0) + \beta V(Y_0))\big] = \mathbb{W}_{\psi,\beta V}(\mu,\nu).$$

Let (X_t, Y_t) be the coupling constructed in the proof of Theorem 6.5.2. By $\langle \mathbf{n}, \nabla V\rangle|_{\partial D} \leq 0$ and (2.1.3), the local time terms does not make any trouble when we apply Itô's formula to $\psi(|X_t - Y_t|)$ or $V(X_t) + V(Y_t)$. So, by repeating step **C** in the proof of Theorem 5.4.1, we derive (6.5.36). □

6.6 Notes

The study of reflected DDSDEs goes back to [Sznitman (1984)] where the coefficients are Lipschitz continuous and the dependence of distribution is of integral type. In a more general framework but with convex domain D, the propagation of chaos and large deviation principle for small noise have been investigated in [Adams *et al.* (2022)] where more references on reflected DDSDEs can be found.

In this chapter, most results in Chapters 3–5 have been extended to the reflected case, except Bismut formula and the Donsker-Varadhan large deviations. Noting that Bismut formulas have been established for reflected diffusion processes, see for instance [Wang (2014)], this type of formulas should also be valid for the reflected DDSDEs. When D is convex, it should be easy to extend the Donsker-Varadhan LDP in Theorem 5.6.1 to the reflected setting.

It is interesting to study the exponential ergodicity and long time LDP for non-convex D.

Chapter 7

Killed DDSDEs

In this chapter, we consider the killed DDSDEs which in turn describe the nonlinear Dirichlet problems, where the distributions are restricted to an open domain and thus might be sub-probability measures. We first introduce the link of killed DDSDE and nonlinear Dirichlet problem with a general well-posedness result, then study the killed DDSDE for several different situations. This part is due to [Wang (2023f)].

7.1 Killed DDSDE for nonlinear Dirichlet problem

Let $D \subset \mathbb{R}^d$ be a connected open domain with closure $\bar{D}$, and let

$$\mathcal{P}^D := \{\mu \text{ is a measure on } D, \ \mu(D) \leq 1\}$$

be the space of sub-probability measures on D equipped with the weak topology.

Consider the following time-distribution dependent second order differential operator on D:

$$L_{t,\mu} := \mathrm{tr}\{(\sigma_t \sigma_t^*)(\cdot, \mu)\nabla^2\} + \nabla_{b_t(\cdot,\mu)}, \quad t \in [0,T], \mu \in \mathcal{P}^D,$$

where $T > 0$ is a fixed constant, σ^* is the transposition of σ, ∇^2 is the Hessian operator, $\nabla_b := b \cdot \nabla$ is the derivative along b, and for some $m \in \mathbb{N}$,

$$b : [0,T] \times D \times \mathcal{P}^D \to \mathbb{R}^d, \quad \sigma : [0,T] \times D \times \mathcal{P}^D \to \mathbb{R}^d \otimes \mathbb{R}^m$$

are measurable such that

$$\int_0^T \mathrm{d}t \int_D \{|b_t(x,\mu_t)| + \|\sigma_t(x,\mu_t)\|^2\}\mu_t(\mathrm{d}x) < \infty, \qquad (7.1.1)$$
$$\mu = (\mu_t)_{t \in [0,T]} \in C([0,T]; \mathcal{P}^D).$$

To introduce the nonlinear Dirichlet problem for $L_{t,\mu}$ on $\mathcal{P}^D$, let $C_D^2(D)$ be the class of $f \in C_b^2(\bar{D})$ with Dirichlet condition $f|_{\partial D} = 0$, where $f \in C_b^2(\bar{D})$ means that f is a bounded C^2 function on $\bar{D}$ with bounded first and second order derivatives. For any $t \in [0,T]$ and $\mu, \nu \in \mathcal{P}^D$ such that

$$\int_D \{|b_t(x,\mu)| + \|\sigma_t(x,\mu)\|^2\}\nu(\mathrm{d}x) < \infty,$$

define the linear functional on $C_D^2(D)$:

$$L_{t,\mu}^{D*}\nu : C_D^2(D) \ni f \mapsto (L_{t,\mu}^{D*}\nu)(f) := \int_D L_{t,\mu} f \mathrm{d}\nu \in \mathbb{R}.$$

The corresponding nonlinear Dirichlet problem for $L_{t,\mu}$ is the equation

$$\partial_t \mu_t = L_{t,\mu_t}^{D*}\mu_t, \quad t \in [0,T] \tag{7.1.2}$$

for $\mu : [0,T] \to \mathcal{P}^D$. We call $\mu. \in C([0,T]; \mathcal{P}^D)$ a solution to (7.1.2), if

$$\mu_t(f) = \mu_0(f) + \int_0^t \mu_s(L_{s,\mu_s} f)\mathrm{d}s, \quad t \in [0,T], f \in C_D^2(D),$$

where $\mu(f) := \int f \mathrm{d}\mu$ for a measure μ and $f \in L^1(\mu)$.

When $\mu_t(\mathrm{d}x) = \rho_t(x)\mathrm{d}x$, (7.1.2) reduces to the nonlinear Dirichlet problem

$$\partial_t \rho_t = L_{t,\rho_t}^{D*}\rho_t, \quad t \in [0,T],$$

where $L_{t,\rho_t} := L_{t,\rho_t(x)\mathrm{d}x}$, in the sense that

$$\int_D (f\rho_t)(x)\mathrm{d}x = \int_D (f\rho_0)(x)\mathrm{d}x + \int_0^t \mathrm{d}s \int_D (\rho_s L_{s,\rho_s} f)(x)\mathrm{d}x,$$
$$t \in [0,T], f \in C_D^2(D).$$

To characterize (7.1.2), we consider the following killed DDSDE on $\bar{D}$:

$$\mathrm{d}X_t = 1_{\{t<\tau(X)\}}\{b_t(X_t, \mathcal{L}_{X_t}^D)\mathrm{d}t + \sigma_t(X_t, \mathcal{L}_{X_t}^D)\mathrm{d}W_t\}, \quad t \in [0,T], \tag{7.1.3}$$

where W_t is the m-dimensional Brownian motion on a complete filtration probability space $(\Omega, \{\mathcal{F}_t\}_{t\geq 0}, \mathbb{P})$,

$$\tau(X) := \inf\{t \in [0,T] : X_t \in \partial D\}$$

with $\inf \emptyset = \infty$ by convention, and for a $\bar{D}$-valued random variable ξ,

$$\mathcal{L}_\xi^D := \mathbb{P}(\xi \in D \cap \cdot)$$

is the distribution of ξ restricted to D, which we call the D-distribution of ξ. When different probability spaces are concerned, we denote $\mathcal{L}_\xi^D$ by $\mathcal{L}_{\xi|\mathbb{P}}^D$ to emphasize the reference probability measure.

Definition 7.1.1. A continuous adapted process $(X_t)_{t\in[0,T]}$ on $\bar{D}$ is called a solution of (7.1.3), if $\mathbb{P}$-a.s.

$$\int_0^{T\wedge\tau(X)} \{|b_t(X_t, \mathcal{L}_{X_t}^D)| + \|\sigma_t(X_t, \mathcal{L}_{X_t}^D)\|^2\} \mathrm{d}t < \infty$$

and

$$X_t = X_0 + \int_0^{t\wedge\tau(X)} \{b_s(X_s, \mathcal{L}_{X_s}^D)\mathrm{d}s + \sigma_s(X_s, \mathcal{L}_{X_s}^D)\mathrm{d}W_s\}, \quad t \in [0,T].$$

We call $(\tilde{X}_t, \tilde{W}_t)$ a weak solution to (7.1.3), if there exists a complete filtration probability space $(\tilde{\Omega}, \{\tilde{\mathcal{F}}_t\}_{t\in[0,T]}, \tilde{\mathbb{P}})$ such that $\tilde{W}_t$ is m-dimensional Brownian motion and $\tilde{X}_t$ solves (7.1.3) for $\tilde{W}_t$ replacing W_t.

Remark 7.1.1. (1) It is easy to see that for any (weak) solution X_t of (7.1.3), $\mu_t := \mathcal{L}_{X_t}^D$ solves the nonlinear Dirichlet problem (7.1.2). Indeed, since $\mathrm{d}X_t = 0$ for $t \geq \tau(X)$, we have

$$X_t = X_{\tau(X)} \in \partial D, \quad t \geq \tau(X),$$

so that

$$X_t = X_{t\wedge\tau(X)}, \quad \mathcal{L}_{X_t}^D(\mathrm{d}x) = \mathbb{P}(t < \tau(X), X_t \in \mathrm{d}x), \quad t \in [0,T].$$

By this and Itô's formula, for any $f \in C_D^2(D)$ we have

$$\mu_t(f) = \mathbb{E}[(1_D f)(X_t)] = \mathbb{E}[f(X_t)]$$

$$= \mathbb{E}[f(X_0)] + \mathbb{E}\int_0^t 1_{\{s<\tau(X)\}} L_{s,\mu_s} f(X_s)\mathrm{d}s$$

$$= \mu_0(f) + \int_0^t \mu_s(L_{s,\mu_s}f)\mathrm{d}s, \quad t \in [0,T].$$

(2) An alternative model to (7.1.3) is

$$\mathrm{d}X_t = 1_D(X_t)\{b_t(X_t, \mathcal{L}_{X_t}^D)\mathrm{d}t + \sigma_t(X_t, \mathcal{L}_{X_t}^D)\mathrm{d}W_t\}, \quad t \in [0,T]. \quad (7.1.4)$$

A solution of (7.1.3) also solves (7.1.4); while for a solution X_t to (7.1.4),

$$\tilde{X}_t := X_{t\wedge\tau(X)}$$

solves (7.1.3). In general, a solution of (7.1.4) does not have to solve (7.1.3). For instance, let $d = m = 1$ and $D = (0,\infty)$, consider $\sigma_t(x,\mu) = 2x$, $b_t(x,\mu) = 2\sqrt{x}$. Let Y_t solve the SDE

$$\mathrm{d}Y_t = Y_t \mathrm{d}W_t + \left(1 - \frac{1}{2}Y_t\right)\mathrm{d}t, \quad Y_0 = 0.$$

Then $X_t := (Y_t)^2$ solves (7.1.4) but does not solve (7.1.3), since $\tau(X) = 0$ and $X_t > 0$ (i.e. $X_t \notin \partial D$) for $t > 0$.

(3) The SDE (7.1.4) can be formulated as the usual DDSDE on $\mathbb{R}^d$, so that the superposition principle in [Barbu and Röckner (2020)] applies. More precisely, let $\mathcal{P}$ be the space of probability measures on $\mathbb{R}^d$, and define

$$\bar{b}_t(x,\mu) := 1_D(x) b_t(x, \mu(D \cap \cdot)), \quad \bar{\sigma}_t(x,\mu) := 1_D(x)\sigma_t(x, \mu(D \cap \cdot))$$

for $(t, x, \mu) \in [0, T] \times \mathbb{R}^d \times \mathcal{P}$. Then (7.1.4) becomes the following DDSDE on $\mathbb{R}^d$:

$$\mathrm{d} X_t = \bar{b}_t(X_t, \mathcal{L}_{X_t})\mathrm{d} t + \bar{\sigma}_t(X_t, \mathcal{L}_{X_t})\mathrm{d} W_t, \quad t \in [0, T].$$

We often solve (7.1.3) for D-distributions in a non-empty sub-space $\hat{\mathcal{P}}^D$ of $\mathcal{P}^D$, which is equipped with a complete metric $\hat{d}$. Let $C^w([0, T]; \hat{\mathcal{P}}^D)$ and $C_b^w([0, T]; \hat{\mathcal{P}}^D)$ be defined as in (3.1.2).

Definition 7.1.2. (1) If for any $\mathcal{F}_0$-measurable random variable X_0 on $\bar{D}$ with $\mathcal{L}_{X_0}^D \in \hat{\mathcal{P}}^D$, (7.1.3) has a unique solution starting at X_0 such that $\mathcal{L}_X^D := (\mathcal{L}_{X_t}^D)_{t \in [0,T]} \in C_b^w([0, T]; \hat{\mathcal{P}}^D)$, we call the SDE strongly well-posed for D-distributions in $\hat{\mathcal{P}}^D$.

(2) We call the SDE weakly unique for D-distributions in $\hat{\mathcal{P}}^D$, if for any two weak solutions (X_t^i, W_t^i) w.r.t. $(\Omega^i, \{\mathcal{F}_t^i\}_{t \in [0,T]}, \mathbb{P}^i)(i = 1, 2)$ with $\mathcal{L}_{X_0^1|\mathbb{P}^1}^D = \mathcal{L}_{X_0^2|\mathbb{P}^2}^D \in \hat{\mathcal{P}}^D$, we have $\mathcal{L}_{X^1|\mathbb{P}^1}^D = \mathcal{L}_{X^2|\mathbb{P}^2}^D$. We call (7.1.3) weakly well-posed for D-distributions in $\hat{\mathcal{P}}^D$, if for any initial D-distribution $\mu_0 \in \hat{\mathcal{P}}^D$, it has a unique weak solution for D-distributions in $\hat{\mathcal{P}}^D$.

(3) The SDE (7.1.3) is called well-posed for D-distributions in $\hat{\mathcal{P}}^D$, if it is both strongly and weakly well-posed for D-distributions in $\hat{\mathcal{P}}^D$.

When (7.1.3) is well-posed for D-distributions in $\hat{\mathcal{P}}^D$, for any $\mu \in \hat{\mathcal{P}}^D$ and $t \in [0, T]$, let

$$P_t^{D*}\mu = \mathcal{L}_{X_t}^D, \quad t \in [0, T], \quad \mathcal{L}_{X_0}^D = \mu.$$

We will study the well-posedness under the following assumption.

$(A^{7.1})$ For any $\mu \in C_b^w([0, T]; \hat{\mathcal{P}}^D)$, the killed SDE

$$\mathrm{d} X_t^\mu = 1_{\{t < \tau(X^\mu)\}}\{b_t(X_t^\mu, \mu_t)\mathrm{d} t + \sigma_t(X_t^\mu, \mu_t)\mathrm{d} W_t\}, \quad t \in [0, T] \tag{7.1.5}$$

is well-posed for initial value X_0^μ with $\mathcal{L}_{X_0^\mu}^D = \mu_0$, and $\mathcal{L}_{X^\mu}^D \in C_b^w([0, T]; \hat{\mathcal{P}}^D)$.

Under this assumption, we define a map

$$C_b^w([0,T]; \hat{\mathcal{P}}^D) \ni \mu \mapsto \Phi.\mu := \mathcal{L}_{X^\mu}^D \in C_b^w([0,T]; \hat{\mathcal{P}}^D). \tag{7.1.6}$$

It is clear that a solution of (7.1.5) solves (7.1.3) if and only if μ is a fixed point of Φ. So, we have the following result.

Theorem 7.1.1. *Assume* $(A^{7.1})$. *If for any* $\gamma \in \hat{\mathcal{P}}^D$, Φ *has a unique fixed point in* $\{\mu \in C_b^w([0,T]; \hat{\mathcal{P}}^D), \mu_0 = \gamma\}$, *then* (7.1.3) *is well-posed for* D-*distributions in* $\hat{\mathcal{P}}^D$.

7.2 Monotone case

In this part, we solve (7.1.3) under monotone conditions with respect to the L^1 or truncated L^1 Wasserstein distances:

$$\begin{aligned}\mathbb{W}_1(\mu,\nu) &:= \inf_{\pi \in \mathcal{P}_D(\mu,\nu)} \int_{D \times D} |x-y| \pi(\mathrm{d}x, \mathrm{d}y), \\ \hat{\mathbb{W}}_1(\mu,\nu) &:= \inf_{\pi \in \mathcal{P}_D(\mu,\nu)} \int_{D \times D} (1 \wedge |x-y|) \pi(\mathrm{d}x, \mathrm{d}y), \quad \mu,\nu \in \mathcal{P}^D,\end{aligned} \tag{7.2.1}$$

where $\pi \in \mathcal{P}_D(\mu,\nu)$ means that π is a probability measure on $\bar{D} \times \bar{D}$ such that

$$\pi(\{\cdot \cap D\} \times \bar{D}) = \mu, \quad \pi(\bar{D} \times \{\cdot \cap D\}) = \nu.$$

7.2.1 Monotonicity in $\hat{\mathbb{W}}_1$

$(A^{7.2})$ For any $\mu \in C^w([0,T]; \mathcal{P}^D)$, $b_t(x,\mu_t)$ and $\sigma_t(x,\mu_t)$ are continuous in $x \in D$ such that for any $N \geq 1$ and $D_N := \{x \in D : |x| \leq N\}$,

$$\int_0^T \sup_{D_N} \{|b_t(\cdot,\mu_t)| + \|\sigma_t(\cdot,\mu_t)\|^2\} \mathrm{d}t < \infty.$$

Moreover, there exists $K \in L^1([0,T];(0,\infty))$ such that for any $x,y \in D$ and $\mu,\nu \in \mathcal{P}^D$,

$$\begin{aligned}2\langle b_t(x,\mu) - b_t(y,\nu), x-y\rangle &+ \|\sigma_t(x,\mu) - \sigma_t(y,\nu)\|_{HS}^2 \\ &\leq K(t)\{|x-y|^2 + \hat{\mathbb{W}}_1(\mu,\nu)^2\}, \\ 2\langle b_t(x,\mu), x\rangle + \|\sigma_t(x,\mu)\|_{HS}^2 &\leq K(t)(1+|x|^2), \quad t \in [0,T].\end{aligned}$$

$(A^{7.3})$ There exists $r_0 \in (0,1]$ such that the distance function ρ_∂ to ∂D is C^2-smooth in
$$\partial_{r_0} D := \{x \in \bar{D} : \rho_\partial(x) \le r_0\},$$
and there exists a constant $\alpha > 0$ such that
$$|\sigma_t(x,\mu)^* \nabla \rho_\partial(x)|^{-2} \le \alpha, \quad L_{t,\mu}\rho_\partial(x) \le \alpha, \quad x \in \partial_{r_0} D, t \in [0,T].$$

Theorem 7.2.1. *Assume $(A^{7.2})$ and $(A^{7.3})$. Then the following assertions hold.*

(1) *(7.1.3) is well-posed for D-distributions in $\mathcal{P}^D$. Moreover, for any $p \ge 1$ there exists a constant $c > 0$ such that for any solution X_t to (7.1.3) for D-distributions in $\mathcal{P}^D$,*
$$\mathbb{E}\Big[\sup_{t \in [0,T]} |X_t|^p \Big| \mathcal{F}_0\Big] \le c\big(1 + |X_0|^p\big). \tag{7.2.2}$$

(2) *There exists a constant $c > 0$ such that*
$$\sup_{t \in [0,T]} \hat{\mathbb{W}}_1(P_t^{D*}\mu, P_t^{D*}\nu) \le c\hat{\mathbb{W}}_1(\mu,\nu), \quad \mu,\nu \in \mathcal{P}^D. \tag{7.2.3}$$

Under assumption $(A^{7.2})$, for any $\mu \in C^w([0,T]; \mathcal{P}^D)$, the SDE (7.1.5) satisfies the semi-Lipschitz condition before the hitting time $\tau(X^\mu)$, hence it is well-posed, and for any $p \ge 1$ there exists a constant $c > 0$ uniformly in μ such that
$$\begin{aligned}\mathbb{E}\Big[\sup_{t \in [0,T]} |X_{t \wedge \tilde{\tau}}^\mu|^p \Big| \mathcal{F}_0\Big] &= \mathbb{E}\Big[\sup_{t \in [0,T]} |X_{t \wedge \tau(X^\mu) \wedge \tilde{\tau}}^\mu|^p \Big| \mathcal{F}_0\Big] \\ &\le c(1 + |X_0^\mu|^p)\end{aligned} \tag{7.2.4}$$
holds for any solution X_t^μ of (7.1.5) and any stopping time $\tilde{\tau}$.

By Theorem 7.1.1, to prove the well-posedness of (7.1.3) for D-distributions in $\mathcal{P}^D$, it remains to show that for any $\gamma \in \mathcal{P}^D$, the map Φ in (7.1.6) for $\hat{\mathcal{P}}^D = \mathcal{P}^D$ has a unique fixed point in
$$\mathcal{P}_\gamma^T(\bar{D}) := \{\mu \in C^w([0,T]; \mathcal{P}^D) : \mu_0 = \gamma\}. \tag{7.2.5}$$
To this end, for $i = 1,2$, let $\mu^i \in C^w([0,T]; \mathcal{P}^D)$, and let X_t^i solve (7.1.5) for μ^i replacing μ with $\mathcal{L}_{X_0^i}^D = \mu_0^i$, i.e.
$$\begin{aligned}\mathrm{d}X_t^i &= \mathbf{1}_{\{t<\tau(X^i)\}}\big\{b_t(X_t^i, \mu_t^i)\mathrm{d}t + \sigma_t(X_t^i, \mu_t^i)\mathrm{d}W_t\big\}, \\ t &\in [0,T], \mathcal{L}_{X_0^i}^D = \mu_0^i.\end{aligned} \tag{7.2.6}$$
Simply denote
$$\tau_i = \tau(X^i) \text{ for } i = 1,2, \quad \tau_{1,2} := \tau_1 \wedge \tau_2.$$

Since
$$\Gamma := \{(x,y) : x \in D, y \in \partial D, |x-y| = \rho_\partial(x)\}$$
is a measurable subset of $D \times \partial D$ and $\Gamma_x := \{y \in \partial D : (x,y) \in \Gamma\} \neq \emptyset$ for any $x \in D$, by the measurable selection theorem, see Theorem 1 in [Evstigneev (1988)], there exists a measurable map $P_\partial : D \to \partial D$ such that
$$|P_\partial x - x| = \rho_\partial(x), \quad x \in D. \tag{7.2.7}$$

We will use the following coupling by projection.

Definition 7.2.1. The *coupling by projection* $(\bar{X}_t^1, \bar{X}_t^2)$ for $(X_t^1, X_t^2) = (X_{t \wedge \tau_1}^1, X_{t \wedge \tau_2}^2)$ is defined as
$$(\bar{X}_t^1, \bar{X}_t^2) := \begin{cases} (X_t^1, X_t^2), & \text{if } t \leq \tau_{1,2}, \\ (X_t^1, P_\partial X_t^1), & \text{if } \tau_2 < t \wedge \tau_1, \\ (P_\partial X_t^2, X_t^2), & \text{otherwise.} \end{cases} \tag{7.2.8}$$

It is easy to see that $\mathcal{L}_{\bar{X}_t^i}^D = \mathcal{L}_{X_t^i}^D = \Phi_t \mu^i$ for $i = 1, 2$; i.e. the distribution $\mathcal{L}_{(\bar{X}_t^1, \bar{X}_t^2)}$ of the coupling by projection $(\bar{X}_t^1, \bar{X}_t^2)$ satisfies
$$\mathcal{L}_{(\bar{X}_t^1, \bar{X}_t^2)} \in \mathcal{P}_D(\Phi_t \mu^1, \Phi_t \mu^2).$$

Thus, by (7.2.1) and Definition 7.2.1,
$$\begin{aligned}\hat{\mathbb{W}}_1(\Phi_t \mu^1, \Phi_t \mu^2) &\leq \mathbb{E}\big[1 \wedge |\bar{X}_t^1 - \bar{X}_t^2|\big] \\ &\leq \mathbb{E}\big[1 \wedge |X_{t \wedge \tau_{1,2}}^1 - \bar{X}_{t \wedge \tau_{1,2}}^2|\big] + r_0^{-1}\mathbb{E}[\{r_0 \wedge \rho_\partial(X_t^1)\}\mathbf{1}_{\{t \wedge \tau_1 \geq \tau_2\}}] \\ &\quad + r_0^{-1}\mathbb{E}[\{r_0 \wedge \rho_\partial(X_t^2)\}\mathbf{1}_{\{t \wedge \tau_2 \geq \tau_1\}}].\end{aligned} \tag{7.2.9}$$

Lemma 7.2.2. *Assume $(A^{7.2})$. Then there exists a constant $c > 1$ such that for any $t \in [0,T]$ and $\mu^1, \mu^2 \in C^w([0,T]; \mathcal{P}^D)$,*
$$\begin{aligned}&\mathbb{E}\big[|X_{t \wedge \tau_{1,2}}^1 - X_{t \wedge \tau_{1,2}}^2|^2 \big| \mathcal{F}_0\big] \\ &\leq c|X_0^1 - X_0^2|^2 + c\int_0^t K(s)\hat{\mathbb{W}}_1(\mu_s^1, \mu_s^2)^2 \mathrm{d}s.\end{aligned} \tag{7.2.10}$$

Consequently, for any $t \in [0,T]$,
$$\begin{aligned}&\mathbb{E}\big[1 \wedge |X_{t \wedge \tau_{1,2}}^1 - X_{t \wedge \tau_{1,2}}^2|\big] \\ &\leq \sqrt{c}\,\mathbb{E}[1 \wedge |X_0^1 - X_0^2|] + \left(c \int_0^t K(s)\hat{\mathbb{W}}_1(\mu_s^1, \mu_s^2)^2 \mathrm{d}s\right)^{\frac{1}{2}}.\end{aligned} \tag{7.2.11}$$

Proof. It suffices to prove (7.2.10), which implies (7.2.11) due to Jensen's inequality.

By $(A^{7.2})$ and Itô's formula, we obtain

$$d|X_t^1 - X_t^2|^2 \le K(t)\{|X_t^1 - X_t^2|^2 + \hat{\mathbb{W}}_1(\mu_t^1, \mu_t^2)^2\}dt + dM_t, \quad t \in [0, T \wedge \tau_{1,2}]$$

for some local martingale M_t. This and (7.2.4) imply that

$$\beta_t := \mathbb{E}\big[|X_{t\wedge\tau_{1,2}}^1 - X_{t\wedge\tau_{1,2}}^2|^2 \big| \mathcal{F}_0\big]$$

is bounded in $t \in [0, T]$ and satisfies

$$\beta_t \le \beta_0 + \int_0^t K(s)\{\beta_s + \hat{\mathbb{W}}_1(\mu_s^1, \mu_s^2)^2\}ds, \quad t \in [0, T].$$

By Gronwall's inequality, we prove (7.2.10). □

Lemma 7.2.3. *Assume $(A^{7.3})$. Then there exists a constant $c > 1$ independent of μ such that for any solution X_t^μ to (7.1.5) and any stopping time $\tilde{\tau}$,*

$$1_{\{t\wedge\tau(X^\mu)\ge\tilde{\tau}\}}\mathbb{E}\big[r_0 \wedge \rho_\partial(X_t^\mu)\big|\mathcal{F}_{\tilde{\tau}}\big] \le c1_{\{t\wedge\tau(X^\mu)\ge\tilde{\tau}\}}\rho_\partial(X_{t\wedge\tilde{\tau}}^\mu), \quad t \in [0, T].$$

Proof. By the strong Markov property of X_t^μ which is implied by the well-posedness of (7.1.5), we may and do assume that $\tilde{\tau} = 0$ and $x = X_0^\mu \in D$, such that the desired estimate becomes

$$\mathbb{E}^x\big[r_0 \wedge \rho_\partial(X_t^\mu)\big] \le c\rho_\partial(x), \quad t \in [0, T], \quad (7.2.12)$$

where $\mathbb{E}^x$ is the expectation under the probability $\mathbb{P}^x$ for X_t^μ starting at x. If $\rho_\partial(x) \ge \frac{r_0}{4}$, this inequality holds for $c := 4$. So, it suffices to prove for $\rho_\partial(x) < \frac{r_0}{4}$.

Let $h \in C^\infty([0, \infty))$ such that

$$h' \ge 0, \ h'' \le 0, \ h(r) = r \text{ for } r \in [0, r_0/2], \ h'(r) = 0 \text{ for } r \ge r_0.$$

By $(A^{7.3})$,

$$dh(\rho_\partial(X_t^\mu)) \le \alpha dt + dM_t, \quad t \in [0, T \wedge \tau(X^\mu)], \quad (7.2.13)$$

where M_t is a martingale with

$$d\langle M \rangle_t \ge \alpha^{-1}dt, \quad t \le \hat{\tau} := \inf\{t \ge 0 : \rho_\partial(X_t^\mu) \ge r_0/2\}. \quad (7.2.14)$$

By (7.2.13) we obtain

$$\begin{aligned}\mathbb{E}^x[r_0 \wedge \rho_\partial(X_t^\mu)] &\le 2\mathbb{E}^x[h(\rho_\partial(X_{t\wedge\tau(X^\mu)}^\mu))] \\ &\le 2\rho_\partial(x) + 2\alpha\mathbb{E}^x[t \wedge \tau(X^\mu)].\end{aligned} \quad (7.2.15)$$

On the other hand, let
$$\eta_t := \int_0^{\rho_\partial(X_t^\mu)} e^{-2\alpha^2 s} ds \int_s^{r_0} e^{2\alpha^2 \theta} d\theta, \quad t \in [0, T \wedge \tau(X^\mu) \wedge \hat{\tau}].$$

Since $h(r) = r$ for $r \leq \frac{r_0}{2}$, by (7.2.13), (7.2.14) and Itô's formula, we find a martingale $\tilde{M}_t$ such that
$$d\eta_t \leq -dt + d\tilde{M}_t, \quad t \in [0, T \wedge \tau(X^\mu) \wedge \hat{\tau}].$$

Consequently,
$$\mathbb{E}^x[t \wedge \tau(X^\mu) \wedge \hat{\tau}] \leq \eta_0 \leq c_1 \rho_\partial(x) \tag{7.2.16}$$

holds for some constant $c_1 > 0$. Therefore,
$$\begin{aligned} \mathbb{E}^x[t \wedge \tau(X^\mu)] &\leq \mathbb{E}^x[t \wedge \tau(X^\mu) \wedge \hat{\tau}] + T\,\mathbb{E}^x\big[1_{\{t \wedge \tau(X^\mu) > \hat{\tau}\}}\big] \\ &\leq c_1 \rho_\partial(x) + T\,\mathbb{P}^x\big(t \wedge \tau(X^\mu) > \hat{\tau}\big), \quad t \in [0, T]. \end{aligned} \tag{7.2.17}$$

To estimate the second term, let
$$\xi_t := \int_0^{\rho_\partial(X_t^\mu)} e^{-2\alpha^2 s} ds, \quad t \in [0, T \wedge \tau(X^\mu) \wedge \hat{\tau}].$$

By $h(r) = r$ for $r \in [0, \frac{r_0}{2}]$, (7.2.13), (7.2.14) and Itô's formula, we see that ξ_t is a sup-martingale, so that
$$\begin{aligned} \rho_\partial(x) \geq \xi_0 &\geq \mathbb{E}^x[\xi_{t \wedge \tau(X^\mu) \wedge \hat{\tau}}] \\ &\geq \mathbb{P}^x\big(t \wedge \tau(X^\mu) \geq \hat{\tau}\big) \int_0^{r_0/2} e^{-2\alpha^2 s} ds. \end{aligned} \tag{7.2.18}$$

Combining this with (7.2.15) and (7.2.17), we prove (7.2.12) for some constant $c > 0$. □

Proof of Theorem 7.2.1. (a) Well-posedness. Let $\gamma := \mathcal{L}_{X_0}^D$, and consider $\mathcal{P}_\gamma^T(\bar{D})$ in (7.2.5). We intend to prove that Φ is contractive in $\mathcal{P}_\gamma^T(\bar{D})$ under the complete metric
$$\hat{\mathbb{W}}_{1,\theta}(\mu^1, \mu^2) := \sup_{t \in [0,T]} e^{-\theta t} \hat{\mathbb{W}}_1(\mu_t^1, \mu_t^2)$$

for large enough $\theta > 0$. Then Φ has a unique fixed point in $\mathcal{P}_\gamma^T(\bar{D})$, so that the well-posedness follows from Theorem 7.1.1.

To this end, let $\mu^i \in \mathcal{C}^\gamma$ and let X_t^i solve (7.1.5) with $\mu = \mu^i$ and $X_0^i = X_0, i = 1, 2$. By $r_0 \leq 1$, Lemma 7.2.3, and noting that
$$1_{\{t \wedge \tau_2 \geq \tau_1\}} \rho_\partial(X_{t \wedge \tau_{1,2}}^2) \leq 1_{\{t \wedge \tau_2 \geq \tau_1\}} |X_{t \wedge \tau_{1,2}}^2 - X_{t \wedge \tau_{1,2}}^1|,$$

we obtain

$$\begin{aligned}
&\mathbb{E}\Big[1_{\{t\wedge\tau_2\geq\tau_1\}}\{r_0\wedge\rho_\partial(X^2_{t\wedge\tau_2})\}\Big]\\
&= \mathbb{E}\Big(1_{\{t\wedge\tau_2\geq\tau_1\}}\mathbb{E}\big[\{r_0\wedge\rho(X^2_{t\wedge\tau_2})\}\big|\mathcal{F}_{\tau_1}\big]\Big)\\
&\leq c\,\mathbb{E}\Big[1_{\{t\wedge\tau_2\geq\tau_1\}}\{r_0\wedge\rho_\partial(X^2_{t\wedge\tau_{1,2}})\}\Big]\\
&\leq c\,\mathbb{E}\Big[1\wedge|X^1_{t\wedge\tau_{1,2}} - X^2_{t\wedge\tau_{1,2}}|\Big].
\end{aligned} \qquad (7.2.19)$$

By symmetry, the same estimate holds for $\mathbb{E}\Big[1_{\{t\wedge\tau_1\geq\tau_2\}}\{r_0\wedge\rho_\partial(X^1_{t\wedge\tau_2})\}\Big]$. Combining these with $X^1_0 = X^2_0 = X_0$, (7.2.9) and (7.2.11), we find a constant $c_1 > 0$ such that

$$\hat{\mathbb{W}}_1(\Phi_t\mu^1, \Phi_t\mu^2) \leq c_1\bigg(\int_0^t K(s)\hat{\mathbb{W}}_1(\mu^1_s, \mu^2_s)^2\mathrm{d}s\bigg)^{\frac{1}{2}}, \quad t\in[0,T].$$

This implies that Φ is contractive in $\hat{\mathbb{W}}_{1,\theta}$ for large enough $\theta > 0$.

(b) Estimate (7.2.2). Let $\mu_t = \mathcal{L}^D_{X_t}$ for the unique solution of (7.1.3), we have $X_t = X^\mu_t$ since μ is a fixed point of Φ. So, (7.2.2) follows from (7.2.4).

(c) Estimate (7.2.3). Take X^1_0, X^2_0 such that

$$\mathcal{L}^D_{X^1_0} = \mu, \quad \mathcal{L}^D_{X^2_0} = \nu, \quad \mathbb{E}[1\wedge|X^1_0 - X^2_0|] = \hat{\mathbb{W}}_1(\mu,\nu). \qquad (7.2.20)$$

Let X^1_t and X^2_t solve (7.1.3). Then they solve (7.2.6) with

$$\mu^1_t := \mathcal{L}^D_{X^1_t} = P^{D*}_t\mu, \quad \mu^2_t := \mathcal{L}^D_{X^2_t} = P^{D*}_t\nu,$$

so that $\mu^i_t = \Phi_t\mu^i$, $t\in[0,T]$, $i = 1,2$. Thus, by (7.2.9), (7.2.10) and Lemma 7.2.3, we find a constant $c_2 > 0$ such that

$$\begin{aligned}
\hat{\mathbb{W}}_1(P^{D*}_t\mu, P^{D*}_t\nu) &= \hat{\mathbb{W}}_1(\Phi_t\mu^1, \Phi_t\mu^2)\\
&\leq c_2\hat{\mathbb{W}}_1(\mu,\nu) + \bigg(c_2\int_0^t K(s)\hat{\mathbb{W}}_1(P^{D*}_s\mu, P^{D*}_s\nu)^2\mathrm{d}s\bigg)^{\frac{1}{2}}, \quad t\in[0,T].
\end{aligned}$$

By Gronwall's inequality, we prove (7.2.3) for some constant $c > 0$. □

7.2.2 Monotonicity in $\mathbb{W}_1$

Let $\mathcal{P}^D_1 = \{\mu\in\mathcal{P}^D, \|\mu\|_1 := \mu(|\cdot|) < \infty\}$. Define

$$\|\mu\|_{1,T} := \sup_{t\in[0,T]}\|\mu_t\|_1, \quad \mu\in C^w_b([0,T];\mathcal{P}^D_1).$$

$(A^{7.4})$ For any $\mu \in C_b^w([0,T]; \mathcal{P}_1^D)$, $b_t(x, \mu_t)$ and $\sigma_t(x, \mu_t)$ are continuous in $x \in D$ such that for any $N \geq 1$ and $D_N := \{x \in D : |x| \leq N\}$,

$$\int_0^T \sup_{D_N} \{|b_t(\cdot, \mu_t)| + \|\sigma_t(\cdot, \mu_t)\|^2\} dt < \infty.$$

Moreover, there exists $K \in L^1([0,T]; (0, \infty))$ such that for any $x, y \in D$ and $\mu, \nu \in \mathcal{P}_1^D$,

$$2\langle b_t(x, \mu) - b_t(y, \nu), x - y \rangle + \|\sigma_t(x, \mu) - \sigma_t(y, \nu)\|_{HS}^2$$
$$\leq K(t)\{|x-y|^2 + \mathbb{W}_1(\mu, \nu)^2\},$$
$$2\langle b_t(x, \mu), x \rangle + \|\sigma_t(x, \mu)\|_{HS}^2 \leq K(t)\{1 + |x|^2 + \|\mu\|_1^2\}, \quad t \in [0, T].$$

$(A^{7.5})$ There exists $r_0 > 0$ such that $\rho_\partial \in C^2(\partial_{r_0} D)$, and there exists an increasing function $\alpha : [0, \infty) \to [1, \infty)$ such that

$$\begin{aligned}|\sigma_t(x, \mu)^* \nabla \rho_\partial|^{-2} &\leq \alpha(\|\mu\|_1),\\ L_{t,\mu}\rho_\partial(x) &\leq \alpha(\|\mu\|_1), \quad x \in \partial_{r_0} D,\end{aligned} \qquad (7.2.21)$$

$$\begin{aligned}&2\langle b_t(x, \mu), x - y \rangle + \|\sigma_t(x, \mu)\|_{HS}^2\\ &\leq K(t)\alpha(\|\mu\|_1)(1 + |x-y|^2), \quad t \in [0, T], y \in \partial D, x \in D.\end{aligned} \qquad (7.2.22)$$

Theorem 7.2.4. *Assume $(A^{7.4})$ and $(A^{7.5})$. Then the following assertions hold.*

(1) (7.1.3) *is well-posed for D-distributions in* $\mathcal{P}_1^D$. *Moreover, for any* $p \geq 1$ *there exists a constant* $c > 0$ *such that for any solution* X_t *to (7.1.3) for D-distributions in* $\mathcal{P}_1^D$,

$$\mathbb{E}\left[\sup_{t \in [0,T]} |X_t|^p \Big| \mathcal{F}_0\right] \leq c\big(1 + |X_0| + \mathbb{E}[1_D(X_0)|X_0|]\big)^p. \qquad (7.2.23)$$

(2) *If α is bounded, then there exists a constant $c > 0$ such that*

$$\sup_{t \in [0,T]} \mathbb{W}_1(P_t^{D*}\mu, P_t^{D*}\nu) \leq c \mathbb{W}_1(\mu, \nu), \quad \mu, \nu \in \mathcal{P}_1^D. \qquad (7.2.24)$$

It is standard that $(A^{7.4})$ and $(A^{7.5})$ imply the well-posedness of (7.1.5) for $\mu \in C_b^w([0,T]; \mathcal{P}_1^D)$, and instead of (7.2.4), for any $p \geq 1$ there exists a constant $c > 0$ such that

$$\mathbb{E}\left[\sup_{t \in [0,T]} |X_t^\mu|^p \Big| \mathcal{F}_0\right] \leq c\big(1 + |X_0^\mu|^p\big) + c\int_0^t K(s)\|\mu_s\|_1^p ds, \qquad (7.2.25)$$
$$t \in [0, T], \mu \in C_b^w([0,T]; \mathcal{P}_1^D).$$

Let $\mu^i \in C_b^w([0,T]; \mathcal{P}_1^D)$, $i = 1, 2$, let X_t^i solve (7.1.5) for μ^i replacing μ with $\mathcal{L}_{X_0^i}^D = \mu_0^i$, and denote as before

$$\tau_i := \tau(X^i) \text{ for } i = 1, 2, \quad \tau_{1,2} := \tau_1 \wedge \tau_2.$$

Using $(A^{7.4})$ replacing $(A^{7.2})$, the proof of (7.2.10) leads to

$$\mathbb{E}\big[|X_{t \wedge \tau_{1,2}}^1 - X_{t \wedge \tau_{1,2}}^2|^2 \big| \mathcal{F}_0\big] \\ \leq c|X_0^1 - X_0^2|^2 + c \int_0^t K(s) \mathbb{W}_1(\mu_s^1, \mu_s^2)^2 \mathrm{d}s, \quad t \in [0, T], \tag{7.2.26}$$

and instead of (7.2.9), Φ defined in (7.1.6) for $\hat{\mathcal{P}}^D = \mathcal{P}_1^D$ satisfies

$$\mathbb{W}_1(\Phi_t \mu^1, \Phi_t \mu^2) \leq \mathbb{E}\big[|\bar{X}_t^1 - \bar{X}_t^2|\big] \leq \mathbb{E}\big[|X_{t \wedge \tau_{1,2}}^1 - \bar{X}_{t \wedge \tau_{1,2}}^2|\big] \\ + \mathbb{E}\big[\rho_\partial(X_t^1) 1_{\{t \wedge \tau_1 \geq \tau_2\}}\big] + \mathbb{E}\big[\rho_\partial(X_t^2) 1_{\{t \wedge \tau_2 \geq \tau_1\}}\big]. \tag{7.2.27}$$

The following lemma is analogous to Lemma 7.2.3.

Lemma 7.2.5. *Assume* $(A^{7.5})$. *Then there exists an increasing function* $\psi : [0, \infty) \to (0, \infty)$ *which is bounded if so is* α, *such that for any* $\mu \in C_b^w([0,T]; \mathcal{P}_1^D)$, *any solution* X_t^μ *to (7.1.5), and any stopping time* $\tilde{\tau}$,

$$1_{\{t \wedge \tau(X^\mu) \geq \tilde{\tau}\}} \mathbb{E}\big[\rho_\partial(X_t^\mu) \big| \mathcal{F}_{\tilde{\tau}}\big] \leq 1_{\{t \wedge \tau(X^\mu) \geq \tilde{\tau}\}} \psi(\|\mu\|_{1,T}) \rho_\partial(X_{\tilde{\tau}}^\mu).$$

Proof. By the strong Markov property, we may assume that $\tilde{\tau} = 0$ and $x = X_0^\mu \in D$, so that it suffices to prove

$$\Gamma_t(x) := \mathbb{E}^x[\rho_\partial(X_t^\mu)] \leq \psi(\|\mu\|_{1,T}) \rho_\partial(x), \quad x \in D, t \in [0, T]. \tag{7.2.28}$$

(a) Let $\rho_\partial(x) \geq \frac{r_0}{2}$ and $y \in \partial D$ such that $\rho_\partial(x) = |y - x|$. By (7.2.22), we have

$$\mathrm{d}|X_t^\mu - y|^2 \leq K(t)\alpha(\|\mu\|_{1,T})\big(1 + |X_t^\mu - y|^2\big)\mathrm{d}t + \mathrm{d}M_t, \quad t \in [0, T \wedge \tau(X^\mu)]$$

for some martingale M_t. Combining this with $|x - y| = \rho_\partial(x)$, we obtain

$$\mathbb{E}^x[|X_t^\mu - y|^2] \leq \rho_\partial(x)^2 + \alpha(\|\mu\|_1) \int_0^t K(s)\mathrm{d}s \\ + \int_0^t K(s)\alpha(\|\mu\|_{1,T}) \mathbb{E}^x[|X_s^\mu - y|^2]\mathrm{d}s, \quad t \in [0, T].$$

By Gronwall's inequality and $\rho_\partial(x) \geq \frac{r_0}{2}$, we find an increasing function $\psi_1 : [0, \infty) \to (0, \infty)$ which is bounded if so is α, such that

$$\mathbb{E}^x[|X_t^\mu - y|^2] \leq \Big\{\rho_\partial(x)^2 + \alpha(\|\mu\|_{1,T}) \int_0^T K(s)\mathrm{d}s\Big\} e^{\alpha(\|\mu\|_{1,T}) \int_0^T K(s)\mathrm{d}s} \\ \leq \{\psi_1(\|\mu\|_{1,T}) \rho_\partial(x)\}^2, \quad t \in [0, T].$$

Combining this with Jensen's inequality, we prove (7.2.28) with $\psi = \psi_1$ for $\rho_\partial(x) \geq \frac{r_0}{2}$.

(b) Let $\rho_\partial(x) < \frac{r_0}{2}$. Simply denote $\alpha = \alpha(\|\mu\|_{1,T})$ and define

$$\hat{\tau} := \inf\{t \geq 0 : \rho_\partial(X_t^\mu) \geq r_0\}.$$

By $(A^{7.5})$ and Itô's formula, we obtain

$$\mathrm{d}\rho_\partial(X_t^\mu) \leq \alpha \mathrm{d}t + \mathrm{d}M_t, \quad t \in [0, T \wedge \tau(X^\mu) \wedge \hat{\tau}]$$

for some martingale satisfying (7.2.14). So,

$$\mathbb{E}^x[\rho_\partial(X_{t\wedge\tau(X^\mu)\wedge\hat{\tau}}^\mu)] \leq \alpha \mathbb{E}^x[t \wedge \tau(X^\mu) \wedge \hat{\tau}].$$

Combining this with step (a) and the strong Markov property, we obtain

$$\mathbb{E}^x[\rho_\partial(X_t^\mu)] = \mathbb{E}^x[\rho_\partial(X_{t\wedge\tau(X^\mu)}^\mu)]$$
$$\leq \mathbb{E}^x[\rho_\partial(X_{t\wedge\tau(X^\mu)\wedge\hat{\tau}}^\mu)] + \mathbb{E}^x\left[\mathbf{1}_{\{t\wedge\tau(X^\mu)\geq\hat{\tau}\}}\Gamma_{t-\hat{\tau}}(X_{\hat{\tau}}^\mu)\right]$$
$$\leq \alpha \mathbb{E}^x[t \wedge \tau(X^\mu) \wedge \hat{\tau}] + \mathbb{P}^x(t \wedge \tau(X^\mu) \geq \hat{\tau})\psi_1(\|\mu\|_{1,T})r_0.$$

Combining this with (7.2.16) and (7.2.18), we prove (7.2.28) for some increasing function $\psi : [0,\infty) \to (0,\infty)$, which is bounded if so is α. □

Proof of Theorem 7.2.4. Let X_t solve (7.1.3) for D-distributions in $\mathcal{P}_1^D$. Then $X_t = X_t^\mu$ for $\mu_t := \mathcal{L}_{X_t}^\nu$, so that

$$\|\mu_s\|_1 = \mathbb{E}[\mathbf{1}_D(X_s)|X_s|] = \mathbb{E}[\mathbf{1}_{\{t<\tau(X)\}}|X_s|], \quad s \in [0,T].$$

Combining this with (7.2.25), we obtain

$$\|\mu_t\|_1^2 \leq \left(\mathbb{E}\sqrt{\mathbb{E}[\mathbf{1}_{\{t<\tau(X)\}}|X_t|^2|\mathcal{F}_0]}\right)^2$$
$$\leq 2\left(\mathbb{E}\sqrt{c(1+\mathbf{1}_D(X_0)|X_0|^2)}\right)^2 + 2c\int_0^t K(s)\|\mu_s\|_1^2 \mathrm{d}s$$
$$\leq 2c(1+\mathbb{E}|\mathbf{1}_D(X_0)|X_0|])^2 + 2c\int_0^t K(s)\|\mu_s\|_1^2 \mathrm{d}s, \quad t \in [0,T].$$

By Gronwall's inequality, we find a constant $c_1 > 0$ such that

$$\sup_{t\in[0,T]} \|\mu_t\|_1^2 \leq c_1(1+\mathbb{E}[\mathbf{1}_D(X_0)|X_0|])^2.$$

This together with (7.2.25) yields (7.2.23) for some different constant $c > 0$. It remains to prove the well-posedness and (7.2.24).

(a) Well-posedness. Let $\gamma := \mathcal{L}_{X_0}^D \in \mathcal{P}_1^D$. For any $N > 0$, let

$$\mathcal{P}_{1,\gamma}^{T,N}(\bar{D}) := \left\{ \mu \in C_b^w([0,T]; \mathcal{P}_1^D) : \mu_0 = \gamma, \sup_{t \in [0,T]} e^{-Nt} \|\mu_t\|_1 \leq N \right\}. \tag{7.2.29}$$

We first observe that for some constant $N_0 > 0$,

$$\Phi \mathcal{P}_{1,\gamma}^{T,N}(\bar{D}) \subset \mathcal{P}_{1,\gamma}^{T,N}(\bar{D}), \quad N \geq N_0, \tag{7.2.30}$$

where Φ is defined in (7.1.6) for $\hat{\mathcal{P}}^D = \mathcal{P}_1^D$. Let $\mu \in \mathcal{P}_{1,\gamma}^{T,N}(\bar{D})$ and let X_t^μ solve (7.1.5) for $X_0^\mu = X_0$. Then $\Phi_t \mu = \mathcal{L}_{X_t^\mu}^D$. By (7.2.25) and

$$\|\Phi_t \mu\|_1 \leq \mathbb{E} \sqrt{\mathbb{E}[1_D(X_0)|X_{t \wedge \tau(X^\mu)}|^2 | \mathcal{F}_0]},$$

we find a constant $c_1 > 0$ such that

$$\|\Phi_t \mu\|_1 \leq c_1(1 + \|\gamma\|_1) + c_1 \left(\int_0^t \|\mu_s\|_1^2 ds \right)^{\frac{1}{2}}, \quad t \in [0, T].$$

Then for any $N \geq N_0 := c_1 + 2c_1(1 + \|\gamma\|_1)$, we have

$$\sup_{t \in [0,T]} e^{-Nt} \|\Phi_t \mu\|_1$$

$$\leq c_1(1 + \|\gamma\|_1) + c_1 \sup_{t \in [0,T]} \left(\int_0^t e^{-2Ns} \|\mu_s\|_1^2 e^{-2N(t-s)} ds \right)^{\frac{1}{2}}$$

$$\leq c_1(1 + \|\gamma\|_1) + c_1 N \sup_{t \in [0,T]} \left(\int_0^t e^{-2N(t-s)} ds \right)^{\frac{1}{2}}$$

$$\leq c_1(1 + \|\gamma\|_1) + c_1 \sqrt{N} \leq N.$$

Next, for any $N \geq N_0$, we intend to prove that Φ is contractive in $\mathcal{P}_{1,\gamma}^{T,N}(\bar{D})$ under the complete metric

$$\mathbb{W}_{1,\theta}(\mu^1, \mu^2) := \sup_{t \in [0,T]} e^{-\theta t} \mathbb{W}_1(\mu_t^1, \mu_t^2)$$

for large enough $\theta > 0$, so that Φ has a unique fixed point in $\mathcal{P}_{1,\gamma}^T(\bar{D}) := \cup_{N \geq N_0} \mathcal{P}_{1,\gamma}^{T,N}(\bar{D})$. Hence the well-posedness follows from Theorem 7.1.1.

To this end, let $\mu^i \in \mathcal{P}_{1,\gamma}^{T,N}(\bar{D})$ and X_t^i solve (7.1.5) for $\mu = \mu^i$ and $X_0^i = X_0, i = 1, 2$. By Lemma 7.2.5 and noting that $\rho_\partial(x) \leq |x - y|$ for $x \in D$ and $y \in \partial D$, we find a constant $c_2 > 0$ depending on N but uniformly in μ^i, such that

$$\mathbb{E}\big[\rho_\partial(X_t^1) 1_{\{t \wedge \tau_1 \geq \tau_2\}} + \rho_\partial(X_t^2) 1_{\{t \wedge \tau_2 \geq \tau_1\}}\big]$$
$$\leq c_2 \mathbb{E}\big[\rho_\partial(X_{t \wedge \tau_{1,2}}^1) 1_{\{t \wedge \tau_1 \geq \tau_2\}} + \rho_\partial(X_{t \wedge \tau_{1,2}}^2) 1_{\{t \wedge \tau_2 \geq \tau_1\}}\big]$$
$$\leq 2c_2 \mathbb{E}[|X_{t \wedge \tau_{1,2}}^1 - X_{t \wedge \tau_{1,2}}^2|], \quad t \in [0, T].$$

Combining this with (7.2.26) and (7.2.27), we find a constant $c_3 > 0$ depending on N such that

$$\mathbb{W}_1(\Phi_t\mu^1, \Phi_t\mu^2) \leq c_3\mathbb{E}[|X_0^1 - X_0^2|]$$
$$+ c_3\left(\int_0^t K(s)\mathbb{W}_1(\mu_s^1, \mu_s^2)^2 \mathrm{d}s\right)^{\frac{1}{2}}, \quad \mu^1, \mu^2 \in \mathcal{P}_{1,\gamma}^{T,N}(\bar{D}), \ t \in [0,T]. \quad (7.2.31)$$

Since $X_0^1 = X_0^2 = X_0$, this implies the contraction of Φ in $\mathbb{W}_{1,\theta}$ for large enough $\theta > 0$.

(b) Estimate (7.2.24). Now, for $\mu_0^1, \mu_0^2 \in \mathcal{P}_1^D$, let X_0^1, X_0^2 be $\mathcal{F}_0$-measurable random variables on $\bar{D}$ such that

$$\mathcal{L}_{X_0^1}^D = \mu_0^1, \quad \mathcal{L}_{X_0^2}^D = \mu_0^2, \quad \mathbb{E}[|X_0^1 - X_0^2|] = \mathbb{W}_1(\mu_0^1, \mu_0^2). \quad (7.2.32)$$

Letting X_t^i solve (7.1.3) with initial value X_0^i, then $\mu^i := (P_t^{D*}\mu_0^i)_{t\in[0,T]}$ is the unique fixed point of Φ in $\mathcal{C}^{\mu_0^i}$, so that

$$\mu_t^i = \mathcal{L}_{X_t^i}^D = \Phi\mu_t^i = P_t^{D*}\mu_0^i, \quad i = 1, 2, t \in [0,T]. \quad (7.2.33)$$

When α is bounded, (7.2.31) holds for some constant $c_3 > 0$ independent of N, which together with (7.2.32) yields

$$\mathbb{W}_1(\mu_t^1, \mu_t^2) = \mathbb{W}_1(\Phi_t\mu^1, \Phi_t\mu^2)$$
$$\leq c_3\mathbb{E}[|X_0^1 - X_0^2|] + c_3\left(\int_0^t K(s)\mathbb{W}_1(\mu_s^1, \mu_s^2)^2 \mathrm{d}s\right)^{\frac{1}{2}}$$
$$= c_3\mathbb{W}_1(\mu_0^1, \mu_0^2) + c_3\left(\int_0^t K(s)\mathbb{W}_1(\mu_s^1, \mu_s^2)^2 \mathrm{d}s\right)^{\frac{1}{2}}, \quad t \in [0,T].$$

By Gronwall's inequality and (7.2.33), we obtain

$$\mathbb{W}_1(P_t^{D*}\mu_0^1, P_t^{D*}\mu_0^2)^2 = \mathbb{W}_1(\mu_t^1, \mu_t^2)^2$$
$$\leq 2c_3^2\mathbb{W}_1(\mu_0^1, \mu_0^2)^2 e^{2c_3^2\int_0^t K(s)\mathrm{d}s}, \quad t \in [0,T].$$

Then the proof is finished. □

7.3 Singular case with distribution dependent noise

In this part, we assume that σ and b are extended to $[0,T] \times \mathbb{R}^d \times \mathcal{P}^D$, but may be singular in the space variable. For any $\mu \in C^w([0,T]; \mathcal{P}^D)$, let

$$b_t^\mu(x) := b_t(x, \mu_t) = b_t^{\mu,0}(x) + b_t^{(1)}(x),$$
$$\sigma_t^\mu(x) := \sigma_t(x, \mu_t), \quad (t,x) \in [0,T] \times \mathbb{R}^d, \quad (7.3.1)$$

where $b_t^{\mu,0}(\cdot)$ is singular and $b_t^{(1)}(\cdot)$ is Lipschitz continuous. As in the last section, we consider (7.1.3) for D-distributions in $\mathcal{P}^D$ and $\mathcal{P}_1^D$ respectively.

7.3.1 For D-distributions in $\mathcal{P}^D$

$(A^{7.6})$ *There exist* $K \in (0, \infty)$, $l \in \mathbb{N}$, $\{(p_i, q_i) : 0 \leq i \leq l\} \subset \mathcal{K}$ *with* $p_i > 2$, *and* $1 \leq f_i \in \tilde{L}_{p_i}^{q_i}(T)$ *for* $0 \leq i \leq l$, *such that* σ^μ *and* b^μ *in* (7.3.1) *satisfy the following conditions.*

(1) *For any* $\mu \in C^w([0,T]; \mathcal{P}^D)$, $a^\mu := \sigma^\mu(\sigma^\mu)^*$ *is invertible with* $\|a^\mu\|_\infty + \|(a^\mu)^{-1}\|_\infty \leq K$ *and*

$$\lim_{\varepsilon \downarrow 0} \sup_{\mu \in C^w([0,T]; \mathcal{P}^D)} \sup_{t \in [0,T], |x-y| \leq \varepsilon} \|a_t^\mu(x) - a_t^\mu(y)\| = 0.$$

(2) $b^{(1)}(0)$ *is bounded on* $[0,T]$, σ_t^μ *is weakly differentiable for* $\mu \in C^w([0,T]; \mathcal{P}^D)$, *and*

$$|b_t^{\mu,0}(x)| \leq f_0(t,x), \quad \|\nabla \sigma_t^\mu(x)\| \leq \sum_{i=1}^l f_i(t,x),$$

$$|b_t^{(1)}(x) - b_t^{(1)}(y)| \leq K|x-y|, \quad t \in [0,T], x,y \in \mathbb{R}^d.$$

(3) *For any* $t \in [0,T], x \in \mathbb{R}^d$ *and* $\mu, \nu \in \mathcal{P}^D$,

$$\|\sigma_t(x,\mu) - \sigma_t(x,\nu)\| + |b_t(x,\mu) - b_t(x,\nu)| \leq \hat{\mathbb{W}}_1(\mu,\nu) \sum_{i=0}^l f_i(t,x).$$

Theorem 7.3.1. *Assume* $(A^{7.6})$ *and* $(A^{7.3})$. *Then the following assertions hold.*

(1) (7.1.3) *is well-posed for D-distributions in* $\mathcal{P}^D$.
(2) *For any* $p \geq 1$, *there exists a constant* $c_p > 0$ *such that for any solution* X_t *to* (7.1.3) *for D-distributions in* $\mathcal{P}^D$,

$$\mathbb{E}\left[\sup_{t \in [0,T]} |X_t|^p \Big| \mathcal{F}_0\right] = \mathbb{E}\left[\sup_{t \in [0,T]} |X_{t \wedge \tau(X)}|^p \Big| \mathcal{F}_0\right] \leq c_p(1 + |X_0|^p). \tag{7.3.2}$$

(3) *There exists a constant* $c > 0$ *such that* (7.2.3) *holds.*

For any $\mu \in C_b^w([0,T]; \mathcal{P}^D)$, instead of (7.1.5) we consider the following SDE on $\mathbb{R}^d$:

$$dX_t^\mu = b_t^\mu(X_t^\mu)dt + \sigma_t^\mu(X_t^\mu)dW_t, \quad t \in [0,T]. \tag{7.3.3}$$

Noting that $\tilde{X}_t^\mu := X_{t \wedge \tau(X^\mu)}^\mu$ solves (7.1.5), the map Φ in (7.1.6) is given by

$$\Phi_t \mu := \mathcal{L}_{X_{t \wedge \tau(X^\mu)}^\mu}^D, \quad t \in [0,T].$$

So, (7.2.9) and (7.2.27) remain true for X_t^i solving (7.3.3) with $\mu = \mu^i \in C^w([0,T]; \mathcal{P}^D), i = 1, 2$.

By Theorem 1.3.1, $(A^{7.6})(1)$ and $(A^{7.6})(2)$ imply that this SDE is well-posed, and for any $p \geq 1$ there exists a constant $c_p > 0$ such that

$$\mathbb{E}\left[\sup_{t\in[0,T]} |X_t^\mu|^p \Big| \mathcal{F}_0\right] \leq c_p(1 + |X_0^\mu|^p), \quad \mu \in C^w([0,T]; \mathcal{P}^D). \tag{7.3.4}$$

We have the following lemma.

Lemma 7.3.2. *Assume* $(A^{7.6})$. *Then for any* $j \geq 1$ *there exists a constant* $c > 0$ *and a function* $\varepsilon : [1, \infty) \to (0, \infty)$ *with* $\varepsilon(\theta) \downarrow 0$ *as* $\theta \uparrow \infty$, *such that for any* $\mu^1, \mu^2 \in C_b^w([0,T]; \mathcal{P}^D)$ *and any* X_t^i *solving (7.3.3) with* $\mu = \mu^i, i = 1, 2$,

$$\mathbb{E}\left[\sup_{s\in[0,t]} |X_s^1 - X_s^2|^j \Big| \mathcal{F}_0\right] \leq c|X_0^1 - X_0^2|^j + \varepsilon(\theta)e^{j\theta t}\hat{\mathbb{W}}_{1,\theta}(\mu^1, \mu^2)^j, \quad \theta \geq 1.$$

Proof. By Lemma 1.2.2, $(A^{7.6})(1)$ and $(A^{7.6})(2)$ imply that for large enough $\lambda \geq 1$, the PDE

$$\left(\partial_t + \frac{1}{2}\mathrm{tr}\{a_t^{\nu^1}\nabla^2\} + b_t^{\mu^1} \cdot \nabla\right)u_t = \lambda u_t - b_t^{\mu^1,0}, \quad t \in [0,T], u_T = 0 \tag{7.3.5}$$

for $u : [0, T] \times \mathbb{R}^d \to \mathbb{R}^d$ has a unique solution such that

$$\|\nabla^2 u\|_{\tilde{L}_{p_0}^{q_0}(T)} \leq c_0, \quad \|u\|_\infty + \|\nabla u\|_\infty \leq \frac{1}{2}. \tag{7.3.6}$$

Let $Y_t^i := \Theta_t(X_t^i), i = 1, 2, \Theta_t := id + u_t$. By Itô's formula we obtain

$$dY_t^1 = \{b_t^{(1)} + \lambda u_t\}(X_t^1)dt + (\{\nabla\Theta_t\}\sigma_t^{\nu^1})(X_t^1) dW_t,$$

$$dY_t^2 = \{\{b_t^{(1)} + \lambda u_t + (\nabla\Theta_t)(b_t^{\mu^2} - b_t^{\mu^1})\}(X_t^2)$$

$$+ \frac{1}{2}[\mathrm{tr}\{(a_t^{\nu^2} - a_t^{\nu^1})\nabla^2 u_t\}](X_t^2)\}dt + (\{\nabla\Theta_t\}\sigma_t^{\nu^2})(X_t^2) dW_t.$$

Let $\eta_t := |X_t^1 - X_t^2|$ and

$$g_r := \sum_{i=0}^{l} f_i(r, X_r^2), \quad \tilde{g}_r := g_r\|\nabla^2 u_r(X_r^2)\|,$$

$$\bar{g}_r := \sum_{i=1}^{2} \|\nabla^2 u_r\|(X_r^i) + \sum_{j=1}^{2}\sum_{i=0}^{l} f_i(r, X_r^j), \quad r \in [0, T].$$

Since $b_t^{(1)} + \lambda u_t$ is Lipschitz continuous uniformly in $t \in [0,T]$, by $(A^{7.6})$ and the maximal functional inequality in Lemma 1.3.4, there exists a constant $c_1 > 0$ such that

$$\left|\{b_r^{(1)} + \lambda u_r\}(X_r^1) - \{b_r^{(1)} + \lambda u_r\}(X_r^2)\right| \leq c_1 \eta_r,$$

$$\left|\{(\nabla \Theta_r)(b_r^{\mu^2} - b_r^{\mu^1})\}(X_r^2)\right| \leq c_1 g_r \hat{\mathbb{W}}_1(\mu_r^1, \mu_r^2),$$

$$\left|[\mathrm{tr}\{(a_r^{\nu^2} - a_r^{\nu^1})\nabla^2 u_r\}](X_r^2)\right| \leq c_1 \tilde{g}_r \hat{\mathbb{W}}_1(\mu_r^1, \mu_r^2),$$

$$\left\|\{(\nabla \Theta_r)\sigma_r^{\nu^1}\}(X_r^1) - \{(\nabla \Theta_r)\sigma_r^{\mu^2}\}(X_r^2)\right\|$$

$$\leq c_1 \bar{g}_r \eta_r + c_1 g_r \hat{\mathbb{W}}_1(\mu_r^1, \mu_r^2), \quad r \in [0,T].$$

So, by Itô's formula, for any $j \geq k$ we find a constant $c_2 > 1$ such that

$$\mathrm{d}|Y_t^1 - Y_t^2|^{2j} \leq c_2 \eta_t^{2j} \mathrm{d}A_t + c_2(g_t^2 + \tilde{g}_t)\hat{\mathbb{W}}_1(\mu_t^1, \mu_t^2)^{2j}\mathrm{d}t + \mathrm{d}M_t \quad (7.3.7)$$

holds for some martingale M_t with $M_0 = 0$ and

$$A_t := \int_0^t \{1 + g_s^2 + \tilde{g}_s + \bar{g}_s^2\}\mathrm{d}s, \quad t \in [0,T].$$

Since $\|\nabla u\|_\infty \leq \frac{1}{2}$ implies $|Y_t^1 - Y_t^2| \geq \frac{1}{2}\eta_t$, this implies

$$\eta_t^{2j} \leq 2^{2j} M_t + 2^{2j} \eta_0^{2j} + 2^{2j} c_2 \int_0^t \eta_r^{2j} \mathrm{d}A_r$$

$$+ 2^{2j} c_2 \int_0^t (g_s^2 + \tilde{g}_s)\hat{\mathbb{W}}_1(\mu_s^1, \mu_s^2)^{2j} \mathrm{d}s, \quad t \in [0,T]$$

(7.3.8)

for some constant $c_2 > 0$. By (7.3.6), $f_i \in \tilde{L}_{p_i}^{q_i}(T)$ for $(p_i, q_i) \in \mathcal{K}$, Krylov's estimate in Theorem 1.2.3 and Khasminskii's estimate in Theorem 1.2.4, we find an increasing function $\psi : (0, \infty) \to (0, \infty)$ and a decreasing function $\varepsilon : (0, \infty) \to (0, \infty)$ with $\varepsilon(\theta) \downarrow 0$ as $\theta \uparrow \infty$, such that

$$\mathbb{E}[e^{rA_T}|\mathcal{F}_0] \leq \psi(r), \quad r > 0,$$

$$\sup_{t \in [0,T]} \mathbb{E}\left(\int_0^t e^{-2k\theta(t-r)}(g_r^2 + \tilde{g}_r)\mathrm{d}r \,\bigg|\, \mathcal{F}_0\right) \leq \varepsilon(\theta), \quad \theta > 0.$$

By the stochastic Gronwall inequality in Lemma 1.3.3 and the maximal inequality in Lemma 1.3.4, we find a constant $c_3 > 0$ depending on N such that (7.3.8) yields

$$\left\{\mathbb{E}\left(\sup_{s \in [0,t]} \eta_s^j \,\bigg|\, \mathcal{F}_0\right)\right\}^2$$

$$\leq c_3 \mathbb{E}\left(\eta_0^{2j} + \int_0^t (g_s^2 + \tilde{g}_s)\hat{\mathbb{W}}_1(\mu_s^1, \mu_s^2)^{2j}\mathrm{d}s \,\bigg|\, \mathcal{F}_0\right)$$

$$\leq c_3 \eta_0^{2j} + c_3 e^{2j\theta t}\varepsilon(\theta)\hat{\mathbb{W}}_1(\mu^1, \mu^2)^{2j}, \quad t \in [0,T], \theta > 0.$$

This finishes the proof. □

Proof of Theorem 7.3.1. Let X_t solve (7.1.3). We have $X_t = X^\mu_{t\wedge\tau(X^\mu)}$ for X^μ_t solving (7.3.3) with

$$X^\mu_0 = X_0, \quad \mu_t := \mathcal{L}^D_{X_t}, \quad t \in [0,T].$$

So, (7.3.2) follows from (7.3.4). It remains to prove the well-posedness and estimate (7.2.3).

(a) Well-posedness. Let X_0 be an $\mathcal{F}_0$-measurable random variable on $\bar{D}$, and let $\mathcal{P}^T_\gamma(\bar{D})$ be in (7.2.5) for $\gamma = \mathcal{L}^D_{X_0}$. By Theorem 7.1.1, it suffices to prove that Φ is contractive in $\mathcal{P}^T_\gamma(\bar{D})$ under $\hat{\mathbb{W}}_{1,\theta}$ for large enough $\theta > 0$.

By (7.2.9), (7.2.19) and Lemma 7.3.2 for $X^1_0 = X^2_0 = X_0$, we find a constant $c_1 > 0$ such that

$$\hat{W}_1(\Phi_t\mu^1, \Phi_t\mu^2) \leq c_1\varepsilon(\theta)\hat{\mathbb{W}}_1(\mu^1,\mu^2), \quad \mu^1,\mu^2 \in \mathcal{C}^\gamma.$$

Since $\varepsilon(\theta) \to 0$ as $\theta \to \infty$, Φ is $\hat{\mathbb{W}}_{1,\theta}$-contractive for large enough $\theta > 0$.

(b) Estimate (7.2.3). Let X^1_t, X^2_t solve (7.1.3) with X^1_0, X^2_0 satisfying (7.2.20). Then

$$\Phi_t\mu^i = \mu^i_t := \mathcal{L}^D_{X^i} = P^{D*}_t\mu^i, \quad i = 1,2,$$

so that (7.2.9), (7.2.19) and Lemma 7.3.2 imply

$$\hat{W}_1(\mu^1,\mu^2) = \hat{W}_1(\Phi_t\mu^1, \Phi_t\mu^2)$$
$$\leq c_1\hat{\mathbb{W}}_1(\mu^1_0,\mu^2_0) + c_1\varepsilon(\theta)\hat{\mathbb{W}}_1(\mu^1,\mu^2), \quad t \in [0,T]$$

for some constant $c_1 > 0$. Taking $\theta > 0$ large enough such that $\varepsilon(\theta) \leq \frac{1}{2c_1}$, we derive (7.2.3) for some constant $c > 0$. □

7.3.2 For D-distributions in $\mathcal{P}^D_1$

$(A^{7.7})$ There exist an increasing function $\alpha : [0,\infty) \to (0,\infty)$, constants $K > 0$, $l \in \mathbb{N}$, $\{(p_i, q_i) : 0 \leq i \leq l\} \subset \mathcal{K}$ with $p_i > 2$, and functions $1 \leq f_i \in \tilde{L}^{q_i}_{p_i}(T)$ for $0 \leq i \leq l$ such that σ^μ and b^μ in (7.3.1) satisfy the following conditions.

(1) For any $\mu \in C^w_b([0,T]; \mathcal{P}^D_1)$, $a^\mu := \sigma^\mu(\sigma^\mu)^*$ is invertible with

$$\|a^\mu\|_\infty + \|(a^\mu)^{-1}\|_\infty \leq \alpha(\|\mu\|_{1,T}),$$
$$\lim_{\varepsilon\downarrow 0} \sup_{\mu \in C^w_b([0,T];\mathcal{P}^D_1)} \sup_{t\in[0,T], |x-y|\leq\varepsilon} \|a^\mu_t(x) - a^\mu_t(y)\| = 0.$$

(2) $b^{(1)}(0)$ is bounded on $[0,T]$, σ_t^μ is weakly differentiable for $\mu \in C_b^w([0,T]; \mathcal{P}_1^D)$, and

$$|b_t^{\mu,0}(x)| \leq f_0(t,x) + \alpha(\|\mu\|_{1,T}), \|\nabla \sigma_t^\mu(x)\| \leq \sum_{i=1}^l f_i(t,x) + \alpha(\|\mu\|_{1,T}),$$

$$|b_t^{(1)}(x) - b_t^{(1)}(y)| \leq K|x-y|, \quad t \in [0,T], x,y \in \mathbb{R}^d.$$

(3) For any $t \in [0,T], x \in \mathbb{R}^d$ and $\mu, \nu \in \mathcal{P}^D$,

$$\|\sigma_t(x,\mu) - \sigma_t(x,\nu)\| + |b_t(x,\mu) - b_t(x,\nu)| \leq \mathbb{W}_1(\mu,\nu) \sum_{i=0}^l f_i(t,x).$$

(4) There exists $r_0 \in (0,1]$ such that $\rho_\partial \in C_b^2(\partial_{r_0} D)$, and for any $\mu \in C_b^w([0,T]; \mathcal{P}_1^D)$,

$$\langle b_t^\mu(x), \nabla \rho_\partial(x) \rangle \leq \alpha(\|\mu\|_1), \quad x \in \partial_{r_0} D, \tag{7.3.9}$$

$$\langle b_t^\mu(x), x-y \rangle \leq \alpha(\|\mu\|_{1,T})(f_0(t,x)^2 + |x-y|^2), \tag{7.3.10}$$
$$x \in D, y \in \partial D, t \in [0,T].$$

Note that when $b^{(1)} = 0$, (7.3.9) is implied by the first condition in $(A^{7.7})(2)$.

Theorem 7.3.3. *Assume $(A^{7.7})$. Then the following assertions hold.*

(1) *(7.1.3) is well-posed for D-distributions in $\mathcal{P}_1^D$.*
(2) *For any $p \geq 1$, there exists a constant $c_p > 0$ such that for any solution X_t to (7.1.3) for D-distributions in $\mathcal{P}^D$,*

$$\mathbb{E}\left[\sup_{t \in [0,T]} |X_t|^p \Big| \mathcal{F}_0\right] \leq c_p\{1 + |X_0|^p + (\mathbb{E}[1_D(X_0)|X_0|])^p\}. \tag{7.3.11}$$

(3) *If α is bounded, then there exists a constant $c > 0$ such that (7.2.24) holds.*

By Theorem 1.3.1, $(A^{7.7})$ implies that for any $\mu \in C_b^w([0,T]; \mathcal{P}_1^D)$, the SDE (7.3.3) is well-posed, and for any $p \geq 1$ there exists a constant $c_p > 0$ such that for any $\mu \in C_b^w([0,T]; \mathcal{P}_1^D)$,

$$\mathbb{E}\left[\sup_{t \in [0,T]} |X_t^\mu|^{2p} \Big| \mathcal{F}_0\right] \leq c_p\left\{1 + |X_0^\mu|^{2p} + \int_0^T \|\mu_s\|_1^{2p} ds\right\}. \tag{7.3.12}$$

For any $\mu^1, \mu^2 \in C_b^w([0,T]; \mathcal{P}_1^D)$, let X_t^i solve (7.3.3) for $\mu = \mu^i, i = 1,2$.

For any $N > 0$ and $\gamma \in \mathcal{P}_1^D$, let $\mathcal{P}_{1,\gamma}^{T,N}(\bar{D})$ be in (7.2.29). Since restricting to $\mu, \nu \in \mathcal{P}_{1,\gamma}^{T,N}(\bar{D})$, the conditions in $(A^{7.7})$ hold for a constant α_N replacing the function α, by repeating the proof of Lemma 7.3.2 with $\tilde{W}$ replacing $\hat{W}$, we prove that the following results.

Lemma 7.3.4. *Assume* $(A^{7.7})$. *For any* $N > 0$ *and* $j \geq 1$, *there exists a constant* $c > 0$ *and a function* $\varepsilon : [1, \infty) \to (0, \infty)$ *with* $\varepsilon(\theta) \downarrow 0$ *as* $\theta \uparrow \infty$, *such that for any* $\mu^1, \mu^2 \in \mathcal{P}_{1,\gamma}^{T,N}(\bar{D})$ *and any* X_t^i *solving* (7.3.3) *with* $\mu = \mu^i$, $i = 1, 2$,

$$\mathbb{E}\Big[\sup_{s \in [0,t]} |X_s^1 - X_s^2|^j \big| \mathcal{F}_0\Big] \leq c|X_0^1 - X_0^2|^j + \varepsilon(\theta) e^{j\theta t} \tilde{W}_{1,\theta}(\mu^1, \mu^2)^j, \quad \theta \geq 1.$$

When α *is bounded, the constant* c *does not depend on* N.

Moreover, we need the following result analogous to Lemma 7.2.5.

Lemma 7.3.5. *Assume* $(A^{7.7})$. *Then the assertion in Lemma 7.2.5 holds.*

Proof. It suffices to prove (7.2.28) for some increasing function ψ which is bounded if so is α.

(a) Let $\rho_\partial(x) \geq \frac{r_0}{2}$ and $y \in \partial D$ such that $\rho_\partial(x) = |y - x|$. By (7.3.10) and $(A^{7.7})(2)$, we find an increasing function $\psi_1 : [0, \infty) \to (0, \infty)$ which is bounded if so is α, such that

$$d|X_t^\mu - y|^2 \leq \psi_1(\|\mu\|_{1,T}) \Big(\sum_{i=0}^{l} f_i(t, X_t^\mu)^2 + |X_t^\mu - y|^2\Big) dt + dM_t$$

holds for some martingale M_t, $t \in [0, T \wedge \tau(X^\mu)]$. Next, by Theorem 1.2.3, $(A^{7.7})$ implies that for some increasing function $\psi_2 : [0, \infty) \to (0, \infty)$ which is bounded if so is α, the following Krylov's estimate holds:

$$\mathbb{E}\Big(\int_0^T f_i(t, X_t^\mu)^2 dt \Big| \mathcal{F}_0\Big) \leq \psi_2(\|\mu\|_{1,T}) \|f_i\|_{\tilde{L}_{q_i}^{p_i}(T)}^2, \quad 0 \leq i \leq l.$$

Combining these with $|x - y| = \rho_\partial(x)$, we derive

$$\mathbb{E}[|X_t^\mu - y|^2 | \mathcal{F}_0] \leq \rho_\partial(x)^2 + \psi_1(\|\mu\|_{1,T}) \psi_2(\|\mu\|_{1,T}) \sum_{i=0}^{l} \|f_i\|_{\tilde{L}_{q_i}^{p_i}(T)}^2$$

$$+ \psi_1(\|\mu\|_{1,T}) \int_0^t \mathbb{E}[|X_s^\mu - y|^2 | \mathcal{F}_0] ds, \quad t \in [0, T].$$

By Gronwall's inequality and $\rho_\partial(x) \geq \frac{r_0}{2}$, we find an increasing function $\psi : [0, \infty) \to (0, \infty)$ which is bounded if so is α, such that

$$\mathbb{E}[|X_t^\mu - y|^2 | \mathcal{F}_0] \leq \psi(\|\mu\|_{1,T}) \rho_\partial(x).$$

Since $\rho_\partial(X_t^\mu) \leq |X_t^\mu - y|$, we prove (7.2.28) for $\rho_\partial(x) \geq \frac{r_0}{2}$.

(b) Let $\rho_\partial(x) < \frac{r_0}{2}$. By $(A^{7.7})(1)$, (7.3.9) and $\rho_\partial \in C_b^2(\partial_{r_0}D)$, (7.2.21) holds for some different increasing function α which is bounded if so is the original one. Then step (b) in proof of Lemma 7.2.5 implies the desired estimate. □

Proof of Theorem 7.3.3. Let X_t solve (7.1.3) for D-distributions in $\mathcal{P}_1^D$. We have $X_t = X_{t\wedge\tau(X^\mu)}^\mu$ for X_t^μ solving (7.3.3) with

$$X_0^\mu = X_0, \quad \mu_t = \mathcal{L}_{X_t}^D, \quad t \in [0,T].$$

So, as explained in the beginning of the proof of Theorem 7.2.4 that (7.3.11) follows from (7.3.12), it suffices to prove the well-posedness and estimate (7.2.24).

Let X_0 be an $\mathcal{F}_0$-measurable random variable with $\mathcal{L}_{X_0}^D \in \mathcal{P}_1^D$, and let $\mathcal{P}_{1,\gamma}^{T,N}(\bar{D})$ be in (7.2.29) for $N > 0$. By the proof of Lemma 3.6.4(1), there exists $N_0 > 0$ such that $\Phi \mathcal{P}_{1,\gamma}^{T,N}(\bar{D}) \subset \mathcal{P}_{1,\gamma}^{T,N}(\bar{D})$ for any $N \geq N_0$. For the well-posedness, it suffices to prove that for any $N \geq N_0$, Φ is contractive in $\mathcal{P}_{1,\gamma}^{T,N}(\bar{D})$ under the metric $\mathbb{W}_{1,\theta}$ for large enough $\theta > 0$. This follows from (7.2.27), Lemma 7.3.4 and Lemma 7.3.5.

Finally, by using $\mathbb{W}_1$ replacing $\hat{\mathbb{W}}_1$ in step (b) in the proof of Theorem 7.3.1, (7.2.24) follows from Lemma 7.3.4 with c independent of N. □

7.4 Singular case with distribution independent noise

In this part, we let $\sigma_t(x,\mu) = \sigma_t(x)$ not depend on μ, so that (7.1.3) becomes

$$dX_t = 1_{\{t < \tau(X)\}}\{h_t(X_t, \mathcal{L}_{X_t}^D)dt + \sigma_t(X_t)dW_t\}, \quad t \in [0,T]. \tag{7.4.1}$$

In this case, we are able to study the well-posedness of the equation on an arbitrary connected open domain D, for which we only need $b_t(x,\cdot)$ to be Lipschitz continuous with respect to a weighted variation distance.

For a measurable function $V: D \to [1,\infty)$, let

$$\mathcal{P}_V^D := \left\{\mu \in \mathcal{P}^D : \mu(V) := \int_D V d\mu < \infty\right\}.$$

This is a Polish space under the weighted variation distance

$$\|\mu - \nu\|_V := \sup_{|f| \leq V} |\mu(f) - \nu(f)|, \quad \mu,\nu \in \mathcal{P}_V^D. \tag{7.4.2}$$

When $V \equiv 1$, $\|\cdot\|_V$ reduces to the total variation norm. We will take V from the class $\mathcal{V}$ defined as follows.

Definition 7.4.1. We denote $V \in \mathcal{V}$, if $1 \leq V \in C^2(\mathbb{R}^d)$ such that the level set $\{V \leq r\}$ for $r > 0$ is compact, and there exist constants $K, \varepsilon > 0$ such that for any $x \in D$,

$$\sup_{y \in B(x,\varepsilon)} \{|\nabla V(y)| + \|\nabla^2 V(y)\|\} \leq KV(x),$$

where $B(x, \varepsilon) := \{y \in \mathbb{R}^d : |y - x| < \varepsilon\}$.

7.4.1 Main result

$(A^{7.8})$ σ has an extension to $[0,T] \times \mathbb{R}^d$ which is weakly differentiable in $x \in \mathbb{R}^d$, and b has a decomposition $b_t(x, \mu) = b_t^{(0)}(x) + b_t^{(1)}(x, \mu)$, such that the following conditions hold.

(1) $a := \sigma \sigma^*$ is invertible with $\|a\|_\infty + \|a^{-1}\|_\infty < \infty$ and

$$\lim_{\varepsilon \to 0} \sup_{|x-y| \leq \varepsilon, t \in [0,T]} \|a_t(x) - a_t(y)\| = 0.$$

(2) There exist $l \in \mathbb{N}$, $\{(p_i, q_i)\}_{0 \leq i \leq l} \subset \mathcal{K}$ with $p_i > 2$, and $1 \leq f_i \in \tilde{L}_{q_i}^{p_i}(T)$, such that

$$|1_D b^{(0)}| \leq f_0, \quad \|\nabla \sigma\| \leq \sum_{i=1}^{l} f_i.$$

(3) There exists $V \in \mathcal{V}$ such that for any $\mu \in \mathcal{P}_V := C_b^w([0,T]; \mathcal{P}_V^D)$, $1_D(x) b_t^{(1)}(x, \mu_t)$ is locally bounded in $(t, x) \in [0, T] \times \mathbb{R}^d$. Moreover, there exist constants $K, \varepsilon > 0$ such that

$$\langle b^{(1)}(x, \mu), \nabla V(x) \rangle + \varepsilon |b^{(1)}(x, \mu)| \sup_{B(x,\varepsilon)} \{|\nabla V| + |\nabla^2 V|\}$$
$$\leq K\{V(x) + \mu(V)\}, \quad x \in D, \mu \in \mathcal{P}_V^D.$$

(4) There exists a constant $\kappa > 0$ such that

$$\sup_{x \in D} |b_t(x, \mu) - b_t(x, \nu)| \leq \kappa \|\mu - \nu\|_V, \quad \mu, \nu \in \mathcal{P}_V^D. \tag{7.4.3}$$

Theorem 7.4.1. *Assume* $(A^{7.8})$. *Then* (7.4.1) *is well-posed for D-distributions in* $\mathcal{P}_V^D$, *and for any* $p \geq 1$, *there exists a constant* $c_p > 0$ *such that any solution* X_t *of* (7.4.1) *for D-distributions in* $\mathcal{P}_V^D$ *satisfies*

$$\mathbb{E}\left[\sup_{t \in [0,T]} V(X_t)^p \Big| \mathcal{F}_0\right] \leq c_p V(X_0)^p. \tag{7.4.4}$$

Proof. Let $\mathcal{P}_V(\bar{D})$ be the space of all probability measures μ on $\bar{D}$ with $\mu(V) < \infty$, see (6.5.32), which is a Polish space under the weighted variation distance defined in (7.4.2) for $\mu, \nu \in \mathcal{P}_V(\bar{D})$. We extend $b_t(x, \cdot)$ from $\mathcal{P}_V^D$ to $\mathcal{P}_V(\bar{D})$ by setting
$$b_t(x, \mu) := b_t(x, \mu(D \cap \cdot)), \quad \mu \in \mathcal{P}_V(\bar{D}).$$
Then $(A^{7.8})$ implies the same assumption for $\mathcal{P}_V(\bar{D})$ replacing $\mathcal{P}_V^D$. So, the desired assertions follow from Theorem 7.4.2 presented in the next subsection. $\square$

7.4.2 An extension of Theorem 7.4.1

Consider the following SDE on $\bar{D}$:
$$\mathrm{d}X_t = 1_{\{t < \tau(X)\}}\{b_t(X_t, \mathcal{L}_{X_t})\mathrm{d}t + \sigma_t(X_t)\mathrm{d}W_t\}, \quad t \in [0, T], \qquad (7.4.5)$$
where $\tau(X) := \inf\{t \geq 0 : X_t \in \partial D\}$ as before, and $\mathcal{L}_{X_t}$ is the distribution of X_t.

The strong/weak solution of (7.4.5) is defined as in Definition 7.1.1 with $\mathcal{L}$ replacing $\mathcal{L}^D$. We call this equation well-posed for distributions in $\mathcal{P}_V(\bar{D})$, if for any $\mathcal{F}_0$-measurable random variable X_0 on $\bar{D}$ with $\mathcal{L}_{X_0} \in \mathcal{P}_V(\bar{D})$ (respectively, any $\mu_0 \in \mathcal{P}_V(\bar{D})$), (7.4.5) has a unique solution starting at X_0 (respectively, a unique weak solution with initial distribution μ_0) such that $\mathcal{L}_X = (\mathcal{L}_{X_t})_{t \in [0,T]} \in C_b^w([0, T]; \mathcal{P}_V(\bar{D}))$.

Theorem 7.4.2. *Assume that $(A^{7.8})$ holds for $\mathcal{P}_V(\bar{D})$ replacing $\mathcal{P}_V^D$. Then (7.4.5) is well-posed for distributions in $\mathcal{P}_V(\bar{D})$ and (7.4.4) holds.*

Proof. Let X_0 be an $\mathcal{F}_0$-measurable random variable on $\bar{D}$ with
$$\gamma := \mathcal{L}_{X_0} \in \mathcal{P}_V(\bar{D}).$$
Let
$$\mathcal{P}_{V,\gamma}^T(\bar{D}) := \{\mu \in C_b^w([0, T]; \mathcal{P}_V(\bar{D})) : \mu_0 = \gamma\}.$$
For any $\mu \in \mathcal{P}_{V,\gamma}^T(\bar{D})$, let X_t^μ solve (7.4.1) with $X_0^\mu = X_0$, i.e.
$$\begin{aligned}\mathrm{d}X_t^\mu &= 1_{\{t < \tau(X^\mu)\}}\{b_t(X_t^\mu, \mu_t)\mathrm{d}t + \sigma_t(X_t^\mu)\mathrm{d}W_t\},\\ X_0^\mu &= X_0, t \in [0, T].\end{aligned} \qquad (7.4.6)$$
Let $\Phi_t \mu := \mathcal{L}_{X_t^\mu}, t \in [0, T]$. Then it suffices to prove that Φ has a unique fixed point in $\mathcal{P}_{V,\gamma}^T(\bar{D})$. To this end, for any $N \geq 1$, let
$$\mathcal{P}_{V,\gamma}^{T,N}(\bar{D}) := \Big\{\mu \in \mathcal{P}_{V,\gamma}^T(\bar{D}) : \sup_{t \in [0,T]} \mathrm{e}^{-Nt} \mu_t(V) \leq N\gamma(V)\Big\}.$$

It suffices to find a constant $N_0 > 0$ such that for any $N \geq N_0$, Φ has a unique fixed point in $\mathcal{P}_{V,\gamma}^{T,N}(\bar{D})$. We finish the proof by three steps.

(a) The Φ-invariance of $\mathcal{P}_{V,\gamma}^{T,N}(\bar{D})$ for large N. For any $\lambda \geq 0$ and $N \geq 1$, $\mathcal{P}_{V,\gamma}^{T,N}(\bar{D})$ is a complete space under the metric

$$\rho_\lambda(\mu,\nu) := \sup_{t \in [0,T]} e^{-\lambda t}\|\mu_t - \nu_t\|_V, \quad \mu, \nu \in \mathcal{P}_{V,\gamma}^{T,N}(\bar{D}).$$

Let $\mu \in \mathcal{P}_{V,\gamma}^{T,N}(\bar{D})$. By (7.4.6), ($A^{7.8}$) with $V \in \mathcal{V}$ and Itô's formula, for any $p \geq 1$ we find a constant $c_1(p) > 0$ such that

$$dV(X_t^\mu)^p \leq 1_{\{t < \tau(X^\mu)\}}\{dM_t + c_1\{V(X_t^\mu)^p + \mu_t(V)^p\}dt\}, \quad t \in [0,T],$$

where M_t is a martingale with

$$d\langle M\rangle_t \leq c_1 V(X_t^\mu)^p dt.$$

By using BDG's and Gronwall's inequality, we find a constant $c_2(p) > 0$ such that

$$\mathbb{E}\Big[\sup_{s \in [0,t]} V(X_s^\mu)^p\Big] - \mathbb{E}\Big[\sup_{s \in [0,t \wedge \tau(X^\mu)]} V(X_s^\mu)^p\Big]$$
$$\leq c_2(p)V(X_0)^p + c_2(p)\int_0^t \mu_s(V)^p ds, \quad t \in [0,T]. \tag{7.4.7}$$

Consequently, for $p = 1$ and $c_2 = c_2(1)$ we derive

$$(\Phi_t\mu)(V) = \mathbb{E}[V(X_t^\mu)] \leq c_2\gamma(V) + c_2\Big(\int_0^t \mu_s(V)^2 ds\Big)^{\frac{1}{2}},$$

so that by $\mu \in \mathcal{P}_{V,\gamma}^{T,N}(\bar{D})$, we obtain

$$\sup_{t \in [0,T]} e^{-Nt}(\Phi_t\mu)(V) \leq c_2\gamma(V) + c_2 \sup_{t \in [0,T]}\Big(\int_0^t e^{-2N(t-s)}N^2\gamma(V)^2 ds\Big)^{\frac{1}{2}},$$
$$c_2(1 + \sqrt{N})\gamma(V) \leq N\gamma(V),$$

provided $N \geq N_0$ for a large enough constant $N_0 \geq 1$. By the continuity of X_t^μ in t, $\Phi_t\mu$ is weakly continuous in t. Therefore,

$$\Phi\mathcal{P}_{V,\gamma}^{T,N}(\bar{D}) \subset \mathcal{P}_{V,\gamma}^{T,N}(\bar{D}), \quad N \geq N_0.$$

(b) Let $N \geq N_0$. It remains to show that Φ has a unique fixed point in $\mathcal{P}_{V,\gamma}^{T,N}(\bar{D})$. By (7.4.7) with $p = 2$ and $V \geq 1$, there exists a constant $c_3 > 0$ such that

$$\mathbb{E}\Big[\sup_{t \in [0,T]} V(X_t^\mu)^2 \Big| \mathcal{F}_0\Big] \leq c_3^2 V(X_0)^2, \quad \mu \in \mathcal{P}_{V,\gamma}^{T,N}(\bar{D}). \tag{7.4.8}$$

For any $\mu^i \in \mathcal{C}_V(\bar{D}), i=1,2$, we estimate $\|\Phi_t\mu^1 - \Phi_t\mu^2\|_V$ by using Girsanov's theorem. Let X_t^1 be the unique solution for the SDE

$$dX_t^1 = 1_{\{t<\tau(X^1)\}}\{b_t(X_t^1,\mu_t^1)dt + \sigma_t(X_t^1)dW_t\}, \quad X_0^1 = X_0. \quad (7.4.9)$$

By the definition of Φ, we have

$$\Phi_t\mu^1 = \mathcal{L}_{X_t^1}, \quad t \in [0,T]. \quad (7.4.10)$$

To construct $\Phi_t\mu^2$ using Girsanov's theorem, let

$$\xi_t := 1_{\{t<\tau(X^1)\}}\{\sigma_t^*(\sigma_t\sigma_t^*)^{-1}\}(X_t^1)\{b_t(X_t^1,\mu_t^2) - b_t(X_t^1,\mu_t^1)\}, \quad t \in [0,T].$$

By $(A^{7.8})$, there exists a constant $k > 0$ such that

$$|\xi_t| \le k\|\mu_t^1 - \mu_t^2\|_V, \quad t \in [0,T]. \quad (7.4.11)$$

So, by Girsanov's theorem,

$$\tilde{W}_t := W_t - \int_0^t \xi_s ds, \quad t \in [0,T]$$

is an m-dimensional Brownian motion under the probability measure $\mathbb{Q} := R_T\mathbb{P}$, where

$$R_s := e^{\int_0^s \langle \xi_t, dW_t \rangle - \frac{1}{2}\int_0^s |\xi_t|^2 dt}, \quad s \in [0,T].$$

Reformulate (7.4.9) as

$$dX_t^1 = 1_{\{t<\tau(X^1)\}}\{b_t(X_t^1,\mu_t^2)dt + \sigma_t(X_t^1)d\tilde{W}_t\}, \quad X_0^1 = \tilde{X}_0.$$

By the weak uniqueness of (7.4.6), we obtain

$$\Phi_t\mu^2 = \mathbb{Q}(X_{t\wedge\tau(X^1)}^1 \in dx) = \mathcal{L}_{X_t^1|\mathbb{Q}}.$$

Combining this with (7.4.8) and (7.4.10), we derive

$$\|\Phi_t\mu^1 - \Phi_t\mu^2\|_V \le \mathbb{E}[V(X_t^1)|R_t - 1|]$$
$$\le \mathbb{E}[\{\mathbb{E}(V(X_t^1)^2|\mathcal{F}_0)\}^{\frac{1}{2}}\{\mathbb{E}(|R_t - 1|^2|\mathcal{F}_0)\}^{\frac{1}{2}}] \quad (7.4.12)$$
$$\le c_3\mathbb{E}[V(X_0)\{\mathbb{E}(|R_t - 1|^2|\mathcal{F}_0)\}^{\frac{1}{2}}].$$

On the other hand, by $\mu^1, \mu^2 \in \mathcal{P}_{V,\gamma}^{T,N}(\bar{D})$, (7.4.11), and noting that $e^r - 1 \le re^r$ for $r \ge 0$, we find a constant $c > 0$ such that

$$\mathbb{E}[|R_t - 1|^2|\mathcal{F}_0] = \mathbb{E}[e^{2\int_0^t \langle \xi_s, dW_s \rangle - \int_0^t |\xi_s|^2 ds} - 1|\mathcal{F}_0]$$
$$\le \mathbb{E}[e^{2\int_0^t \langle \xi_s, dW_s \rangle - 2\int_0^t |\xi_s|^2 dt}|\mathcal{F}_0]e^{k^2\int_0^t \|\mu_s^1 - \mu_s^2\|_V^2 ds} - 1$$
$$= e^{k^2\int_0^t \|\mu_s^1 - \mu_s^2\|_V^2 ds} - 1 \le e^{k^2\int_0^t \|\mu_s^1 - \mu_s^2\|_V^2 ds}\int_0^t k^2\|\mu_s^1 - \mu_s^2\|_V^2 ds$$
$$\le c^2\int_0^t \|\mu_s^1 - \mu_s^2\|_V^2 ds, \quad t \in [0,T].$$

Combining this with (7.4.12) and letting $C = cc_3 \mathbb{E}[V(X_0)]$, we arrive at

$$\rho_\lambda(\Phi\mu^1, \Phi\mu^2) \leq C \sup_{t\in[0,T]} e^{-\lambda t} \left(\int_0^t \|\mu_s^1 - \mu_s^2\|_V^2 ds\right)^{\frac{1}{2}}$$

$$\leq C\rho_\lambda(\mu^1, \mu^2) \left(\int_0^t e^{-2\lambda(t-s)} ds\right)^{\frac{1}{2}}.$$

Thus, when $\lambda > 0$ is large enough, Φ is contractive in ρ_λ and hence has a unique fixed point in $\mathcal{P}_{V,\gamma}^{T,N}(\bar{D})$.

(c) Note that for any (weak) solution X_t of (7.4.5) for distributions in $\mathcal{P}_V(\bar{D})$, $\mu_t := \mathcal{L}_{X_t}$ is a fixed point of Φ in $\mathcal{P}_V^\gamma(\bar{D})$. Since Φ has a unique fixed point, (7.4.1) has the (weak) uniqueness. Moreover, by Gronwall's inequality, (7.4.4) follows from (7.4.8) for $X_t^\mu = X_t$ and $\mu_t := \mathcal{L}_{X_t}$, where μ is the unique fixed point of Φ. □

Bibliography

Adams, D., dos Reis, G., Ravaille, R., Salkeld, W. and Tugaut, J. (2022). Large Deviations and Exit-times for reflected McKean-Vlasov equations with self-stabilizing terms and superlinear drifts, *Stoch. Process. Appl.* **146**, 264–310.

Aida, S. (1998). Uniformly positivity improving property, Sobolev inequalities and spectral gap, *J. Funct. Anal.* **158**, 152–185.

Albeverio, S., Kondratiev, Y. G. and Röckner M. (1996). Differential geometry of Poisson spaces, *C R Acad Sci Paris Sér I Math.* **323**, 1129–1134.

Ambrosio, L., Gigli, N. and Savare, G. (2005). Gradient Flows in Metric Spaces and in the Space of Probability Measures, *Lect. in Math. ETH Zurich*, Birkhäuser Verlag, Basel.

Arnaudon, M., Thalmaier, A. and Wang, F.-Y. (2009). Gradient estimates and Harnack incqualities on non-compact Riemannian manifolds, *Stoch. Proc. Appl.* **119**, 3653–3670.

Aronson, D. G. and Serrin, J. (1967). Local behavior of solutions of quasilinear parabolic equations, *Arch. Rational Mech. Anal.* **25**, 81–122.

Bai, Y. and Huang, X. (2023). Log-Harnack inequality and exponential ergodicity for distribution dependent Chan-Karolyi-Longstaff-Sanders and Vasicek models, *J. Theoret. Probab.* **36**, 1902–1921.

Bakry, D. and Emery, M. (1984). Hypercontractivité de semi-groupes de diffusion, *C. R. Acad. Sci. Paris. Sér. I Math.* **299**, 775–778.

Bakry, D., Gentil, I. and Ledoux, M. (2014) *Analysis and Geometry of Markov Diffusion Operators*, Springer, Berlin.

Baudoin, F., Gordina, M. and Herzog, D. (2021). Gamma calculus beyond Villani and explicit convergence estimates for Langevin dynamics with singular potential, *Arch. Rat. Mech. Anal.* **241**, 765–804.

Bao, J., Ren, P. and Wang, F.-Y. (2021). Bismut formulas for Lions derivative of McKean-Vlasov SDEs with memory, *J. Diff. Equat.* **282**, 285–329.

Bao, J., Wang, F.-Y. and Yuan, C. (2015). Hypercontractivity for functional stochastic partial differential equations, *Comm. Elect. Probab.* **20**, 1–15.

Barbu, V. and Röckner, M. (2018). Probabilistic representation for solutions to nonlinear Fokker-Planck equations, *SIAM J. Math. Anal.* **50**, 4246–4260.

Barbu, V. and Röckner, M. (2020). From nonlinear Fokker-Planck equations to solutions of distribution dependent SDE, *Ann. Probab.* **48**, 1902–1920.

Beck, L., Flandoli, F., Gubinelli, M. and Maurelli, M. (2019). Stochastic ODEs and stochastic linear PDEs with critical drift: regularity, duality and uniqueness, *Electron. J. Probab.*, **24**, 1–72.

Bismut, J. M. (1984). *Large Deviations and the Malliavin Calculus*, Boston: Birkhäuser, MA.

Bobkov, S. G., Gentil, I. and Ledoux, M. (2001). Hypercontractivity of Hamilton-Jacobi equations, *J. Math. Pures Appl.* **80**, 669–696.

Bogachev, V., Krylov, N. V. and Röckner, M. (2001). On regularity of transition probabilities and invariant measures of singular diffusions under minimal conditions, *Comm. Part. Diff. Equat.* **26**, 2037–2080.

Bogachev, V. I., Krylov, N. V., Röckner, M. and Shaposhnikov, S.V. (2015). *Fokker-Planck-Kolmogorov equations,* American Math. Soc.

Bogachev, V. I., Krylov, N. V., Röckner, M. and Wang, F.-Y. (2001). Elliptic equations for invariant measures on finite and infinite dimensional manifolds, *J. Math. Pures Appl.* **80**, 177–221.

Burkholder, D. L. and Gundy, R. F. (1970). Extrapolation and interpolation of quasi-linear operators on martingales, *Acta Math.* **124**, 249–304.

Butkovsky, O. (2014). On ergodic properties of nonlinear Markov chains and stochastic McKean-Vlasov equations, *Theory Probab. Appl.* **58**, 661–674.

Carmona, R. and Delarue, F. (2019). *Probabilistic Theory of Mean Field Games with Applications I,* Springer, Berlin.

Carrapatoso, K. (2015). Exponential convergence to equilibrium for the homogeneous Landau equation with hard potentials, *Bull. Sci. Math.* **139**, 777–805.

Carrillo, J. A., McCann, R. J. and Villani, C. (2003). Kinetic equilibration rates for granular media and related equations: entropy dissipation and mass transportation estimates, *Rev. Mat. Iberoam.* **19**, 971–1018.

Carroni, M. G. and Menald, J. L. (1992). *Green Functions for Second Order Parabolic Integro-Differential Problems,* Chapman & Hall/CRC.

Chaudru de Raynal, P. E. (2017). Strong existence and uniqueness for degenerate SDE with Hölder drifts, *Ann. Inst. H. Pionc. Probab. Stat.* **53**, 259–286.

Chaudru de Raynal, P. E. (2020). Strong well posedness of McKean-Vlasov stochastic differential equations with Hölder drift, *Stoch. Proc. Appl.* **130**, 79–107.

Chaudru de Raynal, P.-E. and Frikha, N. (2022). Well-posedness for some nonlinear SDEs and related PDE on the Wasserstein space, *J. Math. Pures Appl.* **159**, 1–167.

Chen, M.-F. and Li, S.-F. (1989). Coupling methods for multi-dimensional diffusion process, *Ann. Probab.* **17**, 151–177.

Chen, M.-F. and Wang, F.-Y. (1997). Estimates of logarithmic Sobolev constant: an improvement of Bakry-Emery criterion, *J. Funct. Anal.* **144**, 287–300.

Chen, Z., Zhang, X. and Zhao, G. (2021). Supercritical SDEs driven by multiplicative stable-like L??vy processes, *Trans. Amer. Math. Soc.* **374**, 7621–7655.

Cherny, A. S. (2003). On the uniqueness in law and the pathwise uniqueness for stochastic differential equations, *Theory Probab. Appl.* **46**, 406–419.

Criens, D. (2023). Propagation of chaos for weakly interacting mild solutions to stochastic partial differential equations, *J. Stat. Phys.* **190**, 40 pp.

Crisan, D. and McMurray, E. (2018). Smoothing properties of McKean-Vlasov SDEs, *Probab. Theory Relat. Fields.* **171**, 97–148.

Da Prato, G., Flandoli, F., Priola, E. and Röckner, M. (2013). Strong uniqueness for stochastic evolution equations in Hilbert spaces perturbed by a bounded measurable drift, *Ann. Probab.* **41**, 3306–3344.

Da Prato, G., Flandoli, F., Priola, E. and Röckner, M. (2015). Strong uniqueness for stochastic evolution equations with unbounded measurable drift term, *J. Theoret. Probab.* **28**, 1571–1600.

Da Prato, G. and Zabczyk, J. (1996). *Ergodicity for infinite dimensional systems*, Cambridge Uni. Press, Cambridge.

Davis, B. (1970). On the integrability of the martingale square function, *Israel J. Math.* **8**, 187–190.

Dembo, A. and Zeitouni, O. (1998). *Large Deviations Techniques and Applications*, Second Edition, Springer, New York.

Desvillettes, L. and Villani, C. (2000). On the spatially homogeneous Landau equation for hard potentials, Part I: existence, uniqueness and smoothness, *Comm. Part. Diff. Equat.* **25**, 179–259.

Desvillettes, L. and Villani, C. (2000). On the spatially homogeneous Landau equation for hard potentials, Part II: H-Theorem and Applications, *Comm. Part. Diff. Equat.* **25**, 261–298.

Donsker, M. D. and Varadhan, S. R. S. (1975). Asymptotic evaluation of certain Markov process expectations for large time, I-IV, *Comm. Pure Appl. Math.* **28**, 1–47, 279–301; **29**, 389–461; **36**, 183–212.

Doob, J. L. (1948). Asymptotic properties of Markoff transition probabilities, *Trans. Amer. Math. Soc.* **63**, 394–421.

Down, D., Meyn, S. P. and Tweedie, R. L. (1995). Exponential and uniform ergodicity of Markov processes, *Ann. Probab.* **23**, 1671–1691.

Dupuis, P. and Ishii, H. (1990). On oblique derivative problems for fully nonlinear second-order elliptic partial differential equations on nonsmooth domains, *Nonlin. Anal. Theo. Meth. Appl.* **15**, 1123–1138.

Eberle, A., Guillin, A. and Zimmer, R. (2019). Quantitative Harris-type theorems for diffusions and McKean-Vlasov processes, *Trans. Amer. Math. Soc.* **371**, 7135–7173.

Elworthy, K.D. and Li, X.-M. (1994). Formulae for the derivatives of heat semi groups, *J. Funct. Anal.* **125**, 252–286.

Evstigneev, I. V. (1988). Measurable selection theorem and probabilistic control models in general topological spaces, *Mathematics of the USSR-Sbornik* **59**, 25–37.

Fan, X., Yu, T. and Yuan, C. (2023). Asymptotic behaviors for distribution dependent SDEs driven by fractional Brownian motions, *Stoch. Proc. Appl.* **164**, 383–415.

Fang, S. (2004). *Introduction to Malliavin Calculus*, Tsinghua University Press and Springer, Beijing.

Fang, F. and Zhang, T. (2006). A study of a class of stochastic differential equations with non-Lipschitzian coefficients, *Probab. Theory Related Fields* **85**, 580–597.

Fournier, N. and Guillin, A. (2017). From a Kac-like particle system to the Landau equation for hard potentials and Maxwell molecules, *Ann. Sc. De Lecole Norm. Super.* **50**, 157–199.

Funaki, T. (1985). The diffusion approximation of the spatially homogeneous Boltzman equation, *Duke J. Math.* **521**, 1–23.

Girsanov, I. V. (1960). On transforming a certain class of stochastic processes by absolutely continuous substitution of measures, *Theory Probab. Appl.* **5**, 285–301.

Goldys, B. and Maslowski, B. (2006). Exponential ergodicity for stochastic reaction-diffusion equations, *Stochastic Partial Differential Equations and Applications* **XVII**, 115–131, Lect. Notes Pure Appl. Math., 245, Chapman Hall/CRC, Boca Raton, FL.

Grothaus, M. and Stigenbauer, P. (2014). Hypocoercivity for Kolmogorov backward evolution equations and applications, *J. Funct. Anal.* **267**, 3515–3556.

Grothaus, M. and Wang, F.-Y. (2019). Weak Poincaré inequalities for convergence rate of degenerate diffusion processes, *Ann. Probab.* **47**, 2930–2952.

Guérin, H. (2002). Existence and regularity of a weak function-solution for some Landau equations with a stochastic approach, *Stoch. Proc. Appl.* **101**, 303–325.

Guillin, A., Liu, W. and Wu, L. (2022). Uniform Poincaré and logarithmic Sobolev inequalities for mean field particle systems, *Ann. Appl. Probab.* **32**, 1590–1614.

Guillin, A. and Wang, F.-Y. (2012). Degenerate Fokker-Planck equations: Bismut formula, gradient estimate and Harnack inequality, *J. Diff. Equat.* **253**, 20–40.

Hairer, M., Mattingly, J. C. and Scheutzow, M. (2011). Asymptotic coupling and a general form of Harris theorem with applications to stochastic delay equations, *Probab. Theory Relat. Fields* **149**, 223–259.

Hammersley, W., Šiška, D. and Szpruch, L. (2021). McKean-Vlasov SDEs under measure dependent Lyapunov conditions, *Ann. Inst. H. Poinc. Probab. Stat.* **57**, 1032–1057.

Halmos, P. R. (1950). Measure Theory, *Springer*.

Hao, Z., Röckner, M. and Zhang, X. (2021). Euler Scheme for density dependent stochastic differential equations, *J. Diff. Equat.* **274**, 996–1014.

Hao, Z., Zhang, X., Zhu, R. and Zhu, X. (2021). Singular kinetic equations and applications, *arXiv:2108.05042*.

Hino, M., Matsuura, K. and Yonezawa, M. (2021). Pathwise uniqueness and non-explosion property of Skorohod SDEs with a class of non-Lipschitz coefficients and non-smooth domains, *J. Theo. Probab.* **34**, 2166–2191.

Hong, W., Hu, S. and Liu, W. (2024). McKean-Vlasov SDEs and SPDEs with locally monotone coefficients, *Ann. Appl. Probab.* **34**, 2136–2189.

Hong, W. and Liu, W. (2021). Distribution dependent stochastic porous media type equations on general measure spaces, *arXiv:2103.10135v3*.

Huang, X. (2018). Strong solutions for functional SDEs with singular drift, *Stoch. Dyn.* **18**, 1850015.

Huang, X. (2019). Harnack and shift Harnack inequalities for SDEs with integrability drifts, *Stoch. Dyn.* **19**, 1950034.

Huang, H. and Lv, W. (2020). Stochastic functional Hamiltonian systems with singular coefficients, *Comm. Pure Appl. Anal.* **19**, 1257–1273.

Huang, X., Ren, P. and Wang, F.-Y. (2021). Distribution dependent stochastic differential equations, *Front. Math. China* **16**, 257–301.

Huang, X., Ren, P. and Wang, F.-Y. (2023). Probability distance estimates between diffusion processes and applications to singular McKean-Vlasov SDEs, *arXiv:2304.07562*.

Huang, X., Röckner, M. and Wang, F.-Y. (2019). Nonlinear Fokker-Planck equations for probability measures on path space and path-distribution dependent SDEs, *Discrete Contin. Dyn. Syst. A.* **39**, 3017–3035.

Huang, X. and Song, Y. (2021). Well-posedness and regularity for distribution dependent SPDEs with singular drifts, *Nonlin. Anal. Theo. Meth. Appl.* **203**, 112167.

Huang, X., Song, Y. and Wang, F.-Y. (2021). Bismut formula for intrinsic/Lions derivatives of distribution dependent SDEs with singular coefficients, *Disc. Cont. Dyn. Sys.-A* **42**, 4597–4614.

Huang, X. and Wang, F.-Y. (2017). Functional SPDE with multiplicative noise and Dini drift, *Toulouse Sci. Math.* **XXVI**, 519–537.

Huang, X. and Wang, F.-Y. (2018). Degenerate SDEs with singular drift and applications to Heisenberg groups, *J. Diff. Equat.* **265**, 2745–2777.

Huang, X. and Wang, F.-Y. (2019). Distribution dependent SDEs with singular coefficients, *Stoch. Proc. Appl.* **129**, 4747–4770.

Huang, X. and Wang, F.-Y. (2021). McKean-Vlasov SDEs with drifts discontinuous under Wasserstein distance, *Disc. Cont. Dyn. Sys.-A* **41**, 1667–1679.

Huang, X. and Wang, F. Y. (2021). Derivative estimates on distributions of McKean-Vlasov SDEs, *Elect. J. Probab.* **26**, 1–12.

Huang, X. and Wang, F.-Y. (2022). Singular McKean-Vlasov (reflecting) SDEs with distribution dependent noise, *J. Math. Anal. Appl.* **514**, 126301 21pp.

Huang, X. and Wang, F.-Y. (2022). Log-Harnack inequality and Bismut's formula for singular McKean-Vlasov SDEs, *arXiv:2207.11536*.

Huang, X., Wang, S. and Yang, F.-F. (2021). Invariant probability measure for McKean-Vlasov SDEs with singular drifts, *arXiv:2108.05802*.

Huang, X. and Yang, F.-F. (2021). Distribution-dependent SDEs with Hölder continuous drift and α-stable noise, *Numer. Algorithms* **86**, 813–831.

Ikeda, N. and Watanabe, S. (1977). A comparison theorem for solutions of stochastic differential equations and its applications, *Osaka J. Math.* **14**, 619–633.

Issoglio, E. and Russo, F. (2023). McKean SDEs with singular coefficients, *Ann. Inst. Henri Poincar Probab. Stat.* **59**, 1530–1548.

Izydorczyk, L., Oudjane, N. and Russo, F. (2019). McKean Feynman-Kac probabilistic representations of nonlinear partial differential equations, *Geometry and invariance in stochastic dynamics*, **378**, 187–212, Springer Proc. Math. Stat.

Jourdain, B. and Méléard, S. (1998). Propagation of chaos and fluations for a moderate model with smooth initial data, *Ann. Inst. H. Poinc. Probab. Stat.* **34**, 727–766.

Kac, M. (1954). Foundations of kinetic theory, *Proceedings of the Third Berkeley Symposium on Mathematical Statistics and Probability,* vol. III, University of California Press, 171–197.

Kac, M. (1959). *Probability and Related Topics in the Physical Sciences,* New York.

Khasminskii, R. Z. (1959). On positive solutions of the equation $Lu + Vu = 0$ (in Rusian), *Teorija Verojat. i ee Primen.* **4**, 309–318.

Khasminskii, R. Z. (1980). *Stochastic Stability of Differential Equations,* T Sijthoff and Noordhoff.

Kinzebulatov, D. and Semenov, Y. A. (2020). Brownian motion with general drift, *Stoch. Proc. Appl.* **130**, 2737–2750.

Konakov, V. and Mammen, E. (2000). Local limit theorems for transition densities of Markov chains converging to diffusions, *Probab. Theory Relat. Fields* **117**, 551–587.

Krylov, N.V. (1980). *Controlled Diffusion Processes,* Translated from the Russian by A. B. Aries. Applications of Mathematics, 14. Springer, New York/Berlin.

Krylov, N.V. (1985). *Nonlinear Elliptic and Parabolic Equations of Second Order,* Nauka, Moscow.

Krylov, N.V. and Röckner, M. (2005). Strong solutions of stochastic equations with singular time dependent drift, *Probab. Theory Relat. Fields* **131**, 154–196.

Kurtz, T. (2014). Weak and strong solutions of general stochastic models, *Electr. Comm. Probab.* **19**, paper no. 58, 16 pp.

Kunita, H. (1990). *Stochastic flows and stochastic differential equations,* Cambridge University Press, Cambridge.

Lanconelli, E. and Polidoro, S. (1994). On a class of hypoelliptic evolution operator, *Rend. Sem. Mat. Univ. Pol. Torino* **52**, 29–63.

Li, H., Luo, D. and Wang, J. (2015). Harnack inequalities for SDEs with multiplicative noise and non-regular drift, *Stoch. Dyn.* **15**, 1550015.

Liang, M., Majka, M. B. and Wang, J. (2021). Exponential ergodicity for SDEs and McKean-Vlasov processes with Lévy noise, *Ann. Inst. H. Poinc. Probab. Stat.* **57**, 1665–1701.

Lindvall, T. and Rogers, L.C.G. (1986). Coupling of multidimensional diffusions by reflection, *Ann. Probab.* **14**, 460–872.

Lions, P. L. and Sznitman, A. S. (1984). Stochastic differential equations with reflecting boundary conditions, *Comm. Pure Appl. Math.* **37**, 511–537.

Liu, W., Wu, L. and Zhang, C. (2021). Long-time behaviors of mean-field interacting particle systems related to McKean-Vlasov equations, *Comm. Math. Phys.* **387**, 179–214.

Liu, W., Song, Y., Zhai, J. and Zhang, T. (2023). Large and moderate deviation principles for McKean-Vlasov SDEs with jumps, *Potent. Anal.* **59**, 1141–1190.

Lv, W. and Huang, X. (2021). Harnack and shift Harnack inequalities for degenerate (functional) stochastic partial differential equations with singular drifts, *J. Theoret. Probab.* **34**, 827–851.

Malliavin, P. (1978). Stochastic calculus of variation and hypoelliptic operators, *Proceedings of the International Symposium on Stochastic Differential Equations* (Res. Inst. Math. Sci., Kyoto Univ., Ktoto, 1976), Wiley, pp. 195–263.

Marin-Rubio, P. and Real, J. (2004). Some results on stochastic differential equations with reflecting boundary conditions, *J. Theo. Probab.* **17**, 705–716.

McKean, H.P. (1966). A class of Markov processes associated with nonlinear parabolic equations, *Proc. Nat. Acad. Sci. U.S.A.* **56**, 1907–1911.

Menozzi, S., Parametrix techniques and martingale problems for some degenerate Kolmogorov equations, *Electr. Comm. Probab.* **16**, 234–250.

Menozzi, S., Pesce, A. and Zhang, X. (2021). Density and gradient estimates for non degenerate Brownian SDEs with unbounded measurable drift, *J. Diff. Equat.* **272**, 330–369.

Nam, K. (2020). Stochastic differential equations with critical drifts, *Stoch. Proc. Appl.* **130**, 5366–5393.

Oksendal, B. (2014). *Stochastic Differential Equations,* Springer, Berlin.

Otto, F. (2001). The geometry of dissipative evolution equations: The porous medium equation, *Comm. Partial Diff. Equat.* **26**, 101–174.

Otto, F. and Villani, C. (2000). Generalization of an inequality by Talagrand and links with the logarithmic Sobolev inequality, *J. Funct. Anal.* **173**, 361–400.

Portenko, N. I. (1990). *Generalized Diffusion Processes,* Nauka, Moscow, 1982 In Russian; English translation: Amer. Math. Soc. Provdence, Rhode Island.

Priola, E. and Wang, F.-Y. (2006). Gradient estimates for diffusion semigroups with singular coefficients, *J. Funct. Anal.* **236**, 244–264.

Qian, Z., Ren, P. and Wang, F.-Y. (2023). Entropy estimate for degenerate SDEs with applications to nonlinear kinetic Fokker-Planck equations, *arXiv:2308.16373*.

Rehmeier, M. and Röckner, M. (2022). On nonlinear Markov processes in the sense of McKean, *arXiv:2212.12424v2*

Ren, P. (2023). Singular McKean-Vlasov SDEs: Well-posedness, regularities and Wang's Hanrack inequality, *Stoch. Proc. Appl.* **156**, 291–311.

Ren, P., Röckner, M. and Wang, F.-Y. (2022). Linearization of of nonlinear Fokker-Planck equations and applications, *J. Diff. Equat.* **322**, 1–37.

Ren, P., Sturm, K.T. and Wang, F.-Y. (2021). Exponential ergodicity for time-periodic McKean-Vlasov SDEs, *arXiv:2110.06473*.

Ren, P., Tang, H. and Wang, F.-Y. (2020). Distribution-path dependent nonlinear SPDEs with application to stochastic transport type equations, to appear in *Pot. Anal., arXiv:2007.09188*.

Ren, P. and Wang, F.-Y. (2019). Bismut formula for Lions derivative of distribution dependent SDEs and applications, *J. Diff. Equat.* **267**, 4745–4777.

Ren, P. and Wang, F.-Y. (2021). Derivative formulas in measure on Riemannian manifolds, *Bull. Lond. Math. Soc.*

Ren, P. and Wang, F.-Y. (2021). Donsker-Varadhan large deviations for path-distribution dependent SPDEs, *J. Math. Anal. Appl.* **499**, 125000. 32pp.

Ren, P. and Wang, F.-Y. (2021). Exponential convergence in entropy and Wasserstein distance for McKean-Vlasov SDEs, *Nonlinear Analysis-Method and Theory* **206**, 112259.

Ren, P. and Wang, F.-Y. (2023). Entropy estimate between diffusion processes and application to McKean-Vlasov SDEs, *arXiv:2302.13500*.

Röckner, M. and Wang, F.-Y. (2003). Supercontractivity and ultracontractivity for (non-symmetric) diffusion semigroups on manifolds, *Forum Math.* **15**, 893–921.

Röckner, M. and Wang, F.-Y. (2010). Log-harnack inequality for stochastic differential equations in Hilbert spaces and its consequences, *Infin. Dim. Anal. Quat. Probab. Relat. Top.* **13**, 27–37.

Röckner, M., Wang, F.-Y. and Wu, L. (2006). Large deviations for stochastic generalized porous media equations, *Stoch. Proc. Appl.* **116**, 1677–1689.

Röckner, M. and Zhang, X. (2021). Well-posedness of distribution dependent SDEs with singular drifts, *Bernoulli* **27**, 1131–1158.

Röckner, M. and Zhao, G. (2023). SDEs with critical time dependent drifts: weak solutions, *Bernoulli* **29**, 757–784.

Röckner, M. and Zhao, G. (2020). SDEs with critical time dependent drifts: strong solutions, *arXiv:2103.05803*.

Rozkosz, A. and Slominski, L. (1997). On stability and existence of solutions of SDEs with reflection at the boundary, *Stoch. Process. Appl.* **68**, 285–302.

Saisho, Y. (1987). Stochastic differential equations for multidimensional domain with reflecting boundary, *Probab. Theory Relat. Fields* **74**, 455–477.

Shao, J. (2013). Harnack inequalities and heat kernel estimates for SDEs with singular drifts, *Bull. Sci. Math.* **137**, 589–610.

Scheutzow, M. (2013). A stochastic Gronwall's lemma, *Infin. Dimens. Anal. Quant. Probab. RelatṪopics* **16**, 1350019.

Skorohod, A. V. (1961). Stochastic equations for diffusion processes with a boundary, *Teor. Verojatnost. i Primenen.* **6**, 287–298.

Skorohod, A. V. (1962). Stochastic equations for diffusion processes with boundaries II, *Teor. Verojatnost. i Primenen.* **7**, 5–25.

Situ, R. (2005). *Theory of Stochastic Differential Equations with Jumps and Applications*, Springer, Berlin.

Stein, E. M. (1970). *Singular Integrals and Differentiability Properties of Functions*, Princeton Mathematical Series, Vol. 30. Princeton University Press, Princeton, NJ.

Sznitman, A.-S. (1984). Nonlinear reflecting diffusion process, and the propagation of chaos and fluctuations associated, *J. Funct. Anal.* **56**, 311–336.

Sznitman, A.-S. (1991). *Topics in Propagation of Chaos*, Lecture notes in Math. Vol. 1464, pp. 165–251, Springer, Berlin.

Talagrand, M. (1996). Transportation cost for Gaussian and other product measures, *Geom. Funct. Anal.* **6**, 587–600.

Tanaka, H. (1979). Stochastic differential equations with reflecting boundary conditions, *Hiroshima Math. J.* **9**, 163–177.

Trevisan, D. (2016). Well-posedness of multidimensional diffusion processes, *Elect. J. Probab.* **21**, 1–41.

Trudinger, N. S. (1968). Pointwise estimates and quasilinear parabolic equations, *Comm. Pure Appl. Math.* **21**, 205–226.

Veretennikov, A. J. (1981). On strong solutions and explicit formulas for solutions of stochastic integral equations, *Sbornik: Mathematics* **39**, 387–403.

Veretennikov, A. Y. (2006). On ergodic measures for McKean-Vlasov stochastic equations, *Monte Carlo and Quasi-Monte Carlo Methods 2004* (pp. 471–486), Springer, Berlin, Heidelberg.

Villani, C. (2009). *Optimal Transport, Old and New*, Springer-Verlag.

Wang, F.-Y. (1997). Logarithmic Sobolev inequalities on noncompact Riemannian manifolds, *Probab. Theory Related Fields* **109**, 417–424.

Wang, F.-Y. (2004). Probability distance inequalities on Riemannian manifolds and path spaces, *J. Funct. Anal.* **206**, 167–190.

Wang, F.-Y. (2007). Harnack inequality and applications for stochastic generalized porous media equations, *Ann. Probab.* **35**, 1333–1350.

Wang, F.-Y. (2010). Harnack inequalities on manifolds with boundary and applications, *J. Math. Pures Appl.* **94**, 304–321.

Wang, F.-Y. (2011). Harnack inequality for SDE with multiplicative noise and extension to Neumann semigroup on nonconvex manifolds, *Ann. Probab.* **39**, 1449–1467.

Wang, F.-Y. (2013). *Harnack Inequalities and Applications for Stochastic Partial Differential Equations*, Springer, Berlin.

Wang, F.-Y. (2014). *Analysis of Diffusion Processes on Riemannian Manifolds*, World Scientific, Singapore.

Wang, F.-Y. (2014). Criteria on spectral gap of Markov operators, *J. Funct. Anal.* **266**, 2137–2152.

Wang, F.-Y. (2014). Integration by parts formula and shift Harnack inequality for stochastic equations, *Ann. Probab.* **42**, 99–1019.

Wang, F.-Y. (2016). Gradient estimates and applications for SDEs in Hilbert space with multiplicative noise and Dini continuous drift, *J. Diff. Equat.* **260**, 2792–2829.

Wang, F.-Y. (2017). Integrability conditions for SDEs and semi-Linear SPDEs, *Ann. Probab.* **45**, 3223–3265.

Wang, F.-Y. (2017). Hypercontractivity and applications for stochastic Hamiltonian systems, *J. Funct. Anal.* **272**, 5360–5383.

Wang, F.-Y. (2018). Distribution dependent SDEs for Landau type equations, *Stoch. Proc. Appl.* **128**, 595–621.

Wang, F.-Y. (2018). Estimates for invariant probability measures of degenerate SPDEs with singular and path-dependent drifts, *Probab. Theory Relat. Fields* **172**, 1181–1214.

Wang, F.-Y. (2023). Exponential ergodicity for non-dissipative McKean-Vlasov SDEs, *Bernoulli* **29**, 1035–1062.

Wang, F.-Y. (2023). Distribution dependent reflecting stochastic differential equations, to appear in *Sci. Chin. Math.* arXiv:2106.12737.

Wang, F.-Y. (2023). Exponential ergodicity for singular reflecting McKean-Vlasov SDEs, *Stoch. Proc. Appl.* **160**, 265–293.

Wang, F.-Y. (2023). Derivative formula for singular McKean-Vlasov SDEs, *Comm. Pure Appl. Anal.* **22**, 1866–1898.

Wang, F.-Y. (2023). Singular density dependent stochastic differential equations, *J. Diff. Equat.* **361**, 562–589.

Wang, F.-Y. (2023). Killed distribution dependent SDE for nonlinear Dirichlet problem, *Disc. Cont. Dyn. Syst. S.* **16**, 1054–1075.

Wang, F.-Y. and Zhang, T. (2014). Log-Harnack inequalities for semi-linear SPDE with strongly multiplicative noise, *Stoch. Proc. Appl.* **124**, 1261–1274.

Wang, F.-Y. and Zhang, X. (2013). Derivative formula and applications for degenerate diffusion semigroups, *J. Math. Pures Appl.* **99**, 726–740.

Wang, F.-Y. and Zhang, X. (2015). Degenerate SDEs in Hilbert spaces with rough drifts, *Infin. Dim. Anal. Quant. Probab. Related Topics* **18**, 1550026, 25 pp.

Wang, F.-Y. and Zhang, X. (2016). Degenerate SDE with Hölder-Dini drift and non-Lipschitz noise coefficient, *SIAM J. Math. Anal.* **48**, 2189–2226.

Wang, Z. and Zhang, X. (2018). Strong uniqueness of degenerate SDEs with Hölder diffusion coefficients, *arXiv:1805.05526*.

Wu, L. (2000). Some notes on large deviations of Markov processes, *Acta Math. Sin. (English Ser.)* **16**, 369–394.

Wu, L. (2000). Uniformly integrable operators and large deviations for Markov processes, *J. Funct. Anal.* **172**, 301–376.

Xia, P., Xie, L., Zhang, X. and Zhao, G. (2020). $L^q(L^p)$-theory of stochastic differential equations, *Stoch. Proc. Appl.* **130**, 5188–5211.

Xie, L. and Zhang, X. (2016). Sobolev differentiable flows of SDEs with local Sobolev and super-linear growth coefficients, *Ann. Probab.* **44**, 3661–3687.

Xie, L. and Zhang, X. (2020). Ergodicity of stochastic differential equations with jumps and singular coefficients, *Ann. Inst. Henri Poincaré Probab. Stat.* **56**, 175–229.

Yamada, T. and Watanabe, S. (1971). On the uniqueness of solutions of stochastic differential equations, *J. Math. Kyoto Univ.* **2**, 155–167.

Yan, J. (1988). On the existence of diffusions with singular drift coefficient, *Acta Math. Appl. Sin.* **4**, 23–29.

Yang, S. and Zhang, T. (2023). Strong solutions to reflecting stochastic differential equations with singular drift, *Stochastic Process. Appl.* **156**, (2023), 126–155.

Yang, S. and Zhang, T. (2020). Strong existence and uniqueness of solutions of SDEs with time dependent Kato class coefficients, *arXiv:2010.1146*.

Ye, H., Gao, J. and Ding, Y. (2007). A generalized Gronwall inequality and its application to a fractional differential equation, *J. Math. Anal. Appl.* **328**, 1075–1081.

Zhang, S.-Q. and Yuan, C. (2021). A Zvonkin's transformation for stochastic differential equations with singular drift and applications, *J. Diff. Equat.* **297**, 277–319.

Zhang, X. (2011). Stochastic homeomorphism flows of SDEs with singular drifts and Sobolev diffusion coefficients, *Electr. J. Probab.* **16**, 1096–1116.

Zhang, X. (2016). Stochastic differential equations with Sobolev coefficients and applications, *Ann. Appl. Prob.* **26**, 2697–2732.

Zhang, X. (2018). Stochastic Hamiltonian flows with singular coefficients, *Sci. China Math.* **61**, 1353–1384.

Zhang, X. (2024). Weak solutions of McKean-Vlasov SDEs with supercritical drifts, *Comm. Math. Stat.* **12**, 1–14. *arXiv:2010.15330*.

Zhang, X. (2021). Second order McKean-Vlasov SDEs and kinetic Fokker-Planck-Kolmogorov equations, *arXiv:2109.01273*.

Zhao, G. (2020). On distribution dependent SDEs with singular drifts, *arXiv:2003.04829v3*.

Zvonkin, A. K. (1974). A transformation of the phase space of a diffusion process that will remove the drift, *(Russian) Mat. Sb. (N.S.)* **93**, 129–149.

Index

$(A^{1.1})$, 6
$(A^{1.2})$, 14
$(A^{1.3})$, 14
$(A^{1.4})$, 44
$(A^{1.5})$, 49
$(A^{2.1})$, 63
$(A^{2.2})$, 63
$(A^{2.3})$, 65
$(A^{2.4})$, 86
$(A^{2.5})$, 99
$(A^{3.1})$, 109
$(A^{3.2})$, 113
$(A^{3.3})$, 118
$(A^{3.4})$, 119
$(A^{3.5})$, 126
$(A^{3.6})$, 137
$(A^{3.7})$, 150
$(A^{3.8})$, 157
$(A^{4.10})$, 216
$(A^{4.1})$, 164
$(A^{4.2})$, 167
$(A^{4.3})$, 169
$(A^{4.4})$, 172
$(A^{4.5})$, 181
$(A^{4.6})$, 209
$(A^{4.7})$, 211
$(A^{4.8})$, 212
$(A^{4.9})$, 214
$(A^{5.1})$, 229
$(A^{5.2})$, 234
$(A^{5.3})$, 240
$(A^{5.4})$, 246
$(A^{5.5})$, 247
$(A^{5.6})$, 247
$(A^{5.7})$, 258
$(A^{5.8})$, 261
$(A^{5.9})$, 265
$(A^{6.1})$, 280
$(A^{6.2})$, 281
$(A^{6.3})$, 290
$(A^{6.4})$, 294
$(A^{6.5})$, 296
$(A^{6.6})$, 299
$(A^{6.7})$, 309
$(A^{6.7})(3)$, 313
$(A^{6.8})$, 313
$(A^{6.9})$, 314
$(A^{7.1})$, 322
$(A^{7.2})$, 323
$(A^{7.3})$, 324
$(A^{7.4})$, 329
$(A^{7.5})$, 329
$(A^{7.6})$, 334
$(A^{7.7})$, 337
$(A^{7.8})$, 341
$(\Omega, \{\mathcal{J}_t\}_{t\geq 0}, \mathbb{P})$: complete filtration probability space, 1
$(\mathcal{P}_k, \mathbb{W}_k)$: L^k-Wasserstein space, 106
$C(E)$: space of continuous real functions on E, 7
$C([0,T]; \mathbb{R}^d)$: set of continuous maps from $[0,T]$ to $\mathbb{R}^d$, 19
$C^k(\mathbb{R}^d)$: space of real functions on $\mathbb{R}^d$ with continuous derivatives up to

order k, 3
$C_0^2(\bar{D})$: the class of C^2-functions on $\bar{D}$ with compact support, 67
$C_0^k(\mathbb{R}^d)$: space of functions in $C^k(\mathbb{R}^d)$ with compact supports, 3
$C_D^2(D)$, 320
$C_N^2(\bar{D})$: the set of C^2 functions on $\bar{D}$ with compact support and Neumann boundary condition, 307
$C_N^2(\bar{D})$: the set of C_0^2 functions on $\bar{D}$ satisfying Neumann condition, 67
$C_b(E)$: space of bounded continuous real functions on E, 7
$C_b(E)$: the space of bounded continuous functions on E, 47
$C_b^w([0,T];\hat{\mathcal{P}})$, 102
$C_c^\infty(\mathbb{R}^d) = \mathbb{R} + C_0^\infty(\mathbb{R}^d)$, 36
$C_{Lip}(\mathbb{R}^d)$, 31
D-distribution, 320
$D^I f(\mu)$: intrinsic derivative of f at μ, 175
$D^L f(\mu)$: L-derivative of f at μ, 175
D_r, 62
$H^{\epsilon,p}(\mathbb{R}^d)$: domain of $(1-\Delta)^{\frac{\epsilon}{2}}$ in $L^p(\mathbb{R}^d)$, 5
I_d: the $d \times d$ identity matrix, 15, 105
$L^p(\mathbb{R}^d)$: L^p-space for Lebesgue measure on $\mathbb{R}^d$, 4
P_1^D, 328
$P_t f(\mu)$, 163
W_t: m-dimensional Brownian motion, 1
$\langle M \rangle_t$: quadratic variational of M_t, 18
$\langle \cdot, \cdot \rangle$: inner product in $\mathbb{R}^d$, 2
$\mathcal{B}(E)$: Borel σ-algebra of E, 8
$\mathcal{B}_b$: the set of bounded measurable functions on a measurable space, 13
$\mathcal{B}_b(E)$: the space of bounded measurable functions on E, 47
$\mathcal{B}_b^+(\mathbb{R}^d)$: set of positive bounded measurable functions on $\mathbb{R}^d$, 34
$\mathcal{B}_b^+(\bar{D})$: space of bounded strictly positive measurable functions on $\bar{D}$, 85
$\mathcal{C}(\mu_1,\mu_2)$: space of couplings of μ_1 and μ_2, 35
$\mathcal{C}^\gamma(\hat{\mathcal{P}}) := C_b^w([0,T];\hat{\mathcal{P}})$: the set of weakly continuous maps from $[0,T]$ to $\hat{\mathcal{P}}$ with initial value γ, 107
Δ: the Laplace operator, 1
Ent: relative entropy, 35
$\mathbb{R}^d$: d-dimensional Euclidean space, 1
$\mathbb{R}^{d \otimes m}$: the space of $d \times m$-matrices, 1
$\mathbb{W}_\psi$: Wasserstein distance induced by ψ, 258
$\mathbb{W}_p$: L^p-Wasserstein distance, 35
$\mathbb{W}_{\psi,,\beta V}$, 246
ℓ_X: distribution density of X, 135
$\gamma \in \Gamma: \gamma : [0,\infty) \to [1,\infty)$ is increasing, $\int_0^\infty \frac{ds}{\gamma(s)} = \infty$, 78
$\mathbb{N}$: space of natural numbers, 11
$\mathcal{P}_V(\bar{D})$, 314
$\mathbf{n}$: inward normal vector field of ∂D, 62
∇: gradient operator, 3
∇^2: Hessian operator, 3
∇_v: directional derivative along v, 3
$\phi \in \Phi$: increasing, $\int_0^\infty \frac{ds}{1+s\phi(s)} < \infty$, 14
∂D: boundary of D, 61
$\partial D \in C^k$, 63
$\partial D \in C_b^{k,L}$, 63
$\partial_r D$, 62
$\partial_{-r} D$, 62
ρ_∂, 63
$\mathcal{C}_D(\mu,\nu)$, 323
$\mathcal{K} := \{(p,q) : p,q \in (1,\infty), \frac{d}{p} + \frac{2}{q} < 1\}$, 5
$\mathcal{L}_\xi^D$, 320
$\mathcal{M}f$: local Hardy-Littlewood maximal function of f, 18
$\mathcal{N}_{x,r}$, 61
$\mathcal{P}(E)$: space of probability measures on E, 17
$\mathcal{P}$: space of probability measures on $\mathbb{R}^d$, 3
$\hat{\mathcal{P}}$: space of probability measures on the sate space of SDE, 35
$\mathcal{P}^D$, 319
$\mathcal{P}_{\gamma,T}^k$, 138
$\mathcal{P}_V := \{\mu : \mu(V) < \infty\}$, 245

Index

$\mathcal{P}_\gamma^T(\bar{D})$, 324
$\mathcal{P}_k(\bar{D})$, 277
$\mathcal{S}_n(g)$: mollifying approximation of g, 7
σ^*: transposition of σ, 6
$\tilde{H}^{\epsilon,p}$, 5
$\tilde{L}^p$, 5
$\tilde{L}_q^p(s,t,D)$, 64
$\tilde{L}_q^p(t)$, 64
$\tilde{\mathcal{P}}^k$, $\mathcal{P}^k$, 137
$\tilde{\mathcal{P}}_{\gamma,T}^k$, 138
$b_t^\mu(x) := b_t(x,\mu_t).$, 172
$h \in \Lambda$: $h : (0,1] \to (0,\infty)$ is increasing, $\int_0^1 \frac{ds}{h(s)} = \infty$, 78
(strong) solution, 2

Condition (**D**), 62

DDSDE: distribution dependent SDE, 101
dimension-free Harnack inequality, 34

Irreducibility, 48

Linear functional derivative, 156
log-Harnack inequality, 34

Malliavin calculus, 26
Markov operator, 34
Markov transition kernel, 47

pathwise uniqueness, 2
petite or small set, 48
power Harnack inequality, 34

SDE: stochastic differential equation, 1
Stochastic Gronwall inequality, 17
Strong Feller, 47
strong Feller, 13

weak solution, 2
weak well-posedness, 2
well-posedness, 2

Yamada-Watanabe principle, 3, 17

Zvokin's transform, 24